STRATEGY AND TACTICS

INFANTRY
WARFARE

S T R A T E G Y   A N D   T A C T I C S

INFANTRY WARFARE

A N D R E W   W I E S T   A N D   M .   K .   B A R B I E R

SPELLMOUNT
Staplehurst

British Library Cataloguing-in-Publication Data:
A catalogue record for this book is available from the British Library

Copyright © 2002 Amber Books Ltd

ISBN 1-86227-172-0

First published in the UK in 2002 by
Spellmount Limited
The Old Rectory
Staplehurst
Kent TN12 0AZ

Tel: 01580 893730
Fax: 01580 893731
Email: enquiries@spellmount.com
Website: www.spellmount.com

Editorial and design by
Amber Books Ltd
Bradley's Close,
74-77 White Lion Street,
London N1 9PF

Project Editor: Charles Catton
Editor: Charlotte Rundall
Design: Neil Rigby at www.stylus-design.com
Picture Research: Lisa Wren and Chris Bishop

Printed and bound in Italy by: Eurolitho S.p.A., Cesano Boscone (MI)

Picture Credits
All pictures supplied by **TRH Pictures**, except the following:
Aerospace Publishing: 15, 41, 44, 46, 47, 50, 72, 73, 75 (b).
US Department of Defense: 156–157, 158, 159, 162, 164, 165,
166, 167, 170(b), 172 (both), 173.

Artworks:
Peter Harper & Tony Randall

CONTENTS

CHAPTER ONE

INDUSTRIAL WAR

In the killing fields of World War I, the role of the infantry underwent a change that marked the birth of modern war. Troops who had gone into battle in line now had to cope with machine guns, aircraft and artillery.

Organized warfare has been a part of human experience since the dawn of civilization. Indeed the frequency of conflict seems to suggest that war is integral to the very nature of mankind. Political régimes rise and fall, the fortunes of religions ebb and flow, cultural values continually alter, but war remains. From the first war in the desert wastes of ancient Mesopotamia to the present war on terror, conflict has been a constant in the development of humankind.

Within the history of warfare, much has changed over the millennia. When Achilles battled Hector before the gates of Troy, wars were small and personal, involving hand-to-hand combat and individual physical prowess. Modern wars are technological wonders based on the military dominance of the microchip. Nearly invisible stealth aircraft can launch attacks using terminally guided munitions that strike their targets with laser-guided precision. Though technological advances, from the introduction of gunpowder to satellite targeting, have altered the manner in which war is prosecuted, in many ways the battlefield remains unchanged by history. War remains the domain where the violent, and all too often short life of the infantryman takes centre stage.

LEFT: Infantrymen from around the world gather in Peking (now Beijing) after the suppression of the Boxer Rebellion at the turn of the 20th century. At that time, military theorists of all nations were certain that the cold steel and spirit of the infantry would prove decisive in future conflicts.

ABOVE: British troops move forward in a clearing operation in the Brandwater area during the Boer War. Few European military leaders took note of the lessons learned in South Africa when laying their plans for what they believed would be decisive victory in World War I.

Throughout history military pundits have often readied the obituary of infantry warfare, predicting that armoured knights, gunpowder, rifled weapons, machine guns, tanks, aircraft, nuclear weapons and finally computers would make the infantryman obsolete. From the Greek phalanx to modern special forces, though, the infantry has done what it does best and fought on, sometimes against all odds. Though technological advances have increased the role of other services in wartime it remains the infantry that takes and holds land in battle. It also remains the infantry that suffers the bulk of the military casualties during conflicts.

Much has changed, but in the end, it is still the 'grunt' that fights and wins wars. Infantry defended Rome and destroyed Carthage, infantry won and lost the Napoleonic Wars, infantry suffered and died at Verdun, infantry overthrew Berlin, infantry fought a ragged war in the tunnels of the Viet Cong, and infantry routs terrorists from their mountain fastness. The methods of war have changed, but the infantry remains Queen of the Battlefield.

THE ROAD TO WORLD WAR I
Military history in the modern era has focused on the martial fortunes of the powerful nation states of Europe. Reaping the rewards of the Industrial

Revolution and world-wide economic dominance, the nations of Europe towered over the remainder of the world in military strength in the 19th century.

After having spent a good deal of their latent energy fighting each other, in the wake of the Crimean War the nations of Europe began to turn their power upon the remainder of the world, embarking on a period of colonialism that would soon see great swaths of the globe under the suzerainty of European overlords. The typical view of colonial warfare contends that European armies, utilizing the military inventions of the Industrial Revolution, were able to run roughshod over outmatched native forces. Indeed some colonial wars did follow this pattern, including an easy defeat by the British of a massive Dervish force during 1898 at Omdurman.

However, when properly armed and led, native forces were able to score significant victories, including the Ethiopian defeat of the Italians. In perhaps the most revealing colonial conflict, the mighty British Empire had to struggle for years to overcome the stubborn guerrilla resistance of an irregular force in the Boer War. Though warning signs were evident, European nations emerged from the period of colonial warfare confident that their modern militaries could overcome all resistance, making modern warfare quick and decisive.

Several other conflicts also indicated that warfare in the age of the Industrial Revolution would be different from what most nations expected. The American Civil War had been anything but quick and decisive as the outgunned South managed to resist the seemingly inevitable Northern victory for nearly five years. Though the war included manoeuvre and quick victories, such as Lee's masterpiece at Chancellorsville, in the end it got bogged down. It became a war of sieges and a Northern constriction of the Southern will to fight, finally developing into what some historians consider to be the first total war.

The Prussian victories over Austria and France, on the surface, seemed to bode well for the concept of decisive war. Inept Austrian and French commanders should receive much of the blame for the quick defeat of their respective nations. Yet after the German victory over French forces at Sedan in 1870, the surrender of Napoleon III and the seeming collapse of France, the war lingered on. Many within the French population rose up and engaged in a guerrilla war against the increasingly frustrated Prussian military.

It was the Russo-Japanese War, though, that should have alarmed complacent European military theorists. Though the conflict is best known for decisive

Japanese naval victories, it was decided by battle between massed infantry formations. The Battle of Mukden presaged the future, involving over 600,000 soldiers on a front of over 40 miles (64km) in a massive struggle that lasted for two weeks.

TOWARDS TOTAL WAR

Thus several conflicts indicated that future wars would not be decisive in nature, but would instead become attritional. Some military men, led by the Polish war theorist Jan Bloch, predicted that the next European conflict would be long and inconclusive. Most military men, believing implicitly in the dominance of the weaponry provided by the Industrial Revolution, remained confident in their belief that audacious attack would lead to lightning victories. Their faith in the wonders of the new weaponry, though, was misplaced.

The Industrial Revolution had changed the very nature of warfare, ushering in the era of modern, total war. Economic powerhouses, the nations of Europe were now able to raise and equip armies numbering in the millions, and keep them in the field for nearly five years. Thus the coming cataclysm would pit nation against nation rather than army against army, in a war that would achieve

BELOW: A Boer Commando in action in the Transvaal. Through the use of marksmanship and mounted mobility, the Boers were able to hold out against the much more powerful British Imperial forces for nearly three years.

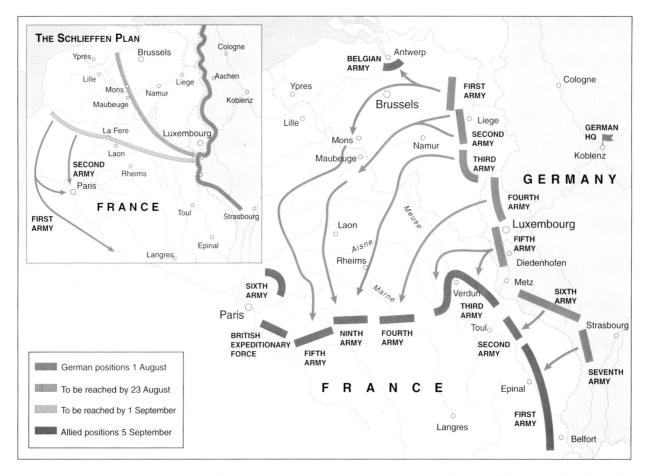

THE SCHLIEFFEN PLAN

German positions 1 August

To be reached by 23 August

To be reached by 1 September

Allied positions 5 September

ABOVE: A German machine gun company moves forward through the Argonne forest early on during World War I. The defensive power of weapons like the MG 08s seen here would become a major factor in infantry tactics to come.

complete victory or utter defeat. It was the factories of Europe and the years of accumulated wealth that made such a conflict possible. The effects of the Industrial Revolution also significantly altered the array of weaponry available to the combatants, having a direct effect on the tactical prosecution of battle. The infantry now carried bolt-action magazine rifles instead of the muzzle-loading smoothbores of old. Though the advance in weaponry made the infantry a more lethal force than ever before, it would be two other new weapons

systems that would come to dominate the battlefields during the Great War, and they were the quick-firing artillery piece and the machine gun.

Perfected by Hiram Maxim in 1883, the machine gun utilized the recoil generated by firing one shot to load the next shell. By the outset of the World War I machine guns were able to fire nearly 600 rounds per minute. War was now mechanized and technological, allowing machine guns to fill the air of the battlefield with deadly steel. Though it is difficult for the modern layman to understand, the generals of the Great War were slow to recognize the revolutionary nature of the machine gun. It had proven deadly in recent conflicts, but most military leaders were unable to admit that the machine gun could replace the fire of carefully trained marksmen. Most believed that in the coming conflict, the machine gun would cause heavy losses, but would prove merely to be an adjunct to the strength of advancing infantry, and they placed a higher value on discipline and military spirit. As it happens, the machine gun would prove more valuable in a defensive role. Working in tandem, the guns, mounted on tripods, could traverse back and forth utilizing interlocking zones of fire to produce a 'beaten zone' that would devastate an infantry force in the open.

DEFENSIVE DOMINANCE

Industrial advances in metallurgy and high explosives also transformed the role of artillery in the Great War. In most previous wars, artillery pieces had aimed over open sights using direct fire. However, stronger barrels and higher muzzle velocities had greatly increased the range of artillery by 1914. In addition, recoil-absorbing devices obviated the need to re-aim artillery pieces after each shot. Finally, the use of high explosive over black powder increased the lethality of each artillery shell. Simply put, artillery could now outrange and outgun the infantry. Firing from miles behind the lines, artillery − including the famous French 75mm (2.95in) field gun, which could fire up to 15 rounds per minute − could decimate attacking infantry forces. Utilizing a frightening array of air-burst shrapnel shells and high explosives, artillery, especially when used in tandem with machine guns, could rain a 'storm of steel' upon any attacker. This new weaponry was destined to make the Great War a struggle in which the defensive dominated, calling for dramatic changes in the tactics used by attacking infantry. To the gentlemen amateur officers of turn-of-the-century militaries, though, such changes were anathema.

For numerous reasons, most military theorists at the turn of the century believed that future conflicts would be quick and decisive in nature. It seemed that modern economies were so interconnected and that industrial war would be so expensive that no nation could stand the strain of war for a long period. Also, the grievous losses inflicted by modern weaponry would force national collapses quicker than ever before. All across Europe military commanders, then, came to believe that the best method to win the coming war was through massive, audacious attack forcing an enemy surrender, and the 'Cult of the Offensive' was born. As war neared, each of the major European powers readied their offensive plans: the Germans had the Schlieffen Plan, the French had Plan 17 and the Russians

had Plan 19. Though they varied in their particulars, the attack plans were all based on the same concept: the war would be won through superior fighting spirit, or what the French called élan; it would be the army that persevered psychologically in the face of heavy losses that would achieve victory; training and discipline would force the infantry through the 'storm of steel'; and only through attacking would their morale be kept high, leading to ultimate victory.

Though much of the planning and staff work that went into the opening phases of World War I was quite detailed − especially the logistical support of the German Schlieffen Plan that called for an massive envelopment of the entire French Army − infantry tactics were rudimentary and essentially linear in nature. Relying in the main on brute force and spirit after a period of manoeuvring, opposing forces planned to meet in battle in thick skirmish lines. With their attendant machine guns and artillery relegated to a supporting role, the massive lines of infantry would blaze away at each other in a great test of wills. The conclusion would come as the morale of one of the forces broke, after which the victors would charge forward with fixed bayonets in order to administer the death blow. Finally, the cavalry forces would gallop in to harry the defeated and retreating foe. In military academies across Europe, officers learned their lessons. World War I would follow a four-step pattern: manoeuvre, locking in battle, the wearing-down fight and, finally, the exploitation.

OPPOSITE: The pre-war Schlieffen Plan for the invasion of Belgium and France, compared with the actual progress of Germany's attack, in which the right wing failed to encircle Paris.

BELOW: After the failure of great offensives in 1914, the combatants on the Western Front began to dig in. The creation of the trench systems would transform the conflict into a bloody four-year stalemate.

Even as the great military minds of Europe dreamed of quick victory, the abilities of modern weaponry available at the time had already tipped the balance of the war to the defensive. There were also some technological shortcomings that worked against the attacker. The battlefields of World War I would be vast, making command and control very difficult. Communications technology, in the form of the telegraph and the telephone, enabled defenders to react with great speed. However, attacking forces had to rely on runners or carrier pigeons to receive their information. Radios were still in their infancy and telephonic communications were too vulnerable. In short, military leaders had little control over their attacking forces once they left their start line. Defenders, though, could react to a changing situation much more quickly. In addition, attackers in the Great War lacked a weapon of exploitation. Cavalry – the weapon of choice in previous wars – proved too vulnerable to the 'storm of steel'. Armour – the weapon of choice in later wars – was not yet up to the task. Thus attackers moved forward only at the pace of marching soldiers. Defenders, though, could utilize lateral rail lines and road communications to shift forces. Simply put, the defenders of World War I could think and react more quickly than the attackers. Thus all of the initial advantages in the war lay with the defender, but every combatant planned to

rely on the strength of the attack. It would be the infantry that paid the price.

THE OUTBREAK OF WAR

In 1914 the nations of Europe stood ready to unleash a cataclysm of war like the world had never seen. It was the titanic clash of arms on the Western Front that would come to represent the epitome of modern, total war and would revolutionize the nature of combat. The French Army, under the command of General Joseph Joffre, numbered some two million men and was augmented by the strength of the 150,000-man British Expeditionary Force (BEF). Following the dictates of the 'Cult of the Offensive' the mighty French force, under the operational orders of Plan 17, attacked eastwards into Alsace and Lorraine. Massed infantry formations, supported by quick-firing 75s and relying on élan, moved forward in linear fashion. Sheer numbers and willpower would be able to overcome the strength of the German defences in the area.

The Germans, in turn, fielded a force of some three million men under the command of General Hemuth von Moltke. German planners realized that the French would attempt to attack into Alsace and Lorraine, states lost to Germany at the close of the Franco-Prussian War, and hoped to use the predictability of the French plan to their advantage. The Schlieffen Plan, developed by Chief of the German General Staff

RIGHT: Soldiers from the Australian and New Zealand Army Corps, or ANZAC, forces rush forward into the attack at Gallipoli. Most military leaders of the time erroneously believed that the spirit, or élan, of attacking infantry could overcome massed defensive firepower, leading to decisive victory.

Alfred von Schlieffen, called for only a
weak defence in Alsace-Lorraine. The
weakness of the defence in this area
would lull the French into complacency
and lure their forces forward away from
their bases of communication. As the
French advanced, the bulk of the German
forces would wheel through Belgium and
into lightly defended northern France.
Thus as the French moved forward, the
Germans would pivot through northern
France, envelop Paris, and attack the
French forces in the south from behind.
Rather like a revolving door, as the
French pushed in Alsace-Lorraine, the
German 'door' would hit them in the
back. Thus Moltke hoped to prosecute a
battle of envelopment, thereby
surrounding and destroying the entire
French Army in less than six weeks.

Initially the Schlieffen Plan worked to
perfection as German infantry thundered
through Belgium and nearly enveloped
the valiant BEF on 23 and 24 August at
the Battle of Mons. Unaware of the
threat, Joffre continued with Plan 17,
against stronger than expected German
resistance. Circumstances and the nature
of infantry warfare in 1914, though, soon
conspired to tip the balance of the
campaign against the Germans. Shaken
and unnerved by the pressure of
command, Moltke committed several
basic military errors. He altered the
Schlieffen Plan and strengthened the
defensive forces in Alsace-Lorraine, both
weakening his attacking forces and

preventing the French from being drawn
forward. Also, frightened by an expected
Russian attack into Germany, Moltke
further weakened his attacking forces by
shifting men to the Eastern Front.
Though German forces were still
advancing on schedule, the weakened
attackers began to slow, and the delicate
logistic support required for such a
massive operation began to falter.

At nearly the same time, General Joffre
belatedly realized the threat posed by the
German advance, and called a halt to Plan
17. Seemingly unperturbed by events, he
began the task of shifting his massive
army northwards to meet the developing

ABOVE: Canadian troops of the 87th Battalion resting in a trench near Willerval. For the ordinary infantryman, trench warfare was rarely better than miserable. In fact, most of the time his life was little more than a squalid and exhausting existence.

quickly than the Germans could advance. The sheer numbers of the French Army and the resilience of a modern nation at war had served to blunt the effect of the continuing string of German victories, successes that might have been decisive in earlier wars. Even with the advantages of surprise and numbers, the German attackers had suffered grievously at the hands of small forces utilizing massed defensive firepower. Thus as September approached, the balance in the north shifted. The Allies now had 41 divisions in the area, while German forces, undersupplied and weakened by exertion, numbered only 25 divisions.

BATTLE OF THE MARNE

On 6 September French and British forces near Paris attacked the flank of the German advance, launching the Battle of the Marne. Far behind the lines and confused by events, Moltke had lost control of the situation, and after a confused, see-saw battle, the Germans chose to retreat to defensive positions along the Aisne River and began to dig in. On 14 November, hoping to score a great victory, Joffre's forces attacked the sketchy German defensive lines along the Chemin des Dames Ridge. Though outnumbered, the Germans, utilizing the defensive prowess of machine guns and artillery, defeated the French advances with relative ease. Searching for an open flank, both the Germans and the Allies began to advance northward in a period of the war called 'the Race to the Sea'. Though the Germans once again attacked the depleted BEF in the savage encounter known as the Battle of First Ypres, the defensive lines once again held. By the end of the year the battle lines had been drawn from the English Channel to Switzerland, and both sides began to construct trench systems. The Great War on the Western Front had become a stalemate, a stalemate running the length of France and involving millions of combatants and the economic outputs of entire nations. The nature of the war had changed. There would be no quick, decisive victory, for the infantry had proven unable to overcome the defensive 'storm of steel'.

German threat to the city of Paris. Utilizing lateral rail lines, Joffre was able to move his forces northwards, reconfiguring his defensive alignments much more quickly than the Germans could advance on foot. In addition, though they had moved from victory to victory, the German infantry had suffered greatly at the hands of the outnumbered Allied forces. Concentrated rifle, machine gun and artillery fire had served to blunt several German assaults, allowing the BEF to escape destruction at Mons and Le Cateau. With their armies weakened and falling behind schedule, the Germans chose to wheel inside of Paris to save time. It was a critical error, offering the flank of the German advance to French forces massing in defence of their capital. The reality of infantry warfare had tipped the tactical balance. The French had been able to react on the defensive more

Across the length of France, both the Germans and the Allies had dug in, hoping to recuperate from their losses, marshal their forces and attack again. Trained for open warfare and yearning for decisive victory, military leaders on both sides originally paid scant attention to their developing defensive lines, but it would be trenches that would come to dominate the remainder of the conflict, and it would be trenches that changed the very nature of warfare.

At first the opposing defensive lines were rudimentary, meant only to hold until the resumption of offensive, manoeuvre warfare. Within a few months, though, the defensive lines were becoming more and more complex, taking on an air of permanence. The trenches themselves, constructed in a defensive zigzag pattern, were usually 8ft (2.4m) deep, honeycombed with large and small underground dugouts that kept their occupants safe from all but a direct hit from a heavy-calibre artillery shell. From the front-line trench, communications trenches led back towards the rear to second- and third-line defensive trenches. Within the trench systems, natural defensive features were converted into redoubts. As a result trench systems in World War I were extensive, often forming confusing rabbit warrens of interlocking defensive emplacements, sometimes of up to 24km (36.8 miles) in extent.

The complicated, nearly invulnerable, trench lines bristled with defensive firepower. Impassable forests of barbed-wire entanglements, thigh-high and sometimes nearly 40m (44yd) in depth, guarded the front-line trench. In addition to the rifle fire of the trench occupants, machine guns – located in highly defended nests and utilizing interlocking zones of fire – stood ready to defend the trenches with countless million rounds of fire. Further to the rear the quick-firing and deadly artillery of the Great War, already sighted in on their targets, were able to cover attackers with a steel rain of high explosive and deadly shrapnel. Thus the defensive works of World War I were quite complicated and quite deadly, much more so than the modern-day military commander, aided by a vast array of new offensive weaponry, could understand.

BELOW: Over the Top! British soldiers emerge from their trenches to move forward into no-man's-land. Survival was often a matter of chance in the face of enemy artillery and machine guns, but the Great War added a new horror, poison gas, which was used on an large scale for the first time.

The Great War had settled down into a massive stalemate on the Western Front, though in Russia and in other far-flung theatres the conflict remained more fluid in nature. Commanders in the West began to wrestle with a problem that seemingly defied solution. Wars are, in the main, won by attacking. How, then, were attackers to overcome the defensive works of their enemy and achieve the long-awaited decisive victory?

The Germans, under their new commander Erich von Falkenhayen, chose to stand on the defensive in the West while concentrating their efforts against the Russians. It was, then, the Allies who would spend 1915 in search of answers to the riddle of trench warfare. In a series of battles spanning much of the Western Front, including the Battle of Neuve Chapelle, it became apparent to most Allied commanders, including Joffre and General Sir Douglas Haig – who would soon take command of the BEF – that it would not be infantry tactics that would reign supreme during the Great War.

The infantryman, armed with his rifle and bayonet, could do but little against an enemy in trenches guarded by barbed wire and defended by machine guns and artillery. The infantry was too vulnerable and carried too little firepower to have a meaningful effect on the outcome of the coming battles. Though some retained hope of its usefulness, the cavalry too proved to be of little value against an entrenched foe. Nor were machine guns, powerful but not yet portable, the answer for those seeking offensive victory. In the end it became obvious to most that the only weapon in the attacker's arsenal that was capable of defeating these defensive trench systems was artillery.

It, too, had a hidden flaw. Firing from miles behind the lines and using indirect fire, for the first time artillery had to be able to hit what it could not see. Trenches, dug-outs and machine gun nests are – especially from 11 miles (17.6km) away – small targets which require direct hits to ensure their destruction. Simply put, the artillerists of the Great War did not yet posses the technical expertise or the communications technology to achieve such accuracy. During this phase of World War I, artillery could destroy large targets, such as massed infantry attempting to cross no man's land, but was not yet able to strike particular targets that it could not see. Even so, it would be artillery that would come to dominate much of the offensive action of the Great War.

THE SOMME

During 1916 on the Western Front both the Germans and the British sought to score decisive victories aimed at shattering the opposing trench systems and thereby restoring a war of movement. Both commanders, Falkenhayen and Haig, relied on the dominance of artillery to achieve their proposed victories, sparing only scant thought for the tactics of their massive, but outmatched, infantry formations. The resulting struggles at Verdun and the Somme would become the signature battles of the Great War, in the main giving the conflict its reputation for futile slaughter. Perhaps the greatest single

BELOW: British soldiers examine the sighting mechanism of an American-built Browning machine gun. Such weapons, able to fire over 750 rounds per minute, meant that the time-honoured tactic of advancing in line abreast had become suicidal. Even so, it took commanders several years and millions of lives to realize that fact.

battle of World War I, Verdun lasted from 21 February to 18 December, claiming over 800,000 casualties in a horrific battle of attrition designed to break the French will to fight. Falkenhayen hoped to pit German steel against French flesh, calling down a hailstorm of artillery fire upon the defenders of the embattled city. In many ways, though, it was the Battle of the Somme that best illustrates the state of warfare existing in 1916.

The British, having raised their first truly mass army, hoped to use their newfound might to rupture the German defensive system near the River Somme, partly in an effort to aid their embattled ally. The commander of the BEF, General Haig, planned to win victory at the Somme through sheer weight of artillery fire. Towards this end, the BEF gathered together some 1400 artillery pieces and countless millions of artillery shells. Victory would rest upon the efforts of but one branch of the armed forces. The artillery had to flatten the German barbed wire, destroy the German trench system as well as its inhabitants, and finally silence the German artillery. The infantry would simply advance in the wake of the pulverizing barrage and occupy the shattered and defenceless German trenches. Advancing shoulder to shoulder in waves, the infantry would walk across no man's land, using discipline, courage and the strength of sheer numbers to overcome any German defensive fire that remained. Thus at this stage of World War I, infantry tactics harked back to the linear tactics used in the days of Frederick the Great. Artillery had become the queen of battle, leaving the infantry as its adjunct.

On 24 June 1916 the British artillery launched its hurricane of fire, beginning a week-long bombardment in which some one-and-a-half million shells rained down on the German trenches. Though the bombardment was quite impressive, and convinced many of the British soldiers that nobody in the German trenches would emerge alive, the barrage failed to achieve any of its main goals. Around 1000 of the guns that took part in the bombardment were field guns, which were not powerful enough either to destroy trenches or duel with German artillery. In addition, nearly a million of the shells fired in the bombardment were

air-burst shrapnel shells, effective against exposed targets but of little value against trenches and dug-outs. Thus only 400 artillery pieces firing 500,000 shells were assigned to do the bulk of the damage, not nearly enough to complete the task. Also, since Britain's industry had only recently begun shell-making, 30 per cent of the shells used in the bombardment were faulty, possibly even duds. Finally, given the inaccuracy of indirect fire, only 2 shells in every 100 were direct hits; the others were loud but ineffective.

ABOVE: British soldiers firing a Lewis Gun, a truly portable machine gun, which would serve to strengthen the role of attacking infantry.

BELOW: German infantry training for trench raids. Clubs, bayonets, pistols and other close-combat weapons were ideal for the vicious man-to-man fighting typical of the trenches.

ABOVE: A British soldier peers warily out into no-mans-land through his trench periscope. Trench raids were a constant threat as both sides sent out intelligence-gathering patrols, whose aim was to capture prisoners who could be grilled during interrogation.

As a result, when the bombardment ceased, though frightened, most of the German defenders were very much alive as the British infantry went 'over the top', dressed their lines and began walking towards their foe. The Germans readied their machine guns and artillery and called down the storm of steel. Though gallant, the British attackers, armed in the main with rifles, could do little against such defensive firepower and an entrenched enemy. Only on the southern portion of the front did the British infantry make substantial gains, while along most other parts of the line, they did not even reach the German front-line trenches. The effort to achieve a crushing victory through weight of artillery fire

had failed, and nearly 57,000 British soldiers had fallen in a single day. As a result, the Battle of the Somme settled down into a six-month struggle of attrition that eventually claimed one million casualties. Though the British made alterations to their offensive schemes, including the introduction of the tank, the reality of the Great War remained unchanged. The defenders were still able to out-think and outperform their attackers. Much would have to change to tip the balance of futility.

TECHNOLOGICAL CHANGES

Many historians and military history aficionados fail to look much past the Somme in their study of the Great War. The remainder of the conflict seems to be a great mass of trench struggles representing the height of military stupidity and a dearth of tactical development. Victory came only after years of attrition, the introduction of fresh American soldiers and the onset of revolution. The truth is, as usual, much more complicated than this.

The years 1917 and 1918 in fact represent a sea change in modern warfare, and a compressed period of tactical innovation rarely seen. During a period that many Great War historians simply call 'the Learning Curve', commanders on both sides of the front lines began to rethink the very nature of modern warfare. As a result, technological changes and tactical innovations would come to solve the riddle of the dominance of the defensive. Warfare shifted dramatically, from being a 'great game', practised by gentlemen amateurs, to being technical and tactical, practised by the first generation of true professionals. War, now industrial and total, had entered the 20th century and had become modern.

Revealing the complete reliance upon the nation in total war, many of the changes that would alter the Great War were advances made by scientists and engineers rather than by military men. Where in 1916 infantry had been nearly powerless against an entrenched foe, new inventions by 1918 had once again made the infantry powerful. The best-known such invention was the tank, but it was, in some ways, at the time the least important. Lightly armed and slow-moving, the tank was in its infancy, and though it would prove a valuable adjunct it was not yet a dominant weapon of war.

It was the more mundane developments that made the infantry powerful, enabling it to work with the other arms of the military and restore balance and movement to the battlefield. New, portable machine guns – including the famous British Lewis gun – added to the infantry's firepower. In addition, the infantry now had its own portable artillery in the form of modern hand grenades and mortars. This new weaponry enabled the infantry to deal with defensive emplacements on its own, rather than having to wait for artillery support. In addition, the infantry also possessed other, more case-specific weaponry, including bangalore torpedoes and the intimidating flamethrowers.

Other technological developments served to help bring the various components of the modern military into closer battlefield harmony. Military aircraft which, like the tank, would come to revolutionize war, saw significant action in the Great War, but had their greatest value in a reconnaissance role. Coupled with advances in aerial photography and radio communications, aircraft were able by 1918 accurately to locate enemy defensive works and serve as artillery spotters, heralding a military communications revolution that is embodied today by the microchip.

Scientific developments more specific to artillery also began to tip the balance in favour of the attacker. In 1916 artillery could not hit what it could not see. By 1918 much had been learned about the effects of barrel wear and weather on shell trajectory. Techniques known as 'flash spotting' and 'sound ranging' were able to pinpoint the location of enemy artillery pieces. Aerial photography had provided artillerists with accurate maps of

OPPOSITE: A German soldier hurls a hand grenade over a barbed-wire entanglement in 1917. Along with the light machine gun, grenades became a vital tool for the World War I infantryman during the bitter, short-range combat in the trenches, helping clear trenches of enemy opposition.

BELOW: An Austrian trench mortar fires against Italian forces in the rugged, mountainous terrain of the Italian Front. This simple weapon gave infantry units their own short-range fire support, and it could be used to drop mortar bombs directly into enemy trenches.

enemy defences. When coupled with relentless on-the-job training such technological advances had transformed the power of artillery. By 1918 artillery was able to hit what it could not see with deadly accuracy, with no preregistration. Thus trenches and dug-outs were no longer safe havens of defence. In barely two years of innovation, artillery had become modern – a science and no longer an art – establishing practices which are still prevalent today.

As the balance of the war began to shift, military leaders in every combatant nation struggled to develop tactics that – utilizing the new weaponry – would bring the longed-for victory. Though all sides experimented in their offensive tactics, it is the Germans who usually receive credit for updating the old, linear system and thus showing the way forward into modern war.

Having witnessed daring new Russian infantry tactics on the Eastern Front, General Oskar Hutier began to codify a new attack formula that would come to bear his name. His tactical scheme called for the newly powerful infantry to advance in wedge formations and in short bursts rather than in slow, ponderous waves. Highly trained élite soldiers, referred to as storm troops, would lead the assault. With their new firepower and communications abilities the storm troops would probe the enemy lines for weak points.

Once located, the troops could use their own weaponry to achieve a breech in the line, with the goal of advancing to tactical depth. No longer was it necessary to attempt to overthrow the entire enemy defensive system utilizing the brute force of great numbers. The quickly advancing storm troops would attempt to disrupt the enemy defensive system by striking at supporting artillery and command centres. In many ways the new style of warfare was *Blitzkrieg* without tanks. The enemy defences were now seen as a system. It was the job of the storm troops to short-circuit the brain of the system rather than to batter the body. Though the Germans were the first to codify 'Hutier Tactics', they had learned much of the way forward from Russian innovations, ones that would become very important in the World War II and the Cold War. At the same time the Western Allies were also moving along the learning curve, readying their own advanced tactics for 1918.

THE BIRTH OF MODERN WARFARE

The fighting on the Western Front in 1918 – all too often ignored by military historians – represents the culmination of a revolutionary period in the history of warfare. A far cry from the blundering, gentlemen amateurs of 1914, the military practitioners of 1918 laid the foundations for every war since. Blitzkrieg, Soviet Deep Battle and even Coalition forces in

the Gulf War trace their origins back to the Western Front at the close of the Great War. In four short years, warfare had come of age.

The Germans, under the command of General Paul von Hindenburg and his Chief of Staff General Erich von Ludendorff, unleashed their new style of warfare on 21 March 1918 in a last, desperate bid to achieve total victory. The surprised men of the British 5th Army suffered greatly in the assault, very nearly breaking. With the aid of an accurate artillery bombardment, dubbed a 'fire waltz', German forces advanced nearly 7 miles (11km) before nightfall, an unheard-of development for the Great War. By 6 April the Germans had advanced nearly 40 miles (64.3km), and victory seemed to beckon.

At this point, however, the realities of the Great War intervened. As Joffre had before, Haig and General Philippe Pétain, the new French commander, were able to rush reinforcements to the scene more quickly than the now-exhausted Germans could advance on foot. Thus the Allied lines finally held, leaving the Germans to defend a vulnerable salient having suffered over 200,000 casualties, a total that they could ill afford after nearly

five years of war. Undaunted, the Germans went on to launch a further series of offensives, leading to very similar results and culminating in the Second Battle of the Marne. Thus the Germans had proven the strength of their new tactics, but were still unable to solve the overall riddle of trench warfare.

The Allies had also been experimenting with new tactics, as evidenced by the first massed use of tanks in the Battle of Cambrai in November 1917. Even as the

ABOVE: The arrival of the US Army in 1918 brought new strength to the Allied cause. These members of the 369th Regiment, 93rd Division – one of two segregated 'Colored' Divisions in France – are seen in the Argonne region on 4 May 1918.

ABOVE: A line of Renault FT-17 light tanks creeps through the French countryside near the end of the Great War. Although the tank was far from being a war-winning weapon at this stage, it had an immense psychological effect on defending infantry, and far-sighted pioneers were already thinking about how it should be used in future. It was the basis for the Russian T-series of tanks, the most widely produced of which was the famous T-34.

Germans attacked, the British were planning to put their new tactics to the test yet again. Even more so than the Germans, the British planned to use unprecedented levels of all-arms coordination during their coming offensive, giving each facet of the military its role to play in a resounding, harmonious symphony of war.

The artillery, which had been called upon to win battles in 1916, was to prepare the way. Accurate and powerful, the artillery had only to keep the Germans under cover during the infantry advance. Using the new infiltration tactics the British infantry would then advance to depth, accompanied by over 400 lumbering tanks, a weapons system that the Germans had virtually ignored. Though the tanks would not be decisive, they could deal with enemy strongpoints and machine-gun nests. Overhead, Allied air superiority would communicate the progress of the battle to the other facets of the military, as well as engage in a ground-attack role in order to interdict any enemy movement. Much had changed since the stalemate trench-warfare tactics of 1916.

The British forces, aided by the arrival of masses of American troops and a slow decline in German morale, on 8 August rushed forward into the Battle of Amiens. Within four days, the BEF had advanced

over 15 miles (24km), but their pace suddenly began to slow. Once again the defenders were able to move laterally more quickly than the attackers could move forward, the same problem the Germans had faced in the spring. At this point, Haig chose to halt the offensive, unlike Ludendorff before him, who had pressed his troops to breaking point. The rules of World War I dictated that there would be no breakthrough, and Haig had learned his lesson. Instead British forces attacked a week later further north, forcing the Germans to react by shifting their dwindling reinforcements in that direction. Once that attack had run its predictable course, it was abandoned, leading to a French attack further south and another German reaction.

The Allies had stumbled onto the operational level of warfare. A tactical victory in a single battle would not lead to a strategic victory in the war. However, a series of related tactical victories – ripping off great chunks of the German defensive network and forcing continued German troop movements – could lead to a strategic victory. In the '100 Days' battles of 1918 each tactical victory led logically to the next tactical victory, forcing the Germans further and further back towards their border and even breaching the vaunted Hindenburg Line in a matter of days.

Battered by continuing military defeat and haunted by famine and revolution at home, on 11 November 1918 the Germans quit World War I. Their defeat was not total, and the deeply flawed Treaty of Versailles would help to assure that the Great War was but the first total war for European and world dominance.

The Great War, far from being a tactical wasteland, had seen the military systems of the world change from lumbering linear forces into the deadly modern forces that would so devastate the globe in the next conflict. In a four-year period of innovation, modern warfare had been born. Commanders now presided over technological, professional forces. Weaponry had been altered for ever by science, making war more efficient and more deadly. Many now saw armies as systems, vulnerable to deep penetrations. Others began to concentrate on the newly discovered operational level of war. Modern, industrialized nations seemed nearly impervious to decisive battlefield defeats. Only a series of coordinated battlefield efforts would suffice to win victory in a total war.

The new weapons, tactics and operational discoveries of the Great War would be the subject of many debates during the interwar period, especially among the vanquished. The victors, who had done so much to revolutionize warfare, would ignore the lessons of the Great War at their peril.

ABOVE: British Mark V heavy tanks of the 8th Tank Battalion move forward in the Battle of the St Quentin Canal. At this stage, tanks were seen primarily as infantry support weapons. Three of the tanks carry fascines, bundles of sticks used to help in crossing enemy trenches.

LEFT: In the US Army's first offensive of the Great War, American troops rush forward in order to launch an attack on the trenches of the veteran German Eighteenth Army at Cantigny. The soldier at bottom centre carries an example of the terrifying flamethrower; new weaponry and tactics had changed the very nature of infantry combat.

CHAPTER TWO

NEW THINKING

The interwar period saw many experiments in strategy and tactics. Among the most far-reaching were those involving mechanization and armour, both of which would impact on the role of the infantryman.

The nature of trench warfare during World War I led to high casualties each time armies clashed on the battlefield. For example, the 1916 Battle of Verdun, which lasted nearly 10 months, resulted in almost 800,000 casualties. On 1 July 1916, the first day of the Battle of the Somme, British casualties numbered over 57,000. By the time the battle ended six months later, the British had suffered 420,000 casualties, while the French lost 195,000 men and the Germans 650,000. The combined losses totalled over one million soldiers. When the war ended, the total casualties – killed, wounded, prisoners and missing – for the four-year conflict exceeded 37 million. While the Allied Powers suffered over 22 million casualties, the Central Powers lost over 15 million troops. These losses would have a huge impact on both the peace process and the nature of the military establishments in 1939 when war broke out again.

On 11 November 1918, an armistice ended the Great War. Following the cessation of hostilities, the leaders of the participating countries met at Versailles to negotiate a peace treaty. President Woodrow Wilson of the United States, Prime Minister David Lloyd George of Great Britain, Premier Georges Clemenceau of France,

LEFT: Technology revolutionized tactics and strategies between the wars. These Guardsmen, training at Pirbright, England, would become part of the reorganized British Army, which was the first to transform infantry units into machine-gun and rifle battalions, and mount them in motor vehicles.

ABOVE: A key to the new tactics was mobility. The British Army tested an all-mechanized force in the 1920s, and by the 1930s, reconnaissance units, like this detachment from the Northumberland Fusiliers, had finally traded in horses for motor vehicles.

and Prime Minister Vittorio Orlando of Italy tried to solve the world's problems at the Versailles Peace Conference, and they hoped that their solution would prevent future wars. At the same time, military leaders began to analyze the previous four years in order to develop new tactics designed to prevent the atrocities of trench warfare from being repeated in future conflicts.

Not all officers were ready to accept change, however. No one wanted a repeat of the horrific losses of the previous four years. Many political leaders and ordinary civilians thought that the solution lay in abolishing war; military leaders looked for a new weapon that would allow the quick defeat of an enemy. Some looked to the development of air power and strategic bombing, and others looked to the formation of tank divisions to reduce casualties, but most did not think that the infantry would become obsolete. Rather the infantry would play a secondary role to the air force or the tanks.

Although the military doctrines and organizations of the major powers

evolved between the world wars, a number of factors hindered the evolution of new weapons and tactics. A spirit of pacifism permeated the interwar period. The citizens of many nations rejected war and everything associated with the military. In addition, many nations felt the need to formalize their condemnation of war. Consequently, 15 countries signed the Kellogg–Briand Pact in 1928. A further 45 nations would add their names to the agreement, which stated that war was acceptable only for the self-defence of a nation.

Furthermore, governments, bowing to pressure both from their own citizens and from abroad, became increasingly reluctant to allocate funds for military research. A series of international conferences took place in the 1920s and 1930s. The purpose of these was to restrict the armaments of the militaries and navies of the major powers. Although not entirely successful, the delegates at the Washington Naval Conference negotiated restrictions on the number of capital ships that the participating countries could maintain. Some countries, such as the United States, downsized their armies by choice; others, like Germany, reduced the size of their armies because of the terms of the Versailles Peace Treaty.

ECONOMIC CONSTRAINTS
The United States and other countries could not justify the allocation of funds to develop new weapons when they had stockpiles of weapons and ammunition left over from World War I. By the time armies had depleted the stockpiles or could no longer use ammunition from the war, economic constraints impeded the production of new ones. While some governments were willing to finance the development and testing of new tactics and weapons, the Great Depression severely limited the availability of funds for tanks, aircraft, and other new equipment. Despite these limitations the mechanization of industry as well as the technological advances of the 1920s and 1930s would affect the conduct of the next war.

During the interwar period, the implementation of mass-production techniques increased industrial output and reduced costs. At the same time, the exploitation of raw materials, the construction of refineries and power plants, and agricultural productivity also grew. Scientific experimentation resulted in improvements in aviation, vehicles and communication. Better vehicles meant better cross-country transport abilities, the increased protection, firepower and speed of tanks and other armoured fighting vehicles, and the possibility of the rapid transportation of troops by means other than rail. All of these improvements would affect the nature of the next conflict.

There was, however, a downside to the rapid changes in technology. Concerned that the speed of these changes would quickly render them outdated, governments hesitated to invest in existing designs. In addition, officials had no way of predicting the effect of these technological advances on military tactics. Financial constraints frequently prevented the testing of new equipment before it was placed in the field. While many of the changes suggested the development of mechanized warfare, the leading military authorities did not always concur with this theory. The proponents of traditional combat arms resisted the push for tactical changes. The evolution of military tactics and weaponry proceeded at a different pace in each country.

By the end of World War I, the participating armies had undergone great changes. None of them were the same as they had been at the beginning of the conflict. By 1918 the most professional force in the field was the British Army, whose officers had been the first to understand the importance of operational thinking and the development of armoured equipment and doctrine. While other military officials had a narrow view of the uses of tanks in the field, British commanders took the lead in testing expanded roles for mechanized formations. Several British officers developed combined-arms tactics, beginning in 1918. Unlike his colleagues, however, Colonel J. F. C. Fuller placed much more emphasis on the role of tanks operating alone, rather than in conjunction with the infantry. In his Plan 1919, Fuller, drawing on German tactics, designed a large-scale armoured offensive, supported by air attacks against supply lines, that would result in a series of breakthroughs followed by the disruption of the enemy's headquarters. Once the enemy's command structure had been disrupted, British forces would launch follow-up attacks against the German front line. Unfortunately for Fuller, the war ended before he could convince his superiors to implement his plan.

BELOW: British experiments in mechanization proved the value of motor vehicles, and the United Kingdom was the first major power whose army switched completely from horse-drawn transport to internal combustion power. These British infantrymen pass in review, mounted in tracked personnel carriers developed from Carden-Loyd tankettes.

emphasis after the war was on the protection of the British Empire, few government officials supported Fuller's expensive plan. In addition, the tank programme, like those of the Royal Navy and the Royal Air Force, suffered from defence-spending cuts during the depression. Resistance to innovation impeded the tactical development of the British Army during the interwar period.

Because of their experiences with trench warfare, British commanders were determined to develop tactics that would return manoeuvrability to the battlefield. Consequently, they recognized the importance of mechanization. The establishment of the Royal Tank Corps in 1923 reflected the desire for change in order to build on the tank successes of World War I. In 1927 the British Army formed the Experimental Mechanical Force (EMF), which was one of the first all-arms mechanized formations, consisting of light and medium tanks and motorized machine-gunners, artillery and engineers. The EMF conducted brigade-level exercises that year and again the following year. In 1928, however, responding to opposition within the military community and budgetary problems, the British War Office disbanded the EMF.

Although both the Germans and the Soviets studied them, Fuller's ideas did not gain wide acceptance in Great Britain. Conventional infantry commanders, although willing to utilize tank and air support, resisted the relegation of their troops to a minor role in future operations. Because the main

British Army officers continued to debate the role of tanks in future

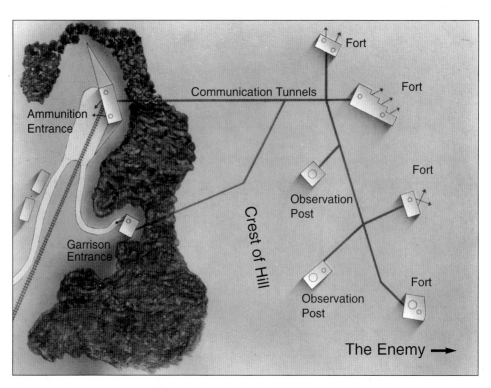

Fort

Fort

Communication Tunnels

Ammunition Entrance

Garrison Entrance

Crest of Hill

Fort

Observation Post

Fort

Fort

Observation Post

The Enemy ⟶

conflicts. Although the EMF had been dissolved, its exercises in 1927 and 1928 had demonstrated the problem that resulted from the formation of a unit consisting of various types of motorized vehicles. Because some vehicles had tracks and others had wheels, they could only advance as a unit at slow speeds, which limited the EMF's ability to move rapidly. As a result, tank proponents began to accept one of Fuller's proposals: the creation of pure-tank units. The Royal Tank Corps began to press for the development of new tanks and new tank tactics.

The British military community did acknowledge, however, that the creation of tank formations would not eliminate the need for infantry units. A tank revolutionary, Percy Hobart, whose floating tanks landed on the Normandy beaches in 1944, advocated a primary role for tanks in future operations. Not all tank advocates went as far as Fuller and Hobart. George Lindsay, like those who supported more moderate change, suggested that tanks participate in combined-arms formations with support from mechanized infantry, artillery and other service. Some senior British military officials endorsed the use of tanks only as infantry support, while others, stressing their vulnerability, objected to the creation of independent armoured units. Despite their First World War experiences, die-hard cavalry supporters did not believe that tanks would replace horses on the battlefield.

As the debate continued within the British Army, another armour theorist, Basil Liddell Hart, began to voice his opinion. An infantry officer during the Great War, Liddell Hart was influenced by Fuller's writings. He left the army in 1925 and pursued a career as a military correspondent, which enabled him to exert influence over the British Government in military matters. Unlike Fuller, however, Liddell Hart advocated a combined-arms force, with a large role for mechanized infantry; but he did not reject all of the theorist's ideas. In fact, he went beyond them. Liddell Hart expanded Fuller's proposal to use tanks in order to break through enemy lines. He envisaged the use of tanks to exploit the breakthrough and to advance deep behind enemy lines in order to disrupt its command centre.

PROPHETS WITHOUT HONOUR?

Although their ideas influenced the British Army, cuts in defence-spending by the late 1920s and 1930s, as well as the push for empire defence and the rejection of future continental conflicts, caused both Fuller and Liddell Hart to reconsider their positions. While Fuller began to focus more on strategic bombing, and less on tanks, to play the major role in future conflicts, Liddell Hart began to question whether or not armoured warfare could succeed against strong, well-prepared fortifications. Despite his uncertainty, Liddell Hart continued to influence those in power throughout the 1930s because of his position as private military adviser to Leslie Hore-Belisha, Secretary of State for War in the Chamberlain government.

LEFT: French soldiers in one of the Maginot Line's underground complexes in 1939. Unfortunately for French planners, in May 1940 the Germans declined to make a frontal assault on the line. Sweeping through Belgium and around the end of the defences, they isolated more than 400,000 French troops in their massively expensive but useless fortifications.

He argued against the deployment of troops on the Continent, advising that the government divert funds from ground troops to the improvement of the country's air defences. Liddell Hart's advice reflected growing concerns within the military community about the threat of strategic bombing.

The debate among infantry and tank commanders over the role of tanks did not prevent the formation of a division-sized independent mechanized force in 1934. Insufficient training, command disputes and other problems resulted in a poor showing by the force in its first armoured exercise. Following a change in command, the British mechanized a large percentage of its cavalry to create a Mobile Division in December 1937. The Mobile Division included two armoured cavalry brigades consisting of light tanks and armoured cars, one previously established tank brigade, two mechanized infantry battalions, and artillery, engineer and support formations. Although not fully equipped, the Mobile Division contained 500, mostly light, tanks. Without adequate guidance from their superiors, however, the cavalry officers in charge of the division decided to use it in the same way that they would use a cavalry division: for reconnaissance, screening and security. Within two years, the newly reorganized Mobile Division became the 1st Armoured Division.

Although the British developed tanks and combined-arms tactics, they failed to commence a study of World War I's lessons until 1932. The Army created a committee, which issued a report of its findings, but the chief of the imperial general staff did not allow it to circulate within the Army. In the late 1920s and 1930s the British Army implemented a series of experiments in order to analyze the potential of armoured warfare. These demonstrated the capability of tanks, as well as their limitations. As they developed tank and infantry tactics to be used in the next conflict, the British failed to incorporate the lessons from these experiments or those that they should have learned from the previous war. They did, however, restore the infantry's firepower and mobility with new equipment and organization, but the improvements imposed a cost. The British infantry battalion, although more mobile, had less firepower and limited antitank capabilities. By 1939, even though the British Army had developed new equipment and technology, it failed to devise tactics for using them effectively in the field. With the exception of the infantry battalion, combat arms and different weapons systems cooperation had improved little over what it had been in 1914.

FRENCH DEVELOPMENTS

During the 1920s and 1930s, while the British Army struggled to change despite economic, government and public constraints, the French faced different problems. According to the terms of the Versailles Treaty, Germany could maintain a 100,000-man professional army. As long as a German Army existed, France could not trust its former enemy, nor could it feel safe. Consequently, the French Army not only had to prepare for future conflicts, but also had to develop strong defensive capabilities to thwart an unexpected invasion. During the interwar period, although the French had a large army, it had little hope of deterring a sudden German attack because it consisted mainly of insufficiently trained reservists. France, like many other countries, did not want to invest in revitalizing its weary army, but believed that a German threat still existed. It was not a question of if Germany would attack again, but when. The French therefore allocated time and resources to the construction of a series of self-contained armed concrete forts: the Maginot Line. Recognizing how long it would take to mobilize their

BELOW: Japanese expansionism on the mainland of Asia meant that her troops saw combat a number of times during the 1930s. Here, gas-masked Japanese Marines fight through the rubble caused by the Imperial Air Force's bombardment of Shanghai, after the outbreak of war with China in 1937.

LEFT: The Soviets were among the pioneers of airborne operations. The Red Army was the first to deliver battalion and regimental-sized formations of infantry from the air, dropping paratroopers from TB-3 aircraft in military exercises through the 1930s. Ignored by most western armies, the exercises were avidly studied by the Germans.

reservists, French officers hoped to force the Germans to attack through Belgium, which would then give the army more time to prepare.

Feeling a sense of responsibility for their ally, French military strategists expected to come to Belgium's aid in the future; they therefore planned to fight the next war in Belgium, not at home. Unfortunately, Belgium's declaration of neutrality in 1936 prevented both the French and the British from deploying troops before the Germans invaded. Therefore, the French, in conjunction with the British, drew up plans for the rapid deployment of forces into Belgium and the Netherlands in the event of a German invasion. The French, responding to an emphasis by the military on mobility, allocated funds for the development of armoured forces. Because they considered them armoured infantry, the French commanders deployed their tanks throughout the army rather than in large independent formations. As a result, the role of tanks would be restricted initially to close infantry support. Because of their subordination to the infantry, the tanks' rate of advance was tied to the infantry.

By the early 1930s Maxime Weygand, the Chief of Staff, pushed for motorization and mechanization, despite limited funding. What with the Depression and the cost of constructing the Maginot Line, the military in general suffered from funding cuts. Because of

Weygand's fervent commitment to modernization, however, the military motorized seven infantry divisions and provided half-tracks and armoured cars for several cavalry divisions. In 1934 Weygand ordered the formation of the first light mechanized division, the *Division Légère Mécanique* (DLM). By the late 1930s, the French had established four DLMs. Because of his background as a cavalryman, Weygand assigned reconnaissance and security missions to the DLMs, rather than battlefield tasks.

In 1934 Lieutenant Charles de Gaulle's writings almost sabotaged the French Army's move toward modernization: he

ABOVE: Japanese control of Manchuria brought them into contact with the USSR. In 1938 and 1939, border disputes grew into wars in all but name, in which the Soviets had rather the better of things. These Japanese troops in Manchuria are about 50km (31 miles) from the combat zone on the Khalkin Gol river.

advocated the creation of an independent tank brigade that would advance in linear formation. A motorized infantry unit would follow the tank brigade for the purposes of mopping up. At a time when French citizens were promoting peace and the idea of citizen soldiers, de Gaulle proposed a large professional standing army that many feared could be used to start a war with Germany. Not willing to accept the slow modernization of the French Army, de Gaulle pushed for the rapid development of armour and tactics. The resulting debate contributed to the lack of French preparation for 1940. Although they had almost as many tanks as the Germans by 1940, the French did not establish their first armoured divisions until after war erupted. At that time the French Army consisted of a large, untrained, poorly organized militia and a small mechanized force. It was not until March 1940 that the French issued a regulation for large armoured-unit tactics. A few weeks later, German troops invaded France.

While pacifism entered the public debate in Great Britain and France, the Bolsheviks in the Soviet Union commenced preparations for armed conflict almost as soon as they came to power. Because of the ideological differences between Communism and capitalism, they believed that war was inevitable. Only total mobilization of their resources would result in a victory for the Soviets.

Despite the willingness to prepare, the actuality was somewhat different. In the 1920s the Soviets slowly developed their military capabilities. Following the industrial expansion plan of 1929, the armed forces, which grew in size, experienced rapid increases in the production of tanks and aircraft.

By 1928 Soviet military leaders had developed the 'Deep Battle' doctrine. According to Soviet military philosophy, a series of connected operational successes would lead ultimately to strategic victory. Despite the development of a doctrine, Soviet leaders did not agree on the nature of the future Red Army. An armour advocate, Mikhail N. Tukhachevsky, persuaded his superiors to create the army's first armoured divisions in 1931. Tukhachevsky, the guiding force behind the 'Deep Battle' doctrine, envisaged the deployment of conventional infantry and cavalry divisions, mechanized formations and aircraft, all working together to achieve the same goal:victory. According to Tukhachevsky's plan, the infantry working in concert with the other arms and weapons systems would launch a two-part battle. In the first phase, a massed force led by tanks would attack the enemy's front line on a narrow front and achieve a breakthrough of his conventional defences. Soviet artillery and mortars would silence the enemy's artillery and antitank guns. During the second phase, the rate of the tanks' advance would not be tied to that of the infantry. The tanks could exploit the hole in the enemy line to advance and scatter his reserves by launching small, deep attacks. The mobile elements of the force would move quickly in order to outflank the enemy or in order to penetrate the enemy's rear areas. Long-range artillery fire, bombing, parachute attacks, smoke and deception operations would support the main operation.

ARMOUR OR CONSCRIPTS

Despite the new doctrine, there was some opposition within the Soviet military leadership to devoting resources to the development of mechanized units at the expense of the infantry. Many still supported the concept of a large conscript army. As a result, the Soviets attempted to accomplish both the maintenance of a large conscript army and the development of a large armoured force.

Josef Stalin dealt the Soviet military a heavy blow in 1937 when he unleashed his secret police against his enemies. Most of the proponents of modernization, including Tukhachevsky, died during this purge. By 1939 the state had begun to dismantle the armoured division, and at a time when German panzer divisions were achieving their first successes.

In the United States, however, pacifism and the desire for isolation had an impact on the evolution of the military during the interwar period. Although some occupation troops remained in Austria until 1919 and in Germany until early 1923, most American servicemen quickly returned to the US and their families. The government authorized the reduction of America's armed forces. In 1920 Congress passed the National Defence Act which, had it been implemented, would have established the framework for the post-war army. In the 1920s the US military, like that in Great Britain, suffered from budget cuts. After the Depression hit in 1929, military spending declined even more. Despite the reduction in funding, US military commanders, like their European counterparts, strove to improve their armed forces.

The United States Army, when it entered the field in 1918, had a unique organization. The infantry had a square division structure. By 1920, a number of American officers, including Generals Fox Conner and John Pershing, recommended the structural reorganization of infantry divisions. Copying the European triangular divisions, Pershing suggested the streamlining of the divisions and the development of new tactics. The majority of Pershing's proposals became the victims of infighting with his colleagues in the War Department; therefore, the structure of the square division remained virtually unchanged throughout the 1920s. Although an emphasis on officer training returned to mobile warfare in the mid-1920s, major alteration of equipment and organization did not occur until the mid-1930s because of budget cuts and the overall neglect of the army.

BELOW: Defeated armies often learn more from a battle than the victors, and by the time Hitler came to power in Germany, the new German Army was ready to revolutionize warfare. Based on their infiltration tactics of 1918, to which was added mobile armour and motorization of the infantry, their new idea of warfare was called *Blitzkrieg*.

SHORTSIGHTED AMERICANS

While the US Army focused on the infantry, it neglected the development of mechanized and armoured forces. The subordination of the tank forces to the infantry resulted in a plan to use the tanks as support for the infantry in breakthrough attacks. By the end of the 1920s, British experiments with mechanized forces began to have an impact on the US Army, which conducted its own tests with Experimental Armoured Forces (EAF) in 1928 and 1929. Although a permanent force was created the following year, the Depression caused budget cuts that forced its elimination. Chief of Staff Douglas MacArthur (1930–35) continued to push for army-wide mechanization and motorization, but the budgetary problems forced him to limit his mechanized experiments to the cavalry units.

In 1935 the situation began to change when General Malin Craig, who agreed with the reforms advocated by Conner and others, became the US Army's Chief of Staff. Craig immediately ordered a General Staff board to reassess the army's combat organization and tactics. He also suggested that the board consider the formation of small, mobile divisions that could use mechanical rather than human power. In its report the board suggested that the divisions be reorganized along the lines used by European armies. On numerous occasions between 1936 and 1939, the 2nd Infantry Division tested

RIGHT: The Germans, like the Soviets, tested many of their weapons as well as their tactics during the Spanish Civil War. Here a Messerschmitt Bf 109 strafes Republican troops during the Battle of Madrid.

BELOW: Disliked by many in the German Army, Hitler's SS bodyguard, the *Leibstandarte Adolf Hitler*, was organized as a motorized regiment. Its first major operation was the occupation of Austria after the *Anschluss*.

the new formation. In 1939 General George Marshall became the chief of staff. Like his predecessors, he was committed to mechanization and the development of new tactics. Consequently, he ordered the army to conduct a series of exercises to test combined-arms operations. The Third Army conducted the first large-scale GHQ manoeuvres in 1940.

LESSONS FROM DEFEAT

In some ways, defeat surprised Germany in 1918. As soon as the war ended, the military launched an investigation to determine what had happened. While the victors for the most part had no compelling reason to race towards change, the Germans welcomed new weapons and tactics with open arms. General Hans von Seeckt played a key role in the evolution of the German Army during the interwar period.

In addition to his job of rebuilding the defeated army, based on his own experiences on the Russian Front, Seeckt concluded that an immobile mass army could be outmanoeuvred by a smaller, highly trained, mobile force. Mobility and surprise would take precedence over firepower. The German commanders studied the writings of Fuller and Liddell Hart. Once Seeckt and his staff had devised a doctrine, they ordered the appropriate organization and equipment. Industries designed technological developments in order to achieve the doctrine.

From the start, Seeckt circumvented the terms of the Versailles Treaty. Forced to reduce the officer corps, he selected the best general staff officers and ordered them to rebuild the military. In addition, he created the basis for Germany's early successes in the next war. In order to do this, Seeckt incorporated traditional German military tactics: a concentrated force on a narrow front to achieve a breakthrough, and the integration of weapons and arms in order to overcome the defences of the enemy.

He also emphasized a decentralized command structure that allowed the commanders to make their own tactical decisions. Although some German commanders disagreed with him, Seeckt stressed the importance of modernization and of tanks. Despite the

restrictions imposed by the Treaty of Versailles, German officers designed an armoured doctrine and secretly tested tanks in the Soviet Union.

When developing the new doctrine, the German commanders had to consider the nature of the next war. Numerous officers believed that tanks should be used only as infantry support; others believed that antitank guns would make tanks far too vulnerable in the field. After Hitler came to power, the armoured theorists gained more and more support, and generally succeeded in implementing some of their best and most ingenious ideas.

One of the most influential supporters of mechanization was Heinz Guderian. Guderian advocated the establishment of new all-arms mechanized formations that would magnify the effectiveness of the tank and facilitate mobile warfare. With Hitler's support, the German General Staff began to execute a new mechanization policy. Various antitank units, engineer companies and infantry divisions received orders to become motorized.

The new German doctrine extended beyond mechanization and tanks to the development of close air support for ground operations. Although ultimately committed to strategic bombing and the concept of air power, the Luftwaffe commanders agreed to allocate resources for ground support. With the integration of air and mechanized formations into ground operations, the Germans moved closer to the development of the famous 'Blitzkrieg', or lightning warfare, tactics. These new tactics would be unleashed for the first time in Poland on 1 September 1939.

ABOVE: Germany's secret plans to re-arm came out into the open in 1935, when Hitler cast aside the shackles of Versailles. Plans to re-equip the *Wehrmacht* were under way, but for the moment the General Staff had to use dummy tanks made from plywood and canvas to try out their new theories of mobile war.

CHAPTER THREE

BLITZKRIEG

The German campaigns of 1939 and 1940 introduced the world to a new word and a new kind of battle: *Blitzkrieg* or 'Lightning war.' Using new tactics, the Wehrmacht swept triumphantly through Europe.

The nature of World War II not only forced armies to incorporate all available weapons and arms into mobile, flexible units, but it also made them adapt to numerous threats, terrain and climates. Several trends emerged during the course of the war.

First, mechanized combined-arms forces matured during this period. Although most armies in 1939 considered armoured divisions as a limited, supported mass of tanks, by 1943 armoured divisions had evolved to become a balance of various combat arms and support services. Second, due to the concentration of mechanized forces into mobile divisions, ordinary infantry units lacked the necessary defensive antitank weapons, as well as armoured vehicles, for the offensive. Third, the desire to thwart *Blitzkrieg* drove these trends. Fourth, both the environments and the fluid tactics of the war fostered the creation of specialized combat units, such as amphibious and airborne divisions. Fifth, a war conducted strictly with ground operations ended with World War II. Finally, major changes in command, control, communications and intelligence resulted from the emergence of the complicated, multidimensional conflicts of this war.

LEFT: Spearheading Germany's invasion of France in May 1940, a Panzer IV of the Wehrmacht bursts through the Ardennes Forest. Rapidly advancing toward the Channel coast, the tanks and their accompanying infantry and artillery demonstrated the main characteristic of a successful *Blitzkrieg*: speed.

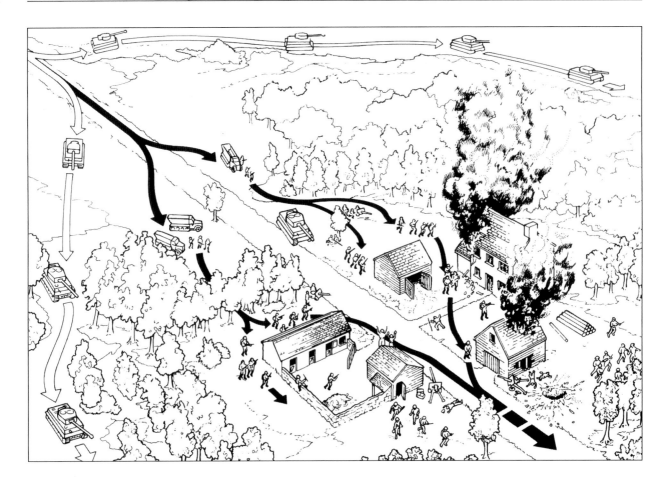

ABOVE: The Germans based their *Blitzkrieg* tactics on speed and the exploitation of the enemy's weaknesses. Armour, supported by air attacks, would burst through the *Schwerpunkt*, the decisive point, usually the weakest area of an enemy's defences. Advancing rapidly, they would cut lines of communication, leaving any advancing infantry to isolate and mop up any resistance.

Although not completely developed until the inter-war period, *Blitzkrieg* tactics can be traced back to the German spring offensive of 1918. The Germans, even though they did not use tanks, succeeded in disrupting and demoralizing their opponents. After the war, General Hans von Seeckt built upon the 1918 German tactics to develop '*Blitzkrieg*', or lightning war. The new doctrine that resulted extended beyond mechanization and tanks to the development of close air support for ground operations. Mobility and surprise would take precedence over firepower. Under Seeckt's direction, the German military developed the capability of implementing offensive campaigns which were always fast-moving.

The *Blitzkrieg* doctrine stressed several important ideas. First, the situation on the battlefield had changed. Mechanization and technological advancements brought modern war. As a result, the situation dictated 'decentralized, mission-oriented orders'. Second, two factors would dominate the field: speed, and the exploitation of the enemy's weaknesses. Therefore, troop commanders, whether operating on the offensive or the defensive, had to adapt to the fluidity of the battle. They could not afford to wait for orders to come from their superior officers. Third, the various combat branches could no longer function alone. The new modern war required closer coordination and cooperation between them. Finally, unlike during the previous war, commanders could not lead from headquarters established well to the rear of the front lines. Because the new tactics were designed to exploit any given situation on the battlefield, the commanders had to be close enough to the front lines to issue appropriate orders. Using air-supported panzer divisions, the Germans hoped to overwhelm the enemy at a particular point and achieve a breakthrough, before finally advancing rapidly and then encircling the defenders.

While Seeckt and other commanders developed the new German military doctrine in the 1930s, Germany began to test the provisions of the Versailles Treaty under the leadership of Adolf Hitler. Despite limits imposed on the size of the

German Army by the treaty, Hitler ordered the implementation of massive rearmament. Between 1933 and 1939 the size of the army and armoured forces greatly increased. In March 1936 German troops occupied the demilitarized Rhineland, breaking another provision of the treaty. Two years later, Germany implemented the *Anschluss*, or union with Austria. Once Austria was controlled, Germany turned to another neighbour, Czechoslovakia. Using the mistreatment of Germans in the Sudetenland as an excuse, German forces moved toward the Czech border.

As tensions increased and Hitler refused to back down, Benito Mussolini suggested a conference in Munich to resolve the situation. Hitler agreed. On 24 September 1938 Hitler and Mussolini met with British Prime Minister Neville Chamberlain and French President Edouard Daladier. No government officials from the Soviet Union or Czechoslovakia received an invitation to participate in the proceedings in Munich. Bowing to pressure from Hitler and Mussolini, and armed with Hitler's assurance that he had no quarrel with the rest of Czechoslovakia, Chamberlain and Daladier agreed to allow German occupation and control of the Sudetenland. Chamberlain returned to Great Britain and announced that he and his colleagues had achieved peace. All could breathe a sigh of relief. The great leaders had averted war, but not for long.

SEIZURE OF CZECHOSLOVAKIA

Following the Munich Conference, the economic situation hindering German rearmament did not improve. What Germany needed was an influx of resources and industry. Control of the rest of Czechoslovakia would fulfil this need. Hitler just needed an excuse to occupy the neighbouring country. The Czechs provided one. In March 1939 a political crisis broke out. Hitler took advantage of the situation and ordered German forces to invade Czechoslovakia and seize the capital, Prague. The Germans hit the jackpot. After entering Czechoslovakia, German troops seized over 1200 aircraft, 800 tanks, almost 2000 antitank guns, 2300 field artillery pieces and 57,000 machine guns, and over 600,000 rifles. Although the Germans could celebrate the addition of these materials to their rearmament stash, their action in Czechoslovakia would bring unforeseen consequences.

BELOW: On 1 September 1939, German troops crossed the Polish frontier to launch the war in Europe. Here, a mixed column of cavalry and infantry move into the Polish corridor towards Danzig (now Gdansk).

A bitter outburst of public outrage in Great Britain forced Chamberlain to reconsidered his country's diplomatic position. Not ready to recognize that war was unavoidable, Neville Chamberlain implemented a diplomatic policy to prevent further German aggression in Europe. At the same time, he had to acknowledge that Britain's armed forces were unprepared should war erupt on the Continent. As a result, the British Government allocated funds for rearmament and admitted the importance of the army's role on the Continent. In addition, the British Government, realizing the possibility of Germany's next goal, guaranteed the independence of Poland. Hitler took two actions in response.

On 3 April 1939 he ordered the OKW (Oberkommando der Wehrmacht, or the Armed Forces High Command) to prepare for the invasion of Poland. The codename of the operation was Case White. In August 1939 the Führer announced that he and Stalin had signed the Nazi–Soviet Non-Aggression Pact, in which Germany and the Soviet Union put aside their differences for the time being in order to accomplish a common goal: the division of Poland. According to the Case White plan, two army groups, carrying out a war of manoeuvre and penetration, would attack and encircle the Polish forces. Commanded by

General Ludwig Beck, Army Group North, which included the Third and Fourth Armies, would advance into East Prussia before turning south and heading for Warsaw. Commanded by General Gerd von Rundstedt, Army Group South, consisting of the Eighth, Tenth and Fourteenth Armies, would advance through Silesia to occupy the Vistula on both sides of Warsaw. Encircling the Polish defenders, the two army groups would then systematically destroy them.

Forty infantry, six panzer, four light and four motorized divisions would carry out Case White. More than 1300 aircraft would supply air support for the offensive. Including 30 divisions, 11 cavalry brigades and two mechanized brigades, the Polish Army, with fewer than 900 tanks at its disposal, would be greatly outnumbered. Only one-third of the army's formations would be at full strength when the German offensive began. The Polish Army also lagged behind the German Army in terms of artillery, mobility and communications. Because they were so badly outnumbered, the Poles had to weaken their defences in their industrial regions in order to station more forces to areas on the border with Germany.

Having set an initial target date of 26 August, the Germans began mobilization on 15 August. Last-ditch efforts to find a peaceful resolution to the tensions on the

RIGHT: As German columns sliced through Poland in two giant pincers, Polish resistance began to collapse. Himmler's armed SS received its baptism of fire during the campaign: here the *Leibstandarte* have attacked enemy forces near Sacharzow.

Polish border resulted in a delay. On 31 August, however, Hitler issued Directive No. 1, ordering the commencement of Case White the next day.

On 1 September 1939 the Nazis unleashed *Blitzkrieg* against Poland. While the Luftwaffe attacked Polish air bases and military targets in Warsaw, German tanks and infantrymen rammed into the defenders. In less than three weeks, German forces occupied over half of the country. German aircraft bombed and strafed the enemy soldiers and overwhelmed Polish aircraft in the air. By 4 September, the Luftwaffe controlled the skies over Poland and could devote its full strength to supporting the ground offensive and to attacking the enemy's railway system. Attacks by the Luftwaffe caused the collapse of the Polish forces concentrating in the region west of Warsaw. As German aircraft disrupted the enemy defenders, Army Group North burst through the Polish Corridor between East Prussia and Pomerania, and Army Group South proceeded to push towards Lodz and Kraków.

COLLAPSE OF POLAND

Under the weight of German attacks, Polish resistance collapsed within a week. On 7 September the Polish High Command withdrew from Warsaw, which caused it to lose control of its forces. Polish resistance forced the advancing Germans to implement a second pincer movement in the area east of the Vistula toward the River Bug. By 16 September the German forces had trapped the defending formations between the Rivers Vistula and Bug. German ground and air forces prepared to demolish the encircled Poles. On 16 September German aircraft dropped more than 700,000 bombs on the trapped enemy. Two days later 120,000 Polish soldiers – one-fourth of the Polish Army – surrendered to the Germans.

Although 100,000 Polish soldiers continued to defend Warsaw, the Germans were close to a decisive victory. Under the threat of massive destruction to the defending troops – as well as to the civilian population – from German air and artillery bombardments, the soldiers in Warsaw decided to surrender. As the Germans thrashed the Polish forces in the west, Soviet troops entered into the conflict from the east. Claiming their desire to 'protect' the population as a motive, Soviet forces advanced into Poland on 17 September.

By the end of the month, Polish resistance had virtually ended. The last surrender of Polish troops occurred on 6 October 1939. The total Polish casualties numbered over 900,000 – 70,000 killed, 133,000 wounded and 700,000 captured. The Germans, on the other hand, suffered fewer than 45,000 total casualties: 11,000 were killed, 30,000 wounded, and 3400 missing.

ABOVE: France and Britain declared war on Germany following the invasion of Poland, but did little more than bring their defences to a high state of alert. Here, French soldiers, seen on 27 October 1939, man an artillery observation post somewhere on the Maginot Line at the German border.

Case White revealed several things about the new German doctrine. The Germans, committed to *Kesselschlacht*, or 'vast cauldron battle', tactics, and the strategy of *Vernichtungsschlacht*, battles of annihilation, expected to launch a series of decisive battles resulting in quick victories. As soon as victory was achieved in Poland, the OKH (*Oberkommando des Heeres*, Army High Command) analyzed the army's performance.

The OKH then instituted a rigorous training programme throughout the army to correct the weaknesses and deficiencies demonstrated by the after-action reports. The speed of the German advance in Poland had revealed several problems that the Germans addressed during the next six months. Although the German doctrine was based on speed, exploitation, combined arms, and decentralized command and control, the Polish campaign revealed the need for improvement in both doctrine and battlefield performance.

Part of the problem faced by the German High Command was an unwillingness to adopt Heinz Guderian's theories about conducting a war of manoeuvre and penetration. Instead of concentrating the panzer and light motorized divisions at a few key points, they divided the formations among the various armies. Generally, the German commanders used the mechanized and motorized forces along a broad front to break through the enemy's defences. Once organized resistance ended, the armoured forces moved into the rear. In

addition, the coordination between ground forces and fire support was not as developed as that between tanks and motorized infantry. Consequently, in the early stages of the attack, German artillery fire came close to hitting Guderian's advancing tanks. The ground forces received little air support because the Luftwaffe's attempts to gain air superiority and to destroy the Polish lines of communication took precedence. Several factors prohibited good cooperation between the Luftwaffe and the army: the complexity of close air support operations; coordination and communication problems between ground and air units; and lack of adequate training in close cooperation between ground and air formations. The German High Command addressed these problems and ordered additional training between October 1939 and April 1940. Following the invasion of Poland, the evolution of the panzer division structure towards balanced combat arms began. By the time they opened their offensive in France, the Germans had perfected combined-arms operations.

WAR IN THE WEST

Although Hitler had planned to attack France as soon as possible after the defeat of Poland, other factors, besides additional training for the army, caused a postponement of the operation until May 1940. While their troops trained, the OKW used the time to analyze the French defences and to perfect their invasion plans. The approach of the French High Command to modern warfare helped the Germans achieve victory quickly. When World War I ended, the French High Command held the belief that cumulative attrition would result in victory. Therefore, a combination of fortifications and firepower, along with attacks into specific areas, called killing zones, would defeat an enemy. Unlike the Germans, the French focused on set-piece assaults with limited objectives, rather than on speed of communications, junior officer initiative or 'high tempo' operations. Emphasizing the importance of the defensive, the French constructed a series of fortifications that constituted the Maginot Line. Built in the 1930s, these fortresses provided the primary defence

BELOW: Firepower was an important component of *Blitzkrieg*. In the attack on Poland, the well-equipped German troops used weapons like the 3.7cm (1.46in) PaK 35/36 antitank gun, whose design had been started by clandestine design teams long before Hitler and the Nazis came to power in 1933.

LEFT: Attacking through the old battlefields of the Somme in May 1940, elements of the 6th and 7th Rifle Regiments and the 7th Motorcycle Battalion, part of Rommel's 7th Panzer Division, advance across the French countryside. Many of the photos of the 7th Panzer Division's advance were taken by Rommel himself.

on the border with Germany. But they did not extend along the French–Belgian border. According to their mobilization plan, the French hoped to station 15 divisions in the Maginot forts and 12 along the Italian border, holding another 58 divisions in reserve or ready to race to Belgium in the event of a German invasion. Britain promised to provide a force to help the French meet a German threat in Belgium.

In constructing a plan, the OKW had to take Hitler's demands into account. Because he did not think that the British and the French would tolerate another great war, Hitler ordered the OKH to capture the Low Countries, as well as Northern France to the River Somme. Then the Luftwaffe and the navy could launch an attack against Great Britain. If the situation progressed as Hitler thought it would, the British would seek terms for an end to their involvement in the conflict. Left alone, the French would have to sue for peace. According to the original plan, *Fall Gelb* or Case Yellow, the main German thrust would come through Belgium. Case Yellow was compromised, though, when the French obtained a copy of the plan following the crash of a German plane.

MANSTEIN'S PLAN

In February 1940 General Erich von Manstein proposed an alternative plan. The general suggested that the main effort, or *Schwerpunkt*, should, for several

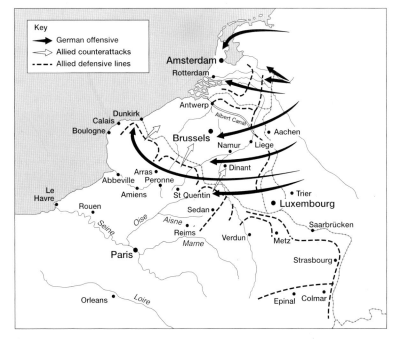

reasons, be through the Ardennes Forest instead of northern Belgium. By advancing through the forest, German forces could divide the Allied defences. Because of the terrain – thick woods, no major roads, and several streams and rivers – the French had not constructed defences in the region. According to Manstein's plan, the Germans would launch a three-pronged attack. Army Group A would initiate a concentrated blow through the Ardennes Forest. The momentum of the attack would carry the army group across the River Meuse and into northern France, where it could manoeuvre into position. Army Group B

ABOVE: The German invasion of France and the Low Countries in May 1940. The British and French advance into Belgium allowed the rapidly advancing Panzers of Army Group A to burst through the Ardennes, sweeping around towards the Channel and cutting off the Allied armies.

ABOVE: German infantry crosses the River Meuse. They established a bridgehead on the other side, allowing the Panzers to cross within hours. Allied planners had expected that it would take the Germans four days.

the OKH opposed Manstein's plan as too risky, Rundstedt supported it. Hitler approved the plan in March.

On 10 May 1940 the Germans unleashed Case Yellow. It began with air attacks. Using its advantage in numbers of modern aircraft, the Luftwaffe planned to achieve air superiority. To accomplish this objective, German air forces attacked Allied air bases and targeted bridges spanning the rivers Lek, Waal and Maas. As a result, the Luftwaffe divided the Netherlands in two and totally disrupted the enemy's defences. The Luftwaffe suffered heavy losses during the first few days of the battle, but it accomplished two goals: German aircraft gained air superiority, and provided support for the effort on the ground. As the Luftwaffe attacked the Netherlands, German paratroopers gained control of the major bridges leading into Holland. Once they held the bridges, the 9th Panzer Division burst through the enemy's defences and advanced into the country. Four days later, the Dutch surrendered.

would attack the Low Countries. After defeating the Dutch and the Belgians, the group would tie down the French and British troops that rushed to the defence of Belgium and prevent them from advancing to stop Army Group A. Army Group C would implement a deceptive attack against the Maginot Line to prevent French formations from moving to the Ardennes Forest. Although

With the Luftwaffe providing cover, German panzer and infantry divisions continued the march to the west. As Army Group B rolled into Belgium, French and British forces raced to stop

the advancing Germans. German glider troops attacked and captured Eben Emael, the Belgian fortress situated at the juncture of the Albert Canal and the River Meuse. After the fortress fell, Army Group B's infantry divisions swiftly marched into the centre of the country. The Belgian Army did not recover from the initial blows launched by the enemy. Convinced that they had been right and that the Germans planned to advance through Belgium and attack France, the French High Command ordered more French and British troops to the country in order to stop the enemy.

ARDENNES SURPRISE

While the French and British focused on Belgium, Army Group A launched the main German attack through the Ardennes Forest. Spearheaded by Panzer Group Kleist, which was protected by air cover, Army Group A easily cracked the weak French defences. Their advance through the forest caught the enemy off guard and proceeded so smoothly that by the evening of 12 May seven panzer divisions, followed by infantry and motorized units, neared the River Meuse. The next day the Germans opened a furious attack against the French defenders. The 7th Panzer Division's infantry, commanded by General Erwin Rommel, crossed the river, drove the enemy back and quickly constructed a bridge across the Meuse. German panzers raced across the bridge and established a bridgehead. Allied forces reached the river too late to stop the German advance; fleeing refugees clogged the roads and impeded their progress. Panicking, some soldiers fled rather than engage the enemy in battle. By 14 May Guderian's 1st and 2nd Panzer Divisions crossed the river further south. Two days later, Army Group A turned north, and then raced towards the English Channel.

The rapid advance through the Ardennes Forest caught the OKW off guard. They had not finalized plans for the next phase of the operation. Despite the opposition of French forces, the spearheads of Panzer Group Kleist pushed them back and drove across open country on 19 May. On 20 May advance elements reached the Channel, with British and French troops retreating in

front of them. Germany infantrymen struggled to keep up with the panzer group, but they succeeded in securing the encirclement of the enemy forces. The French launched a weak counterattack from the south on 19 May, while the British attacked from the north on 21 May. But these counterattacks had little effect on the advancing enemy. Neither the French nor the British had the reserves necessary to stop the Germans.

The situation got progressively worse as German forces moved up the coast. Misreading the situation, Hitler, who feared that the British and the French could mount an effective counterattack, ordered the panzers to stop on 24 May and consolidate their forces with the

infantry. Bowing to pressure from the advancing enemy, the Belgian Army surrendered on 28 May. With the French First Army providing protection, the British Expeditionary Force (BEF) organized the evacuation of troops from Dunkirk: 'Operation Dynamo'. Although they had to abandon their heavy equipment and many of their weapons, most of the BEF avoided capture by the Germans. Between 31 May and 3 June, a total of approximately 340,000 Allied troops managed to escape.

On 5 June, Army Group B launched the final phase of their Western Europe campaign, *Fall Rot*, or Case Red. The French stiffly defended against the German onslaught. The 7th Panzer Division, led by Rommel, seized two railway bridges that spanned the River Somme. After crossing the Somme, the

BELOW: Italy's invasion of the Balkans quickly ran into trouble. In April 1941 Hitler was forced to commit forces to help the Italians, just when they were most needed in the build-up for the invasion of Russia. The spring thaw and muddy road conditions made rapid movement of forces difficult.

OPPOSITE: Mussolini's grandiose plans for empire led him into military adventures for which the Italian Army was unprepared. Poorly equipped and poorly led, the Italians needed Hitler and the Wehrmacht to help them out in Yugoslavia, Greece, and North Africa.

division moved towards the River Seine. Within six weeks from the start of the offensive, over 98,000 French soldiers surrendered to the 7th Panzer Division, which also seized numerous guns, tanks and armoured vehicles. By 9 June, the Germans had pierced the French defences all along the front. Army Group A, pushing aside enemy troops, moved deeper into France. No longer willing to allow the destruction of the country, the French Government surrendered Paris on 14 June and signed an armistice on 22 June 1940. Three and a half million troops fell to the Germans, who only had 45,000 of their soldiers killed in the process. The Germans had again successfully implemented combined-arms tactics in order to achieve their goal. Panzer, infantry and aircraft formations had perfected the *Blitzkrieg* tactics which they had employed in Poland less than a year earlier.

The German victory in France seemed to validate *Blitzkrieg*. Unlike the earlier campaign in Poland, the Germans focused their mechanized forces in large groups in key locations along the front. Supported by five motorized divisions, seven panzer divisions moved through the Ardennes Forest along a 70-km (44-mile) front. British and French forces, on the other hand, were spread thinly along a wide front in a linear defensive position. The German thrust easily broke through a point of the enemy's linear defence. By 14 May German mobile

forces accomplished deep exploitation of the enemy. The French, thrown off guard by the rapid enemy advance, could not react quickly because of their command and control structure. The lack of enemy resistance in the rear enabled the German commanders to use armoured reconnaissance formations to lead their column's advance. The better organization of their armoured divisions gave the Germans another advantage over the British and the French. In addition, the division of the German panzer formation command structure into a series of subordinate headquarters enabled the commanders to operate effectively in the field. Combat engineers made road repairs, which facilitated the Germans' rapid advance over the poor roads of the Ardennes Forest. A coordinated supporting effort by tanks, artillery, and tactical aircraft enabled German infantry and engineers to cross the River Meuse on 13 May. The rapid defeat of France demonstrated the importance of a number of factors: combined-arms mechanized formations, penetration attacks, exploitation of the enemy's rear, and the Germans' advantage in combined-arms training as well as procedures.

Following the defeat of France, however, events did not progress as Hitler had hoped. On 10 June 1940, before the French collapse, Italy declared war on Britain and France. First, the next phase in the German strategy would fail. Then, before Hitler's forces could implement an offensive against another enemy – the Soviet Union – Italian expansionist military actions would force German troops to become involved in the Balkans and in North Africa.

BALKAN ADVENTURE

Jealous of the German acquisition of European territory, Mussolini decided that Italy was entitled to expand its holdings. By the summer of 1940, Mussolini looked to the Balkans as the solution to retrieving Italy's lost glory. Following a series of incidents, Greece refused to accept an ultimatum from Italy. Consequently, on 28 October, Italy attacked Greece; Britain promised Greece support. It did not take long for the Italian offensive to go horribly wrong. On 3 November, Greek forces

BELOW: Waffen SS troops of the *Leibstandarte Adolf Hitler* Division manhandle a motorbike across rough Greek terrain in May 1941. Although administratively separate from the army, in combat SS units were under the command of the Wehrmacht general staff.

captured 5000 Italian soldiers. Ignoring this setback, Italy invaded Albania five days later. As Italy's military position in the Balkans deteriorated, Hitler considered sending troops in to the rescue. Greek forces repeatedly stopped the Italian advances in Northern Greece and Albania. As they forced the Italians to retreat, the Greeks captured several important Albanian cities. Issuing Directive No 20 on 13 December, Hitler ordered preparation for 'Operation Marita': relief of the stalled Italian operation in Albania. At the end of the month, Mussolini appealed to Hitler for help in Albania. A month later, with Directive No 22 Hitler reiterated his intention to supply this. Despite the November and December directives, however, German forces did not enter the fray in the Balkans for several months. They did, though, become committed elsewhere in the Mediterranean region, particularly in North Africa.

YUGOSLAVIA ATTACKED

Following the overthrow of the Yugoslav Government on 25 March, Hitler signed Directive 25, which ordered German troops to destroy the country, and he authorized the bombing of Belgrade. The code name for the operation was 'Punishment'. On 6 April 1941 the Luftwaffe launched bombing missions against Belgrade and virtually destroyed the Yugoslav air force on the ground. While the Luftwaffe bombed the capital, German, Italian, and Hungarian forces invaded Yugoslavia from the north. German infantry quickly over-ran Yugoslavia. Leading the way, German troops quickly ruptured the Yugoslav defences. On 12 April, a day after Italian forces began to advance along the coast, Belgrade surrendered. Pressured by both the Italian and German armies, the Greek Army withdrew from Albania. The Germans occupied Sarajevo on 15 April. Two days later, Yugoslavia surrendered. The Germans seized a total of over 330,000 prisoners.

At the same time as they were gaining ground in Yugoslavia, German forces continued their offensive in Greece, the Wehrmacht's second Balkan operation. In many respects, the German campaign in Greece mirrored the one in

ABOVE: Egypt 1942. A large proportion of Rommel's *Afrika Korps* was actually provided by the Italian Army. In spite of their poor showing against the British in 1940 and 1941, the best Italian units performed well when following good leadership.

Yugoslavia. Attacking southwards along a wide front, XXXX Panzer Corps outflanked Greek defences and forced Greek and British troops to retreat before them. The Luftwaffe's overwhelming air superiority convinced the British to withdraw from Greece. By the end of April, the British had completed their evacuation, and the Germans and Italians occupied Greece.

After the subsequent fall of Crete, the Germans turned their attention to two other regions: North Africa and the Soviet Union. As had been the case in the Balkans, Italian military action in North Africa resulted in German military participation in the region. The Italians' first venture into colonial expansion had occurred before the outbreak of war in Europe. Between 1935 and 1936 they had conquered Ethiopia. By end of the decade, Mussolini was ready to restore the Roman Empire in the Mediterranean. Following the Italian declaration of war on Britain and France in June 1940, a series of raids began in North Africa, which escalated into a full-scale conflict.

In July 1940 Italian air forces attacked Alexandria while ground troops seized British outposts on the Sudanese border. Throughout the month, Italian forces continued to harass the British in Egypt. By early August, after concentrating on the Libya–Egypt border, Italian troops attacked British Somaliland. By the middle of the month, the Italians had forced British troops to vacate British

Somaliland. A few days later the Italian invasion of Egypt began. Following the capture of Sidi Barrani, 97km (60 miles) from the frontier, the Italians began to build their fortified camps.

The small British force stationed in the Middle East had received pre-war training to high standards, but it was poorly equipped. In mid-1940, however, Churchill ordered the transfer of what little resources Britain had to Egypt to counter the threat from Libya. Consequently, a single battalion of 48 heavily armoured infantry support tanks (Mark II or Matilda) travelled to Egypt. The battalion joined two understrength, but well-trained, British divisions. The British spent the next several months solving several logistical problems and also preparing to go on the offensive against the Italian forces.

On 9 December the British launched their attack. The 4th Indian Infantry Division, in conjunction with infantry support tanks, moved against the poorly supplied Italian infantry. The Italians had protected their camps with minefields and obstacles. Recognizing the dangers of a frontal assault, the British advanced between the Italian camps, attacked them from the west, and moved forward along the unmined entrance roads to each

camp. While artillery and mortar fire distracted and pinned down the Italian defenders, two companies of infantry tanks, supported by platoons carrying Bren machine guns, attacked. After the tanks broke through the enemy's defences, trucks drove up, and infantrymen dismounted and proceeded to mop up the defeated Italians.

Advantages in superior training, mobility, and equipment resulted in the British victory in Egypt in December 1940. Following the Italian fiasco in Egypt, Mussolini appealed to Hitler for help. The introduction of German troops into the North African theatre would negate the British advantages there. In addition, because of obligations in other theatres, the British reduced the size of their forces in Egypt in early 1941 and early 1942. Consequently, the Germans were to face a partially trained and poorly equipped enemy when they attacked in March 1941.

In early February 1941 General Erwin Rommel arrived in North Africa to assume control of the German *Afrika Korps*. Because of the vast, open, desert spaces, the fight for North Africa would be characterized by the rapid movement of troops. Spearheaded by tank formations, motorized infantry quickly

BELOW: Not all of the fighting in North Africa occurred in open desert areas. Here, South African troops use grenades and gunfire while fighting their way through the ruined houses of Sollum, Libya.

advanced across the desert. Rommel had orders to protect Tripoli and the remaining Italian troops. Quickly assessing the situation, he realized that in order to succeed, he must go on the offensive. Led by armoured forces, the Germans attacked and disrupted the new, poorly acclimatized British troops. Rommel pushed the British nearly 644km (400 miles), and forced them out of Libya. They managed to hold only Tobruk. Lacking the power needed to destroy the enemy, though, Rommel's force remained between two British strongholds: Tobruk, and the defences along the Egyptian frontier.

BRITISH AMATEURISM

Despite their obligations in Europe and their new offensive in the Soviet Union, the Germans' fight for North Africa had not yet ended. The two British advantages – having broken the German code and possessing air superiority – did not, however, outweigh their weaknesses on the ground. Despite their superiority in numbers of tanks, the British had to divide their tanks between Tobruk and the Egyptian frontier. Due to the German victories and demands for counteroffensives by the British Government, British commanders in the desert had no time to analyze their mistakes and to correct them through additional training. In addition, the constant turnover in command prevented the British troops from learning the lessons of desert warfare. Furthermore, when newly trained formations arrived in Egypt, the British commanders frequently applied them in a piecemeal fashion, which reduced their effectiveness. Unlike the British commanders, who generally had not studied combined-arms tactics, the Germans arrived in North Africa with several tactical advantages: a system of combined-arms formations, flexible commanders, and tactics designed to utilize mass combat power.

When they arrived in North Africa, the Germans also had certain other advantages over the British. Because the British continued to use unsecured communications, the Germans easily intercepted their unencrypted messages. The Germans' medium tanks were better-armed and armoured than the

British light tanks. During the desert battles of 1941 and 1942, the lack of armoured vehicles or effective antitank guns increased the vulnerability of Commonwealth troops. Numerous British armoured units, because they considered the infantry a nuisance, returned to mindset of pure-armour cavalry. In June 1941, the British learned the folly of launching a tank attack without first identifying the location of the enemy's antitank gun line. Following the rapid loss of 17 tanks, the British recognized the importance of firepower over other battlefield elements. In addition, the tendency of armoured units to operate alone and leave infantry formations exposed increased the already growing mistrust and hindered combined-arms cooperation.

Confident, thanks to his numerous successes in North Africa, Rommel wanted to keep the British off balance. He advocated an advance into Egypt, and Hitler agreed. But neither Rommel nor Hitler realized the toll that the previous months had taken on the *Afrika Korps*, which would be unable to drive to Alexandria unless the British defences shattered. The culmination of the *Afrika Korps*' thrust into Egypt was the turning point of the North Africa campaign: the Battle of El Alamein, in July 1942.

ABOVE: As leader of the *Afrika Korps*, Erwin Rommel applied the theories of rapid movement and flexibility he had learned as an infantry officer on the Italian front in 1918 to tank warfare in North Africa. His initial successes made him a national hero in Germany. Following the German capture of Tobruk in June 1942, Hitler promoted Rommel to the rank of Field Marshal.

TOP: When reinforcing a defensive position, all weapons were utilized. Here, a dug-in British Cruiser tank turret was used as an antitank weapon.

ABOVE: Australian troops attack a German position at El Alamein. Commonwealth troops first stopped the overstretched *Afrika Korps* in midsummer, before taking the offensive under General Bernard L. Montgomery.

Before the battle, however, the British Eighth Army would receive a new commander, Field Marshal Bernard Montgomery, who had only three months in which to prepare for the next offensive. From the beginning, British commanders did not always succeed in using the correct tactics to counter the German combined-arms task organization concept. While still in England, Montgomery identified the weaknesses of the British approach in North Africa. The British tendency to form small combined-arms task forces affected the divisions' ability to operate

effectively. The numerous small task forces frequently suffered defeat at the hands of the *Afrika Korps'* concentrated actions. Although some British commanders in Egypt made attempts to adapt the theory of combined-arms armoured divisions, the next German offensive began before the British units could execute the necessary organizational and tactical changes.

Following his arrival in North Africa, Montgomery initiated a retraining programme. Because he could not completely change the existing procedures, he looked for ways in which to adapt traditional command and control methods to desert warfare. When the British stopped the German advance at Alam Haifa in September 1942, they bought time to implement changes. Montgomery used the time to prepare the Eighth Army for the El Alamein offensive of October–November 1942. He returned to the combined-arms tactics of World War I, in which each corps controlled its own artillery. The first step of Montgomery's plan involved a night penetration of enemy defensive positions by engineers, infantry, and artillery. Protected by these forces, armoured units would advance and tempt the Germans to counterattack. Montgomery hoped to take advantage of the Eighth Army's strengths at a time when the enemy was experiencing fuel and equipment shortages that would impede their manoeuvrability. The result of Montgomery's strategy was a battle of attrition, in which the general frequently adapted his plans to suit the actuality of the battlefield in front of him.

THE TURN OF THE TIDE

In late August the Germans attacked the British position at El Alamein, but they were driven back. Instead of exploiting his success, however, Montgomery focused on preparations for the October campaign. According to his plan, his forces would mount several diversionary attacks to make the Germans turn to the south. Then the main thrust would hit the *Afrika Korps*' northern flank. Relying on artillery and engineers to cut a path through the Germans' mined defensive position, the British forces would pierce the enemy's defences and achieve a breakthrough. The offensive did not, however, proceed as planned.

On 23 October the offensive began with a heavy artillery barrage. After the barrage, the infantry attacked. But because the paths cleared through the minefields were too narrow, the infantry advance stalled, and the British armoured force became stuck in the open minefields. Fortunately for the British, Rommel was on sick leave in Germany, and he did not return until the battle had been raging for 48 hours. The German commanders failed to react quickly to the British assault. They committed their forces to the battle in a piecemeal

fashion. Yet although the 15th Panzer Division lost 75 per cent of its tanks on the first day, German resistance did not falter; in fact, it continued.

Realizing that his plan was not succeeding, Montgomery ordered his force to redeploy on 27 October. The next day the Germans took a heavy blow in the north. Although they failed to breach the enemy's defences, the British reduced Rommel's armoured strength to 90 tanks, while they had more than ten times as many. On 2 November the British struck again and slowly advanced through the enemy's minefields. German firepower knocked out 200 British tanks, but the Germans lost 60 tanks that they could ill afford to lose.

The Germans could not hold out much longer. On 3 November, Rommel ordered a retreat, but Hitler issued a counterorder, which forced the Germans to remain in place and suffer even more horrific losses. It would no longer be possible for the *Afrika Korps* to establish a strong defensive position within Libya. On 4 November Hitler finally and reluctantly agreed to a withdrawal, and the *Afrika Korps* rapidly retreated across Egypt and also Libya. But the British Army failed to exploit this retreat, not moving quickly enough, and the Germans succeeded in retreating as far as Tunisia. This was the beginning of the end. On 8 November Allied forces landed in Morocco and Algeria and trapped the famous *Afrika Korps* between them and General Montgomery's advancing Eighth Army.

BELOW: Surrounding Tobruk, the Germans cut off the British supply line and forced the town to surrender in June 1942. The capture of the British fortress was followed by a headlong dash for Egypt, which was only brought to a halt by the Allied defensive lines at El Alamein.

CHAPTER FOUR

EASTERN FRONT

The German attack on the Soviet Union in June 1941 marked the beginning of a long struggle for supremacy in the East. It was a struggle marked by the most ferocious fighting in history, culminating in the fall of Berlin.

The German rescue of the Italian military efforts in the Balkans and North Africa frustrated Hitler. Following the successful operations in Poland, the Low Countries and France, the Führer, despite having to abandon the invasion of Great Britain, was ready to embark upon the next phase of his plan for the creation of a new German Empire: the conquest of the Soviet Union. The commitment of German forces in Greece, Yugoslavia, Albania and North Africa forced a delay in the implementation of Hitler's plan until June 1941. Code-named 'Operation Barbarossa', the German invasion of the Soviet Union was the culmination of Hitler's desire to acquire *Lebensraum*, or 'living space', for the expanding German Reich. Hitler believed that historical precedent existed to justify German imperialism. The United States had expanded westwards across the vast North American continent, and Great Britain had amassed a great colonial empire. In order to realize its own 'manifest destiny', Germany should conquer and settle the huge stretches of land to the east.

According to the Führer's plan, the 120 divisions allocated for the offensive had less than five months in which to destroy the Soviet Army. This would be accomplished in two phases: first, a series of

LEFT: On 22 June 1941, the Nazi Germany unleashed 'Operation Barbarossa', the invasion of the Soviet Union. Although led by the Wehrmacht's Panzers, the bulk of the 120 German divisions were infantry, advancing through the Russian steppes on foot and supported by horse-drawn lines of communication.

53

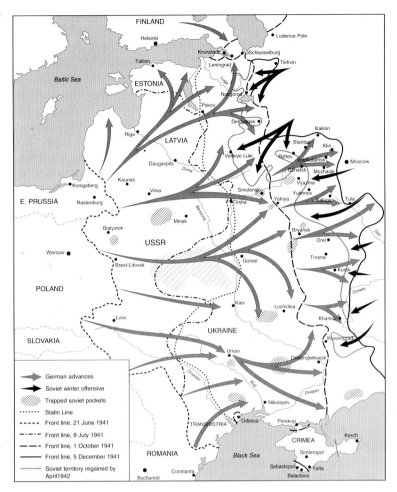

Map labels: FINLAND, Helsinki, Lodenoe Pole, Kronstadt, Schlusselburg, Tallinn, Leningrad, Tikhvin, ESTONIA, Baltic Sea, Luga, Novgorod, Pskov, Demyansk, Riga, LATVIA, Kalinin, Staritsa, Klin, Velikiye Luki, Rzhev, Volokolamsk, Moscow, Daugavpils, Dvina, Gzhatsk, Mozhaisk, Kaunas, Vilna, Vyazma, Konigsberg, Smolensk, Yukhnov, Tula, E. PRUSSIA, Rastenburg, Orsha, Yelnya, Kaluga, Don, Minsk, Bryansk, Bialystok, Berezina, Orel, USSR, Trosna, Warsaw, Gomel, Kursk, Brest-Litovsk, Donets, POLAND, Kiev, Lochvitsa, Lvov, Kharkov, UKRAINE, Krasnograd, SLOVAKIA, Uman, Dnepropetrovsk, Dnieper, Dniester, Bug, Nikolayev, TRANSNISTRIA, Odessa, Perekop, Kerch, ROMANIA, CRIMEA, Simferopol, Black Sea, Sebastopol, Yalta, Constanta, Balaclava, Bucharest

Legend:
- German advances
- Soviet winter offensive
- Trapped soviet pockets
- Stalin Line
- Front line, 21 June 1941
- Front line, 9 July 1941
- Front line, 1 October 1941
- Front line, 5 December 1941
- Soviet territory regained by April 1942

ABOVE: 'Operation Barbarossa' was carried out by three massive army groups, making a three-pronged advance into the Soviet Union. Army Group North aimed for Leningrad, Army Group Centre was directed towards Moscow, while Army Group South pushed through the Ukraine towards Rostov and the Caucasus oilfields.

Kesselschlachten ('cauldron battles') to destroy forces in European Russia, and second, a decisive battle of annihilation for Moscow. Hitler was confident that the German Army's tactical excellence would allow it to achieve strategic military victory in one single campaign, or *Vernichtungsschlacht*. The previous campaigns in Poland, the Low Countries and France seemed to validate the Germany strategy; however, when implemented in 'Operation Barbarossa', *Vernichtungsschlacht* proved unattainable.

Stalin's purge of the officer corps immobilized the Red Army at the same time that Hitler's military achieved a series of victories from September 1939 until the summer of 1941. Logistical problems stemming from the occupation of eastern Poland further adversely affected the effectiveness of the Red Army. In addition, the Red Army's inability to combine formations for offensive operations became apparent during the Russo–Finnish War of 1939–1940. Although the Soviets eventually forced the Finns to agree to an armistice in March 1940, the Red

Army was in bad shape. Consequently, the Soviet Government instituted a series of reforms in military organization, command structure, deployment and equipment. When the Germans attacked in 1941, however, the Red Army had not completed any of the reforms and had managed several of the planned changes incorrectly. As a result, the German forces were able to inflict heavy damages on the Soviet military.

THE RED ARMY

Perhaps one of the most important changes implemented by the Soviets in response to the Germans' 1940 victories was a return to large, combined-arms mechanized formations. In 1940 and 1942 the Soviets formed almost 30 mechanized corps that contained two tank and one motorized rifle division. Manpower and equipment shortages prevented complete implementation of the new plan. There were two causes of concern: tank shortages and the lack of medium or heavily armed and armoured tanks. In addition, by 1941 many of the light tanks were obsolete tactically and worn out mechanically. Although the Soviets had designed medium (T-34) and heavy (KV-1) tanks, there were problems with the designs. Production levels of these tanks lagged because the purges had also affected industrial management. The Soviets also lacked sufficient supplies of trucks for the movement of infantry and artillery and mines, as well as modern fighter planes.

By the summer of 1941, the Red Army still suffered from numerous deficiencies, including the use of a combination of obsolete and new weapons and troops that had not trained with the newer equipment. The 1941 invading German force, on the other hand, was highly trained, as well as being well-equipped. Operationally, the campaign was a demonstration of the height of *Blitzkrieg* tactics, particularly the encirclement battle. The failure of infantry and logistics troops to keep up with advancing panzer units, however, restricted the Germans' exploitation deep into the enemy's rear. As a result, Soviet forces had time to reorganize.

Planning to destroy the Soviet Army in a bloody *Kesselschlacht*, the German Armies unleashed *Blitzkrieg* tactics that had already been tested. According to the plan, the Luftwaffe would create havoc behind enemy lines and render the Soviet air force useless. Armoured and mechanized units would penetrate and encircle the enemy army, which the infantry would then destroy.

The Soviet Army did not, however, collapse when it met the 'superior' German Army in battle. The Germans encountered an enemy army which was commanded by a leadership that was still reeling from the purges of the 1930s, but Josef Stalin expected his military forces to do the impossible: stop the German advance. Although the Soviets' military organization included over 300 divisions, none of the formations equalled a panzer group or a panzer army, which could accomplish large-scale, in-depth penetration to the enemy's rear. The Soviet mechanized divisions were unbalanced and dispersed in a way that made the concentration of these formations difficult. In June 1941 the Soviet military was definitely inferior in terms of tanks and aircraft. The Russian aircraft outnumbered those possessed by the Luftwaffe; however, because of obsolete and worn-out equipment, the Soviet air force did little to hinder German domination of the skies at the beginning of the war.

OPERATION BARBAROSSA

The plan for 'Operation Barbarossa' included a three-pronged assault on the Soviet Union: in the north towards Leningrad, in the centre towards Moscow and in the south towards Rostov and the Caucasus. When the campaign did not end in five months, the Germans began to experience logistical problems that had not been anticipated by Hitler or by the German High Command. Not only did the logistical problems hinder the German advance, but they also became critical during the harsh Russian winter of 1941.

The commencement of 'Operation Barbarossa' brought results beyond Hitler's wildest imagination. The Luftwaffe destroyed 2000 Soviet aircraft on the ground on the first day, which allowed the Germans to achieve air superiority from the beginning. Coordinating air and ground attack, the all-arms cooperation worked extremely well within the panzer formations. The coordination proved less successful in other divisions because the armoured and mechanized formations tended to outdistance the supporting infantry. As a result, the infantry generally bore the brunt of desperate enemy counterattacks. The slow movement of the infantry did not, however, prevent the Germans from advancing 322km (200 miles) in the first five days. Their vast encirclements demolished almost 100 Soviet divisions in the first week alone. By mid-July, advance German formations were less than 160km (100 miles) from Leningrad and 322km (200 miles) from Moscow.

The armies in the north and centre advanced rapidly, while Army Group South (AGS), which encountered stiff Soviet armoured resistance, moved much more slowly. Strong enemy opposition held up the Germans outside Kiev. Then Hitler made a costly mistake. In September he ordered the transfer of armour from Army Group Centre (AGC) to the south. On the one hand, the additional forces facilitated the encirclement of 600,000 Soviet troops at Kiev; on the other hand, however, AGC's advance lost momentum, which prevented it from occupying Moscow on schedule. Although Soviet forces in the centre had been on the run, the weakened AGC missed an opportunity to crush them. The Soviet forces used the time to regroup in the Soviet hinterland.

BELOW: Although the Panzergrenadiers serving alongside the Panzers were motorized, the slower pace of the following infantry divisions reduced the overall speed of the Wehrmacht's advance. The slow movement of the infantry did not, however, prevent the Germans from advancing no less than 322km (200 miles) in the first five days of the campaign.

The Soviet military scrambled to adopt measures to counter their weaknesses as enemy forces advanced deep into the Soviet Union. Two problems became noticeable immediately. First, most Soviet commanders and staff officers did not possess the skills necessary to organize the various arms and weapons effectively for defensive or offensive operations. The commanders who had survived the purges frequently received promotions that far exceeded their skill and training. Many of them failed to take the terrain or approaches that required antitank defences-in-depth into account when they deployed their troops, which caused the Soviet general staff to reprimand them. Second, the Red Army had shortages of specialized units and weapons – engineers, tanks, antitank guns, and artillery – which hindered the commanders' deployment of forces effectively in the field.

Solving these problems required major changes during the desperate struggle to thwart the German offensive. On 15 July 1941, the Soviet Supreme Headquarters (STAVKA) issued Circular No 1, which reassigned control of the specialized units to more experienced commanders and staffs, who could concentrate these limited resources to critical points on the front. The new organization of the military resulted in the formation of field armies that consisted of four to six divisions and specialized – artillery, tank, and antitank weapons – units. The STAVKA directive also reduced the amount of artillery in each division and the size of the mechanized corps.

For the rest of the year, the Red Army struggled in a conflict where the German advantages in equipment and initiative made the traditional doctrines of Deep Battle and large mechanized units ineffective. The few new tanks that the Soviets could place in the field were used as infantry support. The situation would change, however, after December 1941 when the Soviets stopped the Germans from seizing Moscow.

DRIVE FOR MOSCOW

Even though the fighting continued, 'Operation Barbarossa' was basically over by 30 September. Hitler was not, however, ready to call it off. Instead, he shifted the focus of the German offensive back to Moscow. On 2 October 1941 German forces opened a new offensive against the Soviet capital. Commencing 'Operation Typhoon' with great encirclements and tactical victories, German forces encircled most of the four Soviet armies situated west of Viaz'ma within a week, but they experienced difficulty in containing the enemy within the circle. While some enemy troops

BELOW: The Russian winter of 1941 was particularly harsh. Poorly prepared for the cold and lacking adequate winter clothing and equipment, the German Army suffered increasing losses to disease and the weather. Many soldiers literally froze to death.

broke free, others began to destroy heavy weapons and vehicles to keep them out of the Germans' hands.

Overall, the German offensive initially progressed well, but winter came early in 1941 and changed the situation. The first snowfall began on the night of 6 October. Following the snow was a rainy period of mud, or the *rasputiza*, which strikes Russia each spring and autumn as the seasons change. Because of the scarcity of paved roads in those days, travel was extremely difficult during the *rasputiza*. As the attackers discovered, they would lose mobility until the ground froze solid. Experienced with the *rasputiza*, the Soviets took advantage of the enemy's loss of momentum. Although the Germans threatened Leningrad and Rostov, Stalin ordered Marshal Georgi Zhukov to leave Leningrad and organize the defence of Moscow. When he arrived, Zhukov authorized a series of spoiling attacks to divert the enemy while his forces prepared the defences outside the city.

By 27 November, however, the Germans were only a few kilometres from Moscow. By the end of November the Russian winter was in full swing. With the drastic drop in temperatures, the lack of adequate antifreeze for the mechanized formations slowed the Germans' movement even more. Hitler

failed to realize the seriousness of the situation on the Eastern Front.

RED ARMY COUNTEROFFENSIVE

While his forces reinforced their defensive positions around Moscow, Zhukov planned a counteroffensive. The Germans failed several times to regain the momentum of their advance, and each attempt took its toll in dwindling supplies and exhausted soldiers. On 6 December Soviet forces launched a massive counterattack against the overextended, exposed AGC. Although Bock requested permission to withdraw, Hitler denied his request and forced him to resign. By the end of December 1941, despite the continuing battle, the German threat to Moscow had ended. While the Soviet counterattack continued until the spring of 1942, Hitler refused to allow the AGC to surrender. The Soviets' attempt to do too much enabled the German force to survive, although Hitler concluded that his 'stand fast' order was responsible.

After their troops drove the Germans out of Moscow, the Soviet commanders returned to their organization and doctrine. By the spring of 1942, Soviet factories were producing a large number of new weapons, which were used to create new tank corps. By the summer of 1942, the Soviet commanders had

ABOVE: In the early days of 'Operation Barbarossa', the Germans caught the Soviets by surprise. Resistance collapsed in many areas, and the Germans captured large numbers of enemy soldiers. Millions of Soviet POWs were marched westwards to Germany, where they were to be used as forced labourers.

ABOVE: Soviet infantrymen, camouflaged in white to blend in with their snowy surroundings, catch an uphill ride on a T-34 tank. Large numbers of these superb fighting vehicles were used in Zhukhov's counteroffensive in front of Moscow in late 1941, supporting fresh infantry divisions arriving from Siberia.

initiated a new organization of one rifle and three tank brigades, along with supporting elements, which functioned as the mobile exploitation force of a field army. By the autumn, they added mechanized corps, which largely consisted of motorized infantry. The job of the new mobile corps, which were actually division-size or smaller, was the implementation of deep exploitations of 150km (94 miles) or more into the enemy's rear, which had been envisioned in the 1930s when the Soviets formulated a new doctrine: the Deep Battle doctrine. Earlier in 1942, the Soviets created tank armies by combined armoured, cavalry, and infantry units, but these armies did not have a common mobility rate or a common employment doctrine. Because they had not yet trained to function as a unit, the first tank armies suffered heavy losses during the summer of 1942 when they engaged the Germans in battle for the first time.

In January 1943 the Soviets finally fielded a cohesive tank army. During 1943, the Soviets organized six tank armies, which spearheaded Soviet offensives for the rest of the war. Generally, two tank corps, one mechanized corps, and supporting elements formed a tank army, which, although in reality corps size, was tank-heavy for two reasons. First, the terrain of European Russia would support tank-dominated operations. Second, because of the inexperience of Soviet tank crews and junior officers, German tank and antitank formations inflicted heavy

losses. In order to compensate for heavier losses in tanks, the Soviet armoured formations continued to contain more tanks than those of other armies.

Hitler began to plan the next big German summer offensive in the east while the Battle for Moscow still raged. On 5 April he issued Führer Directive 41, which outlined the proposed summer campaign in the Crimea, the Don Steppe and the Caucasus. Several factors, including the spring thaw, forced a delay in the operation until late June 1942. In the meantime, both sides used the time to rest, regroup and prepare for the summer campaign. STAVKA ordered the military to form 'shock' armies, with massed artillery on narrow axes in order to execute deep offensive penetrations, as well as exploitation.

TO THE OILFIELDS

Hitler had come away from the 1941–42 campaign believing that his 'stand fast' order had saved the Wehrmacht. Although Barbarossa had demonstrated the weaknesses of their strategic thinking, the Germans failed to change. They stuck to the concept of a decisive battle, of a quick victory in Russia with a knock-out blow. They also recognized the need to cut the Soviets off from their resources in the Caucasus, particularly the oil. Because of the supply situation in the south, the German High Command shifted troops from AGN and AGC to AGS, but the supply problems remained.

The German summer offensive, *Fall Blau* (Case Blue), included a deception plan. Diversionary forces would threaten Moscow in order to divert enemy troops from the south. AGS would then launch the three phases of Case Blue. During phase one, German troops would penetrate enemy defences around Kursk and conduct a typical encirclement. Prior to the commencement of the second phase, AGS would split into two army groups, 'A' and 'B'. Army Group A would conduct the main thrust by driving to and occupying the oil-rich Caucasus, while Army Group B would advance to the River Volga, north of Stalingrad.

Because the Soviets believed that the enemy would renew the assault against Moscow and Leningrad, the deception plan basically worked. The Soviets expected their formidable enemy to initiate an attempt to cut them off from the resources in the Caucasus after a new Moscow and Leningrad offensive. Despite evidence to the contrary, Stalin believed that the Germans would focus on Moscow. When the Germans opened Case Blue on 28 June with an offensive in the Kursk area, the Soviets were caught completely off-guard. The Germans achieved a quick breakthrough and commenced an offensive that seemed destined to repeat the German military performance of the previous summer. On 30 June, the Sixth Army, which was led by General Friedrich Paulus, commenced its attack south of the Kursk assault. Following Zhukov's advice, Stalin ordered a retreat. To stand fast would probably have resulted in annihilation. The Red Army retreated from the area south of Kharkov, and the Crimea fell to the Germans. Sebastopol fell after an overwhelming air and artillery attack on 4 July.

The German forces continued to achieve success in July, but they failed yet again to accomplish a decisive victory. When Hitler decided to redefine the operation's objectives, the result was that the thrust into the Caucasus came to a grinding halt and failed to result in the control of the oilfields. Then Hitler made another decision that had far-reaching and disastrous consequences: he ordered Army Group B to seize Stalingrad. For the first time, Stalingrad became the primary target for the Germans.

On 19 August 1942, the Sixth Army and the Fourth Panzer Army, under the direction of General Paulus, commenced the assault on Stalingrad. Four days later, some German troops moved into the suburbs while others reached the River Volga north of the city. When citizens from the southern part of the city began to flee to the east, the Soviet defenders panicked. Stalin ordered the Red Army to stand firm. Failure to do so would result in the transgressors' treatment as criminals and deserters. More afraid of Stalin than of the enemy, the defenders slowed the Germans' progress.

Eight days after the first shots had been fired, Zhukov became the Deputy Supreme Commander and was given the job of saving Stalingrad. He quickly

BELOW: The sheer size of the Soviet Union swallowed up each German offensive, and Soviet forces showed surprising resilience in attempting to counter the Wehrmacht's thrusts. The Red Army was still no match for German professionalism, but each battle taught the Russians more.

LEFT: German infantrymen accompany a Panzer III as the 2nd Panzer and Panzergrenadier Divisions push across the Steppes towards the town of Orel, which was taken by the Germans the previous year but lost in the Soviet winter offensive of 1941–2.

ABOVE: Wehrmacht soldiers,
wade across a small river as
Kleist's panzers drive towards
the Caucasus.

BELOW: The autumn of 1942
saw the opening of the battle
for Stalingrad. The fighting
was vicious, and little mercy
was shown by either side.

taking the entire city. In the meantime, the Red Army would amass a huge strategic reserve in order to lauch a well-planned counterattack that would also be well supported logistically.

Two months later the Soviets were ready to mount 'Operation Uranus'. The Sixty-Second and Sixty-Fourth Armies defended the city. General Vasili Chuikov, commander of the Sixty-Second Army, unlike many of his colleagues, demanded good intelligence, which he used to organize and move his decreasing forces when necessary to meet threats posed by the enemy. By the end of October, the battle had degenerated into a stalemate in the factory district. Regrouping his forces, Paulus launched one final attack on 9 November. Despite small successes in some areas, the Germans made little progress. The German advance ground to a halt three days later, and Soviet storm troops slowly began to retake lost territory, particularly in the centre of town and in the factory district.

On 19 November, the Soviets counterattacked north of Stalingrad and surprised the enemy. The next day, Soviet troops in the south of the city hammered the German line, which was held mainly by Romanian and Italian formations.

assessed the situation: the enemy army's limited reserves and long, exposed flanks weakened it; by implementing a pincer movement, the Soviets could surround and isolate the German forces in the city. By 13 September, Zhukov and the Chief of the General Staff, Alexander Vasilevsky, had devised a plan: 'Operation Uranus'. Sufficient forces would reinforce the defenders to prevent the enemy from

Within three days the two mobile forces advanced over 240km (149 miles) and surrounded Paulus's army. When he requested permission to surrender, Hitler ordered Paulus to 'stand fast'. On 12 December, under heavy rainfall, the attempt to rescue the trapped soldiers commenced. Initially the rescuers made steady progress, but on 23 December a tank battle with Soviet reinforcements prevented further penetration. Zhukov and the Soviet General Staff, who had anticipated a rescue effort, had made provisions to stop it. Over 60 Soviet divisions and 1000 tanks moved into place to meet the threat. On 24 December, the German rescuers, in danger of being surrounded, retreated and left Paulus and his army to their fate. The Soviets had greatly underestimated the size of Paulus's force; therefore, they had difficulty in crushing the Germans' resistance to their own destruction. Following orders, Paulus continued the fight, but on 2 September he had no choice but to surrender. The battle for Stalingrad had ended.

SOVIET SHOCK TACTICS

A year earlier when the Soviets had counterattacked a larger defensive force, the results had been disastrous. Consequently, STAVKA had issued Circular No 3, which ordered the creation of shock groups: strong concentrations of combat power that would attack on a narrow front and break through the enemy's defences. Eight months later, in Order No 306, Stalin reinforced this idea. According to Stalin's order, successive waves of infantry could not participate in shock-group attacks. Equipment and firepower shortages convinced the Soviets to maximize the effectiveness of the infantry by concentrating the infantry into a single echelon attack. By 1942 the length of the front line and shortage of troops meant that the Germans could not establish deep defensive positions. As a result, the Soviets would be more likely to break through the enemy's defences if they concentrated their infantry for a single, massive thrust. Later in the war, when both the Germans and Soviets had established defence-in-depth, the Soviets adapted their tactics to the new situation.

Following penetration of the German defences, Soviet mobile units moved through the opening to exploit and surround the enemy. Once they had accomplished a series of shallow-depth encirclements that disrupted the enemy's defences, Soviet forces would link up with other penetrating elements to move deeper into the enemy's rear positions before the Germans could withdraw and create new defensive positions. The Soviets achieved a large, operational-level encirclement for the first time at Stalingrad in November 1942. Although not entirely successful, the penetration tactics served as a model for offensive operations for the rest of the war.

The tide in the East began to turn with the Soviet victory at Stalingrad, but the fight was far from over. Despite the losses incurred at Stalingrad, both the Germans and the Soviets began to plan offensives for the summer of 1943. Although the fighting would be as fierce as it had been in the previous two years, the Soviets would have two advantages that would

BELOW: The Battle for Stalingrad began with a German attack in August 1942. The Soviets refused to relinquish control and launched 'Operation Uranus' on 19 November. Despite the fierceness of the assault, the Soviets did not regain control of the city until early February 1943.

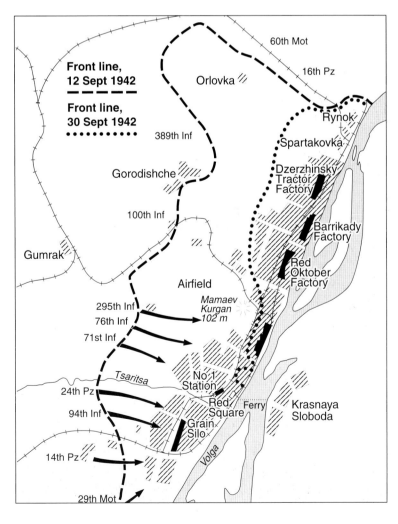

make the difference. First, unlike their enemy, the Soviets could more easily reinforce their armies and build up their supply of tanks, artillery, aircraft, weapons and other supplies. Second, and more importantly, the Soviets accurately determined where the Germans would launch their next offensive, and made the preparations necessary to thwart it. A westward bulge existed in the front lines around Kursk. The German operation 'Citadel', ordered attacks from the north and south against the salient's base to isolate and destroy the Soviet forces.

The first two days of the Battle of Kursk brought huge air and artillery bombardments, fierce fighting and a tank battle. Both sides paid a high cost on the

battlefield. The fighting culminated on 12 July with a huge tank battle near Prokhorovka. All along the front, wave after wave of Soviet tanks pummelled their enemy's forward panzer units. Both the German and the Soviet forces were exhausted after a day of extremely fierce, costly fighting. The Germans had lost almost half of the tanks they had committed to the battle, while the Soviet tank losses numbered almost 200.

THE BATTLE OF KURSK

A tense situation existed throughout the front lines of the Kursk salient. The Soviets launched a major counterattack against the Orel salient. Despite repeated requests, Hitler refused to consider a withdrawal, but finally, on 22 July, he approved an 'elastic defence', which basically allowed Model to begin to retreat. Hitler's decision showed he was willing to accept defeat on the Eastern Front, and signalled the beginning of the Germans' retreat from the Soviet Union.

Once the Soviets had stopped the German attack at Kursk, the momentum on the front began to shift. Quickly taking the initiative, the Red Army commenced attacks along the entire Eastern Front. Preceding each offensive, however, the Soviets implemented elaborate deceptions, which were frequently successful. When they were duped, the Germans concentrated their troops to thwart the fake assaults, which left them exposed in the actual area of the enemy attack.

After the deception was begun and the Germans had moved their forces, the Soviets launched a line-breaking blow against the enemy's defences. Following the creation of a hole in the Germans' line, Soviet tanks and infantry moved up to exploit the breach. Unlike the situation earlier in the war, the Soviets gradually improved their logistical situation so it no longer limited the depth of the exploitation. After the Germans regrouped and repaired their defences, the process would begin again. The Soviets, who maintained the initiative, developed numerous methods of penetration and exploitation, which kept their enemy off-guard.

The Battle of Kursk and the subsequent Soviet counteroffensives severely dented the German campaign in the East. The failure at Kursk and the loss of Kharkov were the culmination of a series of setbacks suffered by the Germans since the Soviet defeat of 'Operation Barbarossa'. Following Kursk and Kharkov, the Germans began to retreat along the entire Eastern Front. The first stage of the German collapse in the east began during the summer and autumn of 1943. The Soviets refused to relax the pressure against their retreating enemy, and the constant battles took their toll on the German infantry.

The tide that had begun to turn at Stalingrad finished turning after Kursk. The Soviets used every tool at their disposal to stop the German advance; they even used partisan forces to sabotage the enemy's efforts. In July 1941 Stalin had appealed to Soviet citizens living in occupied territory to join the effort to defeat the intruders. By January 1942, 30,000 partisans had joined the struggle against the Germans. By the summer this number had grown to 150,000. The Germans' racial policies and their treatment of the Soviets in the occupied regions persuaded many to operate behind enemy lines. By early 1943 almost 250,000 civilians had joined the partisan ranks, and the Soviet military

BELOW: German infantry equipment was often extremely good: the MG-42 general purpose machine gun was probably the best of its kind in the world, and its high rate of fire was feared by all of those who came up against it in battle.

used them whenever possible. Many operated in the forest bordering on Poland and the Ukraine. Divided into groups of 1000 men and women, they worked behind German lines to gather information about enemy troop movements, destroyed bridges, roads, railway lines, and enemy supply depots; they ambushed small enemy forces, and cut telephone lines. Those conducting operations near the front lines worked closely with the military. Beginning in 1943 the partisans engaged in intense activity against the enemy, and cost the German forces nearly 300,000 casualties. They contributed to the exhaustion and over-extension of the enemy armies, and helped hasten their collapse.

More important to their success against the Germans than the breakthrough tactics was the Soviets' numerical superiority. In addition, the Soviets implemented a variety of tactics to pierce the enemy's line, including the use of deception to draw out the Germans, who were then bombarded by Soviet artillery. The Soviet military leadership also devised different roles for heavy and medium tanks. Although the infantry received support from heavy tanks, which attacked enemy strongholds, medium tanks frequently led the assault on the Germans' defensive positions. Combat engineers and trained infantrymen riding on the tanks provided protection against the enemy's antitank weapons. The cost of rapid penetration, however, was a high casualty rate, particularly in the combined-arms – engineers, infantry, and tanks – units. To reduce casualties, the Soviets, by 1944, relied increasingly on deception, concentration of forces, speed, and task organization. The Soviets generally used the forward detachment – a mobile, armed combined-arms unit – to lead a strike. The forward detachment's job was to capture important objectives and prevent the Germans from organizing a defence. The forward detachment did not always slow down its advance to engage the enemy. The unit left that job to the formations advancing behind it.

By 1943 and 1944, certain factors had become glaringly apparent. The Red Army fielded a much better force in

BELOW: Soviet partisans – irregular guerrilla fighters – entered the conflict against the German invaders from the earliest days of the war. Attacks led by partisans like Michail Trakhman, seen here using captured German weapons, were to become a constant thorn in the side of the Wehrmacht. Anti-partisan operations were carried out with great brutality.

1943 than it had in 1941. The army had access to more and better equipment and had accomplished improvements in tactical proficiency and combat ability. The German Army, on the other hand, was a shadow of what it had been several years earlier. Several elements of the German Army suffered decreases by 1943: shortages in manpower, weapons, and other equipment, insufficient overall training, and decline in combat ability. German commanders tended to credit numerical advantage for the Soviets' successes, not their superior breakthrough tactics. Many of the German formations remained below combat strength.

ABOVE: Dressed to blend in with the snow, Soviet guerrilla fighters cautiously move into position. By 1943, there were tens of thousands of partisans operating behind the German lines, their missions being controlled by the STAVKA. By 1944, irregular units were operating in conjunction with regular forces against the Germans.

WEHRMACHT ON THE DEFENSIVE

Troop shortages resulted in the restructuring of German infantry divisions. The Germans provided the infantry divisions with a larger proportion of fire support to counter their declining abilities. By 1944 the Germans increasingly had difficulty in containing the Soviet offensives. They could only hope to slow them down, not stop them. The Germans relied more and more on the defence-in-depth doctrine: absorb the attack, drive a wedge between the enemy's armour and infantry, and attack and destroy each in turn. Improvements in combined-arms cooperation made by the Soviets by 1944 made it increasingly difficult for the Germans to separate attacking units from each other. The decline in morale, improved Soviet intelligence, and a much more professional Red Army made preemptive German withdrawals in the face of an enemy attack much harder.

As the Soviet push depleted the enemy's infantry forces, the Germans focused on the rejuvenation of their panzer divisions. Although the panzer formations remained, the primarily tool the Germans used to counterattack, by 1944 the panzer arm could no longer remain apart from the rest of the German Army, which found itself relying more on combined-arms cooperation, from a necessity rather than a desire. Although the German commanders continued to emphasise offensive tactics, by 1943 and 1944, the German Army was no longer capable of mounting offensive operations on the Eastern

Front. The advancing might of the Soviet military had forced the retreating enemy to assume the defensive.

The Soviet offensive in the summer of 1944 would ultimately decide the fate of Berlin. Larger than any of the Soviets' previous offensives had been, 'Operation Bagration' would result in the destruction of Army Group Centre and the advance of the Red Army to Warsaw. In order to divert enemy troops away from the target area, Soviet troops would attack Finland prior to the beginning of the main offensive.

On 10 June 1944 the assault on Finland began and continued through the next phase of 'Operation Bagration'. Despite German reinforcements, by late August Finland sued for peace. Soviet forces began a two-pronged attack against AGC. As a result of their rapid advance, Soviet forces inflicted 350,000 casualties on the Germans, and destroyed a total of 28 of the enemy's divisions.

The Soviets accomplished even greater successes as the centre of the German front collapsed under the weight of continuous bombardment from tank and infantry forces. Soviet pressure on the Germans continued across the Eastern Front. On 20 August over 900,000 Soviet infantry and tank and mechanized forces burst into Romania. The Romanians quickly decided not to fight, and the German situation rapidly deteriorated. By 2 September Bulgaria had withdrawn from the war against the Western Powers; still the Soviet advance continued, as forward units moved towards Hungary. German resistance in Hungary was much

TOP: Soviet infantrymen enter Berlin in April 1945. Some of the Red Army troops had fought on foot all the way from Stalingrad.

ABOVE: Soviet tank riders hitch a ride on Valentine infantry tanks supplied as Lend-Lease items by the British. For many Red Army soldiers, this was the nearest they got to becoming mechanized infantry.

stronger, though, and Soviet forces did not enter the capital, Budapest, until December. By late December, however, resistance in Hungary was virtually over.

The Soviet military machine rolled on in other areas of the front. On 12 January 1945, when the Red Army attacked along the River Vistula, it met a much weaker enemy force. Hitler had diverted formations westward for the Battle of the Bulge and for the defence of Hungary. Two Soviet offensives began around the same time. In the central part of the Front, the Soviets pitted 5 armies, 2 tank armies and over 1000 tanks against 7 depleted enemy armies. Within a day, they had punctured a hole in the German line. Infantrymen and tanks pushed through the gap and advanced towards the River Oder. A day after the second offensive had begun, the first Soviet troops had pierced the enemy's defences north of Warsaw. By 17 January Warsaw had fallen. By early February they had accomplished the destruction of Army Group A, had crossed Poland and East Prussia, and had reached the River Oder. The Soviets were quick in establishing three bridgeheads across the river. One of these bridgeheads was within a distance of just 70km (43.5 miles) of Berlin.

With their treasured capital city of Berlin in danger, the Germans prepared for the fight of their lives: the defence of their homeland from the Communist hordes. The Germans strengthened their defences along the Oder front. The OKH recommended the implementation of 'defence in depth': the Germans would build several consecutive lines of defence; before the enemy's preliminary artillery bombardment, the troops in the forward-most line would fall back; after the bombardment ended, they would return to their positions and then proceed to stop the advance of the enemy. WIth this aim, Hitler authorized the implementation of defence-in-depth on 30 March.

The Berlin Garrison included several formations: LVI Panzer Corps, which contained 5 or 6 divisions; over 50 *Volkssturm* (home guard) divisions; and several 'Alarm Troops' formations, consisting of clerks, cooks and non-combatants. Three concentric lines of defence extended to the outer suburbs. Each ring consisted of nine defensive sectors, connected by a communications system. The military used the city's subway to move troops, equipment and other supplies without the enemy's knowledge. The heaviest defences and the largest concentration of troops protected the government sector in the centre of town. The situation was grim during the first weeks of April as the final battle neared.

BATTLE FOR BERLIN

Early on 16 April 1945 the Soviets unleashed the largest artillery bombardment of the war. Over 40,000 field guns, mortars and Katyusha rocket-launchers prepared the way for the ground forces. For 30 minutes, wave after wave of Soviet aircraft bombed German defences. Having crossed the River Oder, shouting Soviet infantrymen rushed forward and fired their guns. Soon many were engaged in brutal hand-to-hand combat with their hated enemy. What was to be the last major battle on the Eastern Front had finally, after much preparation, begun.

The rapid Soviet advance quickly bogged down as the searchlights blinded the troops. In addition, the soldiers had difficulty crossing the marshy terrain. Mechanized vehicles got bogged down and caused traffic jams and German artillery bombarded the exposed Soviet tanks. As the Soviets advanced, the Germans' resistance increased. Yet despite the fierce battle, the Soviets inched towards the German capital. After three days of intense fighting, the Soviets prepared to attack Berlin itself.

On the morning of 20 April, Soviet aircraft began bombing Berlin. During the day, Soviet artillery forces moved up. Heavy artillery barrages commenced the

next day. Ground forces moved into the city, and the fighting intensified. On 25 April Soviet forces linked up with American formations on the River Elbe, and on the same day Soviet forces cut Berlin off from the rest of Germany, raining shells on the city. Still the fight for control of the city continued. Rubble clogged up the streets and made the fighting more difficult; most of the city lacked water, gas and electricity.

Although the defenders bitterly opposed their Soviet enemy, Berlin and Germany could not be saved. With American forces nearing the city from the west and Soviet troops nearing his bunker, Hitler made preparations for the end. On 29 April, shortly after 01:00 hours, the Führer married his mistress Eva Braun. Meanwhile, the Red Army moved even closer and would soon overrun Hitler's headquarters.

On the afternoon of 30 April the Soviet assault on the Reichstag began. Despite a counterattack by German tanks, the Soviets gained control. As the red victory banner was hung from a window on the second floor of the Reichstag building, Hitler consumed his last meal. After saying good-bye to his staff, he and his new wife retired to their private rooms. A short time later they ended their lives. Despite Hitler's death, resistance would continue for a few days. On 2 May 1945 the Germans finally agreed to a ceasefire, as well as to the general surrender of all forces remaining in the city. The Soviets had won the battle for Berlin.

ABOVE: As the final collapse of the German Reich nears, Soviet bombers and IS-2 heavy tanks bombard the Reichstag building in Berlin. Following Hitler's death, the defenders of Germany's capital agreed to a ceasefire on 2 May 1945.

CHAPTER FIVE

THE PACIFIC

The war in the Pacific bore little resemblance to the European experience. Fought from island to island, through steaming rainforests and over precipitous mountain ranges, it was a war largely won by the infantryman.

World War II followed a very different path in Asia than in Europe, and historians now often consider the two to be separate, distinct conflicts. While the war in Europe was one of continental dominance fought by traditional adversaries, the war in Asia involved many, sometimes competing, motive forces. Japan, having adopted western technology and military methods in the Meiji Restoration, sought to establish an empire from the lands of rapidly collapsing China. Even as Japan sought to become a colonial power, the traditional colonial nations of Europe succumbed to chaos and defeat in World War II, allowing the Japanese to supplant their authority in much of the Pacific region. Thus the war in the Pacific was also one pitting the rise of Asian power against the legacy of European control, a theme of conflicts in Malaya and Vietnam. Finally, Japanese expansion was seen as a threat by the other great Pacific power: the United States. The resulting conflict was one for dominance over the world's greatest ocean, and the victor would emerge as the next great global superpower. Fought for reasons different to those driving the war in Europe, the conflict in Asia was a war over the future, rather than a repeat of the past.

LEFT: Pre-assault bombardments meant that American forces in the Pacific often had to advance through destroyed terrain. The war in the Pacific involved brutal fighting to winkle out fanatic Japanese defenders from their positions, often only possible through the use of flamethrowers and explosives.

ABOVE: Japanese infantrymen man an improvised armoured train in China in 1937. It was the expansion of Japan's imperial ambitions from mainland China into the rest of Asia that led to conflict with US and Western interests in the Pacific.

The war in the Pacific also followed a different military pattern to that in Europe, and was much more geographically and militarily diverse. In the Pacific, navies and air forces reigned supreme in a war that covered much of the globe, but often involved only tiny land areas. It was a struggle of carrier clashes and amphibious invasions launched against fanatical resistance. Even so, it would be once again the infantry, often in the form of Marines, that took and held terrain in the tropical jungles of innumerable Pacific islands and atolls.

The fighting in China and Southeast Asia, though, was different again. The war in China involved Japanese forces trying to destroy the will to resist of the world's most populous nation. In the jungles of Burma, the numerically small, but élite Japanese forces drove Allied forces back in a war of speed and stealth. These battles saw combat that would give rise to a reimplementation of irregular infantry warfare, presaging the tactics of insurgency that would come to dominate wars later in the century.

JAPAN'S CLASH WITH CHINA
Though Japan was ruled by an emperor and possessed a western form of government, by the 1930s the military was in firm control of the destiny of the country. Infused with the warrior code of *Bushido*, which stressed the martial spirit and preferred death to surrender, the military sought to expand Japan's economic base. It did this by taking control of much of China.

In the wake of the Russo-Japanese War Korea and Manchuria had fallen under informal Japanese rule, becoming chief sources of raw materials for Japanese industry. The Japanese military, though, yearned for more, and by 1930 found even their existing gains threatened by a resurgence of Chinese strength under the Nationalist warlord Chiang Kai-shek. At the same time, though, a Communist insurgency, led by Mao Tse-tung, diverted Chiang's attention and also began to set the tactical rules for nearly every insurgent revolution since. The Japanese military paid little attention to Mao, focusing instead on the threat posed to

Manchuria by Chiang's Nationalist forces. Fearing that negotiations were counterproductive Japanese military leaders in Manchuria took matters into their own hands, a move tolerated and even encouraged by their warrior code.

Without government knowledge, on 18 September 1931 Japanese military forces in Manchuria placed a bomb on an important railway near Muckden. Blaming the 'terrorist attack' on the Nationalist Chinese, the leaders of the Japanese Army in the area, known as the Kwantung Army, attacked. Ignoring subsequent government orders to desist, the Japanese military forces routed Chinese resistance and solidified Japanese control in the area, even leading a punitive expedition against a Chinese boycott of Japan in the important international port of Shanghai, laying waste to much of the city. Facing only faint international condemnation for their actions from western nations who at this point were far more interested in appeasement, the Japanese went on to

drive their forces into the Jehol and Hopei provinces of northern China before calling a halt to the offensive.

In 1937, yearning for glory and certain that their nation had to expand to survive, the Japanese military again moved against China after a minor clash of arms between Japanese and Chinese troops at the Marco Polo Bridge near Beijing. Bragging that they could conquer China in single month, the Kwantung Army moved forward, even arresting a government diplomat sent to defuse the situation.

Undertrained and poorly supplied, the Nationalist Chinese forces stood little chance against the Japanese, and by August had retreated from Beijing. Further south the Japanese laid siege to Shanghai, long a centre of Nationalist Chinese support. After seven weeks of torture the city finally capitulated, and the Japanese made ready to advance up the Yangtze River into the heart of China and towards what seemed to be an easy victory over an outmatched foe.

ABOVE: Japanese troops were very well trained and highly motivated. But their initial successes against the ramshackle Chinese Army in the late 1930s did not lead to ultimate success: indeed, at the height of the Pacific War in the early 1940s, a large part of the Japanese Army was tied down in China.

The fighting in China closely resembled that of other colonial conflicts. Highly trained and motivated Japanese soldiers with the latest equipment including light tanks and plentiful air support, easily defeated poorly trained and poorly armed Chinese troops. Utilizing superior logistic support, air power and tactical mobility, Japanese forces were able to consistently outmanoeuvre their opponents and overwhelm them once battle was joined. The Chinese, though, retained several obvious strategic advantages. Chiang and the Nationalists, though increasingly concerned with their own Communist uprising, could rely on their main weapons of numbers, time and space. If the Chinese people chose to resist, the Japanese Army was not nearly big enough to garrison the entire nation or achieve a total military victory. Thus Chiang chose to protract the war, and fight a war of attrition for which the island nation of Japan was uniquely ill-equipped.

Though the Japanese were somewhat surprised that Chiang and his forces chose to continue fighting after the fall of Shanghai, they remained quite confident, choosing to drive on the Nationalist capital of Nanking to force a decision. Chiang decided to avoid major battle but to contest the Japanese drive up the Yangtze, forcing attritional losses on his enemy. Shocked by continued resistance and taking heavy losses, the Japanese slowly drove toward their goal: supplying their forces in the main by river. In many ways the Japanese war in China was a racial war, replete with Japanese hatred for the supposedly inferior Chinese

people. Casualties taken on the drive to Nanking only served to heighten the racial animus. The Chinese still dared to resist against their rightful overlords, too ignorant to realize that the outcome of the war was a foregone conclusion. In December 1937 Nanking finally fell to the force of Japanese arms, and the pent-up Japanese hatred for their enemy found an immediate outlet. For a month Japanese forces ran riot through the city, engaging in an orgy of looting and slaughter, killing over 200,000 civilians in an atrocity so bad that even Nazi Germany's government complained.

Believing that they had made their point, the Japanese now attempted to enter negotiations with the Nationalist Chinese. Since the Japanese negotiating team demanded near-total surrender Chiang chose to fight on, retreating with his forces to remote Chunking. Here he hoped to continue the long war of attrition and pleaded for international aid. By the end of 1938, the Japanese had seized most of their military goals in China, but Japanese rule did not extend into the great Chinese hinterlands, where hundreds of millions of Chinese remained outside the sphere of Japanese control. As a result, the war in China would drag on inconclusively, with neither side able to secure victory. The Nationalists and the Communists sometimes fought hard against Japanese rule, but often squabbled among themselves. It was the vast Chinese peasantry that suffered most; millions died from famine and want caused by a nation embroiled in constant turmoil and war.

JAPAN TRIUMPHANT

Japan's success in China and obvious designs on areas in Southeast Asia and the Pacific proved quite disquieting for the western Allies. The European nations, though, could devote only scant resources to defence of their colonies in the area due to German threats to their very existence. The United States, under the leadership of President Franklin Roosevelt, recognized the threat posed by Japan, especially to US holdings in the Philippines, but could do little due to powerful forces of isolationism. Even so when Japan took control of French Indo-China in July 1941, Roosevelt took powerful economic action by announcing

BELOW: Japanese infantrymen capture Mandalay Railway Station in Burma as Imperial forces sweep triumphantly through Southeast Asia. British and Commonwealth troops proved no match for the Japanese in jungle warfare, and would not be able to fight on equal terms for at least three years.

an embargo on oil shipments to Japan. The headstrong Japanese, who received some 80 per cent of their oil imports from the United States, saw Roosevelt's act as a declaration of economic war. Threatened with economic collapse the Japanese military made ready to attack the United States. The Japanese plan of action was fairly simple. A surprise attack on Pearl Harbor, condoned by the code of *Bushido*, would cripple the US while Japanese forces seized important islands, including the Philippines, in the Pacific and the countries of Southeast Asia. Such an action would seize important sources of raw materials, making Japan nearly self-sufficient. In addition it would throw up a defensive cordon around the home islands that the US, much less its sorely-pressed British ally, would not have the willpower to test out.

On 7 December, though negotiations with the US continued, Admirals Yamamoto and Nagumo presided over the surprise carrier assault on Pearl Harbor. Though successful, the attack was, in many ways, flawed. All but two of the US battleships lost in the attack were salvageable, and the all-important US carriers were not even present at the time of the attack. Thus the attack on Pearl Harbor was not the crippling blow that Yamamoto had intended, and it only served to waken a sleeping colossus. With its mighty industrial base able to produce

in a month what Japan produced in a year, once America developed the unity required to fight a total war, the outcome was never in doubt. Even so, as the US recovered from its shock and prepared for Pacific war, but chose to concentrate its efforts on Germany, Japanese forces went from victory to victory.

In most areas of the Pacific War the Japanese were able to rely on air power and the almost uncontested might of their fleet to seize and retain the initiative. On most Pacific islands, Allied forces and their indigenous counterparts spread themselves far too thin in a vain attempt to defend everything, allowing the Japanese to achieve victories with relative ease. In two areas, though, Allied forces were deemed strong enough to hold out for extended periods, possibly even dealing the Japanese their first defeat. In the Philippines US forces under General Douglas MacArthur numbered nearly 200,000 and were augmented by a sizable air component. The perceived strength of the US force caused one reporter to comment, 'If the Japs come down here they will be playing in the big leagues for the first time in their lives.' In Singapore and Malaya nearly 60,000 British forces would defend the vaunted 'Gibraltar of the east.' Even though the defensive forces in both cases seemed impressive, in reality the Japanese held all of the advantages.

ABOVE: Using initiative and mobility the Japanese infantry were able to maintain the speed of their advance even in difficult terrain, constantly keeping Allied troops off balance. Here, part of the Japanese force heading for the Indian border makes short work of crossing the Chindwin River in Burma.

MacArthur Caught Out

In a titanic oversight, MacArthur, having been alerted of possible danger by events at Pearl Harbor, fell victim to a surprise Japanese air attack that wiped out over half of the US air strength in the Philippines in a matter of minutes. Now guaranteed air superiority, on 22 December the Japanese invasion force, consisting of two crack divisions under the command of General Homma, landed on the island of Luzon at the Lingayen Gulf. MacArthur, eschewing the plan to use his forces to defend the Bataan Peninsula, decided to meet the invasion head on. The US forces sent to meet the Japanese invasion consisted, in the main, of poorly trained and poorly supplied Filipino soldiers. Having never before witnessed combat the Filipinos found themselves hit by concentrated air, sea and amphibious assaults, and quickly broke. Leaving behind tons of critical supplies, the US and Filipino forces retreated 241km (150 miles) south to Bataan to make a final stand.

The Allied forces drew up defensive positions based on the heights of Mount Natib and awaited the Japanese onslaught. Making matters worse the defenders of Bataan were cut off from all sources of supply and found themselves starving and beset by a host of tropical diseases. In early January Homma's force struck the defences of Mount Natib but was repulsed by heavy artillery fire. Though the terrain was not suitable for tanks and made air power of little value, the Japanese were able to persevere against the staunch American defences by employing their chief strengt: the mobility and resilience of their infantry formations. American forces had thought Mount Natib to be impassable and did little to defend its rugged summit. The Japanese, though, achieved the seemingly impossible, infiltrating through the rough terrain and manoeuvring behind the American defenders. Japanese skill in jungle warfare would become legendary and often followed the pattern of the fighting on Mount Natib. Small, lightly equipped Japanese forces would seek out the path of least resistance – often hacking through trackless jungle, surviving on very little for days on end – to emerge behind an enemy who believed the area impenetrable.

Surprised by the sudden appearance of the Japanese, Allied forces now retreated further down the Bataan Peninsula to the Bagac-Orion Line. Realizing that victory in the Philippines was only a matter of time, General Homma did not make his final effort to seize Bataan until April, by which time US rations were down to 1000 calories a day, and 80 per cent of the soldiers were suffering from malaria. Reinforced by two divisions and aided by amphibious landings to the south of the US defensive line, the Japanese quickly destroyed most American resistance, leaving only the tiny island fortress of Corregidor holding out. In

RIGHT: American Amtracs – Amphibious tractors – rush forward to hit the beach at Iwo Jima. These versatile craft, able to operate on sea and on land, were critical to the success of the American 'Island Hopping' campaigns through the Central and Southwest Pacific.

LEFT: Marines move forward from the beachhead on 'Bloody' Tarawa. After victory at Guadalcanal, US forces were able to keep the Japanese off balance, attacking several important island garrisons, while leaving others to 'wither on the vine'.

the wake of a punishing artillery barrage in early May the starving survivors of Corregidor also surrendered, only to be taken into the nightmare world of Japanese prison camps.

MALAYAN CAMPAIGN

In Malaya events ran a similar course. On 8 January 1941 a daring landing by a single Japanese division pre-empted a planned British move into defensive positions in Thailand. Though the Japanese forces were weak on the ground, usually outnumbered 2 to 1, they made up for their lack of strength with mobility, surprise and command of the air and sea. British and Indian forces, though spread thinly, attempted to use the jungle terrain to their advantage in a defence of northern Malaya. The Japanese infantry, though, was lightly equipped and built for speed, making much better use of the terrain than their adversaries. Using bicycles to advance down jungle trails with great speed, the Japanese soon slipped through and around British defences in the area of Betong. Caught by surprise by the speed with which the Japanese advanced through the jungle's vastness, Allied forces soon began a southward retreat towards the fortress of Singapore. Though nearly surrounded and destroyed at the Battle of Jitra on 15 December, Allied forces attempted to form a series of defensive lines in an effort to involve the Japanese in a slow-moving war of attrition. However, the

Japanese mastery of jungle warfare again proved decisive. British and Indian forces were so thin on the ground that the Japanese never had to launch full-scale offensives into the teeth of the Allied defensive lines. Cautious Japanese probing attacks would locate the weak portions of the defensive system, sometimes only a lone jungle trail, allowing Japanese infantry to infiltrate the British and Indian positions. Again and again the Allies found their defences compromised, sometimes even discovering that the Japanese had reached the next proposed line of defences well in advance of their own retreating forces.

ABOVE: Allied forces cross a river under fire in New Guinea. Fighting in the war against Japan involved soldiers from several nations including the United States, Britain, Australia, the Netherlands and India, often supported by indigenous peoples.

ABOVE: US Marines armed with Thompson submachine guns advance slowly against fanatical Japanese resistance on Iwo Jima. The Japanese fought to the last. Fewer than 1000 of the 21,000 defenders survived to become prisoners.

By 31 January the Allied forces were finally compelled to retreat into the defensive positions of Singapore itself. The Japanese had advanced, in nearly continuous fighting, over 804km (500 miles) through some of the world's harshest terrain in under two months. Cut off from meaningful Allied support, the defence of Singapore was doomed, but the force there remained powerful. Some 45 Allied battalions manned the defences, facing but 31 Japanese battalions making ready to attack. Allied command, under General Percival, spread thin defences along the coastline of the island, holding behind a substantial general reserve that could rush to meet any oncoming threat from the Japanese.

Once again relying on speed and audacity, on 8 February the Japanese flung the majority of their attacking forces in a lightning amphibious landing against only six Australian battalions. Before the reserve forces could arrive, the Japanese were ashore and driving inland, on 15 February capturing the water supply for the city of Singapore. Faced with an urban disaster, General Percival decided to surrender. On the evening of 15 February, Percival arrived at the Japanese lines carrying a Union Jack. He surrendered his 130,000 men to a Japanese force less than half that size. The 'Gibraltar of the East' had fallen in the greatest single military disaster in British history. The Japanese had consistently used their air and naval power, as well as their superior mobility and aggressive style, in order to offset their paucity in numbers. Infantry tactics had, once again, won the day.

ALLIED COUNTERATTACK

Even as the Japanese solidified their massive gains, the United States began to marshal its forces for a Pacific-wide counterattack. Though the economic might of the US would eventually tip the balance of the war, in 1942 American forces were in many ways at a distinct disadvantage. Intelligence, though, served as a great equalizer. Having broken valuable Japanese naval codes, US naval forces surprised the Japanese on 3 June, sinking four carriers and winning the pivotal Battle of Midway. The American victory was one of the true turning points of World War II and shifted the balance of naval power back towards the United States. The Japanese fleet, though, remained powerful, especially in surface vessels, and stood ready to defend its far-flung empire. In a series of battles between near equals, Allied forces began the slow road to the home islands of Japan. US and Allied forces had to secure Australia and stabilize the long supply lines of the Pacific first, necessitating attacks along the periphery of the Japanese Empire. It was then hoped that US forces, led by MacArthur in New Guinea and Admiral Nimitz in the Pacific, could begin a series of 'island hopping' campaigns, each designed to secure naval and air bases and logistic support on the long drive to Japan.

As part of 'Operation Watchtower', designed to destroy the giant Japanese base at Rabaul, US forces began a drive up the Solomon Islands, with landings on the islands of Tulagi and Guadalcanal. Overgrown with thick jungle, Guadalcanal sported a nearly completed Japanese airfield, one that US forces would have to seize to ensure the success of their operations in the Solomons. In an operation that some dubbed 'Operation Shoestring' due to its paucity of logistic support, on 6 August the 1st Marines landed on Guadalcanal and faced no resistance from surprised Japanese construction crews, who retreated into the nearly impenetrable jungle. Quickly the Marines began to construct defensive positions around what they now dubbed Henderson Field to await an expected Japanese counterattack. The Japanese command was unsure of the nature of the unexpected Marine operation, believing that it could only be a raid, and ordered naval units to the area to crush US logistic support for the Marines. In many ways the battle for Guadalcanal would be decided by a series of titanic naval battles that swirled around the island into November as both nations attempted to supply their forces there or ferry new troops into the fray. It was, however, the Marines, undersupplied and facing heavy odds, who conquered Guadalcanal in a series of land battles that would presage the remainder of the war in the Pacific.

By the time that defeats at sea had forced US transports to flee the area, only 10,000 Marines had made it ashore at Guadalcanal with only one month of supplies, no barbed wire, no landmines and no heavy weaponry. The Marines held only a small perimeter around the airfield, leaving much of the island open to Japanese landings and resupply efforts. The existence of the Marines would be day to day and battle to battle, always on the razor's edge of annihilation. For their part, the Japanese ordered Colonel Ichiki and a force of only 1500 men to attack and destroy the Marines. Thinking the Marine force to be small and believing in the martial spirit of his own men, on 20 August Ichiki launched a headlong attack on the Marines defences along the Ilu River. Though the Japanese surged forward in suicidal waves the Marines held firm, and the next day pinned the remaining Japanese attackers onto a beach, crushing them even as Colonel Ichiki committed suicide.

Now more fully aware that the Marines were at Henderson Field in considerable numbers and were there to stay, Admiral Yamamoto ordered Guadalcanal to be retaken at all costs. The Japanese warrior code demanded that the Americans be denied victory. Thus Guadalcanal quickly became a battle that carried great meaning and would mark an even greater turning point in the Pacific war than had Midway. Utilizing night-time destroyer and transport runs dubbed the 'Tokyo Express,' the Japanese massed over 6000 men on the island under the command of General Kawaguchi. American forces, though they could rely on air support from Henderson Field, remained understrength and undersupplied due to continued Japanese naval pressure. Lulled into complacency by nearly continuous victory and convinced that his veteran soldiers could overcome the resistance of the decadent Americans, Kawaguchi

BELOW: US forces made extensive use of superior firepower in the effort to force Japanese defenders from their deeply dug-in positions. Here US Marines fire salvos of high-explosive rockets at the entrenched enemy.

made ready to assault Henderson Field without even taking the time to reconnoitre the its defences. Leaving their base near Tasimboko behind, the Japanese troops traversed the thick jungle with few supplies, sure that they would soon seize the American stores. In the defences surrounding Henderson Field the Marines realized that attack was imminent, for Marine raiding parties had discovered the Japanese build-up.

On the night of 13 September, the attack came as Japanese soldiers poured out of the jungle yelling and howling, throwing themselves at the Marine positions in human waves. Commanded by Colonel Edson, the Marines in the area resisted valiantly on what became

ABOVE: Members of the 6th Marine Division make ready to fire a bazooka at entrenched Japanese defenders on Cemetery Ridge on the island of Okinawa in 1945.

known as 'Bloody Nose Ridge.' The relentless Japanese attack forced the Marines back into their last prepared defences and the struggle became hand-to-hand in many places along the line, but the Marines held. On the next morning, now aided by aircraft from Henderson Field, the Marines put the remaining Japanese to flight, discovering that nearly one half of the Japanese who had taken part in the attack had perished. The Marines had utilized concentrated defensive firepower to hold firm against a brave Japanese attack that exhibited little subtlety. Masters of light infantry tactics and jungle infiltration warfare, the Japanese now found themselves in a war for which they were not prepared, a static war of attrition. Japanese soldiers were

discovering, as had the soldiers of World War I, that spirit had little effect on the firepower of an entrenched foe. Such battles required artillery, armour and great numbers of men, something that the Japanese lacked, but something that the United States would bring to bear in abundance as the Pacific war continued.

Having suffered a bitter defeat on land, and with the naval war slowly turning in the favour of the United States, Yamamoto decided to make one final effort to achieve victory on Guadalcanal. The Tokyo Express, though it took heavy losses, ferried in additional troops. But the United States had also reinforced its tired troops on Guadalcanal for the first time. Having gathered some 20,000 men and 100 artillery pieces,on the night of 24 October the Japanese, now under the local command of General Hyakutake, rushed forward to attack. The bulk of the human wave broke on the defences of the 1st Battalion, 7th Marines under the command of Colonel Lewis 'Chesty' Puller. Though sorely pressed, the Marines relied on the defensive prowess of their small-arms, machine gun and artillery fire to cut the Japanese down in droves. Two days later, having lost over 4000 men, Hyakutake had to admit defeat, leading his men back to their base with little hope for future victory.

Though several naval battles remained and the fighting on Guadalcanal would linger on, the decision on land had been won. By January, American forces had seized the offensive all over the island and the starving and defeated Japanese fled. On 8 February US forces entered the main Japanese base on Guadalcanal only to find it unoccupied. The United States had won its first, and possibly greatest, land campaign of the Pacific War, and the drive on Japan could now proceed in earnest. Casualties, especially for the Japanese, had been high. Of 40,000 Japanese troops sent to Guadalcanal, some 23,000 never returned from the 'Island of Death.' The Japanese had proven in earlier encounters that their light infantry, augmented by air and sea power, could outmanoeuvre and outfight forces of greater size. At Guadalcanal, though, the formula had changed, irrevocably. The US forces now had the initiative and controlled the sea and air. It would be the Japanese that tried to defend everything

against the coming onslaught, thus leaving themselves spread thin. The Americans, not the Japanese, would now mass their forces against immobile foes, using the lessons of speed and mobility that they had learned so well. In addition the Americans could rely on an almost limitless supply of military hardware. Against such odds the Japanese tried in vain to win a crippling naval victory over the Americans. The Japanese soldiers on land, though, could only fight to the death in defence of their islands in the slight hope that American resolve would soon crumble.

ISLAND HOPPING

Along with the climactic naval battles that in many ways decided the war in the Pacific, it was the American island-hopping campaign that came to epitomize the fighting in the theatre. Having seized the initiative, the American forces now began a steady drive towards the home islands of Japan, conquering islands that were needed as bases for further operations, leaving less important island unmolested, and sealing off main Japanese bases by cutting their communications with Japan, in effect leaving them to 'wither on the vine.' In a scenario repeated over and over again in the Pacific War, US forces would pound a target island with relentless air- and sea bombardments. Under cover of this murderous fire, the Marines would hit the beaches in landing craft and in amphibious tractors. The Japanese would sometimes resist the invasion on the beaches, but would more often than not await the advancing US forces in prepared defences further inland. Realizing there was no retreat, Japanese forces would resist to the death, selling their lives dearly for their emperor. In tactics resembling those of World War I the Marines would fight from defensive emplacement to defensive emplacement, often using grenades or flamethrowers to kill the Japanese in their dug-outs and caves. Thus the war in the Pacific was one of attrition, one that the Japanese could not win, and a war that ran an inevitable course. On Beito Island in the Tarawa cluster, only 17 men of the Japanese garrison of 4500 survived the battle. On

ABOVE: Two Marines stand by as a flamethrower fires into a Japanese position during the fighting on Iwo Jima. On some of the larger islands, several Japanese soldiers held out in jungle caves for years after the war, most refusing to believe that their nation had lost.

seemingly countless islands, from Eniwetok to Saipan to Guam, the story remained the same. The battle for Iwo Jima serves to illustrate the tactics used in the island-hopping campaign.

United States forces hoped to seize Iwo Jima for use as an air base from which they could launch attacks on Japan itself. Realizing the strategic nature of the island, over 20,000 Japanese soldiers and sailors waited in defence, commanded by General Kuribayashi. Realizing he could do little but draw the battle out, Kuribayashi chose not to meet the Americans on the beaches, but to rely upon a series of defensive lines to make certain that the Americans paid a dear price in the coming struggle. With a considerable force of artillery and antiaircraft guns at his disposal, Kuribayashi directed his men to construct some 800 pillboxes and over 4.8km (3 miles) of defensive tunnel networks into the volcanic rock of the island. When the Americans came ashore, they would be met by a punishing defence-in-depth system, once again reminiscent of the Great War, and a system which was designed to make the Americans pay for every inch of ground gained.

Against the defensive network the Americans hurled a massive storm of firepower. For 74 days US bombers struck the island, augmented by salvos of naval gunfire. It was the heaviest bombardment yet seen in the Pacific totalling 6800 tons of bombs and 22,000 rounds of heavy shells. The Japanese,

RIGHT: Tanks were not a major factor in most Pacific battles, but in some of the campaigns, they did have a role to play. Here, riflemen from the 29th Marines on Okinawa advance with a flamethrower, using an M4 medium tank as a shield against Japanese fire.

though, lived on in their deep bunkers awaiting the inevitable assault. On 19 February 1945 the Marines, led by the 4th and 5th Divisions, hit the beach near the imposing sight of Mount Suribachi. Initially, though several vehicles bogged down in the black sand beaches, the landing went well. As the Marines neared the higher, inland terraces of Iwo's beaches, weighed down in the clinging volcanic sand, a hail of pre-sighted artillery and mortar fire crashed down, augmented by automatic weapons fire from nearly invisible Japanese defensive works. The Marines advanced slowly against the determined Japanese resistance, but were able to get over 30,000 men ashore by nightfall.

Though the Japanese fought tenaciously the Marines inched forward, by 23 February isolating Mount Suribachi, soon reaching its summit in bitter fighting. Sometimes the Marines resorted to pouring gasoline into defensive ravines along Mount Suribachi's slopes and then setting it ablaze. At the same time the Marines also butted heads with Kuribayashi's main defensive networks surrounding the two airfields on the island. Here US soldiers had to fight their way through hellish, volcanic terrain, flushing determined Japanese from their

hideouts and often using flamethrowers to incinerate the determined defenders. In many ways there was little subtlety in the horrific fighting. Marines slogged their way through defensive features that they dubbed Death Valley and the Meat Grinder. Infantry tactics resembled that of World War I, for the Marines resurrected the artillery tactic known as the creeping barrage. Curtains of artillery fire would keep the Japanese in their pillboxes until

TOP: A flamethrowing tank lashes out at a Japanese bunker on Okinawa. The Japanese resorted to suicide attacks to defend their dwindling positions.

ABOVE: Marines, supported by a bazooka, make ready to assault a Japanese-held ridge on Okinawa.

ABOVE: Exhausted Chindit survivors aboard an aircraft returning to India. Trained to operate in the trackless Burmese jungles, the Chindits were not a notable success militarily, but they did show for the first time that European troops could more than hold their own in the jungle with the Japanese.

the Marines were upon them, blasting and burning them out at close range. Sometimes the Japanese sallied forth, either trying to infiltrate Marine lines, or launching themselves in suicide attacks. Mostly, though, they fought and died inside their defensive networks.

By the end of February the Marines had taken half of the island, and had overthrown much of the most extensive Japanese defences. Realizing that the end was inevitable, Kuribayashi radioed Tokyo apologizing for his defeat, and made ready to resist to the last. Increasingly the Japanese sacrificed themselves in suicide attacks. On 26 March many high-ranking Japanese officers, possibly including Kuribayashi himself, threw themselves at American lines in a last, desperate bid to retain their honour. Iwo Jima had been taken, but at a stunning cost. In the battle some 5931 Marines were killed and over

17,000 were wounded. Nearly 18,000 Japanese soldiers perished in the inferno, and only 216 surrendered, many of whom were gravely wounded. Several Japanese survivors, numbering in the thousands, remained hidden in their defences and doggedly resisting US victory. Some of them remained until several years after the close of the conflict, believing that the war in the Pacific was still raging.

Thus the American ring began to close around Japan, culminating in dramatic naval victories surrounding the Philippines, the invasion of Okinawa and the dropping of the atomic bomb. Though naval battles, from the Coral Sea to Leyte Gulf dominated the story of the Pacific war it was, once again, the infantryman that took and held land, enabling the American naval and air forces to advance. The fighting on the islands of the Pacific was of a grim, attritional nature. Relatively small numbers of Marines and US soldiers advanced from island to island, scoring their first major victory at Guadalcanal. Unlike the mechanized wonders that were European armies, the Marines slogged slowly forward, facing fanatical Japanese resistance and retaliating with brute force. Though the terrain of the Pacific was quite different, the fighting in those islands would have been quite familiar to a soldier who had fought in the Great War.

IRREGULAR WAR IN BURMA

Pushed back to the Indian border, the British sought methods by which to strike back at the Japanese, both to pre-empt any new Japanese offensives and to herald their return to Burma. Such a move would also do much to re-open supply lines to Nationalist forces in China and would receive aid from American and Chinese forces under the command of General 'Vinegar Joe' Stillwell. On 21 September 1942 British and Indian forces, under the overall command of

General Wavell, pressed forward in the Arakan Campaign, but met little success. One aspect of the campaign, though, received much attention and helped to lay the foundations for the jungle fighting which was later seen during the war against counterinsurgents in Vietnam.

General Orde Windgate, having learned the principles of guerrilla warfare from Jewish insurgents in Palestine, led the training of some 3000 British, Ghurka and Burmese soldiers, hoping to lead a

ABOVE: Chinese irregular forces cross the Salween River in Burma. Chindit operations supported the Chinese/American campaign to take northern Burma.

BELOW: Chinese regular troops load an 82mm (3.23in.) mortar during assault on Japanese positions on Pingka Ridge in 1944.

ABOVE: A British infantryman of the Fourteenth Army in Burma. British forces developed jungle fighting skills which would influence insurgency fighting for the next three decades, right through the Vietnam War.

RIGHT: Under the command of General 'Bill' Slim, the British Fourteenth Army in Burma outmanoeuvred and outfought one of the largest Japanese armies outside of Manchuria.

could act in units of brigades or larger, often melted into companies and even squads to facilitate stealthy movement through the enemy-held terrain. On 8 February 1943, with Wavell's reluctant blessing, the Chindits advanced over the Chindwin River into Burma and succeeded in catching the Japanese completely unaware. The Chindits blew up bridges and severed the main Japanese rail route in more than 30 places. Having attracted the attention of the Japanese, Windgate's forces then simply melted into the trackless rainforest. One Japanese general stated, 'If they stay in the jungle, they will starve.' However, the Chindits survived, living off the land and off air drops, and they emerged only to cause yet more havoc. Eventually the Japanese had to devote two whole divisions to the Chindits' destruction.

However, Orde Windgate soon made a critical error. He moved his forces out of their jungle hideaways and out of the range of Allied air drops, in an ill-fated effort to link up with British ground forces further south. Without cover and without supplies, the Chindits found themselves sorely pressed by persistent Japanese attacks. Suffering heavy losses, on 24 March Windgate received orders to retreat. Once again relying on stealth, the Chindits broke down into small units and then exfiltrated through the encircling Japanese forces.

'long-range penetration' raid into Burma which was aimed at disrupting Japanese communications and logistics. Highly trained and motivated, the Chindits, as they came to be known, practised the art of living off of the land in jungle terrain, planning only to be supplied by air drops. Additionally the Chindits, though they

Forced to retreat 241km (150 miles) through rugged terrain, always under the threat of enemy attack and with no supplies, the Chindits lived off the land and had to use all of their jungle skills to survive the trek. Many died on the journey and most considered their mission a failure. However, a small unit had proven that it could, with no vulnerable supply lines, spread havoc deep into enemy territory. With better planning, the Chindits were certain that they could nearly paralyze Japanese movements, and force the Japanese to devote many front-line units to defensive roles in the rear, serving to help bring balance to the war.

IMPHAL AND KOHIMA

In March 1944 the Chindits struck again. Along with a thrust by Merrill's Marauders, American and Chinese soldiers using similar tactics, the Chindits hoped to seize control of northern Burma. At the same time the Japanese went forward into their long-awaited 'U-GO Offensive' aimed at overthrowing British India. Three brigades of Chindits entered Burma by land and by glider landings near the Irrawaddy River at a site dubbed Broadway.

With their airfield connection to the outside world secure, the Chindits set up a series of defensive patrols and also set out to besiege the critical Japanese communications hub of Indaw. At the same time, Merrill's Marauders pressed down the Mogaung River valley towards the major Japanese hub of Myitkyina. Sensing the danger General Honda, the Japanese commander in the area, began to remove troops from the front lines in India to deal with the threat to his lines of communication. The Japanese focused their attacks on the Chindit supply bases, including the Broadway airfield.

In battles that presaged tactics used in Vietnam, the Chindits sent patrols out into the jungle from their secure base area, detecting Japanese movements before they became a definitive threat. In addition, forward air controllers worked to call down punishing air attacks on the unsuspecting Japanese. Both sides manoeuvred for tactical supremacy, but the Chindits were successful in defending their bases and in April they succeeded in

occupying Indaw, where they destroyed massive caches of Japanese supplies.

After Windgate's untimely death, the Chindits, having caused much havoc, moved with Merrill's Marauders to capture Myitkyina and succeeded in their efforts to free northern Burma from Japanese control. Meanwhile, the Japanese offensive against India collapsed, leading to a general Japanese retreat in the area. The Allies now shifted over to the general offensive in Southeast Asia driving towards victory even as naval power in the Pacific broke the back of the Japanese empire. The tactics on the islands of the Pacific, though they had

ABOVE: Fighting through the jungle, often supplied and supported solely by air, British and American Chindits and Marauders developed many of the techniques used by the Special Forces in Vietnam.

involved modern amphibious landings, had closely resembled the brutal infantry advances of the Great War. In Burma, though, the British and Americans had turned the rugged terrain into an advantage. They relied on forward air bases and jungle fighting skills to act as force multipliers for their numerically weak units. The result was a resounding victory. Further to the east, in Vietnam, Ho Chi Minh and his Viet Minh were the only forces now standing against Japanese rule. As World War II ended, the Viet Minh seized control, only to face the return of their French colonial masters. It was in that resulting, unexpected war where the irregular tactics of jungle warfare which had been pioneered by Windgate and Merrill would come to their final fruition.

CHAPTER SIX

AFTER D-DAY

Germany's high-water mark came in 1942, but from that point on, there was nowhere to go but down. Hitler's Third Reich was at war with the world's most powerful nations, and retribution would not be long in coming.

Despite some setbacks, the Germans had amassed a large number of successes from the time of Hitler's ascension to power until the summer of 1942. By that time, the Germans, who were engaged on several fronts, began to suffer a series of defeats. The beginning of the end in North Africa came in October 1942 when the British achieved victory in the Battle of El Alamein. Seven months later the conflict in North Africa ended with the German surrender in Tunisia. In 1943, the Soviets handed the Germans several defeats in the east; the two most devastating defeats occurred at Stalingrad and Kursk. Following their defeat at Kursk in the summer of 1943, German forces began the slow retreat back to Germany.

Sicily became the next target of a joint British–American effort. Before the campaign started, the Allies launched a major air assault against the enemy's island airfields. Code-named 'Operation Husky', the invasion of Sicily began on 10 July 1943 as seven Allied divisions landed on the southeast corner of the island. The largest amphibious landing of the war included two armies, the British Eighth Army, commanded by Montgomery, and the US Seventh Army, led by General George Patton. While more troops landed later on

LEFT: The beginning of the end for Nazi Germany came in Tunisia, soon after the disaster at Stalingrad. British and Commonwealth troops, veterans of years of desert war, linked up with fresh American divisions from Algeria to force Panzer Army Afrika into a final defeat. The next step was to attack the Axis in Europe.

ABOVE: The first Allied landings on Axis territory took place in Sicily. British and American armies swept through the large Italian island, forcing the Germans back to the Italian mainland and dealing Mussolini a shattering blow.

Normandy, on the first day of 'Husky', more soldiers came ashore than was the case on the first day of 'Overlord'. The Allies expected little resistance on the island because of the Italians' low morale and the small number of German defenders and, although they encountered stiff German defensive efforts, the Allies gained control of Sicily by the middle of August. Because of the methodical way in which Montgomery managed the campaign, however, a large number of German troops succeeded in escaping to Italy before they could be captured by the Allies.

ON TO ITALY

The next logical step after the fall of Sicily was Italy. As had been the case in North Africa and Sicily, the toughest resistance in Italy would come from the Germans, not the Italians. Although the Italians accepted the conditions for armistice on 1 September, two days later the Eighth Army came ashore on the toe of Italy. On 9 September a joint British–American force landed south of Naples in the Gulf of Salerno.

Code-named 'Avalanche', the invasion of Salerno nearly failed. Under General Mark Clark's direction, two British

divisions and an American infantry division landed on opposite sides of the River Sele. Because of the steep terrain behind the invasion beaches, the defenders, entrenched on the high ground, had a definite advantage. While the German 16th Panzer Division defended the beach, Allied troops fought their way on to the shore. By 12 September, the Allies had established a narrow beachhead. Because the two forces had not yet linked their positions, they were extremely vulnerable. On 13 September parts of four panzer and panzergrenadier divisions counterattacked the gap between the British and American positions. Despite the Allies' critical situation, stiff fighting and the timely arrival of reinforcements to the beachhead saved the day.

The Germans made the Allies fight for every step they took on Italian soil. Not willing to relinquish control of Italy to the enemy, the Germans constructed three lines of defence. The Gustav Line stretched across the peninsula south of Monte Cassino; the Gothic Line traversed the country north of Arezzo; and the Alpine Line protected the area north of the Verona–Trieste line. Because of the Germans' strong defensive positions south of Rome, the Allies advanced very slowly, which caused them to change their objective. Although the Italian campaign seemed to have become a strategic dead end, the Allies did not abandon it. Instead, they planned to use it to tie down German forces and prevent their transfer to other theatres, particularly the Eastern Front or France.

The Italian campaign did allow the Allies to accomplish another objective: the refinement of the fighting methods that they had developed in North Africa. They recognized the importance of achieving air superiority before launching a land battle, and they worked on improvement of their artillery tactics. In addition, the Allies developed better ground-air liaison and emphasized greater combined-arms cooperation, especially infantry-tank cooperation. Although these changes and improvements would prove useful in the Northwest Europe campaign, numerous factors − slow advances and poorly coordinated, firepower-laden battles − impeded the Allies' progress in Italy.

British and American military officials learned much from the amphibious landings on Sicily and Italy. They would apply the lessons they learned when planning their next major amphibious assault, the invasion of Northwest Europe. The Germans, who were expecting the Allies to come, had begun to construct the Atlantic Wall, a series of defences meant to thwart an enemy assault. The Allies had been planning a cross-Channel invasion for some time, and many factors determined where and when the attack would come. British and American military leaders had no illusions about the cost − in terms of men, weapons, ammunition, equipment, food, and other supplies − of a campaign against German forces. The campaign would have to include an amphibious landing in a well-defended region followed by a thrust to drive them back into Germany and then achieve an unconditional surrender.

PLANNING THE INVASION

By January 1944 the Combined Chiefs of Staff had chosen an invasion site − Normandy − and also established the command structure for the invasion force. While General Dwight Eisenhower was the overall commander of the operation and of Supreme Headquarters Allied Expeditionary Force (SHAEF), Montgomery, the commander of 21st

LEFT: Although the Italians surrendered in September 1943, the Allies did not easily gain control of the peninsula. The Italian terrain is far from easy to fight through, and every major town had the potential to bog down any advance for weeks or even months, as each had to be cleared in fierce house-to-house, or more accurately, ruin-to-ruin battles.

Army Group, would supervise the landings. Eisenhower and Montgomery agreed that the initial force should consist of five divisions. The Allies divided the landing area into five beaches. In the west, American forces would land on Utah and Omaha beaches. British and Canadian troops would land on Gold, Juno, and Sword beaches. Prior to the arrival of the amphibious forces, the British 6th Airborne Division would drop behind the British beaches and capture the bridges across the River Orne. The American 82nd and 101st Airborne Divisions would parachute behind the lines to protect the westernmost beaches, which were the most exposed. It was crucial for the Allies to attack, seize their primary objectives, and land as much men and equipment as possible before the Germans counterattacked. After the forces landed, established beachheads, and began bringing in reinforcements and supplies, it was vital that the Allies break through the enemy's defences and begin a war of mobility, which the Germans would have difficulty containing.

In the months prior to D-Day, the Allies prepared in England and on the Continent. The Allies initiated a bombing campaign designed to disrupt the enemy. In addition to industrial targets in Germany, Allied bombers focused on France and attacked railway lines, marshalling yards, supply depots, and bridges, particularly those across the River Seine. The goal of the Allies' Transportation Plan was to limit the enemy's mobility and access to the Normandy battlefield. By 6 June Allied bombers had damaged or destroyed all the Seine bridges north of Paris. After the landings, the bombers began attacking bridges across the River Loire in order to restrict enemy movement from the south into Normandy.

Determined to limit the enemy's presence in Normandy both before and after the invasion, the Combined Chiefs of Staff approved the implementation of a deception plan, 'Operation Fortitude'. Although there were many facets to the plan, the main goal of 'Fortitude' was to persuade the Germans to continue to focus on the Pas de Calais, not

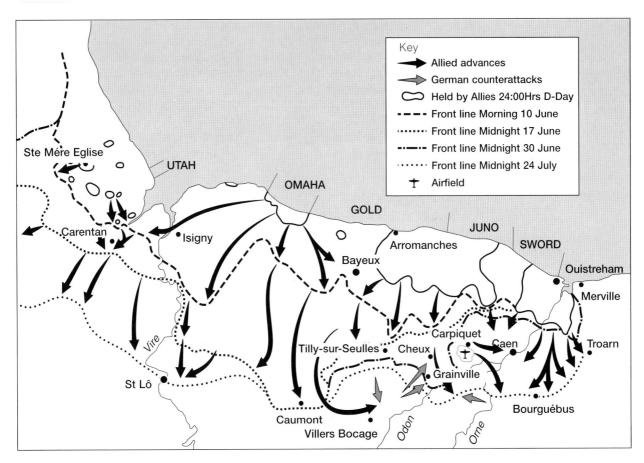

Key
→ Allied advances
⇨ German counterattacks
◠ Held by Allies 24:00Hrs D-Day
– – – Front line Morning 10 June
· · · · Front line Midnight 17 June
–·–·– Front line Midnight 30 June
· · · · · Front line Midnight 24 July
✝ Airfield

Ste Mére Eglise
UTAH
OMAHA
GOLD
JUNO
SWORD
Arromanches
Ouistreham
Carentan
Isigny
Bayeux
Merville
Vire
Carpiquet
Tilly-sur-Seulles
Cheux
Caen
Troarn
St Lô
Grainville
Bourguébus
Caumont
Villers Bocage
Odon
Orne

Normandy, as the invasion site. The Germans, who knew the Allies would come soon, believed that the Pas de Calais, because of its access to the Ruhr industrial area and Berlin, would be the site of the enemy's assault. 'Fortitude' would reinforce that idea.

The Allies planned to use whatever means were at their disposal to ensure the success of 'Operation Overlord'. Although tanks units would begin landing on the first day, the brunt of the attack would be borne by paratroop and infantry divisions. By the end of 1943 the British and Americans, under General Bernard Montgomery's direction, had developed a 'cautious, attritional, firepower-reliant operational' technique. This technique, called 'Colossal Cracks', demanded the meticulous planning of set-piece battles that relied on a heavy concentration of force, massive artillery, and air power in conjunction with integrated tactical air support. Not completely convinced about Montgomery's battle technique, the Americans, while recognizing the limitations which massed firepower on the battlefield placed on mobile operations, believed that mobility was

key to the defeat of the enemy's forces. Concerns about the morale and the casualty levels of their own forces, however, persuaded the Allies to take an attritional approach before implementing sustained mobile offensives. The Allies planned to use mobility to exploit battlefield advantages and achieve success. Their superiority in resources, air power, and manpower would eventually allow the Allies to accomplish their ultimate goal: the defeat of Germany.

ALLIED LANDINGS

Although it was scheduled for 5 June 1944, weather conditions forced the delay of D-Day to 6 June. Planes carrying paratroops and towing gliders took off from several British airfields during the night of 5 June. Around midnight the first paratroop pathfinders were marking drop zones. In the early morning hours of 6 June, parachute and glider forces from the British 6th Airborne Division began landing near the River Orne, north and east of Caen. Their job was to seize bridges across the river and to eliminate the enemy's communications centres and strongholds that could threaten the Sword beach

ABOVE: The Allied invasion of Northwest Europe began when troops landed in Normandy on 6 June 1944. The Allies divided the landing area into five beaches: Utah and Omaha which were to be assaulted by the Americans, and Gold, Juno, and Sword, which were the responsibility of the British and Canadians.

landings. Planes dropped the American 82nd and 101st Infantry Divisons over the marshes of the Carentan estuary. These paratroopers had the job of defending and supporting the landings on the Utah beach.

The invasion took the Germans by surprise. Because of command-structure problems and confusion in the invasion area, the Germans lost their best, and only real, chance to drive the enemy's soldiers back into the English Channel. Shortly before dawn on 6 June, a naval bombardment of the invasion area began and continued periodically throughout the day. At dawn, the first troops arrived on Utah, Omaha, Gold, Juno, and Sword beaches. The troops encountered little trouble on Utah and the British beaches. Despite their initial successes, British and Canadian forces failed to reach their objectives by the end of D-Day. As a result, the 12th SS Panzer Division arrived, established a defensive position, and helped the infantry deny Caen to the Allies for almost six weeks.

The American forces landing on Omaha encountered stiff enemy resistance that hindered their exit from the beach. Several factors contributed to the difficulties encountered by the US 1st and 29th Infantry Divisions. Shortly

before D-Day Allied intelligence indicated the arrival of the German 352nd Division behind Omaha beach. It was too late to make any adjustments in the invasion plan. Then strong currents and poor navigation resulted in the Americans landing in the wrong places. They immediately encountered heavy fire from the enemy's well-placed defensive positions. In addition, the rough seas swamped many of the amphibious trucks that carried much of the essential artillery, and the special 'DD' tanks, which gave armoured support on the other Allied beaches, were launched too far out from Omaha beach, and they tragically sank with their crews still on board.

BLOODY OMAHA

Although General Bradley considered diverting the follow-up divisions to other beaches, he decided to send them where they were most needed: Omaha beach. As the day wore on, small groups of American soldiers braved the withering fire to advance and push the enemy off the high ground. Casualties mounted; by the end of the day, they numbered 3000 on Omaha. As dusk approached, parts of the divisions had begun the move inland. By the next

BELOW: Field Marshal Erwin Rommel examines a section of the coastal defences along Hitler's Atlantic Wall. Barbed wire, mines, and other obstacles protected the areas in front of the German bunkers that lined the coast, all designed to ensure that any Allied landing would never get off the beach.

The liberation of France began when Allied troops landed in Normandy. On 6 June 1944 American soldiers, coming ashore on Omaha beach, experienced withering enemy fire. The first wave of men fired from behind the beach obstacles to cover subsequent waves of men. The rough seas made the landing even more difficult, and because of their heavy packs, soldiers frequently drowned if they could not get their footing in the deep water.

morning, the Allies had managed to secure four beachheads and land a total of over 177,000 men.

The Germans had failed to prevent an enemy foothold in France. They tried to restrict the Allies to a narrow beachhead while quickly moving reinforcements to the front. The Allies, on the other hand, attempted to break through the Germans' defences and move inland. As the first day of the battle progressed, the flow of Allied troops landing on the Normandy beaches continued and served only to make the Germans' counterefforts more difficult.

From the beginning two distinct fights evolved in Normandy: the Americans battling in the west and the British and Canadians fighting in the east. Massive firepower, air power and resources supported both battles. American forces in the west had an important job to accomplish before they began to advance southwards into France. Their goal was the Cotentin Peninsula and Cherbourg. Instead of moving south, they turned to the right and then drove across and then up the peninsula. Stiff enemy resistance kept the Cherbourg port out of American hands until 27 June. Because of the damage inflicted by the Germans on the port facilities, however, the piers remained closed to ships until 7 August.

Once Cherbourg was in their hands, the American forces moved back down the Cotentin peninsula and focused on the enemy's defensive positions in the

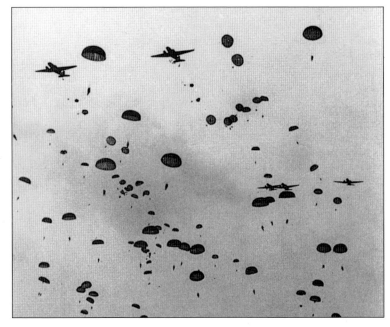

western part of Normandy. Here the Germans used the landscape to their advantage. The defining characteristic of the *bocage* country, the hedgerows, greatly slowed down the American advance. The Americans took advantage of organizational and tactical changes that had been initiated by General Lesley McNair in 1942. Guided by a belief in mobility, McNair pushed for the establishment of triangular divisions, which only contained the minimum forces essential for mobile operations against limited opposition. Triangular divisions included a combination of infantry, artillery, reconnaissance, and

ABOVE: Airborne troops played a vital role in the early stages of the Normandy landings, being dropped ahead of the seaborne assaults in order to seize key bridges intact. Three months later, however, the same idea would lead to a disaster at Arnhem.

ABOVE: Despite heavy enemy fire, American soldiers run down the ramp of a landing craft manned by the US Coast Guard on Omaha Beach on D-Day, 6 June 1944. Although German machine-gun fire pinned down the first waves of troops, air and sea bombardment, plus the landing of additional troops and vehicles by Coast Guard and Navy amphibious landing craft, overwhelmed the defenders.

engineer formations. Group headquarters would control the nondivisional units: armour, antiaircraft, field artillery, mechanized cavalry, and engineer formations. Requirements and missions would dictate the combat power of a division. When applied overseas, these and other concepts stressed by McNair met with limited success.

The Normandy *bocage* demanded close cooperation between engineers, infantry, and mortar units. Because of the hedgerows, the Germans successfully constructed defensive positions in depth. The lack of adequate air and logistical support forced the Germans to utilize the terrain for their defence of Normandy. The thick hedgerows surrounding each field and the villages provided excellent cover for the defenders. Enemy snipers attacked the American soldiers whenever they broke through a hedgerow and entered a field. These snipers frequently either reduced the speed of the Americans' advance through the area or brought it to a halt. Supported by mortars bombarding the enemy, engineers and tank-mounted ploughs cut gaps in the thick hedges. Passing through the holes, tanks and infantry covered each other as they

crossed the enclosed field. Using their ploughs and fierce determination, the Americans slowly forced the Germans to move backwards.

The terrain in the eastern part of Normandy, where the British and Canadians were fighting, was much more open and advantageous for advancement if the Allies could break through the enemy's defences. On 12 June, the British, exploiting a weakness in the German defences, pierced the enemy line. The forward units, which included tanks, soft-skinned vehicles, and supply trucks, came under attack by five Tiger tanks led by Michael Wittmann, a panzer ace. The German tanks caught the British by surprise and almost destroyed the entire column. Lacking weapons capable of knocking out the well-armoured Tigers, the British lost 25 tanks and a greater number of armoured vehicles, and were forced to retreat back to their starting point.

The situation in the Caen area remained the same for the next few weeks. The Germans' II SS Panzer Corps advanced to Normandy in order to counterattack. Learning about the enemy's plan through intercepted intelligence, in an effort to thwart the

Germans, Montgomery decided to launch 'Operation Epsom' on 25 June. Designed to outflank Caen from the west, the offensive did not succeed. It did, however, prevent the Germans' planned assault. At the beginning of July, the Germans still occupied the town. The British mounted major attacks on 8 and 18 July in an effort to capture Caen, which was now largely in ruins.

While their occupation of Caen continued for several weeks, the Germans reassessed their strategy. Realizing that they could not drive the Allies back into the sea, the Germans switched to a containment strategy. If they could restrict the enemy's presence to a small area, the Germans could deny Montgomery's forces access to the open Normandy terrain, which they needed for mobile operations. In response to the enemy's strategy, Montgomery devised plans for an offensive which would break through the German line and reach the open terrain.

ARMOURED ATTACKS

Spearheaded by three British armoured divisions, the 18 July assault, 'Operation Goodwood', attempted to break through the enemy positions south of Caen. Despite early progress, the British advance quickly stalled. Although Montgomery called off 'Goodwood' only 48 hours later, the offensive did net two important results. First, the British seized Caen. Second, they forced the German armour to participate in an attritional battle. Compelled to bring up armoured reserves, the Germans withdrew forces from western Normandy, and this would open the way for the next American offensive, which was named 'Operation Cobra.'

The Allies' advance through the *bocage* country had proven costly; and the infantry bore the brunt of the losses. In the month of July alone, 90 per cent of the US First Army's 40,000 casualties were infantrymen. The Germans' losses were equally heavy. By late July German logistical problems would go a long way to facilitating the Allies' breakout. While the Germans attempted to solve their logistical difficulties and to strengthen their lines before the next enemy attack, the First American Army prepared for a breakout offensive. The adjustments

made by the First Army in its tactics and training would result in improved infantry–tank coordination.

The Americans were able to test changes in the organization of armoured divisions when they reached the open French terrain. General Jacob Devers had instituted a more flexible, functional organization, which included the creation of two Combat Commands headquarters that would, depending upon the mission, control a mixture of subordinate formations. The new divisions combined tank and infantry formations that coordinated their efforts. The armoured divisions that landed in France in 1944 were smaller than previous armoured units and consisted of nine battalions: three each of tanks,

BELOW: The German High Command rushed infantry and panzer divisions to Normandy in the weeks following the invasion. Quadruple Flakvierling 20mm (0.78in) antiaircraft guns provided some protection from fighter bombers, but Allied air superiority was such that movement of troops was soon restricted to night time.

armoured infantry, and armoured field artillery. Supported by aircraft, the tactics of the new armoured divisions revolved around speed and mobility.

Launched on 25 July, 'Operation Cobra' commenced with massive bomber strikes against the German positions the previous day. Following the air strikes, the Allies attacked the German defences. By the next day, the Allies' attack, combined with the Germans' logistical problems, shattered the enemy's resistance, and American forces broke through the line. The Americans swiftly advanced through the open landscape and doubled the size of the territory under their control. The success of

RIGHT: The hedgerows of the French *bocage* country made movement difficult for the Allies. German snipers fired on soldiers attempting to cut their way through the shrubs, and armour like this camouflaged StuG III assault gun could wreak havoc on British or American vehicles trapped in the narrow, sunken lanes.

BELOW: Sherman tanks from the British 4th Armoured Brigade engage a German strongpoint near Cheux on 26 June 1944. Tank for tank, German armour was better than that of the Allies. But the Allies had numbers on their side, and could replace their losses much more easily.

'Operation Cobra' led to key command changes and the activation of General George Patton's Third Army. Despite Montgomery's objections, Eisenhower activated 12th Army Group, commanded by Bradley. General Courtney Hodges became the new commander of First Army, and on 1 August Patton and his Third Army received an active role in the campaign in France.

Although the Germans attempted to penetrate the American line, seize Avranches, and isolate the Third Army in Brittany, staunch resistance by the US 30th Infantry Division prevented the enemy's plan from becoming a reality. Due to concerns about access to the port facilities necessary to sustain the offensive in France, elements of the Third Army's armoured forces raced from Brest, which

it reached by 6 August. Acknowledging Brest's strategic importance, the Germans ordered the defenders to 'stand fast.' The Americans concentrated three infantry divisions, as well as a large store of ammunition, for the attack on Brest. Although the attack commenced on 25 August, the same day that the Germans lost control of Paris, the Americans finally captured the port city on 19 September, suffering a total of 10,000 casualties in the process.

THIRD ARMY BREAK OUT

On 3 August 1944, Generals Eisenhower and Montgomery approved Bradley's decision to turn his forces eastward. The First Canadian, the Second British, and the First US Armies kept the Germans occupied in Normandy while three of the Third Army's corps drove south and then east. Despite the threat of the Allied armoured divisions advancing into the German's rear, instead of allowing a retreat to the River Seine, Hitler ordered an armoured drive from Mortain towards Avranches. He wanted his forces to cut off the US divisions that were now heading south. Ultra intelligence alerted Bradley to the Germans' plan. He ordered reinforcements to Mortain. A combination of bombardment from the air and the resistance by the reinforced units stopped the Germans after they had made only slight gains. As a result of their counterattack at Mortain, the German armoured units were in danger of becoming encircled by enemy forces. Although the Allies missed the chance to capture more enemy forces, they had the Germans on the run.

As the Allies continued the advance to the east while forcing the Germans to retreat, British and American commanders debated their strategy. Montgomery and his advisers advocated a 'single thrust' along the northern route under his command. Many American commanders favoured a 'broad thrust' through the southern route into Germany through the Saar region. By late August Eisenhower decided on a 'broad thrust' on a northern route. Hodges' First Army would take the northernmost route, while Montgomery's forces advanced to take Antwerp before heading into the Ruhr. Patton's Third Army would drive towards Metz. On 1 September Eisenhower took over command of the ground campaign as had been agreed before the D-Day landings had begun.

BELOW: Bocage country, France, 1944. A Bren gunner fires through a hedgerow at a German position. Light machine guns like the Bren had become standard equipment in the infantry platoons of most armies, being used to provide close and accurate support fire.

ABOVE: General Charles De Gaulle, leader of the Free French, triumphantly enters Paris in August 1944. Although the battle for France had not yet ended, the liberation of the capital was a significant milestone.

By early September advance units had reached Belgium and southern Holland. Because of logistical problems, however, the Allies' forward movement had begun to slow down and hindered the development of their strategy. Wanting to restore momentum to the field, Montgomery convinced Eisenhower to support his new offensive plan: 'Operation Market Garden.'

Montgomery constructed a plan that was designed to reach the River Rhine as rapidly as possible before the Germans could construct strong defences and thus open a route to Berlin, ending the war as early as possible. According to the general's plan, armoured units, supported by British, American, and Polish airborne forces, would drive across Holland and the Rhine into Germany. Three airborne divisions would seize the bridges leading up to and across the river. As XXX Corps drove into northern Germany, airborne forces would capture the last, crucial bridge: the one at Arnhem.

There were many problems with Montgomery's plan from the beginning. Although Ultra and Dutch underground intelligence indicated the presence of the 9th and 10th SS Panzer Divisions in the Arnhem area, Montgomery and his staff chose to ignore the evidence. In addition, Montgomery and General F. A. M. 'Boy' Browning, his airborne commander, gave the inexperienced British 1st Airborne Division the most important job of the operation: seizing the bridge at Arnhem. The more experienced American 82nd and 101st Airborne Divisions would capture the bridges at Nijmegen and Eindhoven. To make matters worse, the 1st Airborne Division commanders agreed to a drop zone several kilometres from the bridge. Taken together, these problems would make a successful operation difficult.

Following an air assault, 'Market Garden' commenced on 17 September. Problems arose from the start. The battle did not proceed as planned. After a heroic defence of one end of the bridge for several days by the ill-equipped paratroopers against German armour, the Germans forced them to retreat. The momentum of Montgomery's northern thrust stalled, and provided Hitler with the confidence that he now needed to contemplate a risky countermove.

BATTLE OF THE BULGE

Following the German 'victory' at Arnhem, Hitler ordered the OKW to make plans for a major offensive that would secure a strategic victory. The time had come for the Germans to launch a counterattack. Although the ultimate goal would be Antwerp, the Germans needed to seize Allied fuel dumps in order to mount a successful operation to the coast.

While Eisenhower, Montgomery, and Bradley made plans for a January 1945 offensive, the Germans completed their preparations for what would become known as the Battle of the Bulge. Although the French and Belgian troops had failed to stop the German push through the Ardennes Forest in the summer of 1940, the Germans would not have an easy a time against the American defenders in the winter of 1944. Although thinly spread, the American force was much stronger than the French and Belgians had been four

years earlier. While the 82nd and 101st Airborne Divisions, recuperating after 'Market Garden', were held in reserve, inexperienced divisions from the US forces protected the line where the enemy would attack.

On 16 December the Germans attacked on a broad front between Monschau and Echternach. Because of poor weather, the Allied air forces could not provide protection for the ground forces. Although they were caught by surprise and the poor road infrastructure made movement difficult, the Americans quickly moved up their reserves. Fierce fighting ensued as the Germans breached the American line. The Americans refused to relinquish control of key road junctions, particularly those at St Vith and Bastogne. While US troops slowed the German advance and maintained control of St Vith for six days, a tank detachment in Bastogne repelled the enemy long enough for the 101st Airborne Division to reinforce the American position. Although the forces along the line north and south of Bastogne retreated, the defenders held.

BELOW: British airborne troops taking part in 'Operation Market Garden' fire a 3-inch (76.2mm) mortar. This was the first action picture taken at Arnhem.

TOP: The German assault through the Ardennes Forest in December 1944 came as a complete surprise to the Allies. Spearheading the German offensive was this elite SS *Kampfgruppe*.

ABOVE: The Allies were racing against Soviet troops for the prize of Berlin, little knowing that Eisenhower had already agreed that the Soviets would take the city.

Although cold and running low on ammunition, food, and other supplies, the 101st Division repelled repeated enemy attacks. Until reinforcements arrived, however, the infantry and parachute troops had to draw upon all available resources to stop the enemy's offensive. When the weather finally cleared on 23 December, Allied tactical and strategic air forces went into action against German forces all over the Ardennes. The beleaguered 101st Airborne Division began to receive much-needed supplies. By Christmas Day, the Allies had stopped the German advance and began to push them back. Powerful forces repeatedly hit the northern and southern flanks of the Bulge, supported by ground-attack aircraft. The 4th Armoured Division reached the defenders of Bastogne on 26 December. By the middle of January, Allied troops had forced the Germans to retreat. This was the beginning of a retreat that would not be over until the Germans finally surrendered.

Germany now began to feel pressure from both sides. As the Soviet Red Army drove the Germans out of the Soviet Union, through Poland and the Balkans, and back into their homeland, Allied troops pushed the enemy eastwards. During January and February 1945, the Western Front began to collapse.

A break in Allied communications made coordination of the battle difficult. Consequently, Eisenhower ordered Montgomery to assume control of the fighting north of Bastogne while Bradley did the same in the south. Recognizing the seriousness of the situation at Bastogne, Patton ordered the Third Army to counterattack, break through the German line in the south, and proceed rapidly to relieve the trapped 101st Division. Speed and mobility were once again the key. The Germans, lacking fuel, could not maintain their initial advantage.

By March, Allied forces fought their way into the Rhineland. A week later, the 21st Army Group controlled the river between Düsseldorf and Nijmegen. The US Third and First Armies reached the river by 7 March. One of Hodges' armoured divisions – the 9th – arrived at Remagen as the Germans tried to destroy the railroad bridge. When the Germans' efforts failed, American infantrymen rushed across the damaged bridge and captured the high ground on the other side of the river. Fearing that it might collapse, Hodges sent a large number of units across the bridge.

While the Americans advanced into Germany from the south, Montgomery mounted an offensive in the north. Montgomery's assault began with a massive bombardment by 3300 artillery guns, tactical and strategic air forces, and two airborne divisions. Although the ground attack started slowly, the British advance soon began to pick up momentum. By 1 April British and American forces linked and managed to surround the Ruhr industrial region. American, British, Canadian and French forces eliminated enemy resistance in western and central Germany, while the Soviet troops claimed the ultimate prize of the city of Berlin.

On 21 April the Soviet bombardment of the German capital began. Four days later while the fierce struggle for control of Berlin continued, American and Soviet troops met on the Elbe and divided Germany in half. On 30 April Adolf Hitler took his life. Although resistance continued for a few days, the Germans surrendered. The war in Europe was now over, but the war would continue to rage in the Pacific Theatre for another three months.

Although American and Soviet soldiers shook hands at the River Elbe, distrust between the two countries would soon emerge. At the end of the war, United States and the Soviet Union were the strongest world powers. It would not take long for them to develop military strategies which were aimed at preventing the expansion of each other's influence. As tensions between the two superpowers mounted, a Cold War would break out.

BELOW: With tanks of the 11th Armoured Division leading the way, Third US Army infantry entered Andernach, Germany, on 14 March 1945. An infantryman behind the leading tank spotted the enemy sniper who had been firing on the troops from the building on the left.

CHAPTER SEVEN

INFANTRY AFTER 1945

With the end of World War II, dormant differences between the Allies were free to come into the open. Rivalry between East and West was to lead to a new kind of struggle between two systems of living: the Cold War.

In 1945 World War II finally came to an end. The Germans surrendered in May, and the Japanese officially in September. The cost had been high. Because of the Japanese soldiers' willingness to fight to the death – to surrender was dishonourable – President Harry S. Truman made the tough decision to use the newly perfected atomic bomb. The crew of the Enola Gay dropped the first bomb on Hiroshima on 6 August. Three days later a second atomic bomb fell on Nagasaki. Despite the military's desire to continue fighting, Emperor Hirohito ordered his advisers to accept America's terms for surrender, provided he could retain a ceremonial position. On 2 September 1945, the Japanese signed the surrender. The war was over.

Although the war had ended, peace did not return. A clash between the interests and ideas of East – the Soviet Union – and West – the United States and Britain – quickly dominated international relations. A series of disagreements between the United States, Britain, and the Soviet Union led increasingly to distrust. As tensions grew, the alliance that had existed because of a common enemy began to crack, and a different sort of war broke out between the East and the West: the Cold War. Lasting from 1945 until 1990, the

LEFT: The life of the infantryman has changed more since the end of World War II than in any other period of history. The process of mechanization which began during that conflict has been completed, the foot soldier now being invariably conveyed to battle in armoured personnel carriers like these American M113s.

RIGHT: Participating in the
CARIB-EX military
exercises at the Rio Hato
airstrip in the Republic of
Panama, members of the US
325th Airborne Infantry
direct the air drop of heavy
equipment. US Army, Navy,
Marine and Air Force
personnel took part in
the exercises.

Cold War was characterized by ideological differences, threats, 'brinkmanship', 'containment', hot and cold periods, and a series of crises in Europe, Asia and the Middle East.

The atomic bomb changed the nature of militaries after the war. Both the fear of and the desire to produce atomic bombs drove the leading powers of the world – particularly the United States and the Soviet Union – to embark upon an arms race and to engage in atomic bomb diplomacy. The distrust and disillusionment between East and West resulted in ever-increasing tensions. Questions about the role of conventional troops and the nature of future wars arose. Because both countries wanted to avoid nuclear conflict, they ultimately developed strategies based upon the concept of limited war.

Two trends permeated the major powers' armies in the post-war period and suggested that mechanized armour was not the solution to future combined-arms combat problems. Many strategists believed that the atomic bomb made traditional land warfare obsolete. Other strategists expected any future land combat to be radically different from what had been experienced during the war. In the late 1940s, many in the

United States military believed that in the future the infantry would protect strategic bomber bases and 'mop up' enemy forces dispersed by nuclear attacks. Although they later came to recognize the need for land forces, many strategists believed that nuclear weapons would render the large and conventional armies unnecessary.

Both the United States and the Soviet Union realized that a nuclear conflict between them would be suicide; therefore, the two nations participated in a series of 'proxy wars,' in which they supported smaller countries engaged in conflicts. These proxy, or national liberation wars, because they engaged in guerrilla tactics and stressed political objectives, challenged the use of conventional mechanized armies. Western armies had two ways in which to meet the challenge of national liberation wars. They could either try to use their conventional forces in an unconventional-type conflict, or develop light infantry units at the expense of armoured weapons. While Western armies debated the options and underwent a series of changes, the Soviet Army used the period between 1945 and 1970 to eliminate its technical disadvantages in conventional combat.

After World War II, the Soviet Army underwent four periods of doctrine and organization. Between 1945 and 1953, although they maintained the tactical and operational doctrines and organization used during the war, they disbanded part of their forces. The development of nuclear-equipped arms relegated conventional forces to the background between 1953 and 1967. The desire to construct an armour-heavy force that could survive and exploit a nuclear attack took precedence over the maintenance of a Soviet Army trained in combined-tactics. From the late 1960s until the mid-1980s, however, Soviet attention returned to conventional forces. The Soviet Army prepared a doctrine for a combined-arms mechanized conflict that might or might not be used in conjunction with nuclear weapons. Finally, after the mid-1980s, two factors – the war in Afghanistan and the development of new weapons – dictated that a thorough reorganization of the Soviet Army be conducted.

Because it did not possess any nuclear weapons, in 1945 the Soviet Army developed a doctrine to enable its conventional forces to counter any eventuality in Europe. While they constructed a doctrine based on the reality of the mid-1940s, the Soviets also attempted to develop their own nuclear weapons. By 1948, however, Stalin reduced the size of the military by more than half, but at the same time, he authorized an increase in the number of armoured and mechanized formations. At the same time that the military increased its mechanization, Soviet strategists created a new doctrine. In 1949, after they detonated their first atomic bomb, the Soviets' emphasis on the use of conventional ground forces for national security declined.

SOVIET MODERNIZATION

Following Stalin's death in 1953, the Soviets began a 'Revolution in Military Affairs' and focused on nuclear weapons and electronic and communications improvements. Field Marshal Georgi Zhukov, who desired to adapt Soviet ground forces to the realities of nuclear warfare, received permission to reorganize the military. To improve command and control and provide protection against nuclear weapons, Zhukov reduced unit size and ordered better armour. With the reorganization came a doctrine that dictated a nuclear strike followed by exploitation by mechanized, armour-heavy forces. Combining new equipment and reduced size, all ground units became motorized and, in numerous instances, mechanized. By 1959 the Soviets demonstrated their

BELOW: On exercise in 1952. After helicopter-borne infantry units have established a beachhead on the far side of a river, the remaining force cross the river in assault boats, while engineers set up a pontoon bridge which will allow vehicles to move across the water obstacle.

reliance on nuclear capabilities with the creation of the Strategic Rocket Forces. In the early 1960s, Soviet commanders, driven by the necessity for rapid exploitation and Deep Battle, examined the possibility of limited use of air mobility. In addition, because ground forces had been relegated to a 'mopping-up' role, the infantry experienced a reduction in size, and conventional forces in general took a back seat to the more important nuclear-strike capabilities possessed by the military.

The removal of Nikita Khrushchev from power in 1964 opened the way for changes in the Soviet military. Military leaders debated the future direction of the armed forces, particularly in light of the new American doctrine of flexible response. Recognizing that future wars would not be fought in the United States, the military emphasized forces capable of fighting a wide spectrum of possible conflicts: terrorism, guerrilla warfare, a conventional conflict, or a nuclear war. Acknowledging that it was not realistic to have only one option to an external military threat, Soviet officers returned to the possibility of conventional combined-arms warfare. They analyzed the memoirs of World War II senior commanders and focused on two concepts – mobile group and forward detachment – which were essential to their mechanized exploitation and pursuit methods. By the 1970s the organization of the Soviet military reflected the evolution of military doctrine. Mechanized infantry and conventional artillery battalions became re-attached to tank regiments. The Soviets' doctrine and organization of combined-arms combat had come 'full circle' by the mid-1970s and incorporated improved armoured fighting vehicles and helicopters into the 'Deep Battle' and mechanized combined-arms doctrine.

RIGHT: Although poison gas was not used in World War II, the threat of Nuclear, Biological and Chemical warfare (NBC) was taken very seriously during the Cold War. Seen here in November 1956, members of the US Army Chemical Center in Edgewood, Maryland, wear NBC masks as they make a field demonstration of a new resuscitator that was designed at their facility.

Unlike their Soviet counterparts, American field commanders were not completely satisfied with the organization of the United States military in 1945. An analysis of the performance of the triangular infantry division suggested that each division should have armour that would support infantry attacks and function as the army's primary antitank weapon. The incorporation of this organization, however, would tie the tanks to the infantry and prevent them from attacking and exploiting the vulnerabilities of the enemy. By 1946, the number of armoured infantry battalions in each armoured division increased from a total of three, to four.

DOWNSIZING THE US MILITARY

Two factors prevented the reorganization from having a large impact on the American land forces. First, demobilization after the war reduced the overall size of the military. Second, the emphasis of military doctrine shifted from combined-arms warfare to nuclear warfare. In general, because of the United States' initial monopoly on nuclear weapons, only army commanders envisioned the importance of having combat-ready forces. Despite some changes in the organization and size of the ground forces, the military doctrine emphasized the United States' ability to

launch an air-atomic campaign against Soviet targets to shock and unhinge the Soviet Government. Events would force the United States to rethink its doctrine.

One of the major disagreements between East and West emerged even before Hitler had been defeated: the nature of the post-war world. Stalin proved willing to be the first to challenge the situation by commencing a period of expansion beginning in the spring of 1946. Although the United States and Britain willingly conceded the Soviets' right to a sphere of influence, they feared the spread of world Communism, which did not recognize borders. In response to the apparent Soviet threat, the United States developed the policy of containment. The United States would take a series of steps to 'contain' the spread of Communism and the Soviets' influence. In 1947 President Harry Truman announced the commencement of the Marshall Plan, which would provide the financial means to rebuild Europe. Stalin countered with the Molotov Plan, which would help rebuild Eastern Europe.

The United States planned to use the atomic bomb to 'contain' the Soviet Union. In mid-1947, the Joint Chiefs of Staff developed a war plan – 'Broiler' – which incorporated the bomb for a first strike against Soviet political targets. In the spring of 1948 Truman was able to

ABOVE: During the Cold War the US Army established training units equipped with enemy weapons, initially to establish their capabilities and then train regular units in how to counter them. 'Opfors' members are seen here with an RPG-7 antitank grenade launcher, an AK-47 assault rifle, a flamethrower and a T-54 main battle tank.

test the atomic strategy when the Soviets cut rail and road traffic to West Berlin. Although he did not deploy bombs to air bases in Great Britain, Truman implied that he had. The blockade ended over a year later without the use of nuclear weapons. Although the United States continued to build atomic bombs and develop war plans, such as 'Fleetwood,' the military did not rule out the possibility of a limited conventional campaign. A limited conventional campaign would, however, follow on from an atomic attack.

In 1949, when the Soviets exploded an atomic bomb, the situation changed. By the end of the year, United States' military strategists, placing larger emphasis on the bomb, developed a new war plan: 'Offtackle'. In the early 1950s the US continued to fund the atomic bomb program at the expense of conventional forces. A further complication arose when the Soviet-equipped North Korean Army invaded South Korea in June 1950. America's commitment to the fight in South Korea demonstrated the military's deficiencies. Not only did the troops lack training and combat power, but the Army's force structure also failed to match its doctrine. The need to correct the military's problems and the involvement

on the Korean peninsula resulted in a temporary increase in both the Army's budget, and its size.

THE NUCLEAR SHADOW

A new administration came to power in the United States when Dwight Eisenhower was elected president. Eisenhower and his administration developed a national strategy based on 'massive retaliation' with nuclear weapons. As had been the case for the Soviet Army, the American Army had to devise a doctrine and a structure for ground forces to function on a nuclear battlefield. While maintaining efficient command and control, the doctrine had to demonstrate an ability for greater dispersion and flexibility. The Army had to be prepared for deployment anywhere in the world on short notice. These factors dictated the tactical structure of the army: small, dispersed units that were not nuclear targets, self-sufficient when isolated, and self-supporting without vulnerable supply lines. The result was the 'Pentomic Division', five units within a division that could function on either an atomic or a conventional battlefield. Because of the need for mobility, the 'Pentomic Division' would include a helicopter company and numerous APCs (armoured personnel carriers) in an infantry formation. The pentomic changes had the most effect on infantry units. By 1959 the United States Army, which had undergone major structural and operational changes, was theoretically ready to meet the demands of nuclear warfare.

The early 1960s brought a new administration and a new military doctrine: 'flexible response'. Under the new doctrine, the military would field forces that were capable of fighting a wide range of wars: terrorism to full-scale conventional to nuclear. Because the 'pentomic division' seemed ill-suited to the new doctrine, the Kennedy administration authorized studies into the reorganization of the army. Under the new organization, different types of Reorganization Objectives Army Divisions (ROADs) would operate from the same base. The largest manoeuvre organization with a fixed structure became the battalion. The advantage of the ROAD division was its ability to

BELOW: Post-war Soviet tactics emphasized surprise and speed of movement. Red Army soldiers, wearing the standard leaf-pattern camouflage uniforms of the 1970s, practise deploying from a BTR-152 armoured personnel carrier.

LEFT: Observing the placement of their forces, these Czechoslovakian officers make a final review of their tactical plans before commencing the manoeuvres. Behind them can be seen the Czech-built RM-70 versions of the standard Warsaw Pact 122mm multiple rocket launcher.

adjust its structure to the assigned mission. Consequently, depending upon the threat, the army could deploy either a particular type of division – armoured, mechanized, conventional infantry, airborne, or air mobile – or various kinds of divisions. The drawbacks of using ROAD divisions were some inefficiency and coordination problems.

The United States and Soviet Armies were not the only ones to undergo changes in doctrine and organization during the Cold War. For the 10 years following the formation of the North Atlantic Treaty Organization (NATO), however, the armies of most European nations made few alterations to their existing structure and doctrine. Until 1960 the military policies of the Western European states were similar to those of the post-World War I period. Most European countries did not want to fund new weapons systems. Following 10 years of occupation, West Germany was allowed to re-arm because of the East-West conflict. The Bundeswehr (Federal Armed Forces), because of the inability to mechanize all formations, provided front-line units with different equipment and tactics from other units. Both the French and the British Armies developed three elements. First was a fully mechanized force that had the responsibility of defending Central Europe. The second element was a less well-equipped conscript and reserve force that served at home. The final

element was a lightly equipped, strategically mobile unit that was well-trained and that would participate in conflicts outside the European sphere.

In the 1960s changing circumstances led the British and French armies to shift their attention to the defence of Europe. Because Britain, France, and West Germany focused on the concept of combined arms ('all-arms cooperation') as a principle of tactics, their militaries demonstrated similar large-unit organization with fixed, combined-arms structures. By 1961, with the end of the Algerian War, the French Army turned again to mechanized operations and organizations. In the 1960s and the 1970s the French Army developed the organic combination of different arms within one battalion. The culmination of combined-arms experiments was the mixed or 'tank-infantry' battalion. While

ABOVE: Introduced by the Soviet Army in the 1960s, the BMP was the world's first infantry fighting vehicle. Designed to take a squad into action, it differed from earlier APCs in its ability to provide powerful fire support, as well as its ability to engage enemy tanks with its AT-3 'Sagger' guided missiles.

France took the lead in forming mixed battalions, West Germany began to create mounted infantry integrated with armour. Unlike the Americans who built armoured carriers for transporting infantry who would disembark to fight, the Germans developed infantry fighting vehicles (IFVs) in which the infantry would ride and fight. Other NATO nations, as well as the Soviet Union, copied both the concept and the design of the IFVs. By the late 1970s, however, because of the development of higher-velocity tank guns, infantry formations were even more vulnerable to armoured attack than they had been in 1943.

Following World War II, as the division between East and West widened, both sides formed organizations for collective security. In April 1949 the Western powers, including the United States, the United Kingdom and Canada, signed the North Atlantic Treaty, which created NATO. West Germany became a member in 1955. In response, the Soviet Union, East Germany, and other Eastern European countries signed the Warsaw Treaty of Friendship, Cooperation, and Mutual Assistance in May 1955. The treaty had provisions for a unified military command and for the maintenance of Soviet forces within the territories of the participating states.

NATO DOCTRINE

The main goal of NATO was to unify and strengthen the military response of the Western Allies in the event of an invasion by the Soviet Union and its Warsaw Pact Allies. The purpose of such an invasion would be the spread of Communism. In the early 1950s, because the Soviet Union had much larger ground forces, NATO counted on the possibility of massive United States nuclear retaliation to deter aggression by the Soviets. In 1957, to supplement this policy, NATO deployed US nuclear weapons in Western European bases; many of them were situated in West Germany and pointed to the East. The nuclear weapons would remain under the control of the United States. Throughout the 1950s and 1960s, in addition to increasing its nuclear arsenal, NATO forces systematically developed. Despite deficiencies in size when compared to that of the Soviet military, the sophistication of the NATO units' weaponry and training made them equal in strength to their Soviet adversaries.

As the Soviet Union developed nuclear weapons, it positioned them in Eastern Europe and faced them to the West. Although the situation on the border between Western and Eastern Europe remained tense during the 1950s, 1960s and into the 1970s, other events, such as the Korean War and the Vietnam War, lessened the chance of a nuclear conflict.

The end of the Vietnam War began a period, from 1973 until 1989, that was dominated by confrontation between NATO and the Warsaw Pact. Throughout the 1970s, the Warsaw Pact, which was dominated by the Soviet Union, turned to confrontation because of suspicions of aggression towards the East by the West. In addition to building up its conventional forces, the Warsaw Pact developed and deployed intermediate-range nuclear weapons, including the SS-20 fully mobile missile system. An intermediate-range nuclear

BELOW: On 27 October 1956, a revolt against Soviet control erupted in Hungary. Soviet tanks moved into Budapest as the Red Army quickly quashed the rebellion.

missile force (INF) race was the result, and it was won by NATO, who stationed over 570 new Pershing II and ground-launched cruise missiles in Western Europe, particularly in Western Germany and the United Kingdom. The United Kingdom also began to upgrade its nuclear arsenal with the Trident D5 submarine-launched ballistic missile (SLBM) system. The decision by NATO to increase its nuclear arsenal in Europe sparked a series of protests by various groups, such as the Campaign for Nuclear Disarmament (CND).

The protests did not result in the removal of the missiles or stop the deployment of new missiles. Cruise and Pershing II missiles, which began arriving in Europe in November 1983, re-established NATO's ability to threaten limited nuclear retaliation that was designed to stop short of starting a global nuclear war. The new missiles reduced the time it would take to strike a target. The new NATO strike capabilities had an impact on Soviet military doctrine and strategy. Endeavouring to stress the weaknesses of NATO, the Soviets began to re-emphasize the 'Deep Battle' and 'Deep Operations' doctrines in the 1970s, which they demonstrated publicly. In 1981 the Warsaw Pact conducted the

'ZAPAD 81' manoeuvres in the Baltic States and Poland. The Soviets also developed war plans for multiple 'echelons' of Soviet and Warsaw Pact reserve forces, which would advance through Poland and East Germany. The plans suggested the very real possibility of the Soviets breaking through NATO's Central Front forces in a few days without using nuclear weapons.

The United States led the response by NATO, which was to develop doctrines predicated on technology and skilled manoeuvres, not manpower. In doing so, NATO forces would maintain a credible conventional defence of the Central

TOP: To prevent the escape of its citizens, the East Germans fortified the border with the West and built the Berlin Wall. Here, an East German work party examines the 'Death Strip' near Checkpoint Charlie.

ABOVE: Guarded by US troops, Checkpoint Charlie on the Friedrichstrasse, Berlin, was one of the few points through which people could travel between East and West.

ABOVE: Reconnaissance and
intelligence-gathering are
vital to the success of any
military operation. Originally
performed by the cavalry and
then by reconnaissance
vehicles, like this British
Army Fox, the mission is
now mainly performed by
airborne assets like the
Anglo–French Gazelle utility
helicopter.

Front. First proposed in 1979, the NATO plan was called 'FOFA' (Follow On Forces Attack). Under 'FOFA,' NATO would develop surveillance and reconnaissance systems to 'see deep' into the enemy's rear positions. The goal was to identify ground- and air targets. After the targets were identified, aircraft, missiles and long-range artillery would 'strike deep' against the Warsaw Pact reserve echelons in order to prevent them from reaching the battlefield. As part of their NATO contribution, the United States and British military adopted ideas of manoeuvre, along with 'FOFA', during the 1980s.

During the Cold War, wars occurred more often and lasted longer than they had in the first half of the century. Of the 32 major and 75 minor conflicts that had occurred by the late 1980s, many had nothing to do with the Cold War. Some of the most significant wars of the post-World War II period took place in the Middle East. After 1945, colonial powers lost their hold on this region. Newly independent states and deep-rooted hostilities among diverse religious and ethnic groups emerged. In 1948, with the establishment of the new state of Israel and the neighbouring Arab states' refusal to accept its existence, the instability of the region grew.

International attention focused on the unrest in the Middle East because of the region's oil and its proximity to the Soviet Union. The Arab-Israeli wars came the closest that any war had come to causing a confrontation between the United States and the Soviet Union.

On 14 May 1948, the United Nations declared the creation of the State of Israel, the British withdrew from Palestine, and several Arab states – Egypt, Transjordan, Lebanon, Iraq, and Saudi Arabia – attacked Israel. When the First Arab-Israel War ended in 1949, Israel emerged victorious. Since the foundation of Israel, the Israelis have had to face two threats: an internal one from Arab insurgents and from the 'hit-and-run' attacks by the Palestine Liberation Organization (PLO), and an external threat coming from the regular forces of the nation's Arab neighbours. Until the beginning of the Palestinian *intifada* (uprising) in 1987, the latter was by far the more serious. Between 1948 and 1956, the state of Israel created for itself an effective military force.

Prior to the annexations of territory captured in 1967, Israel's major strategic problems included vulnerable frontiers and economic and demographic weaknesses. Israel, because of its size, could not risk a war of linear or elastic defence or of prolonged attritional struggle. In addition, Israel could not afford to maintain a large standing army. Consequently, the Israelis developed a concept of the 'nation in arms'. The Israeli Defence Forces (IDF) consisted of a small professional component and a large citizen militia. Employing universal conscription of men and women, the Israelis created a reserve force that could be mobilized in 72 hours. In the event of an enemy attack, the regulars' role was to 'hold the ring' until the reserve forces could be mobilized.

ISRAELI MOBILITY

Armour and tactical air power provided the basis for the IDF's striking power. Between 1948 and 1956, the IDF relied on light armour and mechanized infantry instead of tanks. In the 1956 war with Egypt, the Israelis implemented a pre-emptive *Blitzkrieg* in their crushing victory. Following their success, the Israelis modernized and strengthened the IAF (Israeli Armoured Forces). They placed greater emphasis on the tank, which had proven successful in the campaign. Moshe Dayan became an enthusiastic convert. By 1967 tank extremists claimed that tanks could win

battles on their own. Israel's scarce resources purchased tanks rather than artillery or armoured personnel carriers. By 1967 Egypt and Syria had equipped their forces with Soviet kit and trained them in Soviet 'sword and shield' tactics. 'Shields' of infantry, minefields, antitank weapons and armour would hold enemy attacks while armoured/mechanized formations launched counterattacks.

In 1967 the fragile peace that existed in the Middle East ended. Although a United Nations force patrolled the border of Israel on the Sinai and the Gaza Strip, the border with Syria and Jordan was characterized by ambushes, attacks on civilians, and frequent reprisals. In early May, Nasser learned from the Soviets about the Israelis' preparations for an attack against Syria. Although the information later proved to be incorrect, Nasser mobilized his reserves and moved units into the Sinai. Nasser persuaded the United Nations to remove its forces from the Sinai. Egyptian forces occupied the strategic position at the mouth of the Gulf of Aqaba – Sharm el-Sheikh – and cut off Israeli shipping through the Gulf. The Egyptian actions started the Six-Day Arab-Israeli War of 1967.

BELOW: On a routine border patrol, scouts from the Berlin Brigade's Combat Support Company (CSC), drawn from the 4th Battalion, 6th US Infantry Division, check on activity on the Communist side of the Berlin Wall.

Because they realized that they would have to work together to defeat Israel, the Arabs established a unity of command. An Egyptian general commanded the Arab forces on the Jordanian front. The Egyptians also commanded an Iraqi force. Troops from Kuwait and Algeria helped the Arab forces surround Israel. The combined Arab forces outnumbered the Israelis almost two to one. The Arabs had three times as many tanks and almost three times as many aircraft. Because of the diversity of the Arab forces, however, it was inconceivable that there could be any true unity of command.

The Israelis, having learned of the impending attack, quietly mobilized their reserves, and then implemented a pre-emptive air strike on 5 June. Withing the first few hours of the war, the Israelis' crippled the Arabs' air power. Of the 418 Arab aircraft destroyed during the war, almost 300 were lost on the first day. Once the Israelis controlled the skies, the IAF could decisively influence the conduct of the ground war. Their central position would allow the Israelis to move their forces from one front to another. The first stage of the ground campaign began on 5 June when an Israeli armoured division attacked along the coast in the northern part of the Sinai

and captured Bir Gifgafa. Three Israeli divisions engaged and virtually destroyed two Egyptian 'shields' and one 'sword'. Over the next three days, the Israeli divisions exploited their success. By 8 June, advance Israeli forces reached the Suez Canal and exchanged antitank and artillery fire with the Egyptians on the other side of the canal. The IDF succeeded in destroying the Egyptian forces and clearing the Sinai Peninsula.

Jordan entered the battle on the central front around noon on 5 June. Following an artillery barrage, a small Jordanian force crossed the border south of Jerusalem. In response, the Israelis captured Jerusalem and isolated the high ground north of the city that runs parallel to the Jordan River valley. Following attacks on the north and south ends of the high ground, the Israelis controlled all the bridges over the river. Cut off, the Jordanians had no hope of victory. Although the Jordanian forces fought well, the IAF's armour, infantry, and air power overwhelmed them.

The Israelis turned next to the Syrians who had been relatively quiet during the first few days of the war. Early on 9 June, after launching heavy air strikes, the Israeli ground forces attacked. On the Golan Heights, the terrain made Syrian sword and shield defences even more

BELOW: The Israeli Army has seen more combat than most over the last half century, and it trains constantly to keep its edge. Here a field commander with the elite Golani Brigade establishes contact with headquarters during manoeuvres.

formidable. By the end of the first day, however, the Israelis controlled the forward slope of the northern Golan Heights. The next morning, Israeli troops forced their way through the defenders. An Israeli armoured division broke through in the south while paratrooopers hit the Syrians' rear positions. At 18:30 hours on 10 June, a United Nations ceasefire began as resistance by the Syrians fell apart. Although the Syrians held out for 36 hours, the combined use of armour and tactical air power enabled the IDF to achieve victory.

SIX-DAY WAR

The Six-Day War in 1967 was a remarkable victory for Israel. The IDF's use of pre-emptive air strikes against the enemy's air forces, followed by armour-led *Blitzkrieg* supported by tactical air power, proved devastating. The IDF had modified and improved the German version of *Blitzkrieg*. Unlike the Germans, however, the Israelis did not permanently eliminate the Arab threat. The war increased the presence of the United States and the Soviet Union in the Middle East. Although Israel had won the war, the Arabs refused to acknowledge the loss of territory. They demanded the return of the Sinai and the Golan Heights.

For the next three years, the Egyptians would engage in a 'War of Attrition,' in which they launched attacks — cross-border raids, artillery barrages, and air strikes — against positions held by the IDF, who engaged in retaliatory commando raids. Israel, supported by the United States, demanded a negotiated peace settlement in exchange for the occupied territories, as well as an increase in arms shipments to Israel. The Soviets, who consented to rearm Israel and Syria, began sending arms directly to the Arab nations instead of through another country, such as Czechoslovakia. Uprooted Arabs formed the PLO and engaged in terrorist activities against the Israelis. In August 1970, tired of the situation, the Arabs and Israelis agreed to another ceasefire.

Israel's successes in 1967 resulted in a bigger push for 'all-tank' theories. By 1973 the 'tank-heavy' armoured brigades placed less emphasis on all-arms cooperation. The 'armoured shock' of the

frontal assault received the most focus. The available resources purchased tanks and aircraft, not APCs and artillery. The IDF built fixed defences in the Sinai and the Golan Heights. These defences, along with air strikes and mobile defences, were supposed to absorb an initial enemy attack. Recognizing the IDF's superiority in mobile warfare in 1967, the Egyptians planned a limited campaign. They would move into the Sinai and then go on the defence. They would basically invite the IDF to attack. The superior Egyptian numbers and firepower would overcome the superior skill of the IDF. Soviet-supplied antitank and antiair missiles would provide the firepower. At that

point, the Egyptians hoped that the United States along with the Soviet Union would step in and negotiate a compromise peace. This peace, they hoped, would fall in favour the Arab states in the area.

Following Nasser's death in 1970, the more moderate Anwar Sadat became the president of Egypt. Although he did not want to decisively defeat the Israelis, Sadat decided to go to war in October 1973. Believing that a limited victory would result in political gains, Sadat persuaded Syria and Jordan to join the fight so that the Israelis would have to fight a two-front war. Before going to war, Sadat obtained sophisticated weapons from the Soviets and improved the readiness of his forces. Unlike 1967,

ABOVE: The aerial threat remains as dangerous today as it was at the end of World War II. However, the development of sophisticated man-portable surface-to-air missiles like the Short Starstreak (two versions of which are shown here) has given the foot soldier more firepower when confronted by attack helicopters or ground-attack aircraft.

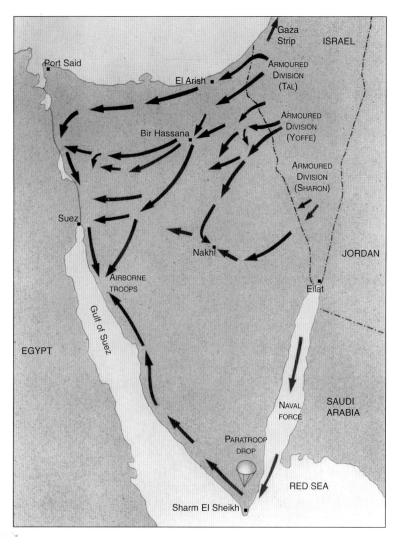

ABOVE: During the 1967 Six-Day War, the Israelis launched a combined-arms air, tank and mobile infantry attack on Egyptian forces in the Sinai desert, prevailing over their more numerous opponents through a classic *Blitzkrieg* series of manoeuvres.

Israeli intelligence did not alert the nation about an impending attack. Consequently, the Arabs succeeded in surprising their enemy.

At 14:00 hours on 6 October 1973, which was the Jewish Sabbath and Yom Kippur, Egyptian and Syrian forces attacked and achieved surprise. While Egyptian forces struck along the entire canal front, the Syrians hit the Golan Heights. Air strikes and a huge artillery barrage prepared the way for five infantry divisions to cross the canal on rubber assault boats. The Egyptians used 50 SAM batteries (SA-6, SA-2, and SA-3) plus conventional cannon and man-portable SA-7 to support the infantry. Once across the canal, Egyptians dug in and waited for the Israelis to attack. By 20:00 hours, the Egyptians had managed to establish several bridgeheads. Egyptian antitank defences included their B-10 and B-11 recoilless guns, AT-3 ('Sagger') and RPG-7s, plus their tanks and various assault guns. The Arabs struck first, with a

series of coordinated blows. With these actions, they forced the IDF to defend two fronts at the same time as they were mobilizing. As a result, the Arabs briefly threatened to defeat Israel.

During the first three days of the war, attacks by the IDF resulted in heavy losses, because most of the IAF had been deployed to the Golan Heights and they had launched a series of unsupported frontal tank attacks. By 10 October, however, the IDF's situation gradually began to change.

On Sunday, 14 October, the Egyptians advanced out of their bridgeheads in six major thrusts, but the cost of their action was high. In the battle that resulted, about 2000 tanks participated. Although the Egyptian forces advanced over 15km (9.3 miles), they lost a total of over 260 tanks, while the IDF only lost 20. By implementing six thrusts, the Egyptians seriously weakened their offensive. The Israeli Air Force inflicted heavy losses. In addition, the IDF had been issued with a new weapon – the TOW (tube-launched, optically tracked, wire-guided) antitank missile – and it was supplied by the United States.

CROSSING THE SUEZ CANAL

At 17:00 hours on 15 October, the Israelis moved on the Suez Canal. Fierce opposition failed to prevent one division from getting across and creating havoc among the surprised Egyptian divisions. Advance units of another two Israeli divisions crossed and destroyed the Egyptian's West Bank air defence sites, which allowed the Israeli Air Force to become more effective. The arrival of the IAF contributed to the IDF's recovery and re-established the balanced 'all-arms' formations that had fought so effectively a few years earlier.

The situation in the north, however, was very different. The Golan Heights was the scene of some of the most critical fighting because the Syrians posed a direct threat to Israel proper. Initially, two infantry divisions, reinforced by two tank battalions, defended the Heights. In late September the arrival of another armoured brigade increased the number of tanks to 175. On 6 October the Syrians attacked with three mechanized and two armoured divisions, which were supported by 32

SAM batteries and antitank missiles. The Syrian force included approximately 1500 tanks. Non-stop fighting lasted for three days and two nights. Superior tank gunnery and IAF support enabled the defenders to survive the first two days of fierce fighting. The arrival of reserves enabled the IDF to switch to the offensive for a two days between the 8th and the 10th of October.

By 10 October, the Israelis had forced the Syrians to retreat behind the 1967–truce line. The Israeli advance continued unchecked for almost two weeks before the United States and the Soviet Union finally urged an end to the conflict. The Soviets agreed to an international peacekeeping force as long as it did not include Soviet or American troops. The Israelis reluctantly agreed to a cessation of hostilities before the intervention of the international peacekeeping force. When the ceasefire began on 24 October the Israelis were not far from Damascus.

Israel had won, despite its being outnumbered. The success of the Israeli *Blitzkrieg* reinforced its importance, but the 'all-tank approach' had been discredited. Although Israel ultimately emerged victorious, Sadat's desire to demonstrate the vulnerability of the Jewish state also succeeded. Despite their losses, the Arabs had achieved a psychological victory over the Israelis.

In the midst of the Cold War, a series of wars occurred in the Middle East, and they were conducted without direct involvement of the United States or the Soviet Union, and without the use of nuclear weapons. The Israelis used conventional forces and conventional weapons to achieve victory, despite the fact that they were greatly outnumbered. During this same period, the United States and the Soviet Union developed military doctrines which were based on nuclear weapons. These doctrines would be put to the test in Southeast Asia; first in Korea, then in Vietnam.

ABOVE: Israel conducted manoeuvres in March 1967 in preparation for the Six-Day War. Although a sandstorm prevented Prime Minister Levi Eshkol from attending, the Army went ahead with a three-day exercise in the Negev desert.

BELOW: The 1967 Jerusalem Battle. Israeli and Arab forces engage in street fighting as each struggle for control of the Holy City.

PROXY WARS

Thankfully, the Cold War never turned hot. However, tension between the superpowers had to find an outlet, which usually took the form of proxy battles between Eastern and Western-influenced movements across the globe.

In the increasingly polarized world of the Cold War, there existed certainties of military thought that most western planners took for granted. The coming conflict between freedom and Communism would take place in western Europe, thus reducing the importance of and interest in other volatile areas of the globe. The war, when it came, would be massive – truly a Third World War – and would quite possibly involve the use of nuclear weapons and cause a global catastrophe. The future seemed quite bleak. The Soviet Union would attempt to expand. The United States would respond in an effort to contain that expansion, and the entire world would pay the price. It was in this atmosphere of 'mutually assured destruction' that both world superpowers began to place checks on their use of military force in attempts to stop the conflicts of the Cold War era from escalating towards totality. Thus the doctrine of Limited Warfare was born, in which the world superpowers would face off on battlefields across the globe, often fighting their murderous conflicts by proxy.

In the wake of World War II the United States, holding to past precedent, had rapidly de-mobilized its ground forces, choosing to rely on the nuclear deterrent and strategic air power in support of

LEFT: During 1950 infantrymen of the US 1st Infantry Division, armed with M-1 Garand rifles, search ruined Korean buildings for Communist infiltrators. The Korean 'Police Action' served as a template for Cold War proxy battles, differing only in the size and scale of the forces involved on both sides.

ABOVE: Communist forces advance in the wake of an artillery barrage in the Korean War. Both the initial assault by North Korean troops and the subsequent Chinese intervention in the conflict took US commanders by surprise.

efforts to contain Communist expansion. In the spring of 1950 there were only 10 active duty army divisions and 11 separate brigades available to respond to a military crisis. At the same time President Truman and his military advisers came to the conclusion that the nuclear deterrent was inherently dangerous and not sufficient to contain all Soviet expansion. In a document entitled NSC 68, the Truman presidency outlined the need for the capacity to meet the Soviet Union in limited war anywhere in the world, resulting in a renewed emphasis on traditional ground forces. Thus the decision had been made to meet Communist threats with military responses of equal intensity.

At the close of World War II, US and Soviet troops had entered Korea to disarm Japanese forces in the area. As the Cold War deepened, the dividing line between US and Soviet forces, at the 38th Parallel, hardened into a tense, international border between North and South Korea. Sidetracked by events in Europe the Americans remained unaware of a North Korean military build-up, and of preparations to invade South Korea. For their part, the Soviets misread US intentions in the area, believing that the Americans had written off South Korea and would not go to war for its

protection. On 23 June 1950 North Korean forces, numbering some 135,000 men, invaded South Korea, which was defended by an army of only 65,000. Facing little effective resistance, the North Koreans quickly seized Seoul and began the process of driving their defeated opponents into the sea.

Determined to save South Korea and contain Communist expansion, Truman turned to the United Nations. The UN Security Council, being boycotted by the Soviets at the time, approved a US-sponsored resolution giving the UN the mandate to defend South Korea from the unprovoked violation of its sovereignty. Control of the war in Korea fell to the United States, but forces from several UN member nations would take part in the fighting. Truman turned to General Douglas MacArthur to command US and UN forces in the conflict. At first the situation looked quite bleak for MacArthur, who could only rely on four understrength American infantry divisions located in Japan. Determined to achieve victory, MacArthur planned to construct a defensive perimeter around and hold firm at the port city of Pusan, while rushing US forces to the front in an effort to delay the North Korean advance. Elements of the 24th and 25th Infantry Divisions, though they were greatly outnumbered, succeeded in slowing the North Koreans, thus gaining valuable time for defensive preparations in the 'Pusan Perimeter'.

By early August the North Koreans had begun attacks on the UN defensive enclave, still holding a numerical advantage over the Eighth Army. Often utilizing human-wave tactics, the North Koreans sorely pressed the US and South Korean defenders. Though the situation was dire, MacArthur held much-needed reinforcements in reserve, waiting to launch an amphibious operation designed to alter the entire war. On 15 September, X Corps, under the command of General Edward Almond, executed a daring amphibious landing in very difficult tidal conditions at Inchon. Though fighting in the area, especially around Seoul, was furious, on 23 September when threatened with envelopment, the North Koreans began a general withdrawal. Aided by punishing air strikes, UN forces quickly drove into North Korea, nearing a complete victory. Truman, though, remained quite conscious that the presence of US forces

along the Yalu River might prompt a formal Chinese entrance into the conflict, an unwelcome escalation towards a true superpower clash. Though steps were taken to placate the Chinese, US military advisers underestimated their reaction to developments in Korea. On 25 November a Chinese force numbering nearly 200,000 men slammed into the exhausted Eighth Army, forcing a hasty UN retreat. Though US air power took a great toll on the Chinese attackers, US and UN forces withdrew to a point below the 38th Parallel and once again Seoul fell to Communist control.

The war in Korea had now become a direct conflict between forces of the United States and China, a situation that rapidly could escalate into a general world conflict. Realizing that the situation had changed dramatically, MacArthur sought more troops and supplies and a new level of national commitment to make war on a much more powerful opponent. For complete victory in Korea the war would have to be taken to China. The Truman government, though, wanted no part of

BELOW: Marines using scaling ladders during the landing at Inchon. The difficult tidal conditions in the area made the amphibious operation perilous, but that also meant that it came as a surprise to the North Koreans who occupied Seoul.

ABOVE: Troops of 3rd Battalion, Royal Australian Regiment, lay down covering fire during operations in Korea. Many battles of the Cold War would have a multi-national component, or were fought entirely by proxy forces.

an all-out war with China. Such a conflict would certainly lead to a clash with the Soviet Union and general war in Europe. Truman instead opted for a more limited war of strategic defence of South Korea. MacArthur bristled at the notion of fighting a war designed to result in a draw, a war that never brought military force to bear on the real enemy: China. By April 1951 differences between Truman and MacArthur had come to a head, resulting in MacArthur's dismissal. The goals of the war and the force used to prosecute the war would

remain limited. Neither the western Allies nor the Soviet bloc desired escalation in Korea; the possible catastrophic results were simply not worth the risk. Thus from 1951 onwards neither side fought for victory and the reunification of Korea. Neither side, though, would accept defeat, and the war in Korea dragged on, developing into a stalemate around the 38th Parallel.

As peace negotiations slowly moved forward, the combatants did battle in prepared defensive emplacements attempting to improve their negotiating position. Thus the war became a trench war aimed at very limited goals, goals more of a diplomatic nature than a military nature. Chinese forces relied on their strength in numbers, sending human waves into attacks that more closely resembled the Somme than battles from World War II. US and UN forces relied on superior defensive firepower and constant air attacks on the Communists' logistic system. Through the tireless efforts of nearly 500,000 labourers, though, the Communist supply lines remained open and effective. The rather primitive Communist logistic network actually proved a very poor target for a modern, technological bombing campaign, a case which would be repeated in the war in Vietnam.

On 27 July 1953 the Korean War ended in a negotiated armistice, ending in many ways where it had begun, along the 38th Parallel. The first major conflict of the Cold War set the rules of modern, limited warfare. US forces had defended South Korea from defeat, thus upholding the doctrine of 'containment'. The amount of force used in Korea had been only enough to stem defeat, avoiding escalation and sublimating the desire for conflict with the real enemy: worldwide Communism. The North Koreans had been denied their goal of conquering South Korea but Chinese forces had at least defended their client state from destruction. Utilizing strength in numbers, the Chinese had prolonged the war long enough to gain what they considered to be an honourable settlement. The parameters of limited warfare had been established, and most military pundits believed that the next limited war which involved the world superpowers would closely resemble the conflict in Korea.

MALAYA AND COUNTERINSURGENCY

Many struggles of the Cold War era were less formal than the Korean War in nature, often taking the form of Communist-inspired insurgencies against governments supported by the western Allies. Insurgent forces, disaffected members of the populace, usually relied on guerrilla warfare and terrorism in their struggles, and often followed the military template set by the Communist revolution in China. Mao Tse-tung had utilized a pattern of protracted war and by 1949 had overthrown the Nationalist régime of Chiang Kai-shek. Using time to create will, the Chinese insurgents had melted into the countryside to indoctrinate the great masses of people in the ideals of revolution.

Fending off government attacks and utilizing guerrilla tactics, the Maoist forces slowly gained adherents while keeping constant pressure on government forces in a slow war of attrition. Time was all-important to the insurgent. In a war that could last for several decades the capitalist government, unwilling and unable to make the necessary concessions to win the true support of the people, would eventually succumb to defeat. Maoist protracted warfare was a recipe for the weak to defeat the strong. Gradually, over a time, the tiny bites would have a cumulative effect and the mosquito would finally defeat the elephant.

During World War II the latent Malayan Communist Party (MCP) had become an armed force standing against Japanese rule. After the close of the conflict the forces of the MCP, led by the charismatic Chin Peng, launched an insurgency designed to overthrow British rule and convert the state to communism. At first glance Malaya seemed to be an ideal place to implicate Maoist protracted war. Covered in dense jungle, the terrain offered the insurgents ample cover. In addition, the insurgents found welcome support from amongst an impoverished Chinese minority population.

The MCP did, however, face several disadvantages in its insurgent campaign. The insurgents, only numbering some 8000 fighters, received little in the way of outside support, and in many ways their message of revolt carried little weight outside the Chinese community. Most Malays remained quite content to live under a rather benign form of British colonialism and were unwilling to risk their lives, realizing that British rule would soon come to a voluntary end. The insurgents often seemed more like thugs than champions of the people, routinely resorting to terrorism and brutality to keep their supporters in line.

OPPOSITE: US Marines engage in house-to-house fighting after the Inchon landings. The surprise of the landings swung the fortunes of war in favour of the UN forces.

BELOW: SAS men in Malaya prepare for a combat jump against Communist guerrillas. Using the experience of the Chindits as a guide, the SAS would set many of the standards for successful counterinsurgency warfare.

Continuing guerrilla raids and terrorist attacks in 1948 forced the British to declare a state of emergency in Malaya. Soon the British had sent a force equalling the strength of two divisions to Malaya to root out the insurgent force. Working in an uneasy and ineffective alliance with the colonial police force the British relied on sizeable military sweeps designed to corner and crush the insurgent fighting forces. However, British intelligence was poor and their enemy was quite elusive, with the result that most of the sweeps met with very little success. At the same time guerrilla activities in Malaya consistently increased. The British, it seemed, were now losing the war.

HEARTS AND MINDS

In 1951 there were some 600 guerrilla attacks throughout Malaya. General Sir Harold Briggs, the new British military commander in the area, though, had developed a plan for victory. The war in Malaya was political in nature and required a political solution. The Briggs Plan called for cooperation between civil and military authorities, designed to separate the insurgents from their base of support. In a 'hearts and minds' campaign of reforms and education, the British colonial authority capitalized on the support of the majority of the Malayan population. The British even promised the Malays independence by 1957, thus crippling any claim the MCP had to being the leaders of a nationalist rebellion. Possibly of the greatest importance, though, the Briggs Plan called for the resettlement of the Chinese squatter population. Construction began on new villages for the Chinese and they were promised farmland and a cash allowance. Most of the Chinese were only too happy to leave the squatter camps behind to benefit from a better life in the new villages. The new British plan severed the MCP from the population, thus denying the insurgents cover, succour and new conscripts.

Though the political aspects of counterinsurgency in many ways took precedence, the British also altered their military philosophy in Malaya. Once isolated from the people, the insurgents were quite vulnerable to attack. The British chose to scrap lumbering sweeps by large units through the jungle in search of their elusive prey. Utilizing superior intelligence gained from supportive locals and MCP defectors, British forces prepared to make use of jungle warfare tactics of the past to bring the conflict to a conclusion. Ex-Chindit Michael Calvert had raised and helped train a special jungle warfare unit of the Special Air Service (SAS). Founded in World War II, the SAS had specialized in sabotage and commando raids, and was the forerunner of modern special forces units. After undergoing extensive training in jungle warfare, the unit – designated the Malayan Scouts – deployed to a series of forts along the fringes of the massive jungle that served as the MCP sanctuary.

The SAS forces led long-range penetration patrols in the best traditions of the Chindits. Utilizing stealth and speed the SAS would drive deep into the jungle – with little in the way of supplies or support – and live off of the land in search of their beleaguered foe. Limiting their use of firepower to minimize collateral damage, the SAS kept the MCP insurgents on the run and

consistently guessing. In some ways the SAS employed guerrilla tactics in fighting the guerrillas. The results were quite successful and led to a collapse of the insurgency. As promised Malaya received its independence in 1957, and by 1960 the stage of emergency within the country had finally ended.

British success in Malaya served as the template for future counterinsurgency campaigns during the Cold War era. Continuing insurgent conflicts from Cuba to Africa would play a key role in the ongoing superpower struggle, becoming much more common than limited war. Western nations often saw insurgencies as only military in nature, requiring a military solution. Though sometimes successful, purely military solutions to insurgency often resulted in disaster. In Malaya the British had recognized that separation of the insurgent from the population through political means was paramount, as it allowed for an eventual military victory utilizing tactics of irregular warfare. In subsequent wars, notably in Vietnam and Afghanistan, counterinsurgency methods

OPPOSITE TOP: British troops practise battlefield insertion from a Whirlwind Mark 10 helicopter in 1965. The speed with which they could deploy troops into difficult terrain would make the helicopter a mainstay of counterinsurgency campaigns around the globe.

OPPOSITE BOTTOM: British troops of the 22nd SAS Regiment patrol the Temenggon area of Northern Malaya in 1954. Innovative SAS tactics coupled with an effective 'hearts and minds' campaign proved to be a war-winning combination in Malaya (now Malaysia).

LEFT: A Wessex helicopter of the Royal Navy delivers Gurkhas to an improvised border landing zone during the confrontation with Indonesia over Borneo.

ABOVE: The French relied on firepower and mobility including combat air assaults in an effort to bring the elusive Viet Minh to battle. Here, the beleaguered garrison at Dien Bien Phu is reinforced from the air.

would sometimes be applied but would often be overlooked or undervalued as both the United States and the Soviet Union emphasized the lessons of conventional limited war.

THE FAILURE OF LIMITED WAR

Though the wars in Korea and Malaya were dissimilar and, in the main, unrelated, the Cold War policy-makers of the United States thought quite differently. Communism was viewed as a monolithic entity, controlled from Moscow, and an entity bent on world domination. Communists in Korea and Malaya – taking part in limited war or revolution – were regarded as but a part of a greater conspiracy against the West. Thus the US considered the Cold War to be a vast, global conflict involving interlocking wars of several types, wars which were directed by the Soviet Union. It was President Truman who gave voice to the evolving policy of the US in the face of what he viewed as Soviet expansionism by stating that the United States would 'assist all free peoples against threats of revolution and attack from without'.

After World War II the French sought to re-assert their authority over Vietnam, which had fallen to the rule of the Japanese. However, the Viet Minh – under the charismatic leadership of Ho Chi Minh – rose up against French rule, sparking a war that would last for nearly 30 years. The Vietnamese insurgents, undersupplied and numbering only 60,000, faced long odds against 200,000 well-armed and highly motivated French troops. Realizing his weakness Ho decided to avoid battle, relying on the attrition of Maoist protracted war to achieve victory and declaring to his French adversaries, 'If we must fight we will fight. You will kill ten of our men and we will kill one of yours. Yet it is you who will tire first.' Unlike the insurgents in Malaya, the Viet Minh enjoyed wide national support in their struggle. Unwilling to reform and unable to offer independence, the French authorities did little to solve the political problems that foment revolution. Thus Ho Chi Minh was able to stand more as a true Nationalist leader rather than a Communist revolutionary. Because the French had done little to win the 'hearts and minds' of the Vietnamese, the Viet Minh insurgency grew in strength, eventually garnering considerable aid and military support from both the Soviet Union and China.

As the conflict in Vietnam became more serious it attracted the notice of the United States. In 1949 the French divided Vietnam in half near the 17th Parallel, proclaiming South Vietnam to be 'free'. The cunning diplomatic move allowed the French to present their colonial war in Vietnam in stark Cold War terminology. South Vietnam – actually a French colonial construct that never requested or desired statehood – was a small nation beset by Communist revolution and invasion. Ho Chi Minh's nationalist uprising was but a well-concealed part of the global Communist conspiracy. South Vietnam was, then, very like South Korea, and could not be allowed to fall, lest 'containment' fail. Indicating new levels of US commitment to the war in Vietnam, in his 1952 inaugural address President Eisenhower remarked, 'the French in Vietnam are fighting the same war we are in Korea'.

By 1954 French national support for the conflict in Vietnam had waned, and French rule in Vietnam came to a disastrous end at the Battle of Dien Bien Phu. Unwilling to see containment fail in Southeast Asia, the United States used diplomacy and threats to assure the continued survival of South Vietnam in the Geneva Accords. Realizing, though, that the war was far from over and that Ho Chi Minh would seek to reunify his country, the United States soon began to send aid as well as advisers to South Vietnam, in the hopes that they could build a nation where none existed.

The South Vietnamese régime, though, was singularly uncooperative. Though it professed to be a democracy South Vietnam was a brutal dictatorship, one that was shot through with graft and inefficiency. The régime had little to offer its people other than poverty and repression, helping the Viet Minh – now called the Viet Cong (VC) – insurgency to grow. By 1965 the Communist insurgents controlled over 60 per cent of the land area of South Vietnam, and the CIA estimated that in a free election

BELOW: French and loyal Vietnamese forces on patrol on 2 February 1954 outside Dien Bien Phu. Such patrols failed to detect a Viet Minh build-up in the area, which led to the later catastrophic French defeat, effectively ending French colonial rule in Vietnam.

some 80 per cent of the population would choose to follow Ho Chi Minh. The US policy of nation-building in South Vietnam had failed, and the régime teetered on the brink of collapse. Seeing containment threatened yet again, the first US combat forces entered South Vietnam on 8 March 1965, confident in their abilities to destroy the insurgent force of a Third World country.

The United States chose to fight the Vietnam War as a limited conflict, seeking only to assure the survival of South Vietnam and the success of containment. As a result US forces would never, in a systematic way, invade North Vietnam or the neighbouring countries of Cambodia and Laos, which provided the Communists with safe base camps and logistic networks. In addition, President Johnson would never fully mobilize US military forces or the popular support of the American people, choosing instead to limit homefront involvement in the conflict. Vietnam became a war that the United States fought with only a small portion of its military and national might, and a war aimed not at specific, obtainable victory but rather at the defence of a rather nebulous status quo. American military planners initially thought little of the difficulties inherent in such limitations, for the problem in Vietnam seemed quite simple. An armed insurgent force, with significant aid from North Vietnam and the Communist bloc, threatened the existence of South Vietnam. The United States military, the most powerful force on the planet, would simply hunt down and destroy the insurgents, putting an end to the problem. Thus most US military planners viewed Vietnam as a rather traditional conflict with a purely military solution, paying little heed to the political and social reasons for the insurgency. Though the US would spearhead a 'hearts and minds' campaign

BELOW: The personification of air mobility. US helicopters hit a landing zone near Bong Son as part of 'Operation Eagle's Claw' in 1966. Ferried into battle by helicopters, US and ARVN (South Vietnamese) forces could strike at Communist troop concentrations across the length and breadth of South Vietnam.

ABOVE: US advisers arrived in Vietnam in the early 1960s, and were soon fighting alongside their South Vietnamese students against the Communist Viet Cong guerrillas. Here, an adviser leads a Vietnamese patrol through the waterlogged Plain of Reeds, 64km (40 miles) from Saigon, in an operation which took place in 1962.

it always stood second to the military effort. Strategically, then, the US viewed Vietnam as another limited war, and chose to limit the application of force to such a degree as to make the war almost unwinnable. Tactically the US prepared to fight a traditional campaign, but this would be to the detriment of true techniques of counterinsurgency. Of course, the mixture was quite volatile.

COMMUNIST REALITY

Though the Americans in many ways fought the wrong war, the North Vietnamese leadership realized that a military showdown with the Americans would be suicidal. Early attempts at conventional battles would prove the point. In the Ia Drang Valley in 1965, substantial US forces clashed with and defeated Viet Cong and North Vietnamese Army (NVA) units for the first time. The battle would set the tactical precedent for the remainder of the conflict. General Vo Nguyen Giap once again chose to rely on Maoist protracted war to even the odds. In command of the American and Allied forces, General Westmoreland adhered to a policy of large unit operations designed to 'find, fix and finish' the enemy.

US forces relied on search-and-destroy missions to locate and neutralize their elusive enemy. Many such missions met with little success, while others led to clashes with Communist forces, usually entrenched and informed of American movements. By late 1956, Giap chose to seek battle against the Americans to get

the measure of his impressive new foe. An NVA build-up in the Central Highlands led to a Communist siege of a US Special Forces camp at Plei Me. Though the attack failed, NVA and VC forces lingered in the area of the Ia Drang Valley. To Westmoreland the situation was perfect. The enemy had been located and had not fled to the sanctuary of Cambodia. Relying on superior firepower, US forces would make the NVA and VC pay.

In the coming battle US forces would rely on their overpowering edge in firepower as a decisive force multiplier. With control of the air, US ground troops could rely on the support of a vast array of aircraft from AC-47 'Spooky' gunships that could pepper targets with 100 rounds per second, to the massive force of B-52 strategic bombers. On the ground US forces crisscrossed South Vietnam with artillery Fire Support Bases (FSBs). Ringed with powerful defences and often the subject of Communist attacks, the FSBs in many ways formed the backbone of the US military effort in Vietnam.

Artillery, usually batteries of 105mm and 155mm guns, would support any search-and-destroy operation in the immediate area. When engaged in combat, US forces would communicate target coordinates to the FSB, which would reply with a devastating barrage of fire. Thus the US hoped to capitalize on its overwhelming firepower edge, enabling even small US forces to pin the enemy into battle with impunity.

Fixing the enemy into battle remained the elusive part of the equation, but Westmoreland believed he had the answer. The helicopter had seen limited military use in Korea, but would be a tactical mainstay in Vietnam. Before 1965 insurgent forces had been able to strike and flee into impenetrable jungle terrain, sure in the knowledge that they could outmanoeuvre any clumsy attempts at pursuit. Aboard helicopters, though, US forces could respond almost instantly to any enemy threat across the length and breadth of South Vietnam, leaving the enemy with no safe haven.

In a tactic which was known as air mobility, US forces, often very small in numbers and reliant upon their firepower edge, would chopper into areas known as Landing Zones (LZ), which were usually chosen for their proximity to a suspected enemy troop concentration. Thus anywhere the enemy operated, the helicopters would soon arrive, carrying troops designed to lock the now located enemy into battle, while massive firepower support assured their destruction. The formula was relatively simple. The superior mobility afforded by helicopters would negate the guerrilla's traditional elusiveness. Denied their primary defence, the application of

ABOVE: Staff Sergeant Collier of the 173rd Airborne amid a burning Viet Cong base camp in War Zone C in 1967. Such search-and-destroy missions were a major part of the US strategy of attrition employed in Vietnam.

RIGHT: US reconnaissance troops leap from a Bell UH-1 'Huey' in a speedy exit over a mountaintop near a suspected Viet Cong base camp. With soldiers standing on the landing skids and leaping from about 19m (6ft) in altitude, 32 men could be landed from six helicopters in less than a minute.

massive firepower would quickly attrit insurgent forces in South Vietnam, leading to total victory.

YEARS OF ATTRITION

On 14 November 1965 the 1st Battalion, 7th Cavalry, commanded by Colonel Harold Moore, flew by helicopter to LZ X-Ray near the Chu Pong Massif in the Ia Drang Valley. Having landed amidst a major NVA staging area, Moore's forces immediately fell under heavy attack. Two NVA regiments, the 33rd and the 66th, attempted to surround and destroy the tiny American force which held a LZ no larger than a football pitch. In savage, often hand-to-hand fighting, the US force grimly held its ground, receiving little by air on the now deadly LZ.

As expected, during the struggle massive US firepower support made the critical difference. Withering artillery barrages and the first tactical use of B-52 strikes, each aircraft laden with 16,329kg (36,000lb) of high explosive, took a terrible toll on NVA and VC forces. After three days of fighting, bloody and beaten, the Communist forces made for their safe havens in Cambodia. In a tragic postscript to the battle, the 2nd Battalion, 7th Cavalry stumbled into a carefully planned Vietnamese ambush and was nearly destroyed while moving to nearby LZ Albany. Even after taking this into account, US forces won a substantial tactical victory in the Ia Drang Valley, losing some 305 dead. In contrast the Communists lost an estimated 3561 dead out of 6000.

Westmoreland seemed vindicated. Helicopter mobility had pinned the enemy into battle, and firepower had prevailed, and this resulted in a favourable rate of attrition of more than 11 to 1. Certainly a few more such successes would ensure victory – a military victory – leaving little reason to rely on the more political and social aspects of a counterinsurgency campaign. Since the military objective had been the destruction of the enemy force, once the NVA and VC fled the area, it no longer held any tactical value. American troops returned to their base camps to await their next opportunity to fix their opponents into battle. Vietnam, then, was not a war about taking and holding ground, rather it was a war about

finding and killing the enemy: attrition at its most basic. There would be no great victories for the public to watch on television, and no pins for anxious relatives to move on campaign maps. There would be only the rough solace of the 'body count' to signify victory. Westmoreland's idea of victory was difficult for the American public to understand. As the war dragged on while the body count never translated into strategic victory, public support for the war slowly began to wane.

The NVA and VC also took several lessons from the Ia Drang. With his forces bloodied by the brush with superior American weaponry, Giap chose to avoid further battle, instead placing an increased reliance on safe havens in Cambodia and Laos in the main beyond the reach of US firepower. In South Vietnam Communist forces would engage in carefully prepared, small unit actions against the Americans, usually taking heavy, but acceptable losses, while

ABOVE: During 1967 members of this US Navy SEAL pause after discovering a Viet Cong booby trap, a Punji Bed, or bed of stakes. The use of booby traps and mines forced American troops to slow their patrols and caused a constant stream of injuries that diverted valuable American resources and sapped American morale.

slowly attriting US strength and eroding American support for the conflict. The strategy closely mirrored the prediction that Ho Chi Minh had made to the French years earlier. Thus even though the Americans held the edge in tactical mobility with the helicopter, the Vietnamese retained the initiative in the war. It was the Communists who chose to fight, or to flee. Tactically, the Communists had learned to neutralize the American firepower edge by 'grabbing American belts'. In battle Communist forces attempted to get as close to American troops as possible, realizing that the Americans would not risk calling down firepower on their own troops. In the end, then, Westmoreland and Giap learned much the same thing from the Battle of the Ia Drang Valley.

Many more such encounters could be lethal to the Communist insurgency. Westmoreland attempted to repeat his success, while Giap chose to fight by different rules, attempting to avoid future disastrous battles of attrition.

ROLLING THUNDER

For the next two years, the Vietnam War progressed through a series of large unit operations designed to wear down enemy strength. At the same time a massive but ill-fated bombing campaign over North Vietnam, 'Operation Rolling Thunder', was intended to force Ho Chi Minh to drop his support of the southern insurgents. From the Mekong Delta to the Demilitarized Zone (DMZ) US forces sought to bring the elusive enemy to battle in operations such as 'Junction City', 'Hickory', and 'Attleboro'. Though the operations often differed in tactics and terrain type – from rice paddies to mountain rainforest – they followed a similar pattern.

US forces, on the ground or by helicopter, would sweep into an enemy area, hoping to attrit enemy forces and 'pacify' the area. The Viet Cong, increasingly augmented by NVA forces, especially near the DMZ, were often aware of the coming sweep and prepared accordingly. Small Communist forces would meet the US or ARVN sweep in prepared positions, before inflicting as much damage as possible and then retreating to safe havens in Cambodia and Laos. In the encounter battles – such as the fight for Hill 875 near Dak To in 1967 – US forces usually inflicted a rate of attrition of over 10 to 1, but failed to destroy Communist units, which would survive to protract the war ever further. Unless they perceived an advantage, the Communists would not stand and fight at all. In 1968 US forces initiated over 1 million sweeps – from multi-divisional search-and-destroy operations to small scout team missions – and only 1 per cent made significant contact with the enemy. Even then US forces often took heavy casualties from mines and booby traps left behind by their elusive foe.

Thus the attritional struggle in Vietnam dragged on, and Westmoreland remained confident that the US was wearing down the Vietnamese will to fight and nearing overall victory. Westmoreland still

BELOW: Viet Cong guerrillas construct defensive tunnel networks by hand in the deep jungle. Seeking shelter the Viet Cong embraced mother earth, constructing thousands of miles of intricate tunnels throughout South Vietnam. Some of the complexes were so extensive that they even contained entire hospitals.

LEFT: US Marines charge up Hill 881 near Khe Sanh. The siege of the Marine base at Khe Sanh could have been another Dien Bien Phu, but unlike the French, the Americans had the air power both to keep the base supplied and to mount round-the-clock operations against the North Vietnamese Army.

believed that the conflict fit traditional patterns of limited war, and always expected a prototypical Communist ground offensive like that waged by China in the Korean War. The attritional rate of 'body count' gave Westmoreland reason for optimism, for in 1968 alone over 100,000 VC and NVA forces perished in battle. However, during the same year over 300,000 Communist troops moved down the Ho Chi Minh Trail, indicating that though the numbers ran heavily in favour of the United States, the war of attrition would indeed be quite lengthy. Even as Westmoreland predicted quick victory for US force of arms, the VC and NVA were preparing for their greatest offensive of the war.

A CHANGING WAR

On the morning of 30 January 1968, during the celebration of the Tet Lunar New Year, Viet Cong forces attacked urban centres across South Vietnam, abandoning the policy of protracted war to seek a decisive military victory. The reasons for the tactical change on the part of the Communists remain quite cloudy. Many historians in the West believe that the level of attrition from 1965 to 1967 made North Vietnamese

leaders unsure of the inevitability of victory in a protracted conflict, forcing a major military gamble in 1968. Others, including General Giap, portray the Tet Offensive as an act of revolutionary war against the Americans. The Viet Cong, fully aware that they could not defeat US military might in a traditional sense, hoped that their audacious offensive would spark a general uprising throughout South Vietnam, thus forcing the Americans to quit the country in

BELOW: A North Vietnamese soldier on alert for an air attack. In the face of the most concentrated bombing in history, Communist forces fought on with a tenacity that surprised American military planners.

ABOVE: Soldiers of the 2nd Battalion, 327th Infantry, fire M-16 rifles and an M-60 machine gun at NVA regulars north of Phu Bai during 1968. The development of lightweight assault rifles like the M16, firing small-calibre ammunition, gave the ordinary infantryman the kind of firepower, fully automatic firepower, previously only available to machine gunners.

defeat. Communist hopes, however, were soon dashed. Though the Viet Cong had attacked sensitive sites across the country, including the American Embassy in Saigon, the South Vietnamese people did not rise up in revolutionary ardour. From Can Tho to Quang Tri City, the Viet Cong attack had taken US and South Vietnamese forces largely by surprise. However, the attack merely served to lock the Communist forces into deadly battle, as Westmoreland had long hoped.

Only in the old imperial capital city of Hue did the fighting linger for long. The struggle for the city involved urban fighting reminiscent of Stalingrad in World War II. US and ARVN soldiers, used to jungle warfare, had quickly to adapt to street-to-street and house-to-house fighting to clear out determined pockets of enemy resistance. Having lost their mobility advantage and with artillery nearly useless in close-quarter fighting, US forces resorted to an all out bombardment of the enemy-held part of the city. During a month of street fighting, much of Hue was destroyed, and 75 per cent of its population made homeless, but the city was saved.

Near the border with Laos the last chapter of the Tet Offensive took place as nearly 40,000 NVA soldiers laid siege a force of 6000 US Marines at Khe Sanh. NVA artillery struck the base with

regularity and ground forces attacked outlying Marine positions using human wave tactics. Cut off from ground communications, the Marines relied on air support for their supplies and their salvation. Having finally located the elusive enemy, Westmoreland unleashed 'Operation Niagara' in defence of the Marines at Khe Sanh.

Devastating B-52 strikes pounded enemy positions in a round-the-clock aerial offensive, killing an estimated 10,000 NVA soldiers. In all American air power dropped the equivalent of 10 Hiroshima-sized atomic bombs in the Khe Sanh area, thus forcing the NVA to call a halt to their siege after some 77 days. During the Tet Offensive the Communist forces had fought bravely, but they had succumbed to superior American firepower, resulting in a staggering loss of life. Of some 85,000 Communist forces involved in the Tet Offensive, nearly 58,000 of them died, a fatality rate of over 70 per cent.

The Tet Offensive was a comprehensive military disaster that nearly destroyed the Viet Cong as an effective fighting force. In full retreat, the remaining VC cadres fled to Cambodia and Laos, thereby abandoning control of much of the Vietnamese countryside. At the same time, though, it became clear that the Tet Offensive had caused a massive shift in

American public opinion concerning the war in Vietnam. Though the war had been controversial, most Americans had been willing to believe their government's claims that the war was nearly over, claims the Tet Offensive seemed to disprove. For many Americans it now seemed that the war in Vietnam would linger indefinitely while US military men and women continued to die for a country that seemingly did not appreciate their presence or their sacrifice. The Tet Offensive, coupled with societal turmoil on the home front, caused American public support for the Vietnam War to plummet, even as Westmoreland sensed victory.

After the Tet Offensive, the Vietnam War changed yet again. Communist forces became more and more reliant on the efforts of the NVA, serving to alter the insurgent nature of the war. In addition Communist forces once again relied on guerrilla tactics in an effort to recover from their massive losses. US forces, too, altered their tactical methods of warfare, especially after General Abrams succeeded Westmoreland in overall command of the war. Partly due to strict orders to avoid taking casualties Abrams put an end to massive search-and-destroy operations, relying instead on the less traditional tactics of counterinsurgency. Thus General Abrams

introduced his 'One War' strategy, which aimed to blend battlefield victories with increased efforts at pacification throughout South Vietnam. Smaller units would patrol the countryside – and work with villagers to ensure their safety – rather than attempting to seek out and destroy elusive enemy main force units. Also, US Special Forces, in tandem with the CIA, undertook efforts to 'neutralize'

LEFT: Amid the rubble of intense urban warfare US Marines stop for a cigarette break. The nearly continuous fighting left many soldiers with lingering psychological problems that often did not become manifest until long after the fighting was over.

BELOW: South Vietnamese Marines return fire during the Tet Offensive in Saigon. Though the surprise Communist attack on the heart of American power was a military failure for the Viet Cong, it did much to convince Americans on the homefront that the war was being run incorrectly.

TOP: South Vietnamese soldiers move into battle during the North Vietnamese offensive of 1972. Though some ARVNs fought well, others were less effective, and there was little hope for victory once American troops had withdrawn.

ABOVE: After quick victories across the country, in 1975 North Vietnamese troops nearing their ultimate goal rush across the tarmac at Tan Son Nhut air base outside Saigon.

the VC infrastructure in the 'Phoenix Program'. Relying on such irregular tactics as political assassinations, the Phoenix Program met with great success, capturing over 34,000 VC operatives and 'neutralizing' thousands more, and in the process, destroying the control which the Communists had until that point held over numerous villages.

Though US tactics in the Vietnam War had changed, these changes came too late to achieve success in the conflict. Unlike the British situation in Malaya, the South Vietnamese Government was unwilling to offer meaningful political or economic reform to accompany the newly effective American military efforts. In addition the North Vietnamese

and the Viet Cong retained the support of the Soviet Union and China, indicating that the conflict could drag on for an indeterminate length of time. With social anarchy at home, and tumbling economic fortunes, the American public increasingly turned against the war in Vietnam. The change in political fortunes was mirrored by a policy of gradual de-escalation of American involvement in the war, conducting a policy that President Nixon would call 'Vietnamization'.

Unwilling to admit defeat or to see the immediate failure of containment, US forces slowly withdrew from Vietnam. Although the North Vietnamese launched a failed ground offensive in 1972, one that was in the main defeated by massive US air strikes, the last American forces exited the war in early 1973 after signing an armistice with the North Vietnamese. Nixon and his supporters realized that the war in Vietnam would continue, and few rational observers were taken by surprise when South Vietnam fell in 1975. The only shock was the speed with which South Vietnam collapsed.

Thus the Vietnam War ended as a victory for the North Vietnamese and a defeat for the American policy of containment. American forces had won every major battle, but had lost the war.

The effects of the first ever American defeat still linger, colouring every subsequent American use of military force. Many blamed the loss on US politicians who had reigned in the power of the military. It seemed, then, that the policy of limited war was to blame, and that the military limitations which were necessary to avoid escalation had made the conflict totally unwinnable.

In addition, the US disaster in Vietnam brought tactics of counterinsurgency into question. It seemed that the British success in Malaya had been an exception born more out of fortunate circumstances than of tactical and systemic success. In the United States the barrage of questions concerning the nature of warfare and the future projection of national power was dubbed the 'Vietnam Syndrome'. The US military had fallen into disarray and disfavour, and though it remained the most powerful nation in the world, the United States questioned its power. In every coming conflict, from Grenada to Kosovo to the War on Terror, the question would now rage, would the new war become 'another Vietnam'?

In the final analysis, limited war has its place and can be effective, as demonstrated in the Falklands and in the Gulf War. However, limited wars prosecuted by and between world superpowers are quite different and are often protracted and inconclusive. In Vietnam the United States tried to fight a limited war on the Korean model, expecting a traditional conflict against a traditional foe. However, the conflict in Vietnam defied western traditions. It was part war of national liberation, part insurgency, part civil war and part Cold War. Even as US forces abandoned their attritional policies in favour of counterinsurgent tactics, the Vietnam War defied labels and controls.

With massive external support, the Vietnamese insurgents and the NVA were able to persevere in the face of tactics that had defeated guerrilla forces elsewhere. Thus in many ways the Vietnam War was a hybrid conflict, part limited war and part insurgency. It was also an integral part of the Cold War.

As such, some military thinkers now see the American effort in Vietnam as a success, even though the country 'fell to Communism'. The US effort succeeded very obviously in one of its major goals, for Vietnam did not escalate into a new world war. Also if seen as one battle of the ongoing Cold War, Vietnam can be seen as something like the Battle of the Somme: a costly setback, but possibly an attritional success, on the way to eventual victory, one which was finally achieved with the fall of the Berlin Wall.

BELOW: North Vietnamese troops fan out through a defeated Saigon, signalling ultimate victory in the Vietnam War. The American failure in Vietnam would eventually redefine the American way of war.

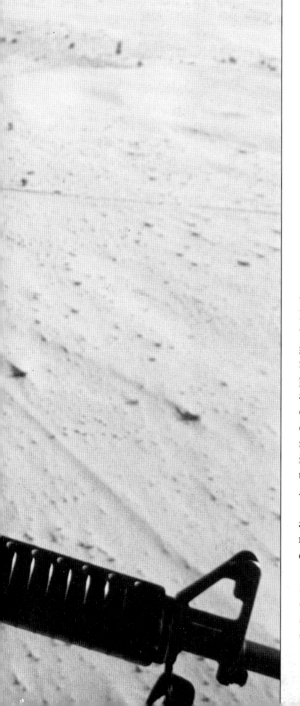

LIMITED CONFLICTS

The end of the Cold War and advances in technology altered the very nature of modern conflict. The infantryman had to be prepared to deploy anywhere in the world, ready to fight limited wars for limited objectives.

In the wake of the Vietnam War military theorists world-wide began to question the effectiveness of limited war in the modern world. After all, if the armed forces of the world's leading superpower had failed to achieve victory against a Third World opponent, how would the militaries of lesser nations fare against similar or even greater threats? The conflict in Vietnam, though, had been unique in many ways and as such only had limited applicability as a new paradigm of modern conflict. Indeed a series of limited wars fought across the globe during the 1980s served to alter military perceptions concerning warfare in general. By the 1990s the success of small, élite forces, utilizing wondrous military technologies provided by the microchip revolution, seemed to indicate that limited war would indeed be the way of the future. The era of total war and threatened total destruction appeared to be over, as a Revolution in Military Affairs transformed conflict and returned decisiveness to war.

The first of the new limited wars pitted the United Kingdom against Argentina over ownership of the Falkland Islands. While neither nation ranked as a world military power, nor expected a major conflict in the Falklands, the war there set important precedents and

LEFT: A Sikorsky UH-60 Blackhawk flies over the Saudi Arabian desert in 1990, during 'Operation Desert Shield'. 'Desert Shield' was to become 'Desert Storm', and the defeat of Saddam Hussein's Iraqi forces seemed to indicate that new technology made it possible to win a decisive victory again, after nearly a century.

139

ABOVE: Following the Argentine invasion of the Falkland Islands (Malvinas) in 1982, the British demonstrated great ingenuity in assembling a force to retake the islands in record time. Here, paratroopers of the 3rd Battalion, the Parachute Regiment head for the shore in Landing Craft as the Task Force made its landings at San Carlos.

was much more revealing than the US conflict in Vietnam. For many years, Argentina had disputed British claims to ownership of the Falklands (Las Islas Malvinas), which was only 600km (373 miles) off of the Argentine coast. On 2 April 1982 Argentine forces, based on the strength of a large conscript army, struck the islands and quickly overcame the resistance of the small Royal Marine garrison, and within months the Argentines had sent some 13,000 reinforcements to the area. Their regional strength made the Argentines confident that Britain, over 12,000km (7458 miles) away, would not contest the loss of its distant dependency. Under the leadership of Prime Minister Margaret Thatcher, though, the British chose to strike back, and thus both countries made ready for an unexpected and unplanned conflict.

Reflecting the reality of the Cold War world, the British armed forces had been considerably downsized. Reliant upon training and technology as force multipliers, the British military had focused much of its attention on continuing troubles in Northern Ireland and did not expect to become involved in a limited national conflict. In the main incapable of prosecuting a large-scale war against Argentina, British planners limited the war to a contest for possession of the

Falkland Islands themselves. Quickly the British cobbled together a sizable force, consisting of over 8000 men and 39 warships, including two aircraft carriers, with which to launch their counterstrike. In preparation for the coming war Britain proclaimed an exclusion zone around the Falklands and by early May began to launch air- and naval bombardments against the Argentine defenders. Later in the month the British task force arrived in the area and an ensuing air battle heralded the beginning of 'Operation Corporate'. The Argentines, using aircraft based on the mainland, were left with little time to linger over the battle area. In addition the most advanced Argentine aircraft, Super Etendards and Mirages, did not fare well against British Sea Harriers. Even so, though the use of advanced Exocet missiles, Argentine pilots sank six British ships during the conflict including the Atlantic Conveyor, demonstrating a British weakness in radar detection.

Though brave Argentine pilots continued their strikes, the British had, in the main, established air superiority over the Falklands and now made ready for their landings on the islands. After a diversionary assault the British 3rd Commando Brigade and two parachute battalions landed against little opposition near San Carlos and Port San Carlos on the island of East Falkland. After consolidating their positions, the 2nd Battalion of the Parachute Regiment on 28 May launched on overland attack on Darwin and Goose Green. Though the fighting was intense and the 2nd Parachute Battalion lost its commanding officer, Colonel H. Jones, the British force quickly subdued a defending force more than three times its own size. British losses in the fighting were substantial – with the 2nd Parachute Battalion losing 10 per cent of its fighting strength – as the fleet at sea continued to suffer debilitating losses to Argentine air attacks. The capture of Darwin and Goose Green, though, served as morale boosters for the British and a signal of coming difficulties for the Argentineans.

TOWARDS PORT STANLEY

Now more wary of British strength the Argentinean commander, General Mario Menendez, stood ready to defend Port Stanley with nearly 9000 men. Convinced that the British had little stomach for real battle, Menendez arrayed his troops into defensive lines in the rugged hills and hoped to prosecute the war to a draw. However, the morale of the Argentine defenders – mostly unwilling conscripts – fell to a dangerous low. Undersupplied, poorly clothed and not believing the war to be worth their sacrifice the Argentinean soldiers, though numerous, stood little chance in the coming battles. Utilizing helicopters whenever possible British forces, now numbering two brigades, slogged their way across the rugged terrain of East Falkland toward their now reluctant foe.

On the night of 11 June, the British force, under the overall command of General Jeremy Moore, struck the first line of Argentinean defences. Though resistance was sharp in areas, the well-trained and equipped British infantry soon infiltrated and compromised the defensive positions. Just two nights later the British struck again, augmented by heavy artillery fire, long before the Argentineans thought they could be ready. Once again the British outmanoeuvred and outfought their dispirited Argentinean foes and quickly seized the remaining high ground dominating Port Stanley. At this point Argentinean morale utterly collapsed, leading General Menendez to sign a surrender document ending the short,

BELOW: Members of the Scots Guards and the SAS move towards a helicopter on Goat Ridge in the Falklands. The highly trained British professionals were more than a match for the numerous, but poorly motivated Argentine conscripts.

but sometimes quite bitter, conflict on 14 June. Losses in the Falklands War were relatively light: 255 British soldiers died as compared to 1000 Argentine fatalities.

The British experience in the Falklands War seemed to revalidate the concept of limited war. With precious little planning the British had made quick work of the conflict, achieving all of their major goals. More importantly, though, the war seemed to prove the efficiency of small, but highly professional armies. Utilizing superior training and morale, the British had been able to overcome a large, outdated conscript force with relative ease. The British also had exhibited high levels of all-arms coordination while the Argentinean forces had struggled to communicate and act as one. Finally in the Falklands War the British possessed a marked technological edge over their Second World opponent. In short, then, the war in the Falklands indicated that the armies of the West would fare well in more traditional wars against conscript armies of much larger size.

In the United States President Ronald Reagan, elected in 1980, sought to rekindle the nation's faith in itself. In 1983 a small Cuban military force and

the outbreak of societal revolution threatened the security of the tiny Caribbean island of Grenada. Not only were the Cubans in the process of constructing a major air base on the island but the safety of hundreds of American medical students also seemed to be in jeopardy. Thus on very short notice Reagan ordered a military operation – dubbed 'Urgent Fury' – to rescue the American students and force the Cubans and their supporters from the island. Though the Cubans were the tangible enemy in many ways the American attack on Grenada was a battle against the legacy of Vietnam.

Initially US forces estimated that they would face some 10 enemy battalions in Grenada, comprising both Cubans and local revolutionary forces. Wildly overestimating enemy strength, the US gathered together a large – if somewhat poorly organized – attacking force that numbered over 6000 men including elements of the Special Forces, the Marines, the 75th Rangers and the 82nd Airborne. With very little time to plan the operation, US forces began to land on Grenada on 25 October 1983, only three days after Reagan's order. Special Forces spearheaded the US assault on Grenada, including operations by Delta Force, Navy Seals and Army Rangers. Suffering from the lack of tactical planning and knowledge, though, most of the Special Forces units failed to achieve their objectives. Only the Seals succeeded by seizing and guarding the British governor of the area and holding out against repeated counterattacks.

AIRBORNE AGAINST GRENADA

Quickly Marine forces and a combat jump by the 1st Battalion, 75th Rangers seized the major airfields on the island, most notably at Point Salines. Here the rangers fell under heavy defensive fire but held the airport and succeeded in rescuing several US medical students. Though the airport was not yet truly secure, elements of the 82nd Airborne soon arrived by plane to augment American strength. Later that day an additional Marine force made an amphibious landing near the major Grenadan city of St George's against little opposition. On 26 and 27 October US units spread out from Point Salines attacking towards St George's and brushing aside Cuban and local defenders thus seizing control of the entire island.

'Operation Urgent Fury' had been a success, securing for the United States a much-needed victory in a limited war. Though the success helped to restore the nation's faltering faith, the seemingly

simple operation had been beset by several problems. The performance of the Special Forces units had been spotty at best, calling their use into question, especially in the wake of their failure of a few years earlier to rescue the Iranian hostages. In addition, the combined US force had exhibited multiple command and control problems. Such failures, especially in the wake of the nearly seamless British success in the Falklands, could not be tolerated. The shortcomings of 'Urgent Fury', together with the greater failure of Vietnam, changes in the

ABOVE: A dug-in and ready for action Royal Marine mortar team prepares to open fire from the wet and windy slopes of Mount Kent during the Falklands War.

BELOW: The attack on Goose Green was the first chance that British Paras had to measure themselves and their strengths against their Argentine opponents.

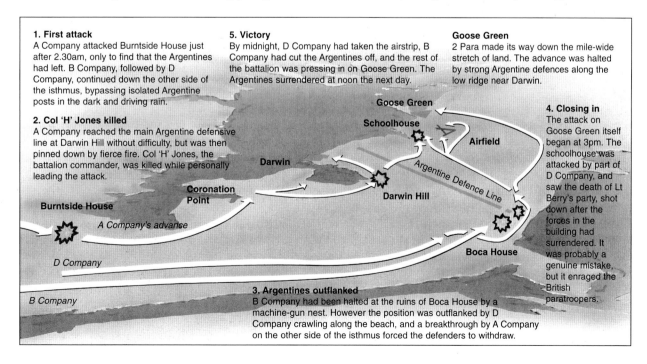

1. First attack
A Company attacked Burntside House just after 2.30am, only to find that the Argentines had left. B Company, followed by D Company, continued down the other side of the isthmus, bypassing isolated Argentine posts in the dark and driving rain.

2. Col 'H' Jones killed
A Company reached the main Argentine defensive line at Darwin Hill without difficulty, but was then pinned down by fierce fire. Col 'H' Jones, the battalion commander, was killed while personally leading the attack.

5. Victory
By midnight, D Company had taken the airstrip, B Company had cut the Argentines off, and the rest of the battalion was pressing in on Goose Green. The Argentines surrendered at noon the next day.

Goose Green
2 Para made its way down the mile-wide stretch of land. The advance was halted by strong Argentine defences along the low ridge near Darwin.

4. Closing in
The attack on Goose Green itself began at 3pm. The schoolhouse was attacked by part of D Company, and saw the death of Lt Berry's party, shot down after the forces in the building had surrendered. It was probably a genuine mistake, but it enraged the British paratroopers.

Goose Green
Schoolhouse
Airfield
Darwin
Argentine Defence Line
Darwin Hill
Coronation Point
Burntside House
A Company's advance
Boca House
D Company
B Company

3. Argentines outflanked
B Company had been halted at the ruins of Boca House by a machine-gun nest. However the position was outflanked by D Company crawling along the beach, and a breakthrough by A Company on the other side of the isthmus forced the defenders to withdraw.

ABOVE: A heavily armed American soldier scans a street in Grenville, Grenada during 'Operation Urgent Fury'. He is armed with an M-203 grenade launcher attached to an M-16A1 rifle and is also carrying an M-72 antitank rocket launcher.

nature of the Cold War and the advent of new technologies, led to a systematic redefinition of the American way of war in the mid-1980s and to what some called a Revolution in Military Affairs.

THE SOVIETS IN AFGHANISTAN

While the nations of the West engaged in victorious limited wars and began to redefine the nature of modern conflict the Soviet Union became bogged down in a war against guerrilla forces in Afghanistan. It was a war that closely resembled the US experience in Vietnam and a war that began the collapse of the massive Soviet empire. Fearing for the security of the pro-Soviet régime in their neighbour to the south, Soviet forces in December 1979 supported an Afghan coup. In tandem with a Special Forces operation in Kabul, four mechanized rifle divisions raced across the border to secure most of the major urban areas of the country and placed Babrak Karmal in power as President of the Democratic Republic of Afghanistan. The massive Soviet force quickly brushed aside resistance from loyal Afghan units and the war appeared to be at an end. Soviet military planners were quite confident that the Afghans – notoriously quarrelsome and ethnically divided – would put up only minor resistance to a

Soviet occupation force. The Soviet occupation, though, caused a great societal upheaval in Afghanistan, leading a total of five million refugees to flee the country. Backed by world opinion and fired by a religious fervour, Afghan rebels took to the rugged mountainous terrain of their nation's hinterland and called for a jihad, or holy war, against the infidel invaders. Though originally small in numbers the guerrillas, termed Mujahedeen, planned to protract their struggle, much as the Viet Cong had done in their war against the United States.

During the first phase of the conflict in Afghanistan, lasting nearly four years, the Soviets expected to crush the Mujahedeen using traditional military tactics dating from World War II and designed for the battlefields of western Europe. Increasingly the Soviet force, raised to a strength of 100,000 men, had to be self-reliant, as the army of their Afghan client government deserted to join the Mujahedeen at a stunning rate. The Mujahedeen for their part relied on traditional guerrilla hit-and-run tactics, launching lightning raids and ambushes on Soviet forces from their mountain hideaways. In addition the Mujahedeen could rely on safe bases of operations in and support from Iran and Pakistan. Like the United States in Vietnam, the Soviets dared not attack the Mujahedeen safe havens for fear of escalating the conflict. The war in Afghanistan, then, would be a limited war on the part of the Soviet Union, following many of the rules of failed efforts in previous such wars.

Initially Soviet military techniques in Afghanistan were clumsy and totally unsuited to the situation at hand. The Soviet theories of modern conflict revolve around their experience in World War II. Expecting any new war to be a total war in Europe, the Soviets were masters of massed armoured movements against an enemy expected to stand and give battle. Though their nation was an armed camp – in constant readiness for the expected massive land war – the Soviets had put little thought into conducting a war against elusive guerrillas, almost ignoring the western experiences of counterinsurgency and limited war. As they would in conflicts in Chechnya later, the Soviets simply attempted to defeat the Mujahedeen by

using the tactics and military doctrine that they had to hand. It was a mistaken attempt to attempt to alter the war rather than alter the fighting methods, a signal and eventually fatal flaw within the Soviet military system. Soviet and loyal Afghan forces, often of divisional strength, would use massive artillery and air support to prepare the battlefield. Following in the wake of the maelstrom, Soviet tanks and mechanized forces would lumber forwards into the countryside in vain attempts to locate and crush the guerrillas, who had already fled to their safe havens. Since much of the Soviet armour and mechanized infantry could not penetrate the rugged terrain, remaining road-bound, the Mujahedeen found it comparatively easy to avoid the clumsy efforts of their enemy. As a result Soviet control of the countryside quickly dwindled and the Soviet Fortieth Army lost the initiative. Reduced to garrisoning major cities and protecting logistic networks, the Soviet military force could only react to Mujahedeen threats.

By 1983 the Soviets and their Afghan allies had altered their military tactics, slowly adapting to the irregular nature of the ongoing conflict. Eschewing their reliance on traditional, large-unit operations, the Soviets began to break their forces down into smaller, tactically more versatile formations. Though these smaller units often launched mechanized conventional attacks, they also began to make more liberal use of air power, especially in the form of the helicopter. Air attacks, based on the strength of MiG-21s and Mil-24 Hind helicopter gunships took a heavy toll on guerrilla strength and mobility. Such Soviet strikes were often followed by infantry assaults from Mi-6 Hook helicopters in tactics that closely resembled US air mobility in Vietnam. The strength of the Soviet small-unit operations served to push the Mujahedeen further into the hills, thus 'pacifying' much of the countryside. The liberal use of Soviet firepower and of force, though, was in many ways counterproductive, as it forced ever greater numbers of Afghans to flee the country and provided the Mujahedeen with a never-ending flow of recruits.

Even though the Soviets had achieved some success, the war in Afghanistan dragged on with no end in sight, and the Mujahedeen, though fractured along ethnic and tribal lines, continued to resist. Unlike the Viet Cong during the war in Southeast Asia, the Mujahedeen had no coherent command structure and often engaged in sporadic, uncoordinated operations that had only attrition as their goal. Thus the war settled down into a

ABOVE: American Airborne Rangers guard local revolutionary fighters taken prisoner in Grenada. Though resistance on the island was slight, US forces encountered several worrisome command and control problems.

period of chaotic Mujahedeen guerrilla raids on Soviet interests, including frequent rocket attacks on the capital city of Kabul. Hoping to press their new-found advantage, the Soviets and their Afghan allies began to prosecute air and airmobile attacks on the Mujahedeen supply lines, especially those emanating from Pakistan. The war had entered a new and a dangerous phase.

The Mujahedeen had long received support from and safe haven in Iran and Pakistan: both nations that could stand behind the Afghans in their jihad against the Soviets. However, the continuing war in Afghanistan also drew the attention of the United States. To the Americans, the Afghan War was a part of the wider global struggle of the Cold War and represented a chance for revenge against Soviet support for the North Vietnamese a decade earlier. Thus the United States, led by the CIA, decided to support the Mujahedeen. However, the support was meant as a detriment to Soviet world power, not as support for an Islamic jihad against the West or even for Afghan independence. Thus two of the motive

geopolitical forces of the late twentieth century met in Afghanistan: the Cold War and Islamic fundamentalism. The mixture proved quite volatile. American support for the Mujahedeen had remained minimal, just enough to keep the revolution alive as not to destroy the slow warming of relations between East and West. However, the Soviet success in Afghanistan helped to lead to a major US change of policy. Partly in an effort to even the odds in the Afghan War, the US in 1987 began to provide the guerrillas with the very latest in air-defence technology, namely the shoulder-fired, 'Stinger' antiaircraft missiles.

The advent of the Stinger in Afghanistan effectively put an end to Soviet air supremacy and airmobile operations. Most of the Soviet air forces, especially their helicopters, proved to be very vulnerable to Stinger attacks. In fact the first 340 Stingers fired in Afghanistan downed an amazing 269 Soviet aircraft. The balance of the war began to shift yet again. Unable to rely on their helicopters for mobility, the Soviets reverted to less successful tactics of mechanized sweeps.

BELOW: A Soviet soldier guards a convoy during the war in Afghanistan. In spite of a major deployment of forces, the Soviets never managed to achieve total control of the countryside, and the already frequent Mujahedeen attacks on Russian convoys became even more so.

Once again the Mujahedeen found their mountain hideaways safe from most attacks and began to spread their control across the countryside. Soviet and loyal Afghan forces found themselves once again confined to cities and major road networks, as up to 80 per cent of the countryside fell under Mujahedeen control. Also as the war drug on and Soviet fortunes began to dwindle, the morale of their army in Afghanistan began to plummet. Mirroring the Vietnam War after 1968, the Soviet population, amid economic turmoil, began to question the worth of continuing the war in Afghanistan.

RUSSIA'S VIETNAM

With no end to the war in sight, the Soviet military leadership faced a difficult choice: significantly increase the military effort in Afghanistan in search of victory, or seek a negotiated end to the conflict. The new Soviet Premier, Mikhail Gorbachev, fully realized that massive economic and world-wide military challenges faced his beleaguered nation. Even as he began to implement policies such as perestroika, Gorbachev came to the conclusion that he had to put an end to the continued drain represented by the war in Afghanistan. At Soviet direction a new Afghan leader, Mohammed

Najibullah, began to offer economic concessions and even a power-sharing arrangement to the Mujahedeen. Sensing victory and without a unified political voice, the Mujahedeen, though, chose to continue their resistance, and pressed for total victory. Unable to arrange an acceptable peace in Afghanistan, the Soviets chose nevertheless to end their conflict there, removing their last troops from the country on 15 February 1989. Soviet support for the Najibullah regime in Afghanistan continued, as did the war in that country. In 1992, as Soviet support died away, Najibullah surrendered power and the Mujahedeen achieved victory.

During the conflict in Afghanistan Soviet forces had suffered some 14,000 killed and 53,000 wounded, while inflicting millions of casualties upon their tactically overmatched foe. In the end the war had proven unwinnable for the Soviets, pitted against an irregular foe with massive international support. Though they had adapted their tactics the Soviets, in the end, chose withdrawal over escalation of the conflict. The war had been in effect the Soviet's Vietnam. The legacies of the Soviet defeat, however, were even more far-reaching than anyone could have ever expected. Facing unrest in Europe and disaffection at home, the Soviet Union quickly crumbled, bringing

ABOVE: Mujahedeen fighters resting in a safe house in Mazar-e-Sharif. Had the Afghans gone toe-to-toe with the Russians, they would have been slaughtered, no matter how high their religious or nationalist fervour. However, Afghans have been fighting invaders for centuries, and they used guerrilla tactics to frustrate the Soviet colossus.

the Cold War to an end. The military and political realities of the era of limited war and revolution suddenly ceased to exist. The contest of wills between the United States and the Soviet Union had been the motive force behind most wars since World War II and had imposed a series of set rules upon the nature of warfare itself. Suddenly those rules were gone, leaving

ABOVE: Soviet forces firing AGS-17 automatic grenade launching during the conflict in Afghanistan. Such weapons provided them with a light and relatively easy to handle source of direct fire support, a kind of short-range artillery capability.

militaries across the world groping towards a redefinition of modern conflict.

The war in Afghanistan also had several, more hidden results. The success of the Mujahedeen once again seemed to indicate that many future conflicts might assume a more asymmetrical form, often involving bitter struggles between ethnic or religious groupings. Presaging the future, following the fall of the Najibullah government in Afghanistan, peace was not restored to the war-torn nation. Instead the Mujahedeen – never a truly unified force – began to squabble among themselves. Tribal warlords vied for control of the stricken nation, resulting in a thoroughgoing humanitarian tragedy. The United States, having achieved victory in the Cold War, paid little heed to the situation in Afghanistan, seeing it as a regional problem. After years of anarchy, a new unifying force came to the area: fundamentalist Islam in the form of the Taliban. Born of the last conflict of the Cold War era, the growing gap between forces of Islamic fundamentalism and the West came to dominate the emerging conflicts of the 21st century.

THE SEARCH FOR TOMORROW

The loss of the Vietnam War had caused the American military to question its doctrine and indeed the entire American way of war. Since the American Civil War the United States had developed a policy of military attrition in major conflicts, often relying on massive industrial strength to achieve victory. However, such attritional policies had failed in Vietnam and also seemed quite ill-suited to the realities of the ongoing Cold War. Against the Soviet military juggernaut an attritional defence of western Europe seemed doomed to failure and overly reliant on nuclear weapons to redress the strategic balance. Indeed, the major US document concerning the defence of western Europe, Field Manual 100-5 Operations 1976, convinced many US and NATO military theorists that the coming conflict would be a catastrophic defeat. Dissatisfied with the legacy of Vietnam and concerned for the future, American military theorists began to move toward a more aggressive approach towards war, one that relied on manoeuvre and overwhelming firepower aimed at a achieving quick victory.

Even as US military thinkers began to reassess the American way of war, groundbreaking technological changes were sweeping the planet, beginning the computer age and causing what many theorists term a Revolution in Military Affairs. The advent of the microchip in the 1980s revolutionized warfare through an exponential leap forwards in the information available to military commanders, affecting both knowledge of the battlefield and the accuracy of existing and new weapons systems. The new technology, in the main a product of US research and development, allowed commanders an unprecedented 'See Deep' capability. Advanced radar systems aboard E-3 AWACS (Airborne Warning and Control Systems) aircraft could detect all enemy air activity, while E-8 JSTARS (Joint Surveillance and Target Attack Radar Systems) aircraft provided information regarding enemy ground movements. With such information – allowing commanders detailed views of enemy movements up to 1000km (622 miles) behind the front lines – US commanders were able to track enemy troops and air assets long before they reached the front lines. In effect they were capable of seeing up to 96 hours into the future.

'STRIKE DEEP'

Technological advances in weapons systems also provided a new 'Strike Deep' capability. Advanced aircraft such as the F-4G Wild Weasel could disable enemy radar systems, while new Stealth aircraft, such as the F-117A, could avoid enemy radar to launch surprise attacks against enemy forces to their strategic depth. Also a variety of 'stand off' weapons, such as the Cruise Missile, now enabled US forces to attack to strategic depth with little risk to themselves. Much of this new weaponry also used 'smart' technology, including television and laser-guidance systems, to strike targets with previously unseen accuracy. To the layman the strength and variety of the new weapons systems is nearly unimaginable. As an example, a new breed of missile could strike an unsuspecting enemy armoured unit far behind the lines with Wide-Area Anti Armor Munitions (WAAM). The missile disperses several terminally guided submunitions above a battlefield, which then use infrared sensors to locate and strike enemy tanks using a depleted uranium charge. Finally technological developments also provided US commanders with a much more powerful 'Strike Shallow' capability. Along with more traditional forms of firepower, such as tactical bombing, US forces could now even rely on 'smart' artillery shells in the battle for the front lines. The advance in firepower, though, is best illustrated by the advent of the Multiple Launch Rocket System (MLRS). One salvo of 12 rockets from a single MLRS could deliver 8000 M-77 submunitions to the battlefield, devastating a 25,083sq km (30,000 square yard) area.

In the late 1970s, from the command centres of NATO in Europe to training centres in the United States, western military theorists struggled to understand how technology had changed war, and struggled to devise new tactical and strategic doctrines. The culmination of these efforts took place under General Donn Starry at the US Army Training and Doctrine Command, resulting in the issuance of Field Manual 100-5 Operations 1982, a new statement of US military doctrine which was revised in 1986. The doctrine, soon known as AirLand Battle, was meant to apply to any possible war, total or limited, anywhere in the world and aimed at achieving a quick, offensive victory through the use of modern technology and manoeuvre. AirLand Battle took a decisively non-linear view of warfare, envisioning enemy forces and their command structure as a

BELOW: A diagram showing a typical Mujahedeen guerrilla attack on a Soviet convoy. Often attacking in narrow mountain passes, the guerrillas usually tried to disable the lead and trailing vehicles in the column, stranding the remainder. Their task was made easier by the fact that Soviet infantry conscripts almost invariably stayed with their vehicles when attacked, rather than trying to engage the attacking guerrillas in an immediate counterattack.

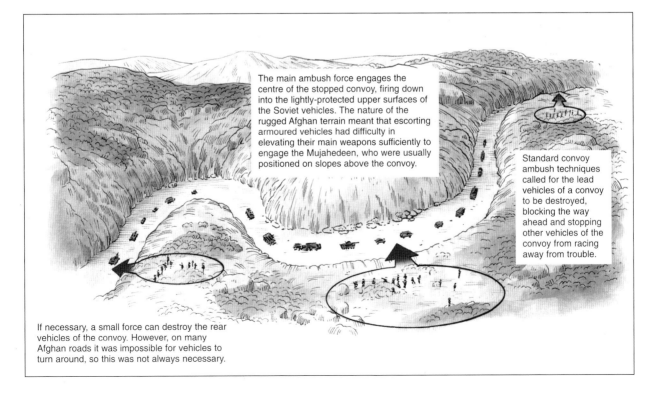

The main ambush force engages the centre of the stopped convoy, firing down into the lightly-protected upper surfaces of the Soviet vehicles. The nature of the rugged Afghan terrain meant that escorting armoured vehicles had difficulty in elevating their main weapons sufficiently to engage the Mujahedeen, who were usually positioned on slopes above the convoy.

Standard convoy ambush techniques called for the lead vehicles of a convoy to be destroyed, blocking the way ahead and stopping other vehicles of the convoy from racing away from trouble.

If necessary, a small force can destroy the rear vehicles of the convoy. However, on many Afghan roads it was impossible for vehicles to turn around, so this was not always necessary.

ABOVE: Soldiers in training for desert warfare during 1991 in Saudi Arabia. Many military pundits thought such training to be of little value, believing that air power alone could win the conflict. No infantryman would have agreed, since history has shown that an enemy was not beaten until he was beaten on the ground.

system. Integrating the abilities of the armed forces to an unprecedented extent – especially a blending of air and ground campaigns – AirLand Battle sought to strike an enemy force to strategic depth in variety of ways. The modern battlefield no longer just encompassed the front lines and possibly a strategic bombing campaign. In AirLand Battle US forces proposed to integrate air, ground, sea, space and cyberspace strikes on a new 'integrated battlefield, stretching from ground level to the edge of the atmosphere, from the front line up to 1000 kilometres into the enemy rear, and from the present moment up to 96 hours into the future.'

In a true cataclysm of modern war, the defender would face attack at all levels, from MLRS salvos on the front lines to destruction of command and control to

cyber warfare. Though the use of their unrivalled technology, US forces proposed to out-think, outgun and outmanoeuvre their rivals. AirLand Battle, then, would be the opposite of World War I, for technology this time provided the attacker with all of the advantages. American planners hoped that such an offensive would return decisiveness to warfare, negating the need for protracted wars of attrition. Continued strikes against enemy command and control and logistics would hopefully 'so disrupt the enemy armed forces as to bring about their collapse, psychological and physical, and disintegration'. In essence the US now proposed to fight wars from the inside out. Instead of battering through an opposing army in years of attrition AirLand Battle aimed at the destruction of the 'brain' of the armed forces and the nation, effectively making the enemy's military might on the front lines now totally redundant.

Many US and NATO military theorists saw AirLand Battle as revolutionary, making all other forms of war and the humble infantryman obsolete. However, there remained high levels of trepidation concerning the new style of war. Reliant on vulnerable and costly technology, AirLand Battle called for an unprecedented and seemingly impossible level of all-arms coordination and information management. Wars involving AirLand Battle would be costly and would require hitherto unseen levels of

RIGHT: Soldiers of the US 101st Air Assault Division in action during 'Operation Desert Storm'. Although tanks and aircraft were the stars of the war against Saddam Hussein, it was the humble infantry who had to take and hold ground until the Iraqis surrendered.

training for and professionalism from soldiers. In addition AirLand Battle was an offensive doctrine, and as such involved great risks for great gains. Thus while some military commanders were lauding AirLand Battle as the wave of the future, others considered it over-ambitious, and they voiced doubts about whether it would work at all.

THE GULF WAR

The initial test of AirLand Battle came in a quite unexpected way. Iraq, under the dictatorial leadership of Saddam Hussein, had long sought more influence in the affairs of the Middle East. In an inconclusive, but bloody struggle with Iran (1980–1988) Iraq had suffered great economic losses and soon sought to recoup those losses at the expense of a lesser foe. Though the Iraqi military was the fourth largest military in the world and was battle tested, it was also fatally flawed. The vast majority of the Iraqi Army consisted of unenthusiastic and poorly trained conscripts who were often quite divided in their ethnic and religious loyalties. As in most dictatorships, Saddam considered his own mass army something of a threat and gave the best of training and equipment only to his most loyal units, including the Republican Guard and the Haras al Ra'is al-Khas. In many ways these units, though by far Saddam's most capable, were too valuable to use in battle, for they also served as his insurance policy against his own people. Though strong on paper – and having learned much from the war with Iran – Iraq's Army was poorly suited for a struggle with a modern western armed force.

The deficiencies of the Iraqi Army mattered little, for their opponent was tiny Kuwait. Totally misreading the situation, Saddam was quite certain that the western Allies would not involve themselves in the coming war. After OPEC failed to placate Iraq by increasing the price of oil, in August 1990 the Iraqi armed forces moved forwards into Kuwait in a daring and unexpected offensive. The tiny sultanate had few defences and quickly fell to the invader, giving Iraq control over 15 per cent of the world's oil reserves. Though Arab nations sought to solve the problem themselves, the nations of the west – dependant on Middle East oil – remained quite frightened that Iraqi forces might push into Saudi Arabia in a bid for control of over 40 per cent of the world's oil. Saddam had, thus, misjudged the level of world interest in his actions and had also chosen the wrong time for such actions. With the end of the Cold War the US and NATO could take military action in the Middle East without fear of Soviet intervention or an unwanted escalation of the conflict. The abatement of Cold War rivalries also made it possible for the United Nations Security Council to adopt a stance calling upon all member nations to oppose the invasion of Kuwait. Finally Saddam had believed that Arab disunity and hatred of Israel would help to serve as a buffer to outside military interference in the region. Once again he was mistaken and soon was quite surprised to witness the formation of a world-wide coalition, which was aimed at the liberation of Kuwait.

Having made the decision to fight, the nations of the coalition now had to gather defensive forces in Saudi Arabia in 'Operation Desert Shield'. In this quite dangerous phase of the war the coalition – including some 30 countries providing naval units, seven providing air units and eight providing ground units – had to mass their forces from all over the globe and transport them to the Middle East. Though the operation in the end went smoothly, it was in many ways a logistical nightmare, one that would have nearly been insurmountable had Saddam chosen to attack. The Iraqi leader, however, chose not to push into Saudi Arabia, instead opting to build massive defensive networks throughout Kuwait, hoping that coalition forces would balk at fighting a war that resembled the defensive struggle of the Iran-Iraq War. Several outside observers agreed that the coming war would be long and bloody. Though estimates varied, it seemed that Iraq had gathered a considerable force in Kuwait numbering some 500,000 men, 4300 tanks (including 500 modern Soviet T-72s), 3000 artillery pieces, 400 aircraft and hundreds of Soviet SS-1 Scud B missiles, possibly tipped with biological or chemical warheads. The force was indeed powerful, but nonetheless poorly motivated, and occupied forward defences which were almost uniquely susceptible to attack by AirLand Battle.

ABOVE: Two US Marine riflemen serving with a reconnaissance platoon keep watch on enemy movements during 'Operation Desert Storm'. The Marine at left is a sniper, armed with an M-40A1 rifle. His companion acts as the spotter, identifying targets for the shooter.

Though several nations took part in the campaign, making command and control uniquely difficult, the United States provided the majority of the forces for the conflict. President George Bush, once the decision for war had been made, left most of the military planning to his trusted military subordinates: General Colin Powell, Chairman of the Joint Chiefs of Staff, and General H. Norman Schwarzkopf, Commander-in-Chief of US Central Command. Both were veterans of the Vietnam War.

While the massive coalition force slowly gathered together in Saudi Arabia Schwarzkopf directed that the Pentagon plan an air campaign against Iraq. The resulting scheme, initially dubbed 'Instant Thunder', called for a massive air campaign prosecuted to enemy strategic depth that would last only nine days, but would cause the immediate defeat of Iraq and the withdrawal of their forces from Kuwait. Thus the faith exhibited by Pentagon planners in the air component of AirLand Battle was so profound that they believed a ground war to be unnecessary. Schwarzkopf and Powell, though, thought differently and overruled their planners. The resultant coalition plan of action called for a six-week sustained air campaign against Iraq, under the control of General Charles Horner. The aerial offensive would use the

massive array of coalition technology to strike at Iraqi command and control, telecommunications and logistic infrastructure and was aimed at decapitating the Iraqi forces in Kuwait. A later goal of the air campaign would be the massed carpet-bombing of Iraqi forces in Kuwait in preparation for the final and huge ground offensive.

Having gathered together some 2600 military aircraft in the Persian Gulf area, on the night of 17 January 1991 the coalition began its air campaign, the first phase of 'Operation Desert Storm'. With air supremacy as a first goal, coalition forces had utilized electronic surveillance to locate the Iraqi 'Kari' air-defence system along with the Air Defence Operations Centre in Baghdad and subsidiary air-defence centres throughout the region. US Special Forces guided AH-64 Apache helicopters to the forward air-defence sights, which destroyed the targets using Hellfire guided missiles. At the same time, Stealth fighters and 54 cruise missiles struck the Air Defence Operations Centre in Baghdad. In addition, EF-111 Raven aircraft eliminated most resistance from Iraqi Surface-to-Air Missile sites.

After a night of fighting, Iraq's air defences had been totally compromised, at the cost of one coalition aircraft lost. Saddam resorted to desperate measures in

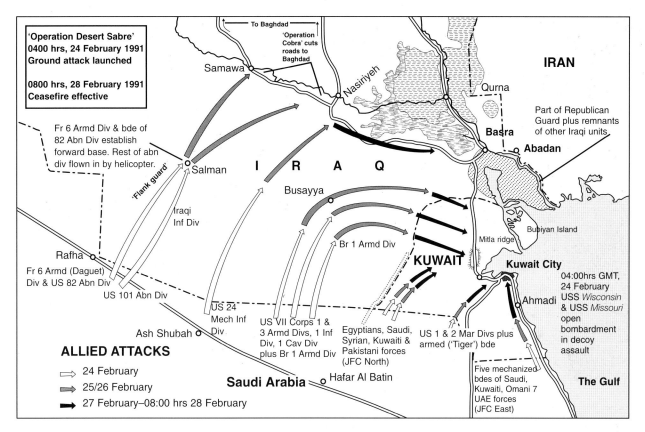

'Operation Desert Sabre'
0400 hrs, 24 February 1991
Ground attack launched

0800 hrs, 28 February 1991
Ceasefire effective

To Baghdad

'Operation Cobra' cuts roads to Baghdad

Samawa

Nasiriyeh

Qurna

IRAN

Fr 6 Armd Div & bde of 82 Abn Div establish forward base. Rest of abn div flown in by helicopter.

Basra

Abadan

Part of Republican Guard plus remnants of other Iraqi units

Salman

I R A Q

'Flank guard'

Busayya

Iraqi Inf Div

Rafha

Br 1 Armd Div

Bubiyan Island

Mitla ridge

KUWAIT

Kuwait City

Fr 6 Armd (Daguet) Div & US 82 Abn Div

US 101 Abn Div

Ahmadi

04:00hrs GMT, 24 February USS *Wisconsin* & USS *Missouri* open bombardment in decoy assault

US 24 Mech Inf Div

Ash Shubah

US VII Corps 1 & 3 Armd Divs, 1 Inf Div, 1 Cav Div plus Br 1 Armd Div

Egyptians, Saudi, Syrian, Kuwaiti & Pakistani forces (JFC North)

US 1 & 2 Mar Divs plus armed ('Tiger') bde

Five mechanized bdes of Saudi, Kuwaiti, Omani 7 UAE forces (JFC East)

The Gulf

ALLIED ATTACKS

⇨ 24 February

⇨ 25/26 February

➡ 27 February–08:00 hrs 28 February

Saudi Arabia

Hafar Al Batin

---

an effort to restore balance, firing Scud missiles blindly at Saudi Arabia and Israel in an effort to rend the coalition asunder. The Scuds, though rather ineffective, caused great worry to coalition planners, for aircraft and Special Forces both had great difficulty in locating and destroying mobile Scud launchers, demonstrating a weakness in the tapestry of technological warfare. Even so, the air war over Iraq was over, as evidenced when many Iraqi fighters flew to Iran for internment.

Next the air campaign moved on to the destruction of Iraqi command, logistics and ability to wage war. The violent, and now virtually unopposed attacks achieved great success, shutting down 88 per cent of the Iraqi national electrical grid, destroying 90 per cent of the Iraqi ability to refine oil, and lowering the flow of supplies coming into Kuwait from 75,000 tons per day to 16,000 tons per day. Air attack also played a vital role in preparing the battlefield, destroying a claimed 32 per cent of Iraq's armoured vehicles and 47 per cent of its artillery before the launch of the ground campaign. Though these numbers remain in doubt, the air assault played a vital role in the coming coalition success in the Gulf War. Air war theorists continued to hope that the bombing campaign would break the back

of Iraq, negating the need for a ground war, but the Iraqis held on stubbornly in the face of heavy losses. Indeed the air war had actually achieved less than AirLand Battle plans had stipulated, for the enemy had not been decapitated or collapsed. Part of this shortfall is due to the fact that western militaries were not yet ready to prosecute a true AirLand campaign. Only some seven per cent of the munitions fired in the Gulf War were precision-guided. Even so, the air campaign cost $1.3 billion, a testament to the cost of modern weapons systems.

ABOVE: After a feint towards Kuwait City, US and British forces attacked in a 'left hook' aimed at cutting off all Iraqi retreat.

BELOW: A US soldier speeds down a desert track during the Gulf War. Coalition superiority in mechanization and technology provided a decisive edge over the numerous but overmatched Iraqi armed forces.

ABOVE: The US Army's Objective Individual Combat Weapon (OICW) combines an advanced sighting system incorporating a thermal imager and a laser rangefinder with an experimental 20mm (0.78in.) grenade launcher and a 5.56mm (0.22in.) assault rifle. It will allow 21st Century soldiers to attack hidden targets more accurately from safer distances.

By 24 February 1991 there were nearly 545,000 coalition forces in the Kuwaiti Theatre of Operations (KTO). After months of planning, Schwarzkopf decided to employ a two-stage strategy. Initially coalition forces would feign a frontal advance into Kuwait and an amphibious landing at Kuwait City. After locking Iraqi attention on the area, over 250,000 coalition forces – mainly the US VII and XVIII Corps – shifted in a westerly direction in order to prosecute a 'left hook', an armoured flank sweep through the Iraqi desert which would be covered by airmobile units.

On the morning of 24 February, coalition forces rolled forward, and even those forces facing the strength of the Iraqi defences made considerable headway at little cost, often finding their greatest difficulty was caring for the masses of prisoners who surrendered to anything that moved, including drone aircraft and journalists. On the extreme left XVIII Corps raced through the Iraqi desert, preceded by an airmobile operation by the 101st Airborne some 275km (171 miles) into Iraq that cut the

road from Kuwait to Baghdad. With supplies often being flown in before their arrival, the remainder of XVIII Corps advanced quickly, finally managing to seal the coalition flank from any oncoming danger from the enemy.

It remained for VII Corps to seal the victory in the Gulf by defeating the Iraqi Republican Guard, which was held in a theatre reserve. VII Corps numbered over 145,000 men, as well as a total of 40,000 vehicles which consumed some 5.6 million gallons of fuel every day. Fighting for VII Corps remained rather sparse, as most Republican Guard units chose to retreat and avoid battle. However, on 25 February, VII Corps encountered several elements of two Iraqi armoured units. The two sides fought a sharp battle amid a blinding, howling sandstorm.

During the fighting, it became quite obvious that the US M1A1 Abrams outclassed all of the Iraqi tanks. Using thermal rangefinders and a substantial advantage in range, the American tanks made short work of their adversaries, destroying hundreds of Iraqi tanks and, in the process, managing to avoid suffering a single loss. On 27 February, with Iraqi resistance on a steady decline, forward elements of VII Corps reached their goal: the Basra-Kuwait City Highway, cutting off the most obvious means of Iraqi retreat, leading to an official cessation of hostilities on the following day.

The Gulf War has its critics, and in many ways it did not represent a true test of AirLand Battle. Many Iraqi units were able to escape the technological trap and escape to the north, where they succeed to this day in maintaining Saddam Hussein as the Iraqi leader. The Gulf War was not, however, intended as a total war, and AirLand Battle had won its quick, resounding victory. It was then time for the politicians to take charge of the peace negotiations.

The casualty rates during the Gulf War were astounding, serving to overcome any criticism and also seeming to herald a true revolution in warfare. During the conflict, the Iraqis had lost a total of some 200,000 dead, 200,000 deserted and 50,000 prisoners. Total coalition casualties stood at less than 1000 killed and wounded. The Iraqis also lost some 3700 tanks and 2000 artillery pieces as a result of the war against the Allies.

To most observers these numbers were conclusive. It seemed that technology had returned decisiveness to war. Air power, smart munitions, stand off weaponry, and quality armour were able to render even the most powerful defences ineffective. The revolutionary nature of the Gulf War victory seemed to indicate that total war was now a thing of the past, and indeed warfare in general might be outdated, for no nation was able to stand against the technological might of the West. Once again, observers proclaimed that the end of infantry warfare was at hand.

FUTURE INFANTRY

Some military theorists, despairing of the further usefulness of infantry in war, advocated bringing the infantryman up to date through further technological developments. One theorist stated that, 'In the end the rifle-bearing infantryman is governed by the same principles that governed the spear hurler and the bowman – first see the target, then try to get your hands to direct your projectile toward it.' This was certainly a quaint anachronism on the modern-day battlefield.

Several US planners began a series of tests and programs to update infantry, including the development of 'The Enhanced Integrated Soldier's System'. The ultra-modern infantryman was to wear body armour as well as a heads-up display, virtual reality helmet, which would allow for full night vision and instantaneous cyber communications. Carrying laser designators and global positioning equipment, the soldier could call in an attack by smart weapons on any designated target.

In the most advanced suggested systems, a soldier would only have to look at a target, before his onboard computer systems, using laser optics, would focus in on the target and then fire a variety of hand-held smart weapons. Such a variety of weaponry might be quite heavy so there was a scheme, code-named PITMAN, to develop an infantry robotic exoskeleton to make the wearer more powerful.

The advocates of technology, though, expected far too much from the Revolution in Military Affairs. Certainly the Gulf War represented a resounding victory for technology in war, but even

it was incomplete. Technological changes have continued, though not at the pace that many had thought, making the armies of the west even more lethal, and expensive. As prices of weapons systems continue to skyrocket, many now wonder whether technological war is not the wave of the future, for only one country can afford to prosecute such war. And even the mighty United States could not prosecute a truly technological war for an extended period of time.

Additionally the success of the Gulf War seems to have convinced 'rogue nations' throughout the world that they should not face the West in such a conflict ever again. Taking the lesson that for every technological advance there is a 'dumb' solution, rogue nations have altered the rules of warfare once again. Rather than becoming more technological, since the Gulf War wars have become less so, involving bloody ethnic conflicts, religious wars and terror strikes. If the combatants hide among the people – or even are the people – technology is foiled, and war becomes, once again, the purview of the infantryman.

BELOW: Though a 'Revolution in Military Affairs' seemed to beckon after the Gulf War, it is still the humble infantryman that takes and holds ground. Indeed, as engineers rush to develop countermeasures against high-tech armoured and aerial wonder weapons, the place of the infantryman will become more, not less, important in future wars.

CHAPTER TEN

PEACE-KEEPING

As the end of the Cold War reduced the threat of global conflict, it released a myriad of new challenges to the world's armies. These ranged from peacemaking and peacekeeping to dealing with the terrorist threat.

The United Nations (UN) was founded in an era of great international insecurity following fast on the heels of World War II. The war had been caused, in many ways, by Hitler's rise to prominence during a period of appeasement, something that many political theorists believed could have been prevented by strong international action. Post-war political and military planners worldwide believed that international collective security, represented in the UN, could limit future conflicts and reign in rogue states, averting the threat of World War III. As the Cold War persisted and the world began to line up in two armed camps, the role of the UN as peacemaker became magnified. Though it had no standing armed force, the UN, under the executive direction of the Security Council, possessed the vast if somewhat nebulous power to intervene in international affairs. However, the inner workings of the Security Council itself served to place a buffer on UN assertion of authority. As permanent members of the Security Council, both the United States and the Soviet Union possessed absolute veto power, something that both nations used with great abandon during the Cold War era. In essence the continued superpower rivalry served to

LEFT: On peacekeeping missions, soldiers are often used to maintain law and order. These paratroopers from the US Army's 504th Parachute Infantry Regiment are helping United Nations' police as they move down an alley in Mitrovica, Kosovo, in February 2000. They are conducting a house-to-house search for weapons.

make the UN ineffective in its major goal. Thus as the Cold War itself and the threat of nuclear destruction placed limits on wars, the UN searched long and hard for its own identity.

Even as the US and the Soviet Union contested for world domination in the Cold War, other forces such as decolonization threatened world peace outside the spheres of direct superpower influence. Though there were no provisions for such actions in its charter, the UN began to take part in peacekeeping missions throughout the world in an effort to monitor several troublespots. The beginnings of what is now known as 'traditional peacekeeping' took place in 1948 when UN forces went to Palestine as impartial observers to monitor the truce between the Palestinians and Israelis. Limited by Cold War rivalry until 1990, the UN would muster some 13 peacekeeping missions across the globe, from the Congo to Lebanon to New Guinea. The nature of the missions varied widely but followed a set pattern. The UN, under the leadership of Dag Hammarskjold, would only send forces into an area with the consent of the belligerent parties involved. Thus the UN would not make peace or force peace on nations, but would only monitor and promote peace agreements that had already been reached. UN peacekeeping forces were small and lightly armed and only used force in self-defence. Hammarskjold also strove to maintain UN impartiality and would not draw peacekeeping forces from any of the five permanent members of the Security Council. Following these guidelines, though UN forces served under difficult circumstances in many of the most dangerous parts of the world, their peacekeeping operations were to achieve several marked successes.

The close of the Cold War brought fundamental change to the peacekeeping process. The lessening of East-West rivalries opened much more of the world to direct United Nations influence, allowing peacekeepers roles in nations that had been Cold War hotspots, including Namibia and Cambodia. Also the Security Council became a 'veto free environment', removing all artificial controls on the expansion of peacekeeping missions. Combined with

BELOW: Peacekeepers often found themselves isolated in the midst of brutal civil struggles. These members of the 709th Military Police Battalion have been surrounded by a hostile crowd in Sevce, Kosovo on 4 April 2000. Several hundred Serbs blocked the road to protest at the arrest earlier in the day of a local suspected of possessing munitions.

the development of a geopolitical atmosphere conducive to peacekeeping, the end of the Cold War created increased the need for peacekeeping missions. Superpower rivalry, often in support of brutal, but ideologically compatible regimes, had done much to control lesser ethnic, tribal and religious rivalries in countries across the globe. With the end of the Cold War, though, the superpowers lost interest in nations of once great strategic importance, including Somalia and Afghanistan. Without artificial controls, several nations overthrew their Cold War leadership only to dissolve into the anarchy of tribal and ethnic warfare. At an alarming rate, nations from Europe to Africa began to implode, leading to brutal civil conflict and humanitarian disasters. As a result, UN peacekeeping activity throughout the world increased greatly, and in 1994 there were 17 UN operations ongoing involving 85,000 personnel from 70 different nations.

UN PEACEKEEPERS

Most importantly, though, the role of UN peacekeepers in the post-Cold War world began to change. The new UN Secretary General, Boutros Boutros-Ghali, sought to seize the moment, seeing the UN as the 'beacon for a new planetary order'. In the new world environment there were few interstate conflicts for the UN to mediate. Instead peacekeeping missions would enter the realm of unstable internal ethnic and civil wars. In such confused, dangerous situations there were often no true local authorities that could offer their consent to peacekeeping missions, thus raising the possibility of attacks on UN forces. Also, continued anarchy in several countries led Boutros-Ghali to believe that UN forces should sometimes intervene in the affairs of sovereign nations without invitation to prevent humanitarian disasters. Such operations, sometimes known as peacemaking missions, were, 'carried out to restore peace between belligerent parties who do not all consent to intervention and may be engaged in combat activities.' Peacemaking forces had to be large and well armed, ready to do combat in the midst of a bewildering swirl of societal violence and decay. The size and nature of such missions also changed the nature of peacekeeping by requiring the involvement of the last remaining superpower. The involvement of American forces in peacekeeping missions represented a major departure from traditional peacekeeping. To many nations in the Third World the presence

ABOVE: Not all contacts with locals are hostile. Staff Sergeant John J. McCarthy of the US Army's 315th Psychological Operations Company, and his interpreter talk to a Kosovar Serb man after giving him a copy of *Dialog* – a KFOR publication printed in Serbo-Croatian and Albanian, on 4 May 2000 – in the Novo Brdo Obstina, Kosovo.

ABOVE: A Czech peacekeeper mans a checkpoint in Bosnia in 1996. The seemingly endless sequence of conflicts in the Balkans marked the first occasion in which former Warsaw Pact forces were under the same command as NATO units.

the Cold War, though, superpower interest in the region lapsed, forcing the government of Somalia to stand on its own. Without controls imposed by the superpower rivalry, disparate Somali factions banded together in opposition to the Barre government, forcing its eventual collapse in the beginning of 1991. In the resulting societal chaos several warlords and ethnic groupings vied for control of the stricken nation. Buying arms on the vast world market Somali warlords – representative of several coming struggles in the post-Cold War era – relied in the main on militia forces for their power. These infantrymen, often including children, received little training or discipline. As a result the warring factions often seemed to be little more than thugs, caring little for life or the rules of war. From Rwanda to Kosovo such armed groups would engage in a seemingly never-ending escalation of slaughter and revenge, usually termed ethnic cleansing. The leaders of such groups, including Mohammed Farah Aydeed, often thought only of their own welfare, and retained only nominal control over their deadly bands. Such situations were fraught with danger for peacemakers, for the lack of law and order or real rulers meant there could never be true consent.

By late 1992 it had become obvious to the outside world that famine had struck Somalia in the wake of the anarchy there, leading to a humanitarian disaster. As the situation worsened and over a million people neared starvation, though the war in Somalia continued, the UN decided to intervene. In September 1992 a small force of UN peacekeepers arrived in Somalia to help secure delivery of international aid shipments to the starving local population. However, Somali warlords, especially Aydeed, viewed the UN intervention with suspicion and as something of an opportunity. Armed bands would regularly steal food shipments for their own uses and the humanitarian disaster worsened. Plainly the UN needed more military punch in Somalia. For the upgraded peacemaking effort, several nations committed troops, but the United States took the lead by offering a force of 30,000 men in an attempt to overawe the recalcitrant Somali warlords.

of American forces did much to diminish the supposed impartiality of the UN, leading to the possibility of interstate conflict with the US. In addition American forces, now trained to 'fight to win' would often see peacemaking in different way to the way the UN leadership would see it.

SOMALIA

The collapse of the East African nation of Somalia offered a stern test for the proponents of peacemaking. During the 1980s the United States had supported the dictatorial Somali regime of Siyad Barre, in part to offset Soviet influence in neighbouring Ethiopia. After the end of

Reflecting the new realities of peacemaking, the UN initially, in most ways, lost control of its mission in Somalia to the world's remaining superpower. Beginning 'Operation Restore Hope' US Marines landed on the beach outside Mogadishu on the morning of 9 March 1993 and were greeted by a phalanx of representatives of the news media. Within days the peacekeepers had taken control of parts of Mogadishu as well as much of the surrounding countryside.

Initially the mission went well and resulted in the effective distribution of aid to the grateful Somalis. However, to many in the UN and the US, such a situation seemed to be only a temporary palliative. Once the peacekeepers left, the anarchy would surely return, bringing with it the humanitarian disaster. At this point the nature of the UN/US mission began to change from one of dispensing humanitarian aid to a mission of disarming Somali militias and bringing an end to the war. The Somali warlords chose to resist, thus leading the peacemaking mission gradually to change into an undeclared urban guerrilla war.

Humanitarian aid and peacemaking are highly involved personal missions and as such are the purview of the infantryman. Even in a situation with the consent of the parties involved, the task of the peacekeeper is quite difficult, requiring tact and forbearance more often than military might. Even in such situations violence often erupts, leaving the lightly armed and usually outnumbered peacekeepers to rely on their personal training and initiative for their very survival. In situations such as that seen in Somalia, though, the task of the infantryman was infinitely more problematic than this. Still needing to maintain the humanitarian mission, small infantry units – often in trucks or all-terrain vehicles – would often find themselves in a sea of people while navigating the narrow streets of Mogadishu. Dealing with such situations is quite difficult, often requiring very specific training in crowd- and riot control. Adding to this problem, Aydeed's militiamen could be anywhere, just waiting to launch a guerrilla attack. Thus some peacekeepers compared their position to that of a 'living target'.

BELOW: A Canadian Coyote reconnaissance vehicle passes a column of British Challenger main battle tanks in Kosovo. NATO forces are used to joint operations, but the multi-national nature of peacekeeping forces can see forces with very different standards of training serving alongside one another in a foreign location, making for a potential command and control nightmare.

On 5 June the simmering war in Mogadishu exploded when Aydeed's forces ambushed a group of Pakistani peacekeepers, killing 23 in a confused and brutal struggle. Dismayed, UN Secretary General Boutros-Ghali called for the arrest of Aydeed and the implied destruction of his militia group. For the first time UN forces had intervened in the affairs of a sovereign nation and chosen sides in a civil war. Frustrated by the development, some of the UN contingents, notably the Italians, expressed their displeasure at what was happening, but operations continued. Several violent skirmishes took place, but UN forces were unable to locate the elusive Somali warlord.

During August élite US forces, including Delta Force and the 75th Rangers, arrived in Mogadishu to prosecute the capture of Aydeed and his chief lieutenants. The UN forces, though, were at a disadvantage that would become typical to peacekeepers. The small size of the elite units and the potential of danger to the population of Mogadishu precluded the use of main force. Also Aydeed was in his element, surrounded by supporters in an insular culture, thus making intelligence-gathering concerning his movements difficult to control. Thus in many ways the UN and US forces struck blindly into

Mogadishu, eventually serving only to unite the population against the now coercive UN and US presence.

On 3 October an impasse was reached in the war in Somalia. Members of Delta Force and the 75th Rangers attempted to capture several Aydeed lieutenants in a 'snatch raid'. The raid was meant to be lighting quick, utilizing the speed and power of helicopters. Though the operation began well, Aydeed's forces used rocket-propelled grenades to down two US helicopters. Hundreds of Aydeed supporters rushed to the scene to attack the stranded, surrounded Americans.

Exhibiting command and control problems common to multinational missions, UN response forces with tanks languished at their base while a company of the US 10th Mountain Division attempted to rush to the aid of their compatriots in trucks, only to stumble into an ambush. Finally the UN relief column, led by Pakistani tanks, reached the scene, but the battle was in the main already over. During the struggle 18 American peacekeepers died and 100 were wounded. Estimates contend that 300 Somalis died in the struggle.

After the battle President Bill Clinton began to withdraw American troops from Somalia, leaving the Somalis to find their own solutions to their problems. By March 1995 all UN forces had

BELOW: The first contingent of US Marines from the 26th Marine Expeditionary Unit load on to a CH-53D Sea Stallion helicopter on the deck of the USS *Kearsarge* (LHD 3) as the ship operates in the Adriatic Sea on 8 June 1999. The Marines are heading to a staging area in Skopje, Macedonia, in support of NATO's 'Operation Allied Force'.

LEFT: Italian peacekeepers man their M109 155mm (6.1in.) howitzers outside Sarajevo. Though the UN had intended its operations in Bosnia to be merely humanitarian in nature, 'mission creep' slowly set in, putting UN troops on the road towards open conflict in the area.

withdrawn from Somalia and the Somali warlords were left to fight a bitter battle over possession of the abandoned UN compound in Mogadishu.

The failure of 'Operation Restore Hope' illustrated the dangers of post-Cold War peacemaking and would and set the precedent for most UN actions to follow. The UN operation had changed from one of humanitarian aid to an undeclared war against one faction in a civil war. The dangers of peacekeeping or peacemaking in still violent situations had caused the UN to take sides in the struggle, thus losing much of its moral authority. The precedent for the infantryman was sobering indeed. Asked to perform as humanitarian aid-workers rather than fighters, the task of the infantry was both frustrating and imminently dangerous. When the task transformed to war, the infantry found itself operating under a fragmented command structure, with little available force, all while becoming the central villain in a brutal civil war.

BELOW: Unlike in many previous UN peacekeeping operations, which were often carried out by lightly equipped troops, the NATO force in Bosnia was heavily armed with state-of-the-art weaponry. NATO made it clear to the Serbs that vehicles like this powerful Italian Centauro armoured car were not for show: they would be used if necessary.

BOSNIA

The collapse of Yugoslavia presented the UN with a critical test of its humanitarian role in the midst of a civil war that resulted in a societal holocaust. In 1992 citizens of Bosnia, following the example set by Slovenia and Croatia, voted to support independence from Yugoslavia, a nation dominated by the state of Serbia. The population of Bosnia, though, was ethnically divided: 44 per cent Muslim Slavs, 31 per cent Serbs and 17 per cent Croats. Quickly the tiny nation fell into a three-sided civil war as the major ethnic groupings vied for control. However, the largest group, the Muslims, soon found themselves at a distinct disadvantage, for the Serbs and the Croats could both rely on support from neighbouring ethnically supportive nations. As the crisis deepened, the UN made the situation worse by declaring an arms embargo on the region, an embargo that only had any real effect on the Muslims. Soon the Serbs gained the upper hand in the civil war, seizing over 70 per cent of the countryside and laying siege to several Muslim cities, including Sarajevo. By 1993 it had become obvious that the Serbs were engaging in an ethnic cleansing of Bosnia, horrifically displacing and killing millions of Muslims.

Against the backdrop of continued slaughter the UN did little, merely hoping that European intervention would solve the problem. Also, as the situation in Somalia worsened, Boutros-Ghali worried that any UN intervention in Bosnia would only serve to add a fourth participant to the ongoing war. Much more so than Somalia, there was no peace to keep in Bosnia, making humanitarian aid or peacemaking extremely difficult. Even so, Boutros-Ghali chose to send a small force to Bosnia to ensure the distribution of humanitarian aid, especially in beleaguered Sarajevo. In 1993 the UN declared a 'no-fly zone' over Bosnia in an effort to halt Serb air strikes.

However, the UN possessed little ability to enforce such a decision and relied on NATO for support. The alliance between the UN and NATO remained quite uneasy as UN forces concentrated on their humanitarian efforts, while NATO forces – often in direct violation of Security Council instructions – began to

BELOW: During the conflict in Sierra Leone peacekeeping forces once again found themselves vulnerable in the face of a societal holocaust, caught between ill-disciplined warring factions who had lost all semblance of restraint.

strike Serb units on the ground. Thus the UN mission in Bosnia had started to shift towards peacemaking, but with NATO as the spearhead and the driving force.

In 1994 the fighting in Bosnia began to centre around several Muslim strongholds that the UN had declared safe havens. During April the Serb refusal to halt attacks on the safe haven of Gorazde resulted in a NATO air attack on Serb positions, forcing compliance. Shortly thereafter NATO aircraft also bombed the area around Sarajevo. Though the war had escalated, involving the first combat operations ever by NATO forces, the bitter ethnic fighting continued. Serb forces sometimes complied with UN regulations, even going to the extent of turning over some of their heavy weaponry to UN control. However, more often than not, the Serbs fought on, even taking UN troops hostage in order to further their harsh struggle.

By 1995 it had become clear that UN humanitarian missions and sporadic NATO strikes would have little effect on the situation as a whole. Serb pressure on the safe havens increased to the point where in August 1995 UN forces

abandoned Srebrenica, which shortly fell to advancing Serbs. The resultant slaughter of civilians in the area forced a UN and NATO change of strategy.

Determined to bring the war in Bosnia under control, NATO had already begun to bring combat ground forces into the area. Led by the British 24th Airmobile Brigade, NATO troops hoped to be able to act as a rapid response force that could quickly come to the aid of any UN peacekeeping units that had fallen under Serb attack. Some NATO leaders argued that the commitment of troops would cross the 'Mogadishu Line', moving from a peacekeeping role to participation in an escalating civil war.

The dynamic in Bosnia, though, was quite different, for UN forces worked hard to maintain their neutrality, while NATO forces represented a mailed fist and an escalation of the conflict. After the fall of Srebrenica the regional dynamics shifted dramatically. NATO forces, pushing aside UN desires and control, embarked upon a sustained air and artillery offensive against the Serbs, dubbed 'Operation Deliberate Force'. The façade of peacekeeping had been

ABOVE: Unlike the situation in the Balkans, the UN force in Sierra Leone is largely composed of units from armies like India, Bangladesh and Nigeria. Command is divided, though the force is the only thing preventing a further descent into anarchy. Several hundred British troops have provided training and leadership for government forces, but they do not come under United Nations' command.

dropped, and forceful peacemaking was now the international goal. The operation was quite unlike that seen in Somalia, and represented a more standard form of warfare. Using techniques perfected in the Gulf War, NATO aircraft punished the static Serb positions. With few restrictions, NATO was able to put its massive edge in firepower to good use, again unlike the cramped streets of Mogadishu. Reeling from years of civil war and taking a relentless pounding from NATO attacks, the Serbs finally relented, resulting in the December 1995 signing of the Dayton Peace Accord. Bosnia would settle down into an uneasy

ABOVE: Rather than being a technological wonder war, the 'War Against Terror' blends the ultra-modern, with the old. Here, US Special Forces troops ride on horseback as they work with members of the Northern Alliance in Afghanistan during the early stages of 'Operation Enduring Freedom' in 2001.

period of peace. For their part, UN forces would now assume the task of monitoring the agreement, thereby returning to a much more successful pattern of traditional peacekeeping.

As the world enters the 21st century, it seems that the implosion of nations and resulting ethnic violence and humanitarian disasters will continue without abatement. In response, though the UN has attempted to return to a more traditional role of peacekeeping, forceful peacemaking missions will become more common. The new paradigm of peacemaking was demonstrated once again in the recent conflict in Sierra Leone. After a period of anarchy and civil war, the warring factions in Sierra Leone – again ill-controlled, ruthless groups of civilian

solders – reached a tentative peace to be monitored by UN peacekeepers. Once again, though, the clarity of the UN mission began to blur, leading to 'mission creep' towards open conflict with the main rebel faction, the RUF, and resulting in a British intervention in the growing conflict. In such situations, where lack of consent regarding peacekeeping leads to conflict, the UN has come to rely on outside intervention, ranging from US to NATO to British support. This shift in policy implementation indicates that the UN and the nations of the West will have to work ever more closely across the planet. However, such an alliance is often counterproductive, convincing the nations of the Third World that the UN is controlled by, and works for, US and western interests. In the resulting struggles it will be the infantryman – in the role of peacekeeper or peacemaker – who bears the burden of the fighting.

THE WAR AGAINST TERROR

After the victory in the Cold War the United States, the lone world superpower, sought to create a 'new world order', one alternately viewed as peaceful or hegemonic, depending on regional geopolitical realities. In the wake of events in Afghanistan and the Middle East – and partly dependent upon the supposed control of the UN by the West – leaders of ethnic, religious and regional struggles began to identify the US and its western Allies as legitimate military targets. As nations continued to implode and regional strife escalated, it seemed quite logical to turn against the leaders of the new world order and to view the system as a failure and a threat.

The crushing western victory in the Gulf War, though, indicated that it would be foolhardy to stand against the might of the United States in open battle. Thus new, disaffected groups – often without the organized support of a national unit – chose to strike at western interests using asymmetrical warfare and terrorism. Often cloaked in the language of fundamentalist Islam, these new military groupings sought to change the reality of modern warfare yet again by using guerrilla and terror tactics to defeat the western edge in firepower and technology. The coming conflict would once again focus on the initiative and perseverance of individual soldiers: it would be an infantryman's war.

Terrorism is usually defined as the use of violence against civilians by revolutionary organizations in an effort to coerce those civilians or their governments. Though terrorism as such has existed throughout the history of armed conflict, it only became an international phenomenon after the establishment of a communist revolutionary ideology. Believing in the revolutionary 'propaganda of the deed' Marxists from Lenin to Che Guevara believed that violent attacks against the government or its supporters could help facilitate the onset of widespread revolutions. Arguably Carlos Marighela stands as the most important modern terrorist theorist. In his *Minimanual of Urban Guerrilla Warfare*, Marighela contends that 'unbridled violence by a fanatical few, regardless of the apparent likelihood of more general support' might result in forcing a government over-reaction, thereby increasing revolutionary support among the people. Though Marighela achieved little success in his native Brazil, his teachings gave hope to other marginalized fringe revolutionary groups around the world. Through the use of seemingly senseless violence against civilians, groups as diverse as the Red Brigades of Italy, the Irish Republican Army and the Palestine Liberation Organization sought to achieve wider political success.

From terror bombings on the streets of London to the hijacking of a French airliner to Entebee airport in Uganda, terrorists made their presence felt across the globe in the early 1970s. Working in small, tight-knit groups – often known as cells – the terrorists proved to be difficult intelligence targets. As the information age dawned, though, the job of tracking terrorists became infinitely more difficult as terrorist organizations began to multiply, expand and work together. Adding to the mounting anti-terrorist difficulties, several 'rogue' states – including Libya, Syria and Iran – began to offer terrorist groups shelter and supplies. With the protection of an established government, terrorists proved to be almost impossible to track down until they had already carried out their acts of violence. To counter terrorism, western nations have developed close ties between their law enforcement agencies and have sought more accurate intelligence regarding terrorist activities, often based on the pattern of success demonstrated by the Israeli Mossad. In addition, western

BELOW: Though the bombing campaign at the beginning of the war in Afghanistan provided television drama, the wonder weapons were not truly effective until infantry forces arrived in the area to act as the eyes and ears of the conflict.

nations have constructed quick-reaction specialist anti-terrorist forces, including the Counter Revolutionary Warfare Team of the British SAS and the American Delta Force. Essentially infantry-driven, these élite formations rely on training and initiative to overcome oftentimes desperate situations involving hundreds of hostages. Even so, as the struggle against terror continues, it has become apparent that the response to terror must be more broad based, and must involve more than simply a military solution.

Fuelled by fundamentalist Islam, several organizations in the Middle East, including Hezbollah and Islamic Jihad, sought to use terror to destroy Israel or aid Palestine, while other such organizations sought change within Arab nations more accurately to reflect their religious and societal views. Especially as the situation in Afghanistan worsened and American troops lingered in the Persian Gulf after the close of the conflict there, many fundamentalist Islamic groups began to isolate and pinpoint the United States as being the source which lurked behind many regional problems.

Utilizing modern, technological weaponry and communications, one terrorist group, known as al-Qaeda, began to construct an international terrorist network of previously unimagined size and sophistication. Headed by Saudi-born millionaire Osama bin Laden, who had fought in the conflict against the Soviets in Afghanistan, al-Qaeda worked patiently behind the scenes in several Arab nations, making use of societal implosion and anarchy to provide cover and recruits. The countries of Somalia and Afghanistan offered the best hope, and after Afghanistan fell under the rule of the fundamentalist Taliban régime, that nation even offered al-Qaeda overt support for the coming terrorist campaign against the United States.

'SEPTEMBER 11'

Though some of the al-Qaeda plots were discovered by western intelligence agencies, several more were not, illustrating the difficulties of fighting international terrorists who enjoy support from a nation. Several smaller strikes against US forces, from Mogadishu to an attack on the destroyer *Cole* in Yemen brought al-Qaeda and bin Laden into focus for the American intelligence community. Even so, US intelligence efforts were not enough to stop the most infamous terrorist attack ever. On 11 September 2001, al-Qaeda operatives, after years of training and planning,

BELOW: US Marines aboard Light Armoured Vehicles pass through a village near Kandahar. The Marines are searching for Taliban fighters and fugitive members of Osama bin Laden's al-Quaeda organization.

LEFT: The US Department of Defense made a conscious decision in the 1970s to turn night into day. American fighting men possess a marked advantage over their less well-equipped foes, since infrared scopes, night-vision goggles and other low-light sensors enable them to fight effectively in pitch darkness.

hijacked four civilian airliners and used them as weapons against the US Pentagon and the twin towers of the World Trade Center in New York City. Succeeding beyond their wildest dreams, the attackers managed to collapse both of the World Trade Center towers and kill an estimated total of over 4000 people, at the same time causing untold economic damage that reverberated across the entire planet. Terrorism had reached a deadly new level of technical and technological expertise that induced worldwide horror, at the same time calling for a massive and innovative response.

AFGHAN INTERVENTION

With world backing and even somewhat reluctant support from the majority of Arab countries, the United States moved to attack the forces of al-Qaeda and their Taliban partners in Afghanistan. In the conflict, US forces had to overthrow the Taliban, thus defeating a nation at war, while also fighting an asymmetrical war against the elusive terrorists both in Afghanistan and in their bases around the world. The delicate situation called for US forces to wage war against an Islamic state and an Islamic terror group, and to retain the support of the Arab world lest the conflict spread and become a religious or ethnic war. Almost immediately, President George W. Bush, on the advice of Secretary of Defense Donald Rumsfeld, ruled out the use of significant US ground forces as too provocative, choosing instead to rely on Afghan resistance fighters and a close alliance with Pakistan. While gathering air power with which to strike the Taliban government, the Bush administration also sent Special Forces units to the region to create and cement alliances with the disparate elements that formed the Afghan resistance, most notably the Northern Alliance. At the same time, diplomats and CIA operatives began close work with the Pakistanis in an effort to gather badly needed intelligence regarding the elusive network of al-Qaeda operatives.

BELOW: A Marine sniper stands guard at the major US base at Kandahar. Constant vigilance is essential in an anti-terror operation like 'Enduring Freedom', since there is very little difference between the appearance of Allied Afghans and those responsible for sheltering the organizers of the 11 September atrocities.

RIGHT: As the nature of warfare has become more complex, the training regimen of the modern infantryman has become more demanding, teaching soldiers to react and think quickly in the most confusing and dangerous situations. Here US soldiers use video cameras on their rifles to investigate a simulated urban environment.

ABOVE: Entering service with the US Army in 2001, the Land Warrior system is designed to use technology to enhance the fighting ability of the infantry soldier.

The military effort against the Taliban and al-Qaeda, dubbed 'Operation Enduring Freedom', began on 7 October 2001 and represented an unexpected mixture of military techniques. In part, the war was a traditional war against the forces of the Taliban government, involving fixed air and ground combat. However, the US in the main chose to fight the conflict by proxy.

In addition, though, the war represented an asymmetrical war against al-Qaeda terrorists, one which involved élite Special Forces and methods of counterinsurgency. The confusing military mix called for unprecedented levels of integration on the part of the branches of the US military and intelligence community. At first it seemed that the War Against Terrorism would closely resemble the Gulf War, relying on the abilities of US air power to achieve victory. Bombers, ranging from high-flying B-52s to stealth aircraft, pounded Taliban and al-Qaeda positions with seemingly unerring accuracy. However, with a singular lack of intelligence concerning targeting in Afghanistan, the US air raids actually accomplished little.

It was not until élite Special Forces units entered Afghanistan that the war there began to change. In small groups, the Special Forces operatives fanned out into the Afghan countryside, adapting to their new situation by any means possible. Secretary Donald Rumsfeld in a recent speech stated that the Special Forces were operating in strained circumstances, that they 'rode horses – horses that had been trained to run into machine-gun fire, atop saddles that had been fashioned from wood and saddle bags that had been crafted from Afghan carpets. They used pack mules to transport equipment along some of the roughest terrain in the world, riding at night, in darkness, often near minefields and along narrow mountain trails with drops so sheer that, as one soldier put it, it took him a week to ease the death grip on his saddle.'

Operating in primitive conditions, the Special Forces relied on their most basic infantry skills: initiative reminiscent of troops on Omaha Beach and spirit and resiliency reminiscent of soldiers in World War I. The ultra-modern war had become a war reliant on well-trained soldiers on the ground, indicating that technology had changed little except the lethality of modern war. Linking up with Northern Alliance fighters, the Special Forces and their allies began to formulate their various plans of attack.

Using stealth, the Special Forces located the Taliban defensive emplacements near the critical town of Mazar-e-Sharif. On the appointed day of attack in early November the Special Forces neared the Taliban lines and designated targets for destruction by precision-guided munitions. Within two minutes the bombs began to strike with deadly accuracy, compromising the Taliban positions. In Rumsfeld's words next, 'hundreds of Afghan horsemen literally came riding out of the smoke, coming down on the enemy in clouds of dust and flying shrapnel. A few carried RPGs. Some had as little as 10 rounds for their weapons. And they rode boldly – Americans, Afghans, towards the Taliban and al-Qaeda fighters. It was the first cavalry attack of the 21st century.'

TALIBAN IN RETREAT

By 9 November Taliban resistance at Mazar-e-Sharif ceased and Northern Alliance troops, with their Special Forces advisers, seized the offensive across the country. Quickly the Taliban edifice crumbled, with their leadership admitting defeat in early December, leading to the formation of an interim Afghan Government under Hamid Karza. Though the struggle against terrorism continues and threatens to engulf other nations, the war in Afghanistan had come to a quick conclusion. The war in Afghanistan had represented a new blend of military techniques and technological levels of sophistication. Much reliance had been placed on the very latest western technology, from smart bombs to new earth-penetrating and thermobaric weapons for use against al-Qaeda cave networks. The coalition also made use of B-52 aircraft, a weapons system over 40 years old. However, the greatest reliance

in the war was placed upon men on the ground, and men on horseback. The blend of old and new styles of warfare was the key element of the campaign in Afghanistan, resulting in a devastating victory in an unexpected war.

TOWARDS THE FUTURE OF WAR

The conflicts of the post-Cold War era have left behind a confusing legacy. Will future conflicts closely resemble the Gulf War, as limited traditional wars seemingly dominated by technological weaponry? The experience of peacekeeping and peacemaking, though, might indicate that

BELOW: Land Warrior is the US Army's first fully integrated soldier fighting system. Through the helmet-mounted display, the soldier can view computer-generated graphical data, digital maps, intelligence information, troop locations and imagery from his weapon-mounted Thermal Weapon Sight (TWS) and video camera.

ABOVE: The role of infantry has changed much since 1900: today's infantryman must be a master of combat, peacemaking and technology.

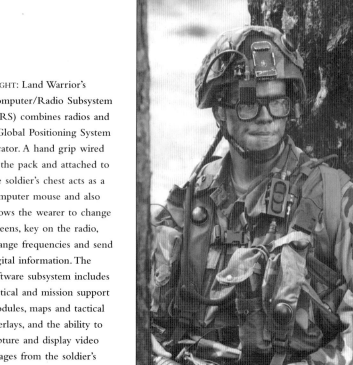

RIGHT: Land Warrior's Computer/Radio Subsystem (CRS) combines radios and a Global Positioning System locator. A hand grip wired to the pack and attached to the soldier's chest acts as a computer mouse and also allows the wearer to change screens, key on the radio, change frequencies and send digital information. The software subsystem includes tactical and mission support modules, maps and tactical overlays, and the ability to capture and display video images from the soldier's weapon sight.

future wars will often take the form of brutal civil conflicts, struggles that pose great dangers to those who dare intervene, but possibly greater dangers if the world community simply stands aside.

Most recently terrorist attacks and rogue nations – with a lingering threat of a wider ethnic and religious conflict – seem to indicate an even more confusing military path towards the future.

The United States and its NATO Allies are now undergoing a series of military reforms, dubbed 'transformation' to make ready for the conflict of the future. Realizing that war in the 21st century can take a number of forms, the watchwords for transformation are 'flexibility' and 'integration'. The military must think ahead in an atmosphere of intellectual honesty towards a war that might take place in caves, in space, in a hostile rogue state, through renewed terrorist attacks on the home front, or in cyberspace. The US has made a number of changes in its defence policy, since the close of the Cold War demanding that its military be able to prosecute wars against two enemies simultaneously.

INTEGRATED WAR

Recent events, though, indicate that such a capability is rather redundant, and that the US should ready itself for a greater variety of conflicts rather than two traditional ones. As a result the US has

now dropped its two-war policy, and is in the process of reducing redundant weapons stockpiles, including reducing its nuclear capability. For American planners the era of Total War is over, as is the era of Limited War. Instead the world is making ready to embark on a period of what is termed as Integrated War.

In the new era of warfare brute strength and technological superiority will remain important but will represent only one facet of combat. US military planners believe, 'The ability of forces to communicate and operate seamlessly on the battlefield will be critical to our success. In Afghanistan, we saw composite teams of US Special Forces on the ground, working with Navy, Air Force and Marine pilots in the sky, to identify targets, communicate targeting information and coordinate the timing of strikes with devastating consequences for the enemy. The change between what we were able to do before US forces, Special Forces, were on the ground and after they were on the ground is absolutely dramatic. The lesson of this war is that effectiveness in combat will depend heavily on 'jointness', how well the different branches of our military can communicate and coordinate their efforts on the battlefield.'

Thus it seems that integration of forces, once called all-arms coordination, will be key to the multidimensional battlefield of the future. The infantryman, once nearly written off as a tool of future warfare unless fitted with a robotic exoskeleton, will continue to play the central role. The foot soldier's death knell has sounded many times, from the beginning of trench warfare to the introduction of the microchip. Still western armies will remain reliant upon the strength of their citizen soldiers. Highly trained and motivated, the infantryman has the most demanding job in the world. Trained in the use of brute force, he must also stand ready to show forbearance and patience as a humanitarian peacekeeper. Skilled in the most modern technology, he must be prepared to jettison the accoutrements of the technological age in favour of cold steel. While ready to call in precision-guided air strikes, the infantryman must be prepared to ride horseback through the Afghan desert. Much has changed in modern warfare, but battles are still won and lost by soldiers on the ground, and remain in the hands of the infantry.

ABOVE: Though some thought that technology had sounded the death knell for the foot soldier, recent conflicts shows that he still has an important role to play. Although his missions and equipment have become vastly more complex, it is still the infantryman who takes terrain and wins wars.

Index

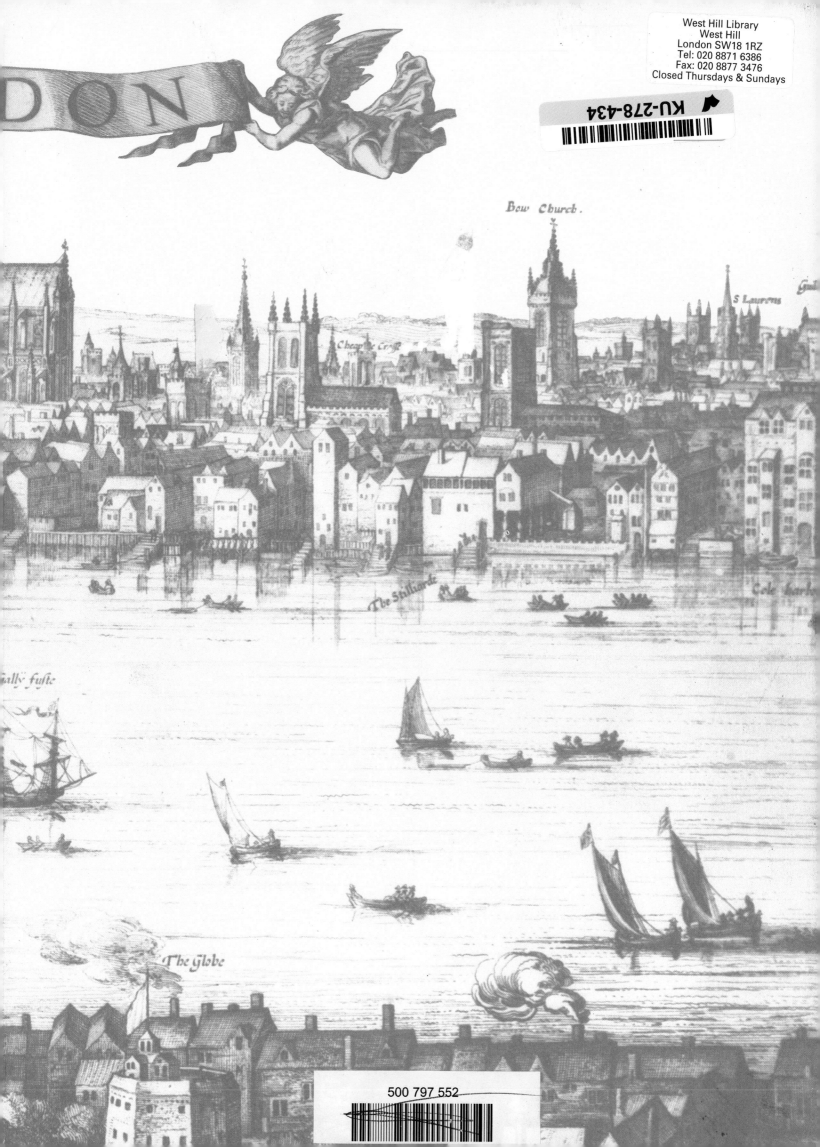

Bow Church.

S Laurens

Cheap e Cro e

The Stillierde

Cole har

ally fufte

The globe

THE TIMES HISTORY OF LONDON

OF

NEW EDITION

THE TIMES

TIMES

HISTORY
OF
LONDON

NEW EDITION

EDITED BY HUGH CLOUT

TIMES BOOKS

CONTENTS

CONTRIBUTORS

GENERAL EDITOR:

Dr Hugh Clout FBA, Professor of Geography, University College London, Dean of Social and Historical Sciences

CONTRIBUTORS:

John Clark, Curator, Medieval Collections, Museum of London

Jonathan Cotton, Curator, Prehistory, Museum of London

Dr Richard Dennis, Reader in Geography, University College London

Jenny M N Hall, Curator, Roman Collections, Museum of London

Dr Vanessa Harding, Lecturer in London History, Birkbeck College, University of London

Gustav Milne, Department of Urban Archaeology, Museum of London and University College London

Dr Richard Overy, Professor of Modern History, Kings College London

Dr Michael Power, Lecturer in Economic and Social History, University of Liverpool

Dr Hugh Prince, Reader Emeritus in Geography, University College London

Dr Eric Robinson, President of The Geologists' Association, London and University College London

John Schofield, Archaeology Officer (City of London), Museum of London

Dr Peter Wood, Professor of Geography, University College London

CONSULTANTS:

Dr John Adams, Professor of Geography, University College London

Dr Clive Agnew, Senior Lecturer in Geography, University College London

Dr Andrew Gurr, Professor of English, University of Reading

Dr Carolyn Harrison, Senior Lecturer in Geography, University College London

Dr Ted Hollis, Late Reader in Geography, University College London

Dr Peter Jackson, Professor of Geography, University of Sheffield

Dr John Salt, Professor of Geography, University College London

ETYMOLOGY COMPILED BY:

Dr Robert Ilson, Honorary Research Fellow, University College London

HUGH CLOUT WISHES TO THANK:

Anne Oxenham of the Department of Geography, University College London, for help in exploring many cartographic sources

Michael Jahn of the London History Library, University College London, for identifying many historic maps and books on London

Malcolm Ward for help in checking many details of London's history on the ground

New Edition
Published in 1999 by
TIMES BOOKS
HarperCollins*Publishers*
77-85 Fulham Palace Road
London W6 8JB

The HarperCollins website address is:
www.**fire**and**water**.com

First published by Times Books in 1991
Reprinted in 1994
Second edition published in 1997

Artwork and typesetting
Swanston Graphics, Derby
Ralph Orme
Andrew Bright
Advanced Illustration

Colour processing
Colourscan, Singapore

Design
Ivan Dodd
Tracey Enever
Mabel Chan

Editorial direction
Elizabeth Wyse
Malcolm Swanston
Isobel Willetts
Caroline Lucas
Ailsa Heritage
Sarah Allen
Philip Parker

Picture research
Jane Parsons

Place names consultant, Index
P J M Geelan

Chronology
Sid Holyland and Len Phillips

Endpapers
Map of London by Nicholas John Visscher, published in Amsterdam in 1616

The publisher wishes to thank Gavin Morgan of the Museum of London, Marrianne Behm, Ian Smith and Lesley Branscombe

Printed and bound in Italy by Rotolito Lombarda spa

British Library Cataloguing in Publication Data
A catalogue reconrd for this book is available from the British Library

ISBN 0 7230 1030 7

INTRODUCTION

THIS BOOK is a visual celebration of two thousand years of urban settlement. The history of London is traced from Roman times, through many centuries of political and commercial greatness, reaching their pinnacle in the Victorian era, to the present when London is confronted by new challenges in a rapidly changing world. After an introduction on the physical resources of the London region, successive chapters are arranged in a conventional chronological way, with the exception of the last two, which concentrate on themes and places in the metropolis. Our interpretation of 'history' is liberal, encompassing social and economic conditions and the processes that led to the formation of London's landscape, as well as key events in the city's past. Great buildings have their place, but so do the poor and deprived whose plight so shocked Victorian observers. Urban history is viewed in a comprehensive way, complemented by the special perspectives of the archaeologists, geologists, geographers and professional historians who have contributed to the book. New evidence on the pattern of London in Roman and medieval times is converted into maps, while images of the city in later periods are created from contemporary statistics and other records.

The book contains not only specially-drawn maps and diagrams but also many reproductions of paintings, facsimiles of historic maps, engravings and photographs. Meticulous reconstructions of parts of London at various times in the past form a unique element. A great deal of imagination and draughtsman's skill have gone into interpreting fragmentary historic records in order to produce convincng and visually attractive images. Last but by no means least are the texts and captions which complement the hundreds of illustrations. These essays provide a succinct digest of our knowledge of the metropolis in successive periods and raise questions about London in the future.

The Museum of London, located in the Barbican, offers a superb series of displays on the capital's social, economic and political life from the earliest days of settlement to present times. London's many specialist museums and local history collections provide an unrivalled range of information on particular topics and neighbourhoods. But urban history is not just about books and museums. It is to be found in the streets we walk, the trains we take and the buildings we use every day. Indeed, the best way to study London's past is to interrogate its present with an inquisitive eye, armed with a detailed map and a good guidebook. Every building, square, street and monument was created at some time past, often according to a logic that was different from our own. Ancient property boundaries are reflected in the alignment of roads and in differences in house type, long-lost trades and activities are recorded in street names, ancient villages form the cores of sprawling suburbs, while historic churches recall the past vibrancy of community life in parts of the city that are now depopulated. The list is endless. A lifetime spent exploring the metropolis convinces me that there is always another street corner to be turned and a fresh townscape to interpret. Continuity and change are vital themes in urban history. New forces have transformed the size, shape, appearance and functioning of our city, but the legacy of the past remains. An appreciation of history helps us understand landscape features that may seem odd or illogical.

But what of the future? Controversy rages every time historic features are threatened with demolition to make way for a new building or a motorway. Some battles for conservation are won and others lost, but every excavation offers the chance of unearthing new evidence of London's past. Sadder still are the hundreds of listed buildings in Greater London that are literally crumbling away through neglect and lack of capital investment. A major survey by English Heritage shows that of 34,000 listed buildings in the capital, 945 are under immediate threat. Most of those at risk are in inner boroughs and include warehouses, redundant churches, hospitals and neglected public monuments. In recent years 'gentrification' has saved many of London's Georgian terraced houses, albeit at the cost of social change. Many more are in a dilapidated state. Saving the city's crumbling buildings will cost vast sums and raises important questions about the wisest use of scarce public resources in a city where public transport, housing, schools and so many other facilities need massive investment if they are to remain viable in the future.

I was born in what is now called the 'inner city' and London is my home town. My life-long fascination with London remains and has been strengthened each weekend over recent years as I have 'checked out' information and visited (and revisited) places that appear in the book. I can do no better than quote Dr Johnson's assertion: 'when a man is tired of London, he is tired of life.' I hope this book will help preserve and extend the reader's fascination with our great city.

I am delighted to have had the opportunity of producing a new edition, which allows new information to be included on the prehistory of the London area and on the capital at the Millennium, as well as some minor modifications.

Hugh Clout

History, Politics and Society

AD
43 Romans invade England.
50 Foundation of *Londinium*.
60 Sack of *Londinium* by Boudicca.
61 *Londinium* rebuilt and designated capital of province.
125 *Londinium* destroyed by fire.
200 *Londinium* designated as capital of Britannia Superior.

457 Britons defeated by mercenaries. *Londinium* disappears from historical record.
604 Mellitus appointed Bishop of London.

842 Viking attack on London: 'Great Slaughter'.
c.871 Danes occupy London. Recaptured by King Alfred (878).
911 Edward takes control of London after Alfred's death.
1016 Cnut captures London, becomes king.
1042 Edward the Confessor becomes king. London made capital of England.
1066 Harold killed at Hastings; William crowned at Westminster Abbey.
1085 Population c. 10-15,000.

1180 Population inside walls c.40,000.

1192 Permission granted for mayor and aldermen with own court.
1207 Archbishop of Canterbury takes up residence at Lambeth.
1215 Magna Carta: Mayor of London one of signatories.

1290 Expulsion of Jews from London ghetto in Old Jewry.

1327 First Common Council of City of London.
1348-9 Black Death epidemic: c.10,000 buried at West Smithfield.

1377 Population c.40,000.
1381 Peasants revolt led by Wat Tyler.

1397 First of Richard Whittington's four terms as Lord Mayor.

Commerce, Industry and Infrastructure

AD
50 Road network begun; Thames bridged; provision of port facilities.

125 New waterfront built.

c.290 London Mint established.

c.640 Gold coins minted in London: the first since Roman times.

899 First mention of Queenhithe.
949 First mention of Billingsgate.

1066 William grants London Charter.

c.1130 Charters establishing liberties.

c.1155 Vintner's Company granted Charter.
1170 Weekly horse fair at Smithfield.
1180 First mention of Goldsmith's Company.

c.1199 First building regulations introduced in the City.

1214 City Charter awarded by King John.

1272 First Craft Guild.
1274 First mention of 'Flete Strete'.
c.1290 The Hop and Grapes in Aldgate High St: London's oldest Licensed House.

1358 138 shops on London Bridge.
1358 First Goldsmith's Hall.

1380 First Skinner's Hall.

1382 First Custom House.

1389 River wall built at Tower.
1394 Farringdon wards formed outside the wall.

Building and Architecture

BC
600 Middle Iron Age domestic sites at Rainham, Dawley, Bedfont, Heathrow.

AD
80-125 Building of basilica and forum, governor's palace, public baths, fort.

200 City wall built.
240 Mithraic temple built

c.600 Saxon London built mainly outside walls.
604 St Paul's Cathedral built.
606 St Mary Overie Nunnery established on the site of present Southwark Cathedral.

898 Conference of King's Council re restoration of London.

c.1000 Earliest reference to London Bridge.

1067 Building of Tower of London and other castles began.

1089 Bermondsey Abbey founded.
1123 Building of St Bartholomew's Priory and Hospital begun.
1140 Priory of St John established at Clerkenwell.
1140 Nunnery of St Mary's established at Clerkenwell.

1176 Old London Bridge begun.

1185 Temple Church consecrated.
1205 St Helen's Nunnery, Bishopsgate.

1212 Southwark Cathedral.
1220 Wakefield Tower at Tower of London begun.
1250 First Gothic arch (at St Bartholomew the Great).
1256 St Paul's Cathedral extended in Gothic style.

1290 Wall extended from Ludgate to Fleet River.

1375 First mention of Staple Inn.

Institutions and Popular Culture

c.1050 St James's Leper Hospital.

1148 St Katharine's Hospital founded by Queen Matilda.

1180 Leisure activities include cock fights, archery, wrestling, skating on Thames when frozen.

1213 St Thomas's Hospital established at Southwark.
1247 Bethlem Hospital for insane established.
1253 Elephant given to King and kept in Tower of London.
1272 Baynard's Castle handed over to Dominican Friars.

1297 Pig-styes banned from streets.

1371 Foundation of Charterhouse at Clerkenwell.

Science and the Arts

c.1173 Fitz Stephen, London historian, gave first description of the city.
1180 'Miracle Plays' performed at Clerkenwell.

1245 Westminster Abbey began to acquire art treasures.
1253 Sculptured bosses carved in Westminster Abbey.

1349-52 Stained glass windows made for St Stephen's Chapel, Westminster Palace.

1377 Effigy of Edward III by John Orchard placed in Westminster Abbey.

1396 Portrait of Richard II painted for Westminster Abbey.

History, Politics and Society

1415 Henry V leads victory parade after Agincourt (London Bridge to St Paul's).
1422 First records of the Honourable Society of Lincoln's Inn.

1440 First reference to the Honourable Society of the Inner Temple.

1535 Sir Thomas Moore executed at the Tower of London.

1550 Population c.80,000.

1580 Proclamation forbidding housebuilding within three miles of any London gate.
1583 Population c.120,000.

1603 Outbreak of plague: c.25,000 deaths.
1605 Guy Fawkes' 'Gunpowder Plot'.

1630 Population c.200,000.

1642 Royalist Army defeated at Turnham Green.
1649-60 Charles I executed in Whitehall; the Commonwealth declared.
1662 Royal Society founded.
1665 Bubonic Plague: c.70,000 deaths.
1666 Great Fire of London.

1700 Population over 500,000.
1702-5 Buckingham House built.

Commerce, Industry and Infrastructure

1400 Billingsgate Market granted its charter.

1422 111 crafts recorded in London.
1425 First Draper's Company Hall.

1479 Billingsgate Market rebuilt by Hanseatic merchants.
1501 First printing press set up in Fleet Street.

1513 Foundation of Royal Dockyard at Woolwich and Deptford.

1554 First mention of The George Inn, Southwark.
1566 Royal Exchange instituted by Thomas Gresham.

1584 First mention of Ye Old Cheshire Cheese Inn, Fleet St.
1593 Horse-driven water pump installed near Queenhithe.
1599 First dry dock built at Rotherhithe.

1613 Opening of 38-mile canal (from Herts to Clerkenwell) – London's main water supply.
1614 East India Docks built at Blackwall.

1651 Hay's Wharf opened.
1656 1,153 taverns in the city.
1663 First toll roads.

1667 Fleet Canal and Thames Quay project.
1669 Regular 'Flying Coaches' – London-Oxford-Cambridge.

1680 'Penny Post' introduced.

1694 Foundation of the Bank of England.

Building and Architecture

1411 First Guildhall built.
1414 Sheen Palace built by Henry V.

1490 Gatehouse, Lambeth Palace.

1512-19 Henry VII Chapel at Westminster Abbey.
1523 Bridewell Palace.

1547 Somerset Palace begun.

1571 Middle Temple Hall.

1616-35 Queen's House, Greenwich.
1619-25 Banqueting House, Whitehall.
1631 Kew Palace.
1635 Piazza, Covent Garden.

1665 Southampton Square, now Bloomsbury Square.

1670-7 College of Arms.
1670-1700 Rebuilding of City churches: 50 by Wren, one by Hawksmoor.
1671-77 The Monument.
1675-1711 St Paul's Cathedral rebuilt.

1698 Berkeley Square.
c.1700 Bedford Row.
1701 The Synagogue of Spanish and Portuguese Jews.
1711 Marlborough House.
1712 St Paul's Chapter House.
1714 St Alphage, Greenwich.

Institutions and Popular Culture

1509 St Paul's School founded by John Colet.

1539 St Bartholomew Hospital refounded after dissolution of monasteries.
c.1550 First use of private coaches.
1552 Christ's Hospital founded.
1558 First proper map of London by Ralph Agas.

1572 Harrow School founded.

1579 Gresham College founded.

1598 John Stow complains of 'terrible number of coaches, world run on wheels'.
1608 Great Frost Fair on Thames; taverns and football.

1611 Charterhouse School founded by Thomas Sutton.

1625 First Hackney carriages permitted to ply for hire.
1634 Sedan chairs for hire.
1637 Hyde Park opened for public use.
1637 50 licensed coaches (by 1652, 200).
1648-9 Frost Fair on Thames; printing press set up on ice.

1676 Foundation of Chelsea Physic Garden.

1682 Foundation of Royal Hospital, Chelsea (Wren)
c.1685 Sadler's Music House Theatre (now Sadler's Wells).

Science and the Arts

1476 First printing press set up in Westminster by William Caxton.

1510 Birth of Thomas Tallis, composer (died 1585).

1552 Birth of Edmund Spencer, poet and writer (died 1599).
1564 Birth of William Shakespeare, poet and dramatist (died 1616).
1573 Birth of John Dunne, poet and Dean of St Paul's (died 1631).
1577 First London theatre built at Shoreditch by James Burbage.
1587 Rose Theatre built in Southwark.
1598 John Stow's 'Survey of London' published.
1599 Globe Theatre built in Southwark.

1620 Birth of John Evelyn, diarist and writer (died 1706).

1633 Birth of Samuel Pepys, civil servant and diarist (died 1703).

1658 Birth of Henry Purcell, musician (died 1695).

1675-6 Foundation of the Royal Observatory, Greenwich.

1685 Birth of George Frederick Handel, composer (died 1759).
1697 Birth of William Hogarth, artist (died 1764).

1705 Her Majesty's Theatre.

History, Politics and Society	Commerce, Industry and Infrastructure	Building and Architecture	Institutions and Popular Culture	Science and the Arts
			1715 Geffrye Museum (built as almshouses)	
		1717 Cavendish Square.	**1718** Maypole removed from front of Somerset House.	**1717** Birth of David Garrick, actor (died 1779).
1720 The 'South Sea Bubble'.	**1720** Charters granted to the Royal Exchange Assurance and London Assurance companies. **1729** 2,484 private coaches, and 1,100 coaches for hire; 22,636 horses in London.	**1720** Hanover Square. **1729** Chiswick House. **1729** Marble Hill House, Twickenham. **1730** St George, Bloomsbury. **1730** St Paul's, Deptford. **1735** The Treasury. **1739-53** The Mansion House.	**1720** Westminster Hospital opened. **1722** Guy's Hospital founded.	**1720** Theatre Royal, Haymarket. **1728** Birth of Oliver Goldsmith, writer (died 1774).
1732 Sir Robert Walpole offered 10 Downing St as official residence.	**1734** *Lloyd's List* established as regular weekly publication.		**1733** The Serpentine created in Hyde Park. **1739** Foundling Hospital, founded by Thomas Coram. **1741** London Hospital founded.	**1732** Covent Garden Theatre (destroyed by fire 1808).
1751 Licensing Act.	**1747** Coal Exchange opened.	**1750-8** Horse Guards. **1750** Westminster Bridge.		**1746** Visit of Antonio Canaletto, Italian artist and painter of London views. **1755** Dr Samuel Johnson published his 'Great Dictionary'. **1757** Birth of William Blake, poet and mystic (died 1827).
1762 Westminster Paving and Lighting Act. **1767** Houses in city numbered for first time.	**1756** 600 stage coaches licensed to towns within 19 miles of London; fixed routes from 123 stations in London. **1761** New Road from Paddington to Islington (first London bypass).	**1758** Kew Palace. **1758** Horse Guards, Whitehall. **1766** City Wall demolished and removal of gates begun. **1769** Kenwood House. **1772-4** Royal Society of Arts. **1775** Boodle's Club, St James's. **1776-86** Somerset House.	**1759** The Royal Botanic Gardens, Kew founded.	**1759** British Museum opened. **1763** James Boswell, biographer, meets Dr Johnson. **1768** Foundation of Royal Academy. **1776** Birth of John Constable, artist (buried Hampstead 1837). **1778** Birth of William Hazlitt, writer (died 1830).
1780 Gordon Riots: c.850 killed.	**1780** London bankers issued their own notes. **1785** First publication of *The Times*. **1794** Grand-Junction Canal opened. **1798** 7,000 watchmakers listed in Clerkenwell. **1802** West India Docks opened. **1802** Stock Exchange opened on new site. **1803** Surrey Iron Railway (Wandsworth-Croydon, horse-drawn): first public railway. **1803** Dickens and Jones department store. **1803** Commercial Road opened: improved access to docks. **1807** Installation of gas lighting in Pall Mall. **1807-9** Royal Mint opened.	**1778** Brook's Club, St James's **1786** Osterley Park House. **1788** White's Club, St. James's. **1789** The facade of Guildhall. **1802-3** Albany Chambers, Piccadilly.	**1784** Balloon ascent from Artillery Ground, Finsbury.	**1784** Birth of James Leigh Hunt, poet (died 1859). **1792** Birth of George Cruickshank, engraver for Charles Dickens (died 1878). **1795** Birth of Thomas Carlyle, writer (died 1881). **1802** 'On Westminster Bridge' sonnet by William Wordsworth.
1811 Population c.1,000,000. **1812** Prime Minister Spencer Perceval assassinated at House of Commons.	**1815** First steamboat service on Thames. **1817** New Custom House. **1819** Burlington shopping arcade, Piccadilly. **1820** Regent's Canal opened.	**1804** Russell Square. **1806** Sir John Soane's House, Lincoln's Inn Fields.	**1805** Moorfields Eye Hospital founded. **1818** Charing Cross Hospital founded. **1819** Brixton Prison opened. **1819** Bedford College for Women founded.	**1806** Birth of John Stuart Mill, philosopher (died 1873). **1809** Covent Garden Theatre rebuilt after fire (Smirke). **1816** Keats' house built, Hampstead. **1821** Haymarket Theatre opened. **1823** Royal Academy of Music opened. **1824** National Gallery founded.
1825 Gallows and turnpike removed from Tyburn. **1829** Metropolitan Police Act.	**1827** First publication of *Evening Standard*. **1828** Covent Garden Market. **1829** General Post Office opened. **1829** First omnibus service: Paddington-City.	**1824** Royal College of Physicians. **1824-31** London Bridge rebuilt. **1825** All Soul's, Langham Place. **1827-33** Carlton House Terrace. **1828** Marble Arch. **1829** Constitution Arch. **1829** Travellers' Club.	**1828** University College, Gower Street founded.	**1827-8** Zoological Gardens in Regent's Park opened. **1829** Cruickshank's 'March of Bricks' cartoon illustrates London's growth.

History, Politics and Society

1832 Cholera epidemic.

1835 Animal fighting made illegal.

1837 Buckingham Palace becomes permanent London residence of the Court.
1837 Typhus epidemic.

1845 Mass meeting of Chartists on Kennington Common.
1848-9 Major cholera epidemic.

1850 Board of Health report on cholera epidemic of 1848-9 and supply of water to metropolis.
1853 Smoke Abatement Act.
1853-4 Cholera epidemic.

1857 Thames Conservancy Act.

1858 'The Great Stink': pollution on the Thames.
1859 Metropolitan Drinking Fountain Association founded.
1860 Metropolis Gas Act.
1860 London Trades Council founded.

1865 Foundation of the Salvation Army in East End.
1866 Last major outbreak of cholera, 5,915 deaths in Poplar.
1866 Sanitation Act.

1868 Toll gates abolished.
1868 Last public execution at Newgate Prison.

1870 School Board of London established.

1878 Epping Forest acquired by City of London Corporation.

1885 Highgate Woods acquired by City of London Corporation.

1888 London County Council created.
1890 Housing Act enabling the LCC to clear slums.

Commerce, Industry and Infrastructure

1834 Hansom Cabs introduced.

1836 First passenger railway in London: London-Greenwich.
1837 Euston Railway Station opened.

1838 Paddington Railway Station opened.
1840 Penny Post introduced.
1841 Fenchurch Street Railway Station opened.
1843 Thames Tunnel opened.

1848 Waterloo Railway Station opened.
1851-2 King's Cross Railway Station opened.
1852 Poplar Docks opened.
1853 Harrod's store opened.

1855 Royal Victoria Docks opened.
1855 Metropolitan board of Works created.

1863 Metropolitan Railway opened first Underground.
1863-9 Holborn Viaduct.
1864 First London bus with stairs.
1864 Charing Cross Station opened.
1864-70 Victoria Embankment.

1866 Cannon St Railway Station opened.

1867-72 St Pancras Station.
1868 Millwall Docks opened.
1868 New Smithfield Market opened.
1868 Abbey Mills Pumping Station opened (Bazalgette).
1869 First Sainsbury's opened in Drury Lane.
1869 Last warship built at Royal Navy Dockyard, Woolwich.
1871 Lloyds Incorporated by Act of Parliament.
1874 Liverpool Street Railway Station opened.
1876 First arrival of refrigerated meat from abroad (America).
1879 First Telephone Exchange, Lombard St (ten subscribers).
1880 Royal Albert Docks opened.

1886 Tilbury Docks opened.
1886 Shaftesbury Avenue opened.
1888 First issue of the *Financial Times*.
1890 First 'Tube' railway: City and South London.

Building and Architecture

1837-52 Houses of Parliament rebuilt after fire.

1841 St George, Roman Catholic Cathedral, Southwark.
1843 Trafalgar Square.

1848-51 Army and Navy Club.

1851-96 Public Record Office.

1853 Brompton Oratory.

1859 Floral Hall Covent Garden.

1862-4 First Peabody Trust buildings erected.

1866 Leighton House.

1868 Royal Albert Hall.

1871-82 Royal Courts of Justice, Strand.
1875-81 Bedford Park Garden Suburb.

1878 'Cleopatra's Needle' erected on Victoria Embankment.

Institutions and Popular Culture

1831 King's College founded.
1833 London Fire Brigade established.
1834 University College hospital founded.
1835 Madam Tussaud's Waxworks opened.
1836 University of London founded.
1837 King's College Hospital founded.

1839 River Police formed.
1839 Highgate Cemetery opened.
1841 Last frost fair on Thames.
1842 Pentonville Prison opened.

1845 Victoria Park opened.
1845 Surrey County Cricket Club founded.
1849 Wandsworth Prison opened.

1852 Holloway Prison opened.
1853 Battersea Park opened.

1860 Battersea Dogs Home founded.
1863 Middlesex County Cricket Club formed.

1865 Metropolitan Fire Brigade formed.

1869 Southwark Park and Finsbury Park opened.

1873 Alexandra Palace opened.

1874 Wormwood Scrubs Prison opened.

1877 First Wimbledon Tennis tournament.

1880 First ever Cricket Test, England v. Australia, at the Oval.

1882 Central London Polytechnic, Regent St founded.

Science and the Arts

1833 Soane Museum founded.

1834 Birth of William Morris, artist, writer, designer (died 1896).

1836 Birth of Walter Besant, London historian (died 1901).
1837 Royal College of Art founded.

1838 National Gallery completed (Wilkins).

1842 British Museum new building begun (opened 1847).

1851 Great Exhibition held in the 'Crystal Palace', Hyde Park.

1858 Alhambra Theatre opened.

1859 National Portrait Gallery opened.

1866 Birth of H G Wells, writer (died 1946).

1867 Birth of John Galsworthy, writer (died 1933).

1870 Royal Albert Hall opened.

1875 Bethnal Green Museum opened.
1876 Albert Memorial completed.

1881 Greenwich recognised as meridian.
1881 Natural History Museum opened (Waterhouse).
1883 Royal College of Music founded.

1888 Shaftesbury Theatre opened.

History, Politics and Society

1899 London Government Act: 28 new metropolitan boroughs created.

1902 Metropolitan Water Board created.

1908-33 London County Hall built.
1908 Port of London Authority created.

1911 Population of Greater London c.7,252,000.

1915-18 German zeppelins bomb London.
1919 'The Cenotaph' war memorial unveiled in Whitehall.
1920 'Unknown soldier' buried in Westminster Abbey.
1922 First Queen Charlotte Ball for debutantes.

1926 The General Strike.

1929 Local Government Act: LCC takes over hospitals and schools.

1931 Population of Greater London c.8,203,000.

1933 London Transport Act (Board formed).

1935 'Greenbelt' established by LCC.
1936 Jarrow unemployed march to London.

1939 Population of Greater London c.8,700,000.

Commerce, Industry and Infrastructure

1897 Queen Victoria's Diamond Jubilee procession.
1899 Savoy Hotel and Theatre built.
1901 First electric trams in London.
1902 Spitalfields Market rebuilt.

1904 First double decker bus running in London.

1905 First telephone box in London.

1906 Bakerloo Line opened.
1906 Piccadilly Line opened.
1907 Northern Line opened.
1907 First Taxicabs in London.
1908 Kingsway Tram Tunnel opened.
1908 Rotherhithe Rd Tunnel opened.
1909 Selfridge's department store, Oxford St opened.

1911-22 Port of London Authority Headquarters erected (Cooper).
1912 Whiteley's department store, Bayswater, opened.

1921 King George V Docks opened.

1924 First Woolworths store in London, Oxford St.
1924 British Empire Exhibition at Wembley.
1925 Great West Rd opened.
1926 London's first traffic roundabout at Parliament Square.
1927 Park Lane Hotel opened.

1930 Dorchester Hotel, Park Lane opened.
1930-4 Battersea Power Station.

1931 Liberty's department store, Regent St opened.
1931 Shell Mex Offices, Strand.
1932 Cockfosters, Arnos Grove, and Manor House Underground Stations (Holden).

1937 Earls Court Exhibition Hall.

Building and Architecture

1891 New Scotland Yard.

1899-1906 The War Office, Whitehall.

1903 Westminster Cathedral.

1905 Kingsway & Aldwych opened.
1905-8 The Quadrant, Regent St.

1907 Central Criminal Court (The Old Bailey).
1908 Rhodesia House, Strand.

c.1910 Ducane Housing Estate, Hammersmith (LCC).
1910 Admiralty Arch.

1930 Y.W.C.A. Hostel, Great Russell St.

1931 *Daily Express* Offices, Fleet St.

1932 R.I.B.A., Portland Place.

1935 South Africa House, Trafalgar Square.
1936 Senate House, University of London.
1937 Bow Street Police Court.
1937 LCC Fire Brigade Headquarters.

Institutions and Popular Culture

1895 First motor bus in London.
1897 Blackwall Tunnel opened.

1904 London Fire Brigade formed.

1908 Twickenham Stadium opened.
1908 Olympic Games held at Shepherds Bush.

1910 London Palladium, Argyll St opened.
1911 'Pearly King' Association formed.

1914 Cinemas in LCC area total 266.

1921 Last horse-drawn fire engine in London.
1922 BBC begins broadcasting from Savoy Hill.
1923 First F.A. Cup Final at Wembley Stadium.

1927 First London greyhound track, White City Stadium.
1929 Dominion Theatre opened (became cinema in 1932).
1929 Tower Pier opened.
1930 Finsbury Park Astoria Cinema opened.
1930 First Chelsea Flower Show.
1930 Leicester Square Theatre opened (became cinema 1968).
1931 London public buildings floodlit for first time.
1932 Arsenal Stadium, Highbury, built.
1932 BBC moved to new offices in Portland Place.

1937 Empress Hall Ice Rink opened.

Science and the Arts

1893 Statue of Eros unveiled at Piccadilly Circus.
1895 First Promenade concert.
1897 Tate Gallery opened.

1900 Wallace Collection opened.
1901 Horniman Museum, Forest Hill opened.
1902 Life and Labour of People of London published by Charles Booth.
1904 London Symphony Orchestra founded.
1904 London Coliseum Theatre opened.
1905 Strand Theatre opened.
1905 Aldwych Theatre opened.

1907 Queen's Theatre opened.

1909 Science Museum founded.

1911 Queen Victoria Memorial unveiled by George V in the Mall.
1911 London Museum founded.
1914 Opening of King Edward VII Galleries at British Museum.

1926 J C Baird gave first demonstration of television in Frith Street, Soho.

1928 Discovery of penicillin by Alexander Fleming at St Mary's Hospital, Paddington.
1930 Whitehall Theatre opened.

1931 Windmill Theatre opened.

1932 London Philharmonic Orchestra founded by Sir Thomas Beecham.
1933 Open-air theatre, Regent's Park opened.
1934 National Maritime Museum, Greenwich founded.

1935 Geological Museum, South Kensington founded.
1936 First regular television service from Alexander Palace.

History, Politics and Society

1940 Air attacks on London docks.
1940 Second great fire of London: 30,000 incendiary bombs.
1941 National Fire Service formed.
1944 First flying bomb hits London.
1946 New Towns Act: eight new towns around London.

1948 First immigrants arrive from Jamaica.

1951 King George VI opens Festival of Britain.
1952 4,000 deaths attributed to 'smog' lasting several days.
1953 Coronation of Elizabeth II at Westminster.

1955 City of London declared 'smokeless zone'.

1957 Survey shows no fish in Thames from Richmond toTilbury (40 miles).

1962 Commonwealth Institute, Kensington opened.
1965 Formation of Greater London Council to replace LCC.

1968 Large anti-Vietnam war demonstration in London.

1973 Statue of Sir Winston Churchill unveiled.
1974 First salmon caught in Thames for 100 years.
1976 Population of Greater London c.7,000,000.

1981 London Docklands Development Corporation formed.
1981 London Wildlife Trust formed.
1981 Greater London Enterprise Board formed.

1986 Greater London Council abolished.
1987 Fire at King's Cross underground station.

1989 *Marchioness* pleasure boat disaster on Thames.

1990 Population of Greater London c.6,500,000,

1992 IRA bombs Baltic Exchange.

1995 Aldwych bus bombed by IRA.
1996 Canary Wharf tower bombed by IRA.

Commerce, Industry and Infrastructure

1944 Port of London used as base for invasion of Europe.

1947 Last horse-drawn cab licence given up.
1947 King George VI Reservoir, Staines inaugurated.

1951 London Foreign Exchange Market reopened after 12 years.
1952 Last tram journey in London.
1953 London Airport, Heathrow, opened.

1956 London Gold Market reopened after 15 years.

1960-64 Post Office Tower.
1961-3 Hilton Hotel, Park Lane, opened.

1966 Carnaby St Market.

1968 Closure of London and St Katharine Docks.
1968 Euston Station opened.
1968 Victoria Line opened.

1974 Covent Garden Market moved to Nine Elms.
1976 Brent Cross Shopping Centre opened.
1979 Jubilee Line opened.
1981 Royal Docks closed, last of London's docks to close.
1982 Enterprise Zone established in London docks.
1982 Billingsgate Market moved to Isle of Dogs.
1982 Thames Flood Barrier at Woolwich completed.
1986 Big Bang in City.
1986 Terminal Four completed at Heathrow Airport.
1987 Docklands Light Railway opened.
1987 London City Airport opened at Docklands.
1991 Spitalfields Market moved to Leyton.

1991 Channel Tunnel Rail terminal under construction.
1991 Queen Elizabeth II Bridge opened.

1994 Eurostar terminal completed at Waterloo station.

1999 Jubilee Line extension completed.

Building and Architecture

1940 The Citadel, the Mall.

1955 Trade Union Congress HQ, Great Russell St (sculpture by Epstein)
1957-79 The Barbican Complex.

1961 United States Embassy, Grosvenor Square.
1962 Shell Centre, South Bank.
1963 Vickers Tower, Millbank.

1968 'Roman Point' Tower Block collapses. Fatalities.

1978 Central London Mosque, Regent's Park.

1982 New British Library building founded.

1986 Lloyd's Building opened.

1987 Princess of Wales Conservatory, Kew Gardens.
1988 New Chapter House, Southwark Cathedral.
1989 Great storm causes much damage in London.
1990 Canary Wharf tower completed.
1991 Broadgate Centre completed at Liverpool Street station.

1995 Conversion of County Hall into flats.

1999 Ferris wheel being built on the South Bank.
2000 Millennium Dome opened, Greenwich

Institutions and Popular Culture

1940 Savoy Cinema, Holloway Road, opened.

1948 First jazz club opened by Ronnie Scott and John Dankworth.
1948 Olympic Games held at Wembley Stadium.
1951 National Film Theatre, South Bank.

1953 First 'coffee bar', 'the Mika', opened on Frith St.

1955 'Bazaar': the first boutique, Kings Rd, Chelsea.

1958 The London Planetarium opened.

1961 First Notting Hill Carnival.

1964 The Beatles recorded at EMI Studios, St John's Wood.

1968 Rolling Stones gave first open-air concert in Hyde Park
1968 London Weekend Television started.

1970 Radio London started.

1973 LBC began.
1973 Capital Radio began.

1985 'Band Aid' Concert, Wembley Stadium; 72,000 attend.

1991 Open-air concert at Hyde Park.

Science and the Arts

1948 'Eros' reinstated at Piccadilly Circus.

1951 Royal Festival Hall, South Bank opened.

1954 'Temple of Mithras' excavation at Bucklersbury.
1955 BBC TV Centre opened at White City.

1957 Imperial College of Science building, Kensington opened.
1959 Mermaid Theatre, Puddle Dock opened.
1961-2 Royal College of Art, Kensington Gore opened.

1967 Queen Elizabeth Hall, South Bank opened.
1968 Hayward Gallery, South Bank opened.

1969 Greenwich Theatre opened.
1970 The Young Vic Theatre opened.
1973 British Library formed.

1975 New Museum of London opened.
1976 National Theatre, South Bank opened.
1980 London Transport Museum opened at Covent Garden.

1982 Barbican Arts Centre opened.

1988 Museum of the Moving Image opened on South Bank.
1989 Design Museum opened on Butlers Wharf.

1990 Courtauld Gallery moved to Somerset House.
1991 Sainsbury Wing opened at National Gallery.

1993 Quaglino's opened.
1993 Buckingham Palace opened to general public.

1996 Millennium site confirmed at Greenwich.
1997 New Globe Theatre opened on Bankside.
1999 New British Library building completed.

MAPPING AND DEPICTING LONDON

Despite the great wealth and political power of London during many centuries, the first detailed depictions of the city did not appear until the 1550s. At that time intricate panoramic views of the city started to be produced, and in 1559 the first detailed map of London was published, a full century after the earliest European printed maps had been engraved in Italy *(page 142)*. Three quarters of the span of London's history had passed without accurate cartographic record being made. A surge of interest in history and geography took place in the Elizabethan Age and gave rise to numerous county maps and to some London maps, which were then reissued in the early seventeenth century. But at this time the science of map-making was new and survey techniques were poorly understood. In Restoration England growing interest in scientific knowledge encouraged cartographers to produce new maps. This trend was further stimulated by the Great Fire of London (1666), which destroyed supplies of existing charts and maps. Damage to the city was so extensive that publishers were dissuaded from reprinting from out-dated copper plates. New surveys were made to assist reconstruction, with John Ogilby's vast map of 1676 (8 ft x 4 ft) giving a reasonably accurate view of the City. Other late seventeenth-century maps embrace Westminster and Southwark, and were sometimes printed in reduced form as pocket maps for visitors.

By contrast, John Rocque's map of 1746 (16 ft x 6 ft) showed the churches, streets, squares and numerous other features of the capital in enormous detail (26 inches to the mile). It depicted the pattern of urban development across 10,000 acres and displayed 5,000 place names. Despite a lack of absolute mathematical rigour, its 24 sheets provided an unparalleled source for tracing the appearance of the rapidly growing city. Methods of survey improved in the late eighteenth century and were employed by Richard Horwood, whose great map of London (17ft x 7 ft) appeared in the 1790s *(page 75)*. The Ordnance Survey was founded in 1791 with its roots in both military and civilian cartography. It continued to improve map-making techniques during the nineteenth century to produce standard topographic maps (one inch to one mile) and a variety of detailed urban plans which chart the vigorous expansion of Victorian London. Specialized maps were also published in the last century to show land use, canals, roads and railways, and many aspects of social life, including poverty, disease, education and religious observance. Twentieth-century maps have been particularly concerned with matters related to urban planning, especially after the widespread destruction incurred in World War II, and have incorporated data from new sources such as aerial photographs and imagery from earth-observation satellites.

This enormous wealth of historical maps and other information has been used in many ways to produce the plates in this Atlas. The most direct method involves the straightforward reproduction of sections of old maps which show how London was depicted by map-makers in the past. For about three centuries these maps were printed in black, with a small proportion being coloured by hand. Not until the 1860s did maps of London start to be printed in colour. A second approach involves abstracting information on specific themes from historic maps in order to create new specialized images. Contemporary verbal descriptions, censuses and statistical enquiries provide other important evidence which has been converted into maps. The most creative cartography involves making maps to show archaeological and documentary evidence which pre-dates London's first printed maps. Last, but emphatically not least, the most imaginative technique in the Atlas is the combining of cartographic, pictorial, statistical and verbal descriptions in order to "reconstruct" buildings and urban scenes in the past, thereby conveying three-dimensional views of London, in contrast with the two-dimensional images of conventional cartography.

Modern mapping *Aerial photographs are of great help in updating maps. The example (far right) shows how the docks were laid out on the Isle of Dogs and identifies the impact of changes in Docklands in the 1980s before the construction of Canary Wharf. The illustration (left) shows a three-dimensional impression of a 1989 proposal for redeveloping the Spitalfields site and replacing the old fruit and vegetable market with offices.*

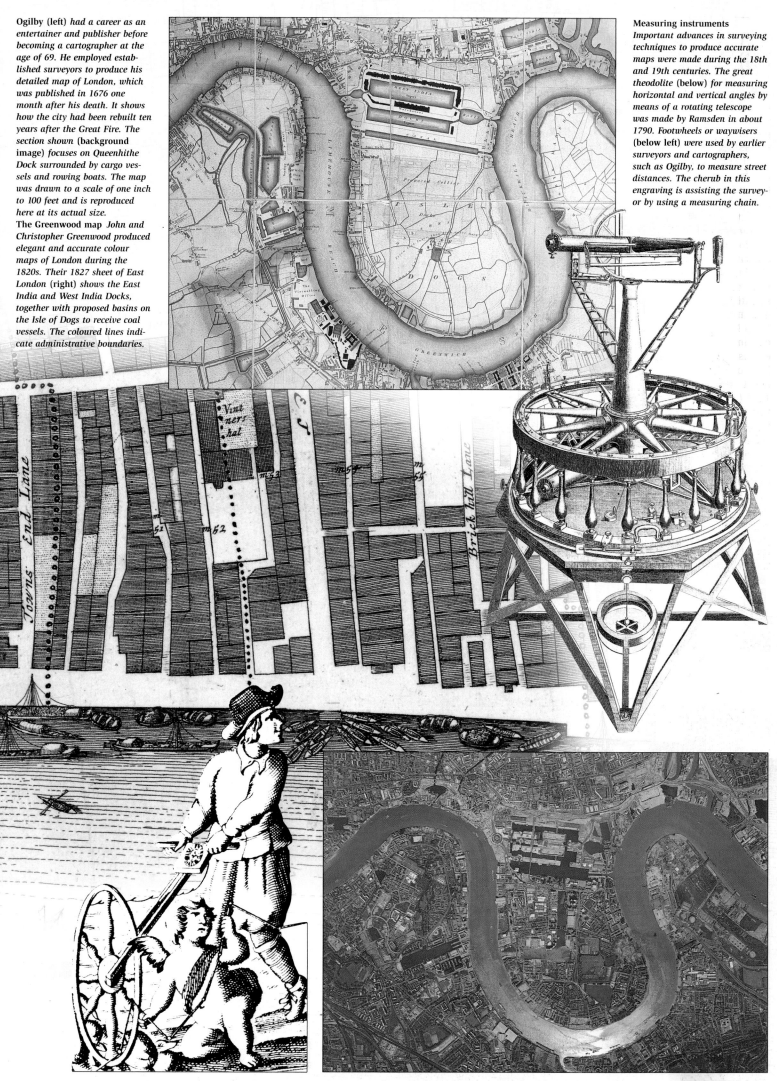

Ogilby (left) *had a career as an entertainer and publisher before becoming a cartographer at the age of 69. He employed established surveyors to produce his detailed map of London, which was published in 1676 one month after his death. It shows how the city had been rebuilt ten years after the Great Fire. The section shown (background image) focuses on Queenhithe Dock surrounded by cargo vessels and rowing boats. The map was drawn to a scale of one inch to 100 feet and is reproduced here at its actual size.*
The Greenwood map *John and Christopher Greenwood produced elegant and accurate colour maps of London during the 1820s. Their 1827 sheet of East London (right) shows the East India and West India Docks, together with proposed basins on the Isle of Dogs to receive coal vessels. The coloured lines indicate administrative boundaries.*

Measuring instruments
Important advances in surveying techniques to produce accurate maps were made during the 18th and 19th centuries. The great theodolite (below) for measuring horizontal and vertical angles by means of a rotating telescope was made by Ramsden in about 1790. Footwheels or waywisers (below left) were used by earlier surveyors and cartographers, such as Ogilby, to measure street distances. The cherub in this engraving is assisting the surveyor by using a measuring chain.

CHAPTER 1
LAND UNDER LONDON

OVER THE CENTURIES, the debris of past generations – known as 'made ground' – and the bricks and mortar of present-day London, have masked the natural land surface. If we are to understand the complete history of the city, then it is with geology that we must begin. Geology is concerned with deposits, such as sands, clays and – further back in time – Chalk, which were laid down by ancient river systems or accumulated on the beds of ancient seas. These sediments consolidated in time to become rocks which were either 'hard' or 'soft'. Erosion by streams fashioned the hills and ridges and hollow plains of the London Basin. 'Deposition' occurred over millions of years; by contrast, the shaping of the present landscape may be measured in mere thousands of years.

London is located in the middle stretch of the Thames valley, near the heart of the London Basin. Moving north from Hyde Park, the suburbs present a switchback of hills before giving way to the slow haul into the Chilterns. These hills are fringed to the north-west by a prominent Chalk scarp edge which overlooks an older plain of clay. Similar changes in topography occur to the west or south of central London. When these traverses are connected together, they constitute the bowl-like shape of the Chalk, known as the London Basin. Clays and sands of younger geological ages – the marine London Clay and the sands of the Bagshot Beds – are found inside the Basin. These materials make up the subsurface geology of the London area.

The Bagshot Beds date back to 40 million years ago. Geologists interpret them as sediments laid down by the flooding of a great river the size of the modern Ganges. Rising in the hills of south-west England, it deposited most of its load of sediment in a broad sheet of sand extending over present-day Middlesex and northern Surrey. 50 million years ago, southern England lay beneath a warm tropical sea which, when it receded, left behind the London Clay which can still be glimpsed in deep excavations at, for instance, building sites. The layer of soft clay facilitated the excavation of the Underground railway, but is notorious for causing the subsidence of housing foundations.

Running water and the sediments it carried, formed the main agent for shaping the natural landscape of south-east England. Early rivers, forerunners of the Thames, steadily removed much of the cover of sand and clay, reducing ancient ridges to the isolated hills of what are now suburban areas. Two or three million years before the present, the early Thames ran through what is now the Vale of Aylesbury, continuing east towards the North Sea. The main river had its source well to the west in Wessex, and was joined by tributaries rising on Chalk slopes to the north and south. At this time, southern tributaries flowed northwards from what is now Surrey, across Middlesex into Hertfordshire. This pattern persisted until 500,000 years ago, when the glaciers of the Ice Age blocked the drainage system, forcing the river to find a route further to the south. In this way the Thames evolved its present course. What were originally floodplain deposits along the Thames were cut into by streams, and deposits of sediment were carved into terraces which flanked the old river courses, the highest terraces being the oldest and the lowest the youngest. Archaeological remains, including the tools of early Man who colonised the area during the past 500,000 years, provide the key to dating these terraces.

The melting of large ice caps during warm climatic periods resulted in a rise in sea level. During cold climatic periods this process was reversed, with vast amounts of water locked up in glaciers. Today, the water levels under London are rising. There are also fears that south-eastern England is sinking in relation to the North Sea. The consequences of economic changes in London itself over the last few decades have also played their part in this process. The decline of manufacturing, especially brewing, in the city means that less groundwater is being pumped up, and hence subsurface water levels have risen. In addition, the land surface is being depressed by new building schemes in what was previously open ground, such as Docklands. The challenges posed by problems of 'global' warming and local concerns over ground stability, all demonstrate the importance of London's geology and topography in the city's past, present and possible future.

MILLION YEARS	EON	ERA	PERIOD	
				Quat.
1·6				Pl
5·3				M
23·7		CENOZOIC	Tertiary	O
36·8				E
57·8				Pa
66·4	PHANEROZOIC		Cretaceous	
144		MESOZOIC	Jurassic	
208			Triassic	
245			Permian	
286			Carboniferous	
360		PALEOZOIC	Devonian	
408			Silurian	
438			Ordovician	
505			Cambrian	
570				
2500	PROTEROZOIC			
	ARCHEON			

Pl = Pliocene
M = Miocene
O = Oligocene
E = Eocene
Pa = Paleocene

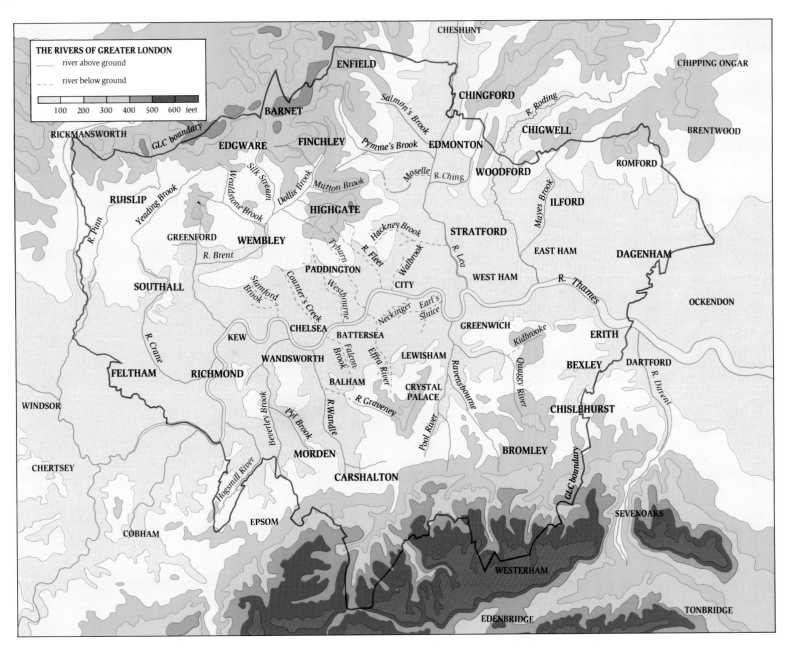

The Thames Basin *The map (above) traces the drainage pattern of the Thames tributaries formed by the sands and clays of the London Basin. The dotted lines indicate what are often called the 'Lost Rivers of London', now largely converted into the sewer system of drainage. Prolonged or heavy rain can result in the sewers flooding in Gospel Oak, Wandsworth and Battersea.*

THE THAMES BASIN

In c.120,000 BP sea level and the level of the Thames in London were much higher than they are now (*below left*). For example, the pavement outside the National Gallery in Trafalgar Square would have been lapped by tidal waters, and the junction of Whitehall and Trafalgar Square would have been underwater, even at low tide. The highest points of land were Ludgate Hill and Cornhill. Bones and fossils found in the Trafalgar Square area show that animals such as hippopotamus and elephant thrived in the warm climate.

By 8,000 BP the level of the Thames was considerably lower than it had been in 120,000 BP (*below right*). At high tide it would have reached the southern end of Northumberland Avenue, some 6½ feet above its present level at the Embankment. Recent development around Upper Thames Street has exposed deposits of alluvium which had built up on the much earlier sands and gravels of the Thames terraces. At this stage, the Fleet River made a significant inlet where it joined the Thames close to Blackfriars. This inlet was an integral part of the port of Roman London.

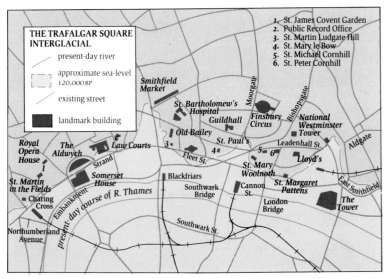

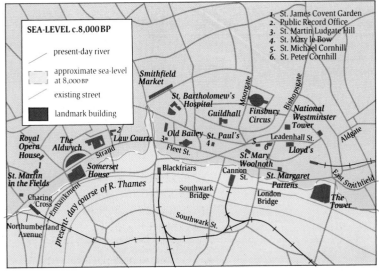

LONDON'S GEOLOGY

The geology of the London area can be best understood by the image of a structural basin, with the Chalk forming the rim and the younger Tertiary beds forming the filling. The green strips on the map (*right*), the Chilterns in the top left corner, the North Downs the broader strip to the bottom right, define that basin.

There is, however, an aspect of London geology which is less well-known, yet at the same time, more striking. This is best revealed by any cross-section of the rocks that are found deep beneath the surface. Such probing reveals a structure which is in complete contrast with the structural basin already identified. The oldest rocks (Silurian age, c. 425 million years) form a central core with younger beds to north and south – an arrangement which has the structure of an anticline (an arch-shaped fold system) contrasting directly with the higher level basin-shaped fold system, or syncline. From the manner in which the younger rocks wedge against the older core, it is possible to surmise that the older core may have been a shallow ridge, or even dry land until it was overwhelmed by the seas in which the Chalk was formed some 90 million years ago. The wedges – which thin out northwards – are best seen to the south of London beneath the North Downs, and are revealed in the contrasts between the deep boreholes at Warlingham and those which probe below central London.

When we examine the solid geology of the London Basin we find that erosion during glacial periods stripped away younger Tertiary deposits within the ancient drainage system of the Thames. This has left broad tracts where London Clay is exposed at the surface (indicated by the chocolate-brown tone on the map). Only small cappings of Claygate Beds and Bagshot Sands remain on some hills. The map shows that these are the well-known high-points of London: Harrow on the Hill, Hampstead Heath and Shooters Hill are all relics of once continuous layers.

The pale buff tones of river terrace deposits, and alluvium impressed upon the deeper tones of the solid geology, demonstrate the great magnitude of the ancient drainage system of the Thames. During glacial episodes, the volume of river flow fluctuated according to the supply of water from the northern ice sheets that were gradually melting. At times then, the Thames was almost Ganges-like in size; at other times, it shrank considerably. The terrace gravels, referred to below, result from these fluctuations.

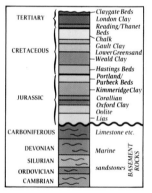

London's geology *This column (right) attempts to summarise the rock units which might be encountered in a borehole beneath London, listing the names used universally for geological time periods. A distinction is drawn between younger rocks and so-called 'basement' rocks, which make up the deeper structure. The cross-section (below) is taken from the north-west of London to the south-east. The line of the cross-section is annotated on the map (right).*

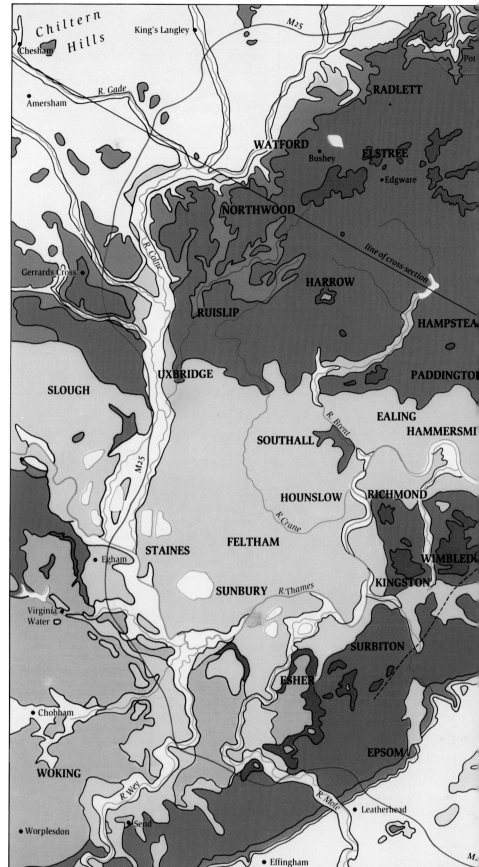

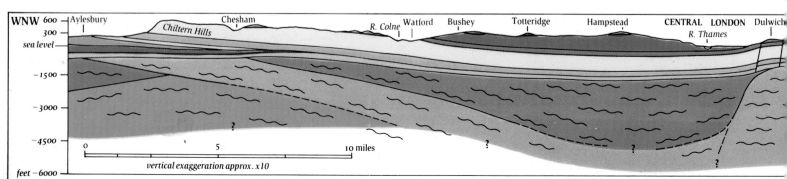

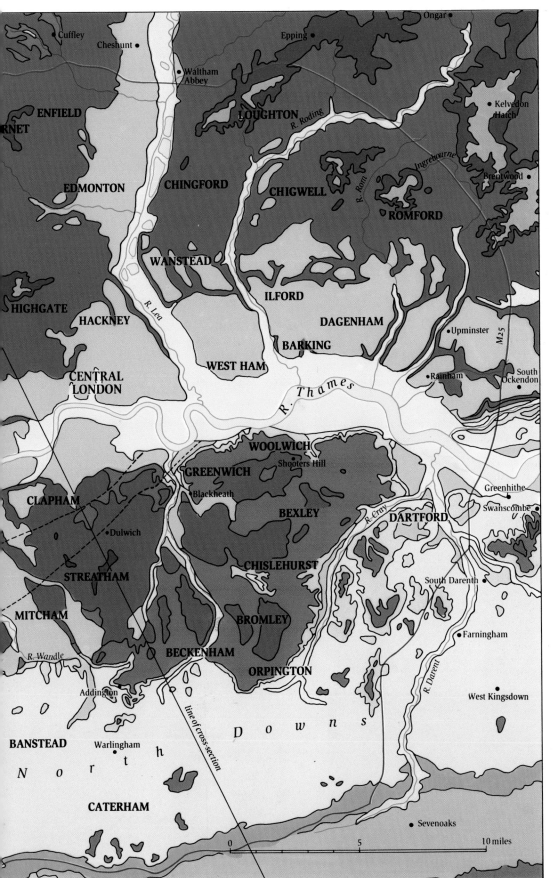

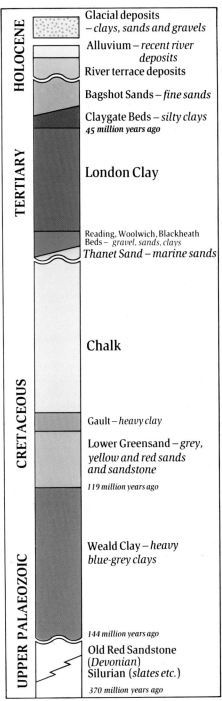

HOLOCENE		Glacial deposits – *clays, sands and gravels*
		Alluvium – *recent river deposits*
		River terrace deposits
TERTIARY		Bagshot Sands – *fine sands*
		Claygate Beds – *silty clays* **45 million years ago**
		London Clay
		Reading, Woolwich, Blackheath Beds – *gravel, sands, clays*
		Thanet Sand – *marine sands*
CRETACEOUS		Chalk
		Gault – *heavy clay*
		Lower Greensand – *grey, yellow and red sands and sandstone* **119 million years ago**
		Weald Clay – *heavy blue-grey clays*
UPPER PALAEOZOIC		**144 million years ago** Old Red Sandstone (*Devonian*) Silurian (*slates etc.*) **370 million years ago**

River terraces of the ancient Thames create and underlie the slopes of London's streets (for example Haymarket to Piccadilly; Whitehall to Trafalgar Square, below). Older terraces were eaten into by later river flow which always tended to bite down to lower levels. In simple terms, the oldest terraces are highest, and those of most recent date lie close to the present river. In fact, the Thames today occupies a route beneath which a super-deepened channel exists. This has always complicated the excavation of tunnels beneath the Thames. The infilled deep channel provides evidence of rising sea levels over the past 10,000 years (page 19).

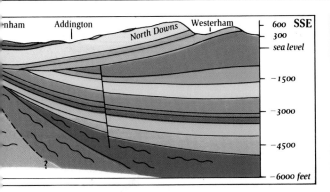

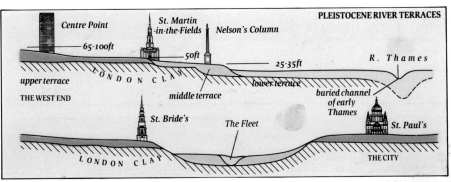

PLEISTOCENE RIVER TERRACES

19

PEOPLE BEFORE LONDON

Groups of people belonging to early forms of the genus *Homo* roamed the London area intermittently from around 450,000 years ago. They scavenged, hunted and foraged across the landscape, and learnt to cope with climatic conditions which varied from the warmth of interglacials to the sub-arctic cold of full glacials. They left behind many thousands of flint tools, principally hand axes, swept from long vanished land surfaces by the ancient Thames and redeposited in the gravel terraces which underlie modern London. Finds of later flint tools associated with the remains of mammoth and woolly rhinoceros sealed beneath brickearths at Southall and Crayford mark undisturbed kill sites.

The emergence of biologically modern people in Europe some 40,000 years ago coincided with a phase of particularly severe climate. Our best evidence for their presence in the London area comes from the very end of the last glacial, as the climate began to warm up. Excavations in the Colne valley at Uxbridge and Staines have revealed butchery sites dating to 10,000 BC, where tundra-loving reindeer and horses were killed and their carcasses processed. A later phase of activity at Uxbridge is dominated by woodland species, including red and roe deer, showing that trees had colonised the landscape.

The melting of the ice sheets led to the final separation of Britain from the continent around 6500 BC. Traditional hunting grounds in the lower-lying valleys were flooded and people forced back up the valleys and onto the valley slopes. From 3900 BC scattered communities constructed earthern enclosures in clearings carved out of the lime- and oak-dominated woodlands fringing the west London Thames. Repeated clearances here led to the creation of a locally open landscape dotted with 'ritual' monuments, one of which, an embanked avenue at Stanwell, is nearly four kilometres long. Special offerings, including human and animal remains, pottery vessels and stone and antler tools, were deposited on land and in the Thames.

Major changes occurred after about 1600 BC: large tracts of land were divided up between farming communities, whose dead were buried in small cremation cemeteries; field systems were laid out in areas of suitable subsoil, while traces of criss-cross ploughing have even been found on the valley floor at Bermondsey. Wooden

Gifts to earth and water Special offerings were frequently deposited on land and in the Thames during later prehistory. The dismembered remains of an early Bronze Age wild cow or aurochs (right), accompanied by six flint arrowheads, were carefully arranged in a deep pit at Harmondsworth, while the splendid bronze oval shield (below), dated to between 400 and 250 BC, was recovered from a former river channel at Chertsey.

An Iron Age tribal centre Occupied during the last two centuries BC, Uphall Camp, Ilford (far right) could be a local oppidum. Recent excavations within its 48-acre interior have produced evidence of weaving and metal-working.

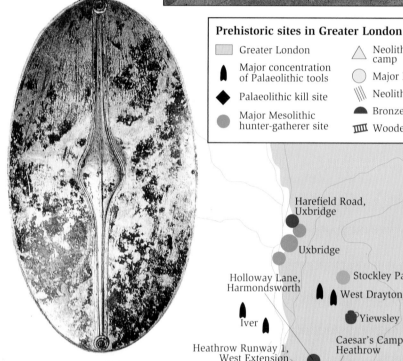

Prehistoric sites in Greater London

▭ Greater London	△ Neolithic causewayed camp
♠ Major concentration of Palaeolithic tools	○ Major Neolithic site
◆ Palaeolithic kill site	⦚ Neolithic 'cursus'
● Major Mesolithic hunter-gatherer site	⬟ Bronze Age barrow
	⬛ Wooden trackway

Harefield Road, Uxbridge

Uxbridge

Holloway Lane, Harmondsworth

Stockley Park, Dawley

West Drayton/Yiewsley

Iver

Yiewsley

Bos Ma

Heathrow Runway 1, West Extension

Caesar's Camp, Heathrow

Southall

Cranford Lane

Old Eng

Manor Farm, Lower Horton

Stanwell cursus

Mayfield Farm, East Bedfont

Lower Mill Farm

West Bedfont

River Crane

Staines Moor

Yeoveney Lodge, Staines

Stanwell Barrow cemetery (9/10 barrows)

Har

Staines

Runnymede Bridge

Littleton

Ashford

Sandy L. Tedding

Thorpe Lea

St Ann's Hill, Chertsey

Staines Road Farm, Shepperton

Hurst Park

Muckhatch Farm

Prehistoric sites in Greater London (c 450,000 BC–AD 43) *Despite the obvious difficulties of undertaking fieldwork in the area, excavations across a range of geological strata have revealed much that is new and unexpected (map above right). Work on the free-draining and fertile gravel terraces has been especially productive and has* located sites such as the *Neolithic circular earthern enclosure at Yeoveney Lodge, Staines (artist's impression, left). More recent has been the realisation that wooden trackways, such as the Bronze Age example from Beckton (far right), survive intact beneath later alluvial deposits on the Thames floodplain.*

trackways constructed with the aid of metal tools were thrown across the riverside marshes, affording people and animals seasonal access to lush water meadows.

Agricultural expansion fuelled competition for the best land, and social relations were underpinned both by feasting and by offerings to the gods, principally metalwork – much of which found its way into the Thames. The occupants of prestigious sites housed within circular 'ring-forts' or on islands in the Thames may have controlled the far-flung bronze exchange networks, although when locally produced iron was introduced around 700 BC, crisis ensued. The collapse of long established contacts hit the lower Thames valley especially hard, so that the area hardly recovered its former pre-eminence until *Londinium* was founded in the mid-first century AD.

Few hillforts were constructed in the region; instead most people lived on small mixed farms. Some, like those below the hillfort on St George's Hill, Weybridge, specialised in the production of iron. The London area appears to have lain outside the mainstream of tribal politics in the last century BC. It could be that the Thames, formerly a highway, now marked the boundary between neighbouring tribes whose main centres (*oppida*) were set well back from the river. On present evidence, London was not itself a tribal centre; better local candidates include Uphall Camp in Ilford or at Woolwich further downstream.

Legend

- ● Bronze Age settlement/field system
- ● Bronze Age cremation cemetery
- ○ Bronze Age ringfort
- ◗ Bronze Age island site
- ⚑ Iron Age hill fort
- ● Iron Age settlement
- ■ Local oppidum
- □ Possible oppidum

Palaeolithic: c 450,000-8300 BC
Mesolithic: c 8300-4500 BC
Neolithic: c 4500-2300 BC
Bronze Age: c 2300-650 BC
Iron Age: c 650 BC-AD43

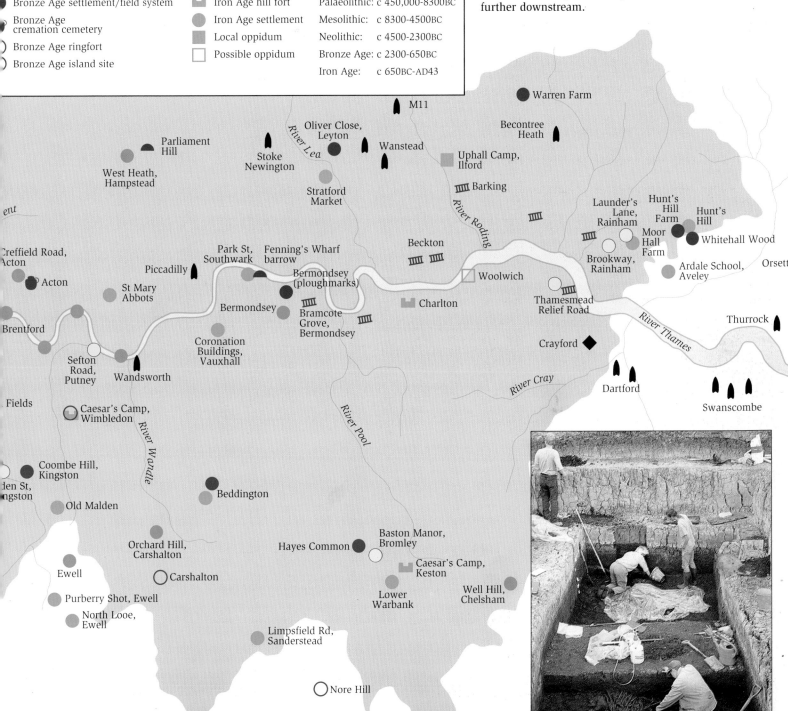

Waltham Abbey
Ambresbury Banks
High Beach
Loughton
Warren Farm
Becontree Heath
M11
Oliver Close, Leyton
Parliament Hill
Stoke Newington
Wanstead
Uphall Camp, Ilford
West Heath, Hampstead
Stratford Market
Barking
Launder's Lane, Rainham
Hunt's Hill Farm
Hunt's Hill
Moor Hall Farm
Whitehall Wood
Beckton
Creffield Road, Acton
Park St, Southwark
Fenning's Wharf barrow
Brookway, Rainham
Ardale School, Aveley
Orsett
Piccadilly
Acton
Bermondsey (ploughmarks)
Woolwich
St Mary Abbots
Bermondsey
Charlton
Thamesmead Relief Road
Thurrock
Brentford
Coronation Buildings, Vauxhall
Bramcote Grove, Bermondsey
River Thames
Sefton Road, Putney
Wandsworth
Crayford
River Cray
Dartford
Swanscombe
Fields
Caesar's Camp, Wimbledon
River Wandle
River Pool
Coombe Hill, Kingston
Beddington
Old Malden
Orchard Hill, Carshalton
Hayes Common
Baston Manor, Bromley
Ewell
Carshalton
Caesar's Camp, Keston
Purberry Shot, Ewell
Lower Warbank
Well Hill, Chelsham
North Looe, Ewell
Limpsfield Rd, Sanderstead
Nore Hill
Banstead Heath
River Lea
River Roding
den St, ngston

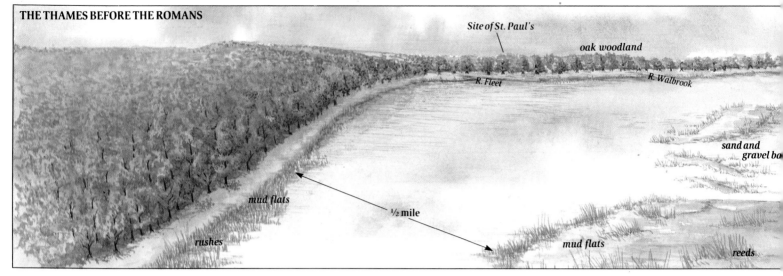

Site of St. Paul's

oak woodland

R. Fleet

R. Walbrook

sand and gravel ba[r]

mud flats

½ mile

rushes

mud flats

reeds

Chapter 1 Land under London

THE TOWN AND THE THAMES

The 'Thames' is one of the oldest documented place names in British history, for it is recorded in the account of Julius Caesar's invasion of these islands in c. 55 BC. The town of London did not exist then — it made its first entry into the history books in AD 60. The river carved a broad and sinuous route through the fertile lands of south-east England, while its estuary opened out into the North Sea beyond which was the Baltic and the mouth of the River Rhine, one of the great waterways of continental Europe. A town on the Thames would therefore share the twin advantages of a prosperous hinterland and ready access to and from the wider world. London, then, was the gift of the Thames. Londoners drank, washed and fished in it; it powered their mills and provided a ready highway in times when more people owned boats than horses and carts. But above all, London prospered by water-borne trade, by the traffic swept upstream from the estuary on incoming tides, and sent out into the North Sea when the tide turned. However, the Thames was not always a passive, benevolent force; riverside settlements were particularly vulnerable to catastrophic floods and such un-welcome visitations as Viking raiders. Other unwanted imports such as plagues were brought in by foreign merchants and sailors.

Nevertheless, Londoners learned to live with their river and eventually saw their harbour become the busiest in the world. In the process, the townscape was transformed on a spectacular scale. The Romans who built *Londinium* would not recognise the vast conurbation we now call Greater London; nor would they recognise the Thames, for the river they had to cross, with its islands, marshes and mudflats, was much wider than the deeper, dirtier Thames we see today. The level of the sea and the Thames relative to the land has been gradually rising. High tide in the Roman period was some nine feet below the present-day level, and the tidal range (the difference between high and low tides) in the 1st century was only some six feet, compared with over twenty feet today. While the Thames has been rising, its banks have been progressively pushed riverwards, narrowing the channel. In the City, this process began with the Romans, but such reclamation schemes continued throughout the medieval period. In other parts of the Thames valley, the inter-tidal marshes were embanked

LONDON'S ADVANCING WATERFRONT

Thames Street

warehouse *quay* AD 100 AD 125 *riverside wall* AD 270 AD 250 AD 1000 AD [...]

natural river bank AD 150

The changing river *The view (top) illustrates the Thames riverscape before the arrival of the Romans. In the foreground is the low-lying Southwark shore; across the untamed Thames is the high ground on the north bank where the Romans were to build* Londinium. *Throughout the city's history, Londoners have changed the shape of the Thames by reclaiming land on the banks of the river. This north-south section (above) shows how the north bank was gradually advanced into the river from c. AD 100 to the present day. In parts of the City, over 110 yards of land were won in this way. The series of plans (right) show the Thames at high tide and low tide in the 1st century AD, compared with today's constricted river. The islands and inter-tidal marshes on the south bank of the Thames in the Roman period are clearly visible – they are now all built over.*

Not only has the Thames changed: so has the site of London and its port. These plans (right) show the location of the Roman settlement, replaced by a Saxon market town on a new site well to the west. The medieval town shared the Roman wall, and its new bridge effectively prevented most larger ships sailing further upstream. Increased trade saw the port expanding far to the east of London Bridge with the development of the enclosed docks from 1800.

THE THAMES AT HIGH TIDE c. AD 50

basilica & forum

R. Walbrook

LONDINIUM

bridge

R. Thames

Stane St. *Watling St.*

causeway

SOUTHWARK

THE THAMES AT LOW TIDE c. AD 50

bas[ilica] & f[orum]

R. Thames

Stane St. *Watling St.*

SOUTHWARK

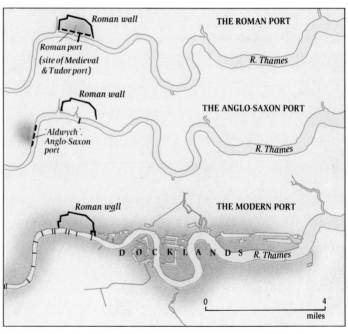

Roman wall

THE ROMAN PORT

Roman port (site of Medieval & Tudor port)

R. Thames

Roman wall

THE ANGLO-SAXON PORT

'Aldwych', Anglo-Saxon port

R. Thames

Roman wall

THE MODERN PORT

D O C K L A N D S *R. Thames*

0 4

miles

and drained: for example the monks of Lesnes Abbey were largely responsible for reclaiming the land now occupied by the Thamesmead estate.

The history of the Port of London is a history of four harbours. First, the Roman centre of *Londinium* which was founded in AD 50 at a major junction of road, river and sea-going traffic. Initially, this port prospered, becoming the pro-vincial capital, but then dramatically collapsed and was abandoned by the 5th century. 200 years later, Saxons developed a quite different port on a new site to the west of the old city. The sprawling riverside market town of *Lundenwic* also pros-pered, but was to attract the attentions of Viking

raiders who sacked it in 842, 851 and 871.

The Londoners then moved inside the area protected by the ancient Roman wall, abandon-ing the old market (Ald-wych) by c. 900. This new settlement saw slow but sure growth in spite of Viking and Norman invasions. The medieval harbour was well placed to lead the nation's all-important trade in wool and cloth, expanding dramatically in the 16th century. But the 18th century saw it choked with its own success. The length of the legal quays upon which all cargoes had to be discharged was only 470 yards, quite insufficient to accommodate the large contem-porary vessels bringing in cargoes from Europe,

Site of Roman City   Site of Tower of London

River   Thames

mud flats

SOUTHWARK

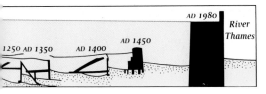

AD 1980
River Thames
1250 AD 1350   AD 1400   AD 1450

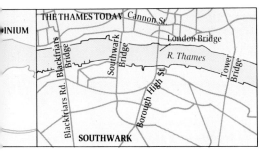

THE THAMES TODAY   Cannon St.

INIUM

London Bridge

Blackfriars Rd.   Blackfriars Bridge   Southwark Bridge   R. Thames   Tower Bridge

Borough High St.

SOUTHWARK

Africa, the Americas and the East. A series of enclosed docks built to the east of the City from 1800 to 1920 remedied the situation. As the Empire grew, so the Port expanded downstream, together with the suburbs and the many industries which relied upon it. London survived the Blitz, and trade reached new records in 1959-62. However, this was the age of supertankers and containerisation and London's docks, some of which were built in the age of sail, were unable to adopt new cargo-handling facilities. From 1967-81, the entire docklands complex from the Tower to Woolwich was closed; the unthinkable had happened — the city was a port no more.

Flooding and reclamation.
Reclamation of land on the Thames banks started with the Roman period and continued in the 19th century with elaborate schemes such as the Victoria Embankment, a broad carriageway incorporating an underground railway line and a new sewer. The map (below left) shows the line of indented waterfront in the Strand area with its projecting river stairs before the scheme began in 1866, and (superimposed) the outline of the new Embankment. The painting (below) shows the view towards the City along the newly-completed Embankment in c. 1873.

London has always been vulnerable to flooding, and the low-lying areas to the east and south of the City are particularly at risk (map top right); records of such disasters go back to 1294, when Southwark suffered severe damage. The situation is made more serious by geophysical changes: since all of south-east England is slowly sinking, the level of the highest Thames tides is therefore rising. The £435 million Thames barrier was completed in 1982. Its flood gates are raised 90 degrees from their normal lowered position to hold back incoming tides, forming a steel wall when needed (right).

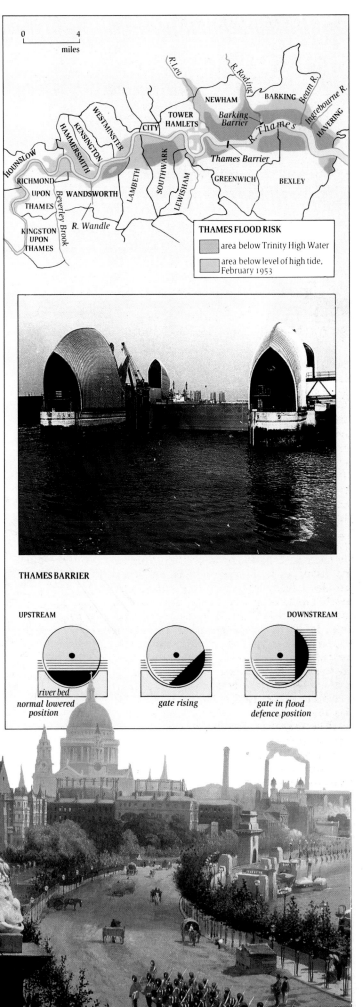

THAMES FLOOD RISK
area below Trinity High Water
area below level of high tide, February 1953

THAMES BARRIER

UPSTREAM   DOWNSTREAM

river bed

normal lowered position   gate rising   gate in flood defence position

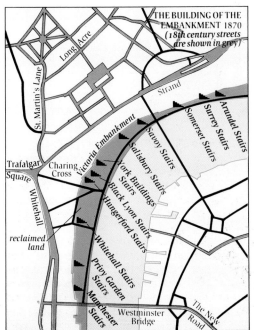

THE BUILDING OF THE EMBANKMENT 1870 (18th century streets are shown in grey)

Long Acre
St. Martin's Lane
Strand
Arundel Stairs
Surrey Stairs
Somerset Stairs
Savoy Stairs
Salisbury Stairs
York Buildings Stairs
Black Lyon Stairs
Hungerford Stairs
Victoria Embankment
Charing Cross
Trafalgar Square
Whitehall
reclaimed land
Whitehall Stairs
Privy Garden Stairs
Manchester Stairs
Westminster Bridge
The New Road

CHAPTER 2
ROMAN LONDON

THE THAMES VALLEY has attracted settlers since the time of the first hunter-gatherers, nearly half a million years ago. It provided everything necessary for successful habitation: water, fertile soils and abundant timber. However, there is no evidence as yet for any settlement of importance in the London area before the advent of the Romans, although isolated homesteads existed on the higher ground of Southwark, south of the Thames, and were scattered throughout the Thames area.

For a more detailed discussion of the physical geography of the London region, see Chapter 1 Land under London, pp. 18-23.

The physical geography of the region was a major influence on the origin and development of *Londinium*. The north bank of the Thames was elevated and well-drained, consisting of a gravel terrace capped by brickearth. Although bisected by a stream (later called the Walbrook), it was a relatively flat platform and ideal for building. Protected to the west by the River Fleet, the steep slopes down to the water's edge provided an additional defence. There was a constant fresh water supply provided by the Thames and the Walbrook and by springs at the base of the gravel terrace that could be exploited by shallow wells.

For political reasons the Emperor Claudius needed a conquest to safeguard his position in Rome and to deploy surplus legions throughout the Empire. In Britain, strong native leaders were emerging, uniting the various tribes in lowland Britain, interfering with those tribes who were allied to or had peaceful relations with Rome. As a state of open hostility with Rome was an ever-increasing threat, Claudius decided to dispatch his formidable army to Britain in AD 43. The force, comprising four legions and auxiliary units from Europe, must have totalled some 40,000 men. Under the command of Aulus Plautius, the troops landed at the natural harbour of *Rutupiae* (Richborough), unopposed by the Britons. From this coastal base, the army gained control of much of the south-east of England. A decisive battle with British tribes took place at the River Medway and from that point on only local pockets of resistance were encountered. The invasion force then advanced to the Thames which the troops crossed successfully in two places, probably by means of a floating pontoon bridge constructed by the military engineers and by swimming across. The Britons retreated eastwards and the pursuing Roman army lost many men in the Essex marshes.

After the crossing of the Thames, the campaign halted on the pretext that reinforcements were needed. Word was sent to the Emperor in Rome and after a delay, Claudius took command of his troops near the Thames and took over the campaign. He marched on the tribal centre, *Camulodunum* (Colchester), where he received the surrender of eleven British kings. After a stay of just sixteen days, Claudius returned triumphant to Rome. He left the army behind to establish *Camulodunum* as the capital of the Province of *Britannia*, and to

Left *The Bucklersbury pavement. This large Roman mosaic dates to the 3rd century AD and was discovered in 1869 close to the Mansion House.*

Above *Head of the Emperor Hadrian. This large bronze head was found in the Thames. He visited London in AD 122.*

impose a single firm control over the various tribal groups on both sides of the Thames.

The Thames must have been finally crossed by a bridge, although its method of construction and material are uncertain. The new roads made by the Roman army leading from the Kent coast to *Camulodunum* and to the rest of the new province beyond had to converge on this bridge, which became strategically important. The earliest occupation of London, therefore, was probably by Roman soldiers defending the new crossing and consisted of temporary fortified camps. It must have taken some years after the invasion of AD 43 to build the permanent bridge and road network around *Londinium*. Archaeological work in Southwark and on the north bank has uncovered evidence of a bridge which could not have been built earlier than AD 50.

Traces of the earliest phase of occupation have been found most clearly in a small area north of an east-west road running parallel with the river, underlying parts of Lombard Street where a regular lay-out of timber buildings and roads is evident. This first *Londinium* was destroyed by fire in AD 60 by Queen Boudicca (Boadicea), who led her tribe, the Iceni, from East Anglia, joined by the Trinobantes of Essex, in revolt against the Romans. The Roman army at that time was scattered throughout Britain and the Governor of Britain, Suetonius Paulinus, was in Anglesey fighting the Druids with a large part of the army. The Governor was unable to save *Camulodunum*, the first target of Queen Boudicca's army, from destruction but attempted to rescue *Londinium*, reaching it with his cavalry before Boudicca's arrival. There were not enough troops to defend the town and he therefore evacuated anyone willing to leave. The inhabitants who remained were massacred by the rebels and the town destroyed. It is possible that some of the many human skulls since recovered from the stream-bed of the Walbrook were ritually deposited there by the tribesmen, although no human remains can be positively attributed to the massacre. *Verulamium* then met the same fate. Suetonius Paulinus finally confronted Boudicca in the Midlands and, although vastly outnumbered, the Romans proved victorious. For some months after the revolt they inflicted retribution on the native Britons.

Nero sent a new *procurator* to administer the province's financial and economic affairs, Julius Alpinus Classicianus. He was a fair and just administrator, putting right the injustices that had caused the rebellion. The office of *procurator* continued to be based in *Londinium*, emphasising the town's importance. The Governor also had his permanent headquarters in *Londinium*, where he maintained a civil service which kept him in touch with the rest of the province and Rome. *Londinium* arose from the ashes of its destruction and underwent substantial expansion and development. A large public building, probably the first basilica and forum, was built on the high ground to the east of the Walbrook. Tacitus, writing in the early 2nd century, described *Londinium* at the time of the Boudiccan revolt as a place teeming with businessmen and a famous centre of commerce. Surviving wooden writing tablets record financial transactions, the repayment of loans and the selling of commodities such as slaves. The port soon revived after AD 60 and when *Londinium* became the capital of the province there was an upsurge in commercial vitality and much upgrading of the port facilities. Indeed, the port seems to have reached its peak of activity during the late 1st century.

Londinium's street system was laid out and many parts of the city had become densely built up by about AD 100. Shops and commercial buildings had been built along the main road leading from the forum. Residential areas and public baths had been developed away from the main streets and the

The new town which grew up on the Thames was named Londinium *by the Romans. The name is of pre-Roman origin, possibly connected with the original British name for the area.*

For a full reconstruction of the 2nd-century city, see pp. 30-31.

For a detailed discussion of the Roman port and London's commercial activities, see pp. 32-33.

For a reconstruction of Roman fort and detailed map of London's defensive walls, see pp. 34-35.

amphitheatre and fort built to the west of the town. The Emperor Hadrian visited *Londinium* in AD 122 and many public buildings were either built or rebuilt for his visit. However, very soon after, in about AD 125, *Londinium* was again severely damaged by fire, a date fixed by the quantities of burnt pottery found on various sites.

The reigns of the Emperors Antoninus Pius and Marcus Aurelius in the mid-2nd century were not a stable period in general for Roman Britain, and *Londinium* in the mid-2nd century was beginning to undergo changes. Houses and workshops in the Walbrook valley and elsewhere in the city and Southwark were no longer required and were deliberately demolished. In certain areas such as Milk Street and Newgate Street to the west, they were overlaid by a dark earth, which initially must have been brought from outside the town. In addition, the large public baths at Huggin Hill and the smaller baths at Cheapside were demolished towards the end of the 2nd century perhaps because their maintenance for a much reduced population could no longer be justified. Despite this, *Londinium* was still regarded as important by the Roman authorities, and this is testified to by the construction of a new timber waterfront and a massive city wall which enclosed 330 acres, making *Londinium* the largest city in Britain. At a period when the working population seems to have abandoned large areas of the city, the wall extended beyond the existing inhabited areas and stretched from modern-day Blackfriars to the Tower where the ends of the wall must have had special fortifications to prevent attackers reaching the city. The tributaries of the Walbrook were allowed to flow through the wall in brick culverts.

The city wall was perhaps built to protect the capital during the civil wars at the end of the 2nd century. Clodius Albinus, Governor of Britain, had declared himself Emperor in AD 193 and had taken most of Britain's troops to fight on the Continent. He may well have realised the importance of protecting *Londinium* against any advances the true Emperor, Septimius Severus, might have chosen to make. The dating of the wall has been made possible by finds of coins: one coin set a date for its construction not earlier than AD 190; and other coins and clay coin moulds belonging to a forger indicate that it could not have been built later than about AD 210.

In AD 200 Britain was divided into two separate provinces. *Londinium* was made capital of *Britannia Superior* (Upper Britain) and *Eburacum* (York) capital of *Britannia Inferior* (Lower Britain). While large numbers of sculptures and inscriptions belonging to the 3rd century indicate that *Londinium* was still a town of considerable importance and wealth, its population was reduced and large areas within the wall were perhaps used for horticulture. The Walbrook area was re-occupied before the mid-3rd century, and the marshy ground was stabilised and substantial buildings were built, many with finely patterned mosaic floors. A large area in the south-western corner of the walled city was occupied during the 3rd century by a pre-

Right *Statue of British hunter-god found in Southwark, 3rd century AD. Statues and inscriptions of the period indicate* **Londinium's** *continuing importance.*

cinct of public buildings, probably including temples, which was possibly entered by a monumental arch, surviving fragments of which were found re-used in the later riverside wall. The temple of Mithras was built in about AD 240 on the east bank of the Walbrook.

In AD 286 Carausius, the admiral of the Roman fleet based in the English Channel, was accused of taking the treasure of Saxon and Frankish pirates for his own use. Crossing to Britain, he declared himself Emperor and ruled Britain until he was murdered by his financial adviser, Allectus, in AD 293. This was a period of political and economic isolation for Britain. To compensate for the loss of official coinage reaching Britain, Carausius in AD 288 created, for the first time, a coin mint in *Londinium*, which continued to produce coins spasmodically into the 4th century.

In AD 296, the army of the Roman Emperor Constantius Chlorus landed in Hampshire and defeated Allectus near *Calleva* (Silchester). Constantius sailed up the Thames and arrived in time to save *Londinium* from being sacked by rebels from the defeated army of Allectus, restoring Britain to the Empire. In the early 4th century the Emperor Diocletian completely reformed the whole imperial system of government with a policy of devolution which was to alter the political status of *Londinium*, and this move perhaps coincided with the demolition of the forum and basilica complex. Britain was sub-divided into four smaller provinces. Despite this division and the political upheavals, *Londinium* retained its place as capital of the new province of *Maxima Caesariensis*. Following Constantine the Great's conversion to Christianity in the early 4th century, the Christians became powerful in Britain. Records show that a Bishop of *Londinium* went to the Council of Arles in AD 314. The Mithraic temple was attacked on several occasions, and sculptures were broken up and discarded. It seems that the Mithraists forestalled one of the Christian attacks by burying the more precious marble sculptures. The temple was to continue in use for pagan (but not necessarily Mithraic) worship for a few more years until the building finally fell into ruin in the mid-4th century.

Political unrest throughout the Empire was affecting Britain. In AD 350 the usurpers Magnentius and Decentius withdrew troops from Britain but failed in their attempt to win the Empire. When Hadrian's Wall was overrun in AD 367, the Emperor Valentinian sent his general, Theodosius, to regain control of Britain. The little archaeological evidence that there is for *Londinium* at this period shows that the city's defences were improved and strengthened, with a riverside wall completing the defensive circuit. For the last half of the 4th century, the main occupied areas in the city were to the east of the Walbrook. Some areas of the city had not been inhabited since the late 2nd and 3rd centuries.

In AD 410 the Emperor Honorius refused to defend Britain further. Britain was now independent from the rest of the Empire and British towns were forced to recruit mercenaries in order to protect themselves against the threat of increased attacks from Saxon and German tribes. Finds of military buckles and belt sets in burials in the extramural cemeteries testify to a military presence in late-Roman *Londinium*. Such adornments, made in the Rhineland area, would have been worn by government officials and army officers, and may indicate that German mercenaries were employed by the army. Nevertheless, life seems to have continued in much the same way as before for some years. Excavations revealed that a fine 3rd-century house in Lower Thames Street continued in use into the 5th century. A fragment of pottery amphora of that date, imported from the eastern Mediterranean, was found there, indicating that supplies were still reaching the city from some distance. However, when the building slowly fell into disrepair and was eventually demolished in the late 5th century nothing was built to replace it. The Romans had left *Londinium* to its fate but they left strong foundations and traditions that were to leave a lasting impression on the later city.

Above Gold medallion from Arras, France, showing the city of Londinium *welcoming Constantius Chlorus who rescued the city from plunder in* AD 296.

Map below *The extent of Roman London at its peak is contrasted to the size of the city in 1820.*

EXTENT OF LONDON 2ND CENTURY AD

LONDON 1820

✝ St. Paul's

✝ Westminster Abbey

LONDON 1820

THE GROWTH OF LONDINIUM

Road systems, quickly established after the Roman conquest, facilitated the movement of troops, and many road alignments laid out by the Romans are still in use today. Watling Street, leading from the Kent coast, presumably crossed the Thames in the Lambeth/Westminster area, prior to *Londinium*'s foundation, and continued its route to *Verulamium* (St. Alban's) and the north-west. With the building of the bridge across the Thames, most traffic would have passed through *Londinium*, with roads leaving the city for Silchester, York, Colchester and Chichester.

An essential part of the officially organised communication system were posting stations (*mansiones*) along the road network, at set distances apart, where fresh horses were available for the imperial post and inns were provided for refreshments. Several known Romano-British centres near to *Londinium* originated as *mansiones*: Ewell on Stane Street; Crayford and Brockley Hill on Watling Street; and Staines on the Silchester road.

For the ordinary farmer in the countryside around *Londinium* the Roman conquest probably meant little more than exchanging a native, tribal landlord for a foreign Roman tax official. The nearest known villas, or more elaborate farm-houses, lie mainly to the south of *Londinium*, for example at Beddington, Keston and Orpington, and further still in the Darenth valley and beyond Ewell.

Early *Londinium* mainly occupied the area around modern Lombard Street and Gracechurch Street. The majority of the early buildings were constructed of timber, clay and mud brick with thatched roofs, the more important buildings were tiled. Boudicca's total destruction of the town in AD 60 left layers of burnt red clay which are all that remain of the clay and timber houses and give clear evidence of *Londinium*'s size. In about AD 120 *Londinium* was again destroyed by fire. The burnt debris shows that the area of habitation had grown considerably over the intervening years, spreading to the south-east and west of the Walbrook stream with major public buildings (as befitted the capital of the province) now in place. *Londinium* was rebuilt once again and continued to prosper. New waterfronts were constructed, shops were rebuilt and temples built and repaired.

The nature of *Londinium* in the 3rd century, however, was changing. The city wall provided a defensive barrier; large town houses replaced areas of local industry; some areas of the town were deliberately demolished and the population was decreasing. Archaeological evidence shows that in the 4th century the main area of occupation was

The London area in the Roman period: *The natural advantages of the surrounding area helped to determine Londinium's position. The Thames was tidal in the city area and being navigable it enabled ships to travel direct from the heart of Europe via the Rhine. After the conquest in AD 43 the Romans quickly established the road network to allow the army and merchants alike fast passage. Towns grew up along the roadsides either originating as official resting posts or at river-crossing points. Other settlements centred on areas where raw products such as clay for pottery were readily available or where shrines or temple worship were established. Much of the countryside around Londinium was thickly wooded, a useful source of timber, and the rural areas probably altered little after the conquest, with many farming communities being subsumed in the Roman villa estates that clustered around Londinium.*

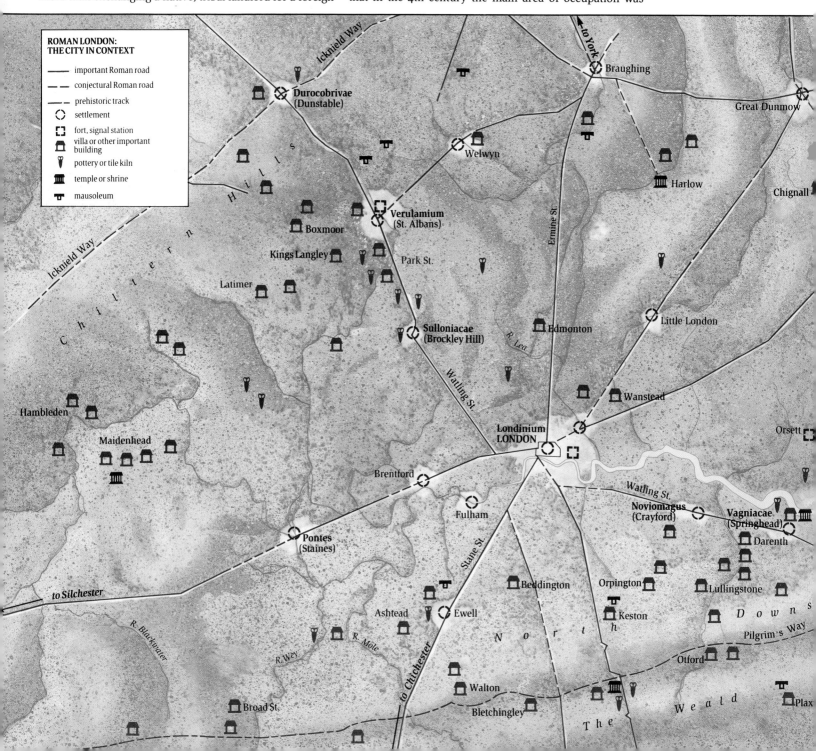

ROMAN LONDON:
THE CITY IN CONTEXT

— important Roman road
-- conjectural Roman road
-· prehistoric track
◯ settlement
▢ fort, signal station
🏠 villa or other important building
▼ pottery or tile kiln
🏛 temple or shrine
⊤ mausoleum

Icknield Way
to York
Braughing
Durocobrivae (Dunstable)
Great Dunmow
Welwyn
Harlow
Chignall
Chilterns Hills
Verulamium (St. Albans)
Boxmoor
Kings Langley
Park St.
Ermine St.
Icknield Way
Latimer
Sulloniacae (Brockley Hill)
R. Lea
Edmonton
Little London
Watling St.
Wanstead
Hambleden
Maidenhead
Orsett
Londinium LONDON
Brentford
Watling St.
Noviomagus (Crayford)
Vagniacae (Springhead)
Fulham
Darenth
Pontes (Staines)
Stane St.
Orpington
Lullingstone
to Silchester
Ashtead
Ewell
Keston
Downs
R. Blackwater
Beddington
Pilgrim's Way
R. Wey
R. Mole
North
Otford
Walton
Weald
Broad St.
Bletchley
The
Plax

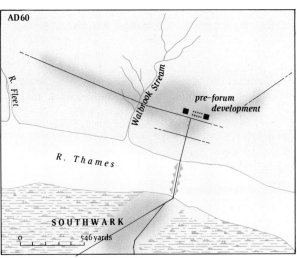

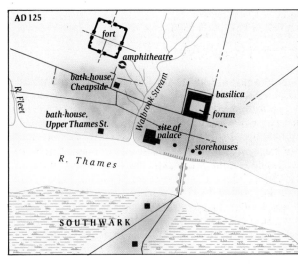

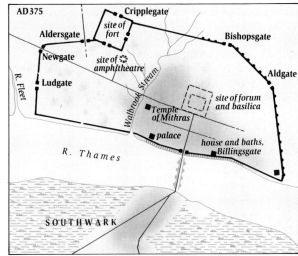

The growth of Londinium (maps right) *Most of the early town lay east of the Walbrook stream. The town was completely destroyed during the Boudiccan rebellion, AD 60/61. The rebuilt town became the thriving provincial capital in the late 1st century AD with many major public buildings. A wall was built around the city c. AD 200 to enclose a dwindling population. Additional fortifications had been added by AD 375 to protect the city from attack.*

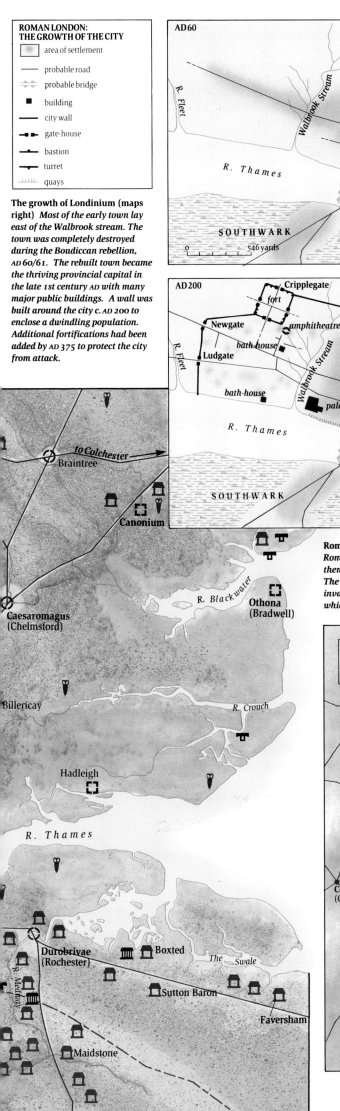

Roman invasion AD 43 *When the Romans invaded they brought with them their skill in building roads. The map (below) shows both their invasion routes and the network which began to take shape.*

centred on the east side of the Walbrook. People were still living in *Londinium* in Roman style, but houses were finally abandoned in the early 5th century, coinciding with the refusal of the Emperor Honorius in AD 410 to send more troops to the province to defend Britain.

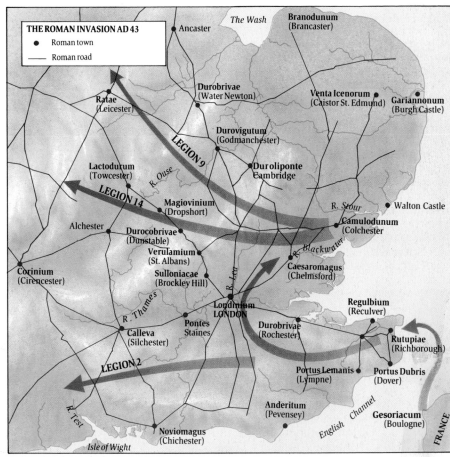

THE ROMAN INVASION AD 43
● Roman town
— Roman road

LONDINIUM: THE ROMAN CITY

In the last 30 years of the 1st century AD, *Londinium* became the capital of the province and was transformed into a Roman city. The timber-framed houses with their walls of wattle and daub and thatched roofs were upgraded and the walls given coverings of painted plaster. Public buildings were built of Kentish ragstone and flat tile-like bricks. Of *Londinium*'s public buildings, the great *basilica* and *forum* was the most important. Built about AD 120 to 125, and situated on the highest ground in the city, it would have dominated the skyline. The *basilica* served as town hall and law courts, indicating that *Londinium* must have now had a constitution for local government. The new *basilica* was more than 164 yards long, longer than any other north of the Alps. It had a great hall with a nave and northern aisle, with a double row of offices to one side. At the eastern end was a raised platform for judges. The *forum* enclosed a great central courtyard on three sides with the *basilica* on the fourth. Its three wings formed the business centre of the city, containing shops and offices, and the central courtyard provided an open market and assembly area. It seems that the whole complex was demolished in about AD 300.

An imposing residence was also sited at the junction of the Thames and Walbrook. Built between AD 80 and 100, it may have been the governor's headquarters. Areas of the building are now buried under Cannon Street Railway Station and recent excavations there have added more information. The great building covered a wide area of at least three acres and consisted of state rooms, reception halls and offices, as well as baths and living accommodation, all built around a central garden court. To the north was a great hall. The east wing consisted of small rooms and may have been the quarters for visiting officials or administrative offices, while other areas probably contained the residential quarters.

For their entertainment, Roman Londoners would have congregated at the amphitheatre in which animal baiting and possibly gladiatorial combat took place. It may also have served to stage plays, house state ceremonials and as a training ground for the army. It was situated to the south-east of the fort and its site has only recently been uncovered. It consisted of an oval arena of packed sand with an under-floor drainage system. A main entrance-way was discovered with the remains of two ante-rooms on either side suitable as waiting rooms for the participants or combatants in the arena events.

As befits the capital of any Roman province, sanitation and hygiene were given a high priority. The Britons were encouraged to adopt the pleasures of civilisation, including daily bathing. Public bath-houses in Upper Thames Street and Cheapside were built in the late 1st century and enlarged in the 2nd. The Cheapside bath-house was sited near to the Roman fort, and because of its small size, was perhaps intended for the military rather than the public. The baths in Upper Thames Street, however, were built on a grander scale, overlooking the Thames in a position where natural spring water fed the baths. In later Roman times, finer houses also had private bath-houses.

Religion in Roman Britain was a combination of Roman and native Celtic ideas, and cults from the east, such as Mithraism, were also popular. The only temple identified in London is dedicated to the mystery cult of Mithras, the god of heavenly light. Mithraism emphasised honesty, purity and courage and was favoured by soldiers, officials and merchants. The temple, on the east bank of the Walbrook, was built about AD 240 and had a central nave and side aisles. Fine marble sculptures of Mithras and other Roman gods and goddesses were found deliberately hidden. It seems that when Christianity finally overcame Mithraism in the 4th century the sculptures from the temple were buried to protect them from destruction.

NEWGATE

LUDGATE

religious precinct

River Fleet

River Thames

govern
pal

SOUTHWARK

GOVERNOR'S PALACE

ALDERSGATE

CRIPPLEGATE

fort

Cheapside Baths

amphitheatre

Walbrook Stream

Temple of Mithras

forum & basilica

BISHOPSGATE

cemetery

ALDGATE

CHEAPSIDE BATHS

COMMERCIAL LONDINIUM

In the wake of the Roman invasion, merchants and traders followed closely after the army. They exploited the fact that ocean-going ships could reach the city, and the Thames provided deep-water anchorage. The town's commercial vitality and growth after AD 60 depended on the construction of adequate port facilities along the river. In the mid 1st-century AD the riverbank lay to the north of modern Thames Street, and the waterfronts were built out into the river on reclaimed land. Their construction and their successive replacements throughout the Roman and later periods, have resulted in today's river lying about 100 yards or more to the south of its original shoreline. Excavations near Pudding Lane, downstream of the Roman bridge, have revealed a series of timber quays dating from the mid-1st–early 3rd centuries. Here goods were unloaded from ships and moved into rectangular storehouses at the back of the quay.

Trade flourished in Roman London, where a prosperous community of differing nationalities demanded the same goods as the rest of the Empire. Many of *Londinium*'s inhabitants would have been British or Gaulish in origin; others came from Germany, Greece and beyond. Although *Londinium* was politically and culturally Roman it must have been a constantly changing racial mixture. While the province was developing, imports were essential, since there were not the native craftsmen to produce the high-quality goods required. However, in time, British goods were also to flood into *Londinium* and the province gradually became self-sufficient. The cosmopolitan population of *Londinium* also had a taste for Mediterranean foods such as olives, olive oil, wine, grape juice, dates, figs and salted fish products, which were imported in large pottery containers (*amphorae*) to meet this demand. Glossy red pottery (samian), standard Roman tableware, was imported in great quantities from regions of France. Bronze tableware, glass and lamps came from Italy; bronze, pottery, glass, millstones and wine from Germany and fish sauce and olive oil from Spain.

The archaeological evidence for shops in *Londinium* is limited, restricted to surviving ground plans or occupational debris. Carbonized grain recovered from Pudding Lane and Fenchurch Street suggests a bakery or mill was sited there, and part of a donkey-driven mill found in Princes Street suggests that the milling of flour on a large-scale must have taken place there. Few traces have survived of the numerous merchants of Roman London, apart from the balances and steelyards which were probably used for weighing out foodstuffs in their shops.

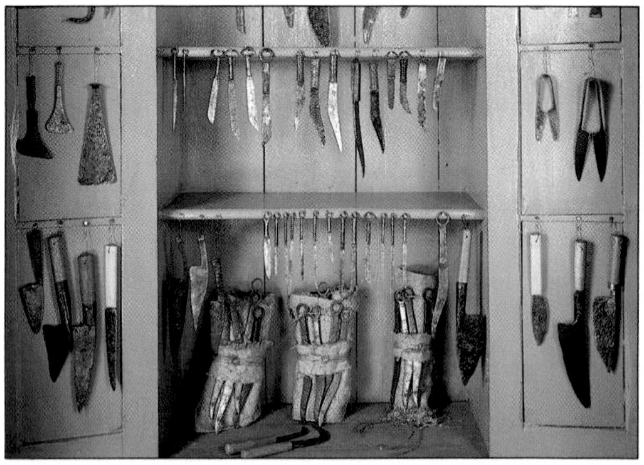

Londinium's ***commercial vitality and growth depended on the rapidly expanding port situated to the west of The Tower*** **(right)*. New quays were constantly being built out into the river, using varying methods of construction: the 3rd-century quays*** **(below right)** ***used small timbers in complicated box-like structures. The busy waterfronts received sea-going ships laden with imports from all over the empire*** **(map below)*. Amphorae were used as containers for olive oil, wine or fish products. The example*** **(below left)** ***comes from Southwark and bears an ink inscription testifying to the fish sauce's quality. Emeralds were imported from Egypt, and were used in this emerald and gold necklace found in Cannon Street. Lamps, such as the example*** **(below centre)** ***found on the bank of the Thames, were imported from Gaul. The bustling port supplied the town's needs through shops and workshops. Excavations in Newgate Street revealed two such buildings of 2nd-century date*** **(plan left)*, while this reconstruction of a cutler's stall*** **(left)*, gives a vivid impression of the appearance of a Roman shop. During the late 3rd century,*** **Londinium** ***produced minted coins, and this bronze*** **antoninianus** ***(see obverse and reverse*** **below)** ***of Allectus (AD 293–296) has the mint mark*** **ML** ***(Moneta Londinii).***

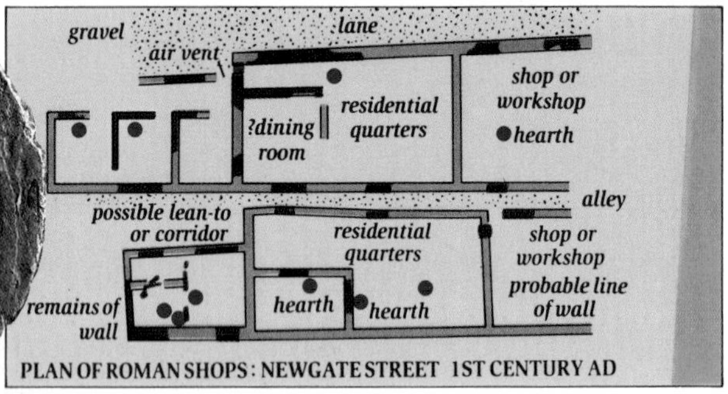

gravel
air vent
lane
?dining room
residential quarters
shop or workshop
hearth
possible lean-to or corridor
alley
residential quarters
shop or workshop
remains of wall
hearth
hearth
probable line of wall
PLAN OF ROMAN SHOPS: NEWGATE STREET 1ST CENTURY AD

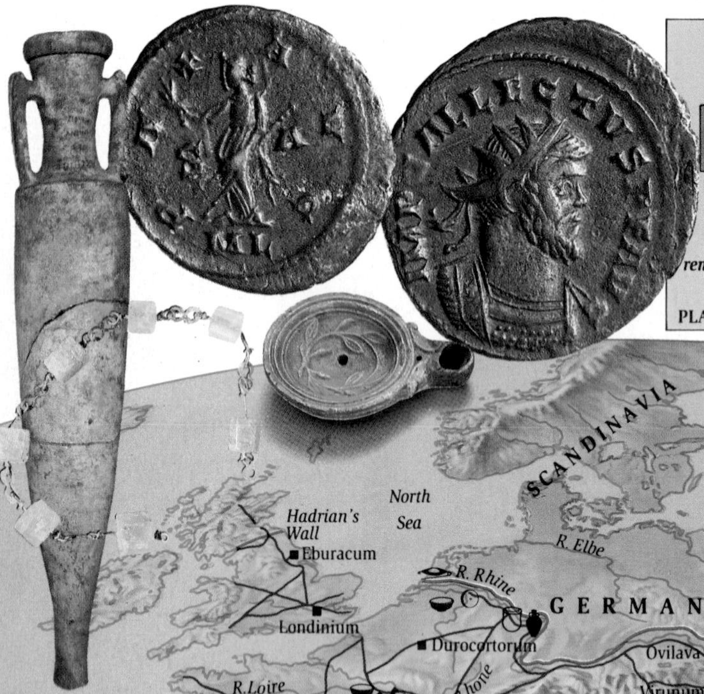

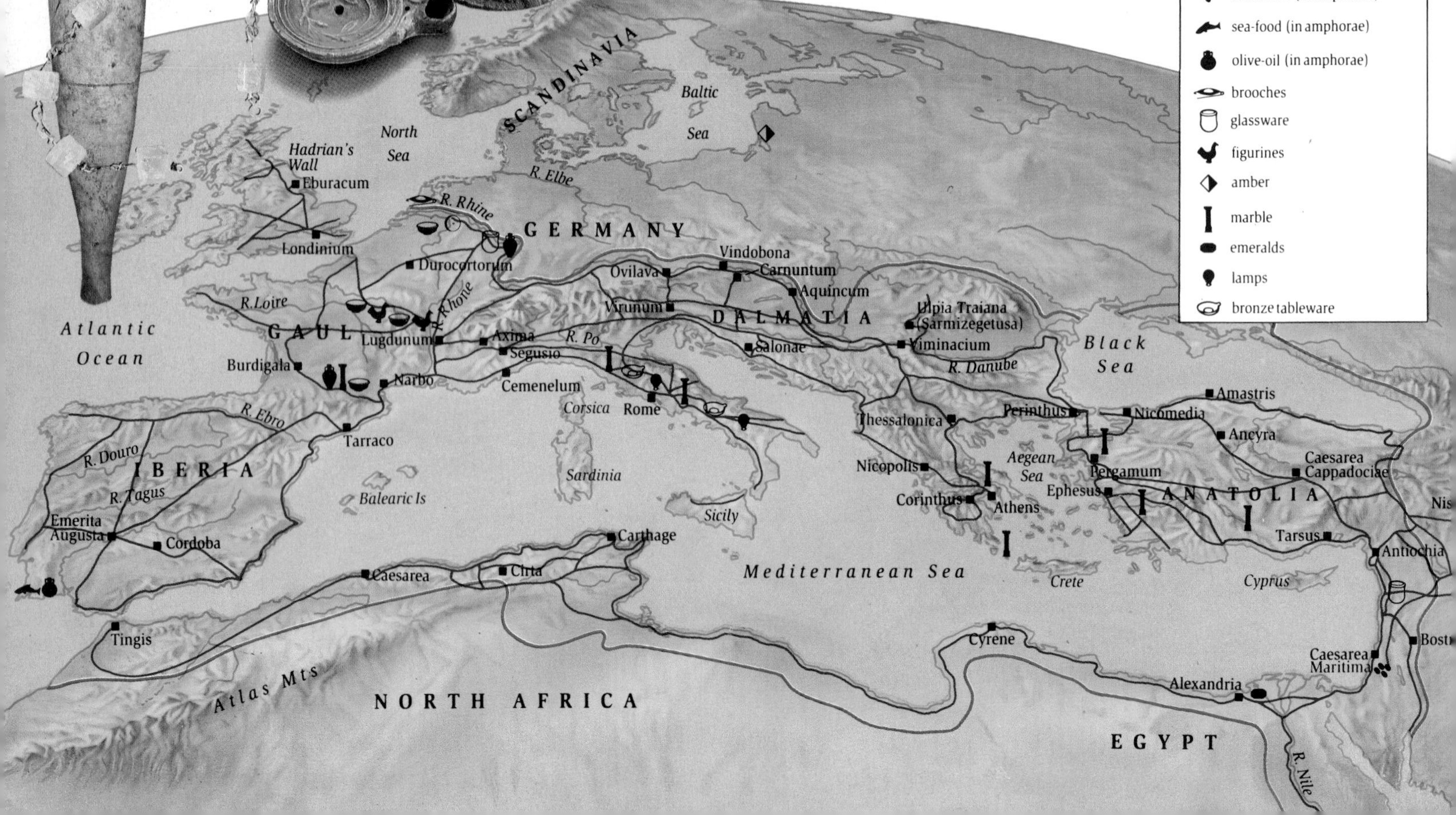

LONDON'S TRADE WITH THE EMPIRE
extent of Roman empire, 2nd century AD
Roman road
provincial capital
commodities imported to London:
wine (in amphorae)
pottery
millstones
dried fruit (in amphorae)
sea-food (in amphorae)
olive-oil (in amphorae)
brooches
glassware
figurines
amber
marble
emeralds
lamps
bronze tableware
SCANDINAVIA
Baltic Sea
North Sea
Hadrian's Wall
Eburacum
R. Elbe
R. Rhine
GERMANY
Londinium
Durocortorum
Vindobona
Carnuntum
Ovilava
Aquincum
Virunum
DALMATIA
Ulpia Traiana (Sarmizegetusa)
R. Loire
R. Rhone
Atlantic Ocean
GAUL
Lugdunum
Axima
Segusio
R. Po
Salonae
Viminacium
R. Danube
Black Sea
Burdigala
Narbo
Cemenelum
Corsica
Rome
Amastris
R. Ebro
Tarraco
Thessalonica
Perinthus
Nicomedia
Ancyra
R. Douro
IBERIA
Sardinia
Nicopolis
Aegean Sea
Pergamum
Caesarea Cappadociae
R. Tagus
Balearic Is
Corinthus
Athens
Ephesus
ANATOLIA
Emerita Augusta
Cordoba
Sicily
Tarsus
Antiochia
Carthage
Caesarea
Cirta
Mediterranean Sea
Crete
Cyprus
Tingis
Cyrene
Bostra
Caesarea Maritima
Atlas Mts
NORTH AFRICA
Alexandria
EGYPT
R. Nile

multiple-box
framework
of jointed beams

horizontally-
laid timbers

CROSS-SECTION OF 3RD-CENTURY QUAY AT
CUSTOM HOUSE SHOWING METHOD OF CONSTRUCTION

THE ROMAN RIVERSIDE

LONDINIUM'S DEFENCES

The early 2nd-century fort was built away from the centre of the town. It covered an area of nearly 12 acres and was constructed on the standard Roman rectangular pattern, with rounded corners and internal towers. There were also intermediate towers containing stairs to the rampart walk. The fort's function was probably not primarily defensive but to provide a suitable barracks for troops stationed in the capital of the province. Soldiers, who acted as military escorts and performed guard and ceremonial duties, would have been based there. Members of the governor's staff, probably seconded from the legions based in Britain, would have assumed the duties of civil servants. A tombstone of a legionary soldier depicts him in military attire, but also carrying a case of writing tablets, to denote his clerical function.

Up until AD 200, the town boundaries were marked by a bank and ditch, but then

Londinium was enclosed by a great wall on the landward side. The wall stretched over two miles from modern Blackfriars to the Tower of London. It was probably at least twenty feet high when built and was nine feet thick on ground level narrowing to eight feet above the sandstone plinth. The wall was made of ragstone quarried at Maidstone and brought, via the River Medway, up the Thames by boat. The outer faces of the city wall, like the fort, were built of squared blocks of ragstone, skilfully cut to shape and the intervening space was filled with irregular lumps of ragstone and mortar. At yard intervals two or three courses of tiles were laid across the wall to provide stability.

The outside of the wall was defended by a ditch about four yards away. The earth dug from both the ditch and the wall's foundation trench formed a great reinforcing bank against the inner face of the wall. Where the existing Roman roads left the city, gate-houses were built, later to be called Aldgate, Bishopsgate, Newgate, Ludgate and in the north of the fort, Cripplegate. The west gate was blocked up in the

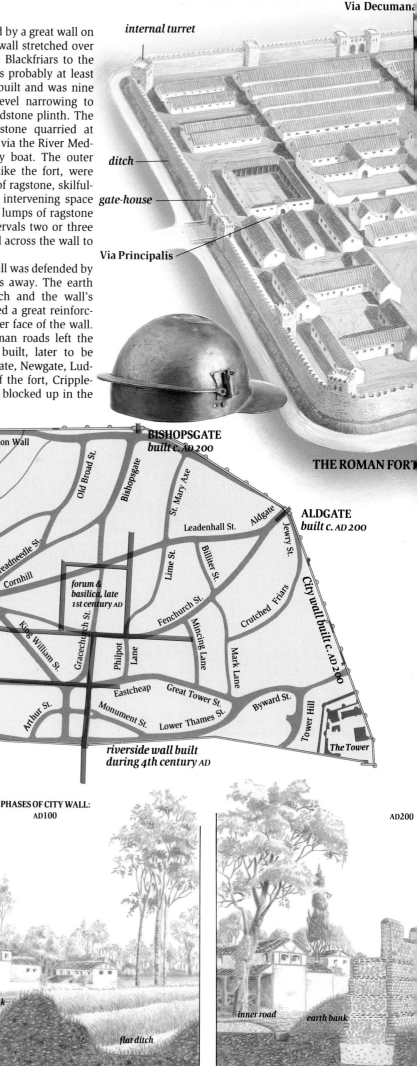

Via Decumana

internal turret

ditch

gate-house

Via Principalis

THE ROMAN FORT

The City wall *was built to a uniform standard in c. AD 200 (below). More than a million ragstone blocks were used for the two faces of the wall, with horizontal tile courses at regular intervals.*

city wall built c. AD 200

fort built *c. AD 120*

London Wall

London Wall

Moorgate

Old Broad St.

Bishopsgate

St. Mary Axe

BISHOPSGATE *built c. AD 200*

Coleman St.

Gresham St.

Lothbury

Leadenhall St.

Aldgate

Jewry St.

ALDGATE *built c. AD 200*

ALDERSGATE *late Roman addition*

King Edward St.

Angel St.

St. Martin's Le Grand

Wood St.

King St.

Princes St.

Threadneedle St.

Cornhill

Lime St.

Billiter St.

NEWGATE *built c. AD 200*

Warwick Lane

St Paul's Cathedral

Cheapside

forum & basilica, late 1st century AD

Gracechurch St.

Fenchurch St.

Mincing Lane

Crutched Friars

Mark Lane

City wall built c. AD 200

Watling St.

Cannon St.

Walbrook

Walbrook

King William St.

Philpot Lane

LUDGATE *built c. AD 200*

St Paul's Church Yard

Godliman St.

Queen Victoria St.

Queen St.

Arthur St.

Monument St.

Eastcheap

Great Tower St.

Lower Thames St.

Byward St.

Tower Hill

The Tower

riverside wall built during 4th century AD

THE DEFENCES OF ROMAN LONDON

— wall
— bastion
— gateway
— turret

— presumed
— recorded
— surviving
— Roman road
— modern road

PHASES OF CITY WALL: AD100

earth bank

flat ditch

AD200

inner road

earth bank

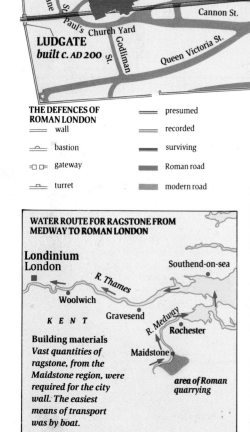

WATER ROUTE FOR RAGSTONE FROM MEDWAY TO ROMAN LONDON

Londinium London

Southend-on-sea

R. Thames

Woolwich

Gravesend

K E N T

R. Medway

Rochester

Maidstone

area of Roman quarrying

Building materials *Vast quantities of ragstone, from the Maidstone region, were required for the city wall. The easiest means of transport was by boat.*

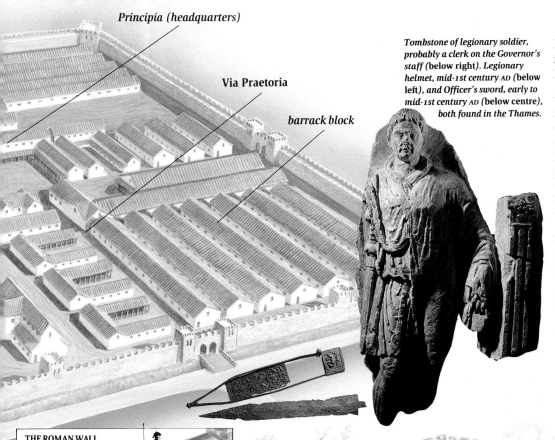

Principia (headquarters)

Via Praetoria

barrack block

Tombstone of legionary soldier, probably a clerk on the Governor's staff (below right). Legionary helmet, mid-1st century AD (below left), and Officer's sword, early to mid-1st century AD (below centre), both found in the Thames.

later Roman period and a new gate, Aldersgate, was inserted just beyond the fort's perimeter to allow traffic to by-pass it.

As *Londinium's* fortunes became more troubled in the later 4th century some 20 semi-circular bastions were added to the eastern side of the city wall as platforms for catapult machines. These towers were probably about 26 feet high and regularly spaced. Their outer walls were constructed like the wall itself and the infill consisted of re-used stonework from neighbouring Roman cemeteries.

In the late 3rd and 4th centuries a riverside wall finally completed the defensive circuit. Excavations near Blackfriars revealed varying styles of wall construction, some taking into account the waterlogged position, with a wooden pile and chalk-raft foundation. Elsewhere large ragstone blocks, including re-used sculptures, were simply rammed into clay. This lack of consistency in the riverside wall's construction suggests it was built at different times in the 4th century with varying degrees of urgency.

THE LATE ROMAN GATE AT ALDERSGATE

THE ROMAN WALL

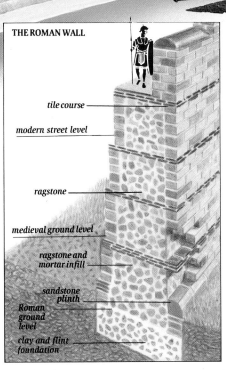

tile course

modern street level

ragstone

medieval ground level

ragstone and mortar infill

sandstone plinth

Roman ground level

clay and flint foundation

Londinium's Defences The city wall, built c. AD 200, was further strengthened by eastern bastions and the riverside wall in the later 3rd and 4th centuries AD when Londinium was increasingly subject to attack.

tower for catapult machine

AD375

...-shaped ditch

ditch, refilled and repositioned beyond tower

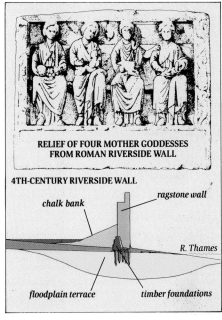

RELIEF OF FOUR MOTHER GODDESSES FROM ROMAN RIVERSIDE WALL

4TH-CENTURY RIVERSIDE WALL

chalk bank

ragstone wall

R. Thames

floodplain terrace

timber foundations

CHAPTER 3
SAXON AND NORMAN LONDON

THE CENTURY IN which Roman Britain was transformed into Anglo-Saxon England and Roman *Londinium* died, remains the most puzzling and hotly-debated period in our history. After AD 410, when the Emperor Honorius instructed the provincial authorities of Britain to look to their own defences, Britain was effectively outside the Roman empire; yet it remained within the Roman sphere of influence. When in 429 Germanus, Bishop of Auxerre, travelled from Gaul to Britain, he found people apparently living a recognisably 'Roman' lifestyle in the south-east of the country, and a flourishing Church. Raids by Picts and Saxons were a problem, but were meeting British resistance. On a second visit a few years later he seems to have found little change.

However, the middle of the 5th century saw dramatic events. In about 446 the Britons sent a vain plea to Rome for support against barbarian invaders. By the end of the 5th century most of the east of Britain was under Anglo-Saxon control. The last evidence for Roman *Londinium* is of a shrunken settlement such as Germanus might have seen in the early 5th century. One site, the Roman bathhouse near Billingsgate, has produced the earliest trace of Saxon presence in the city: a Saxon brooch dropped among the ruins of the abandoned building by a scavenger or squatter. By the end of the 5th century London must have been largely deserted. Whether the inhabitants were British or Saxon is of little account; the place no longer functioned as an urban centre. To the Anglo-Saxon settlers it was irrelevant; as self-sufficient farmers belonging to small family or tribal groupings, they had no use for towns. Few actual habitation sites have been excavated in the London area; the pattern of isolated farmsteads and small village communities seen elsewhere in the country seems to be repeated, some growing up at, or adjacent to, Roman sites, some on new sites – though perhaps farming the old fields.

By the end of the 6th century, the London area was part of the kingdom of the East Saxons which included the later counties of Essex and Middlesex and much, probably originally all, of Hertfordshire, stretching as far as the Chiltern Hills. Even part or all of Surrey ('*Sutherge*', or the 'southern district'), south of the Thames, was at one stage part of the kingdom. No 'princely' burials like those of Sutton Hoo or Taplow (in Buckinghamshire) can confidently be assigned to East Saxon rulers, and they often appear to have been in some sense subordinate to more powerful neighbours. Thus, King Saeberht of Essex was nephew of Ethelbert of Kent and accepted his authority when he received Christian missionaries in 604. He allowed Ethelbert to have a church, dedicated to St.Paul, erected in London as the cathedral of a bishop for the East Saxon people. Other early churches were founded in Roman towns, as at Canterbury and Rochester, and it is likely that it was the former existence of *Londinium*, rather than the presence of any major centre of Saxon population, which ensured the establishment of the church in London. East Saxon conversion to Christianity was, however, only superficial. On the death of Saeberht his people reverted to paganism, and it was the middle of the century before Christianity once more took hold. In the meantime London's function as an '*emporium*' was developing; as early as the 670s a royal charter referred in passing to the 'port of London where

Saxon settlers moved up the Thames and the valleys leading off it, meeting little resistance from former British landowners. Mitcham cemetery (page 40) close to the River Wandle is evidence of this movement.

For a detailed map of the Saxon kingdoms, please see page 40.

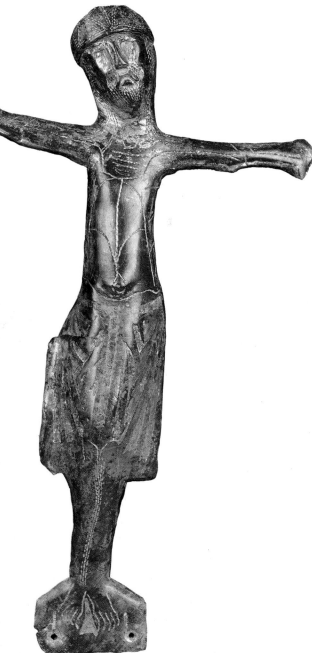

Below *A 13th-century crucifix figure, found at the Cluniac Priory of Bermondsey, founded 1089. It is gilded bronze with enamel decoration.*

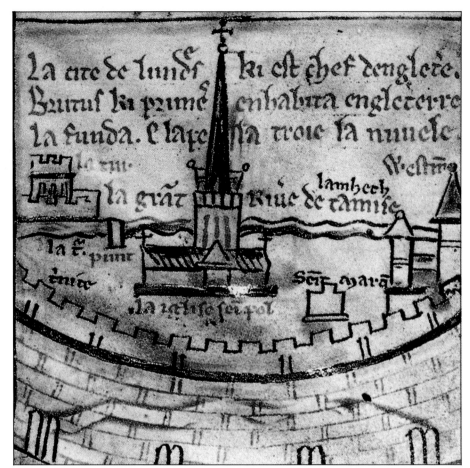

Left *The first known view of London, a section from a road map (from London to Rome) drawn by the monk, Matthew Paris, in about 1252. The view shows three recognisable monuments: on the left, the Tower, St. Paul's in the centre, and to the right, Westminster Abbey. The wall, with its gates, is visible in the foreground.*

ships land'. Since 1985, a growing number of sites excavated north of the Strand, between Trafalgar Square and Aldwych, have produced traces of a flourishing settlement of the 7th to early-9th centuries, and work near Charing Cross in 1988 revealed a river embankment which may have served for ships to beach for unloading. In contrast, the area to the east within the Roman city walls remained largely empty, apart from the cathedral and, perhaps, an as yet unlocated royal palace. In contemporary documents, London is referred to as *Lundenwic*; the Old English ending *-wic*, found also in names like Ipswich and Sandwich, usually denotes a port or trading town. International trade in northern Europe depended on a network of such ports, subject to royal control and taxation. Most trade would have been local – the buying and selling of agricultural produce. Yet finds of imported pottery on sites in the Strand area indicate more extensive trade links. Silver coins minted in London found their way to the Continent; the port of London probably handled export of English woollen cloth, famous throughout Europe.

At the end of the 8th century Vikings from Scandinavia began to raid the towns and monasteries of northern Europe. English chroniclers record 'great slaughter' in London in 842, and a force of 350 Viking ships stormed London and Canterbury in 851. Such direct attacks, together with the total disruption of trade, sealed the fate of *Lundenwic*. In 865 a 'great army' of Danes gathered in East Anglia; in a series of summer campaigns they destroyed the power of the Anglo-Saxon kingdoms in the east and north of England. In 871-72 they made London their winter quarters. Effectively, the only Anglo-Saxon kingdom to survive was the Wessex of King Alfred. Alfred was campaigning in the London area in the early 870s, but hard fighting was to follow before he occupied London, probably a town in ruins. In 886 Alfred formally re-established London, the walled Roman city, as a fortified town – a *burgh*.

The site on the Strand was abandoned and reverted to fields, remembered only as Aldwych, the 'old wic', and marked by old churches stretching from St. Bride's westwards to St. Martin in the Fields. Alfred placed London, together with what was left of old Mercia, in the charge of his son-in-law, the Mercian Ealdorman Ethelred. A programme for the defence and resettlement of the town was put in hand. Grids of new streets were laid out in the centre of the Roman walled area, between Cheapside and the Thames to the west of the

Alfred ceded to the Danes the part of England later known as the Danelaw (see map, page 41), which comprised most of the country east and north of London. He was accepted as king by all the English not under Danish rule.

Walbrook and around Cannon Street to the east. Massive ruins of Roman buildings must have remained a feature of the townscape. The great Roman amphitheatre to the north of Cheapside survived long enough to influence the layout of later medieval buildings. The presence on this site of the medieval Guildhall, centre of the city's government, inspires the suggestion that the open-air *folkmoot*, at which the business of the Anglo-Saxon town was discussed, had once met on the banked seats of the amphitheatre.

Alfred and his successors established a defensive network of forts and walled towns; an early list of such *burghs* includes Southwark (the 'work of the men of Surrey'), which with London itself could guard the Thames and the Thames crossing. The progress made by Alfred's successors in winning back England from Danish rule culminated in the reign of Athelstan (925-39), who claimed on his coins the title 'King of All Britain'. Athelstan's law-codes reflect the importance of London, assigning to it more royal moneyers (mint-masters) than any other town. Yet London did not yet have the massive pre-eminence over other towns it achieved later, and was in no sense the capital city. Royal councils met in London; equally they could meet in any other royal town or convenient centre.

A new series of attacks from Scandinavia began towards the end of the 10th century. In 994, a joint force led by Swein Forkbeard, son of the king of Denmark, and the Norwegian Olaf Tryggvason attacked London; London held out, and the attackers 'suffered greater loss and injury than they ever expected' – according to the English chronicler. For the next twenty years London was the centre of resistance – 'praise God, still [in 1009] it stands safe and sound' – but also contributed vast sums in silver to the *danegeld* collected to buy off the attackers. By the end of 1016 the English King Ethelred and his eldest son Edmund were dead and Cnut, son of Swein Forkbeard, was accepted as king of all England. When his son died in 1042 there was no obvious Danish successor; by popular decision the crown was offered to Edward, surviving son of the English King Ethelred.

Edward the Confessor died and was buried at Westminster in January 1066 (see page 42). **Below** *A scene from the 11th-century 231-foot Bayeux Tapestry shows Edward greeting Harold, Earl of Wessex, inside the royal hall at Westminster.*

Monasteries were established in the late 11th and early 12th centuries on open land on the fringes of the city (Aldgate and Smithfield) and further out (Southwark, Bermondsey Abbey). Nunneries were founded in Clerkenwell and Shoreditch in the early 12th century. Special hospitals or refuges (for example St. Giles', Bloomsbury) were founded on the roads leading out of London, and sheltered sufferers from leprosy who had been driven out of the town (see map, page 43).

A reconstruction of the Tower of London can be found on page 139.

The period from Alfred's resettlement of London in 886 to the death in 1066 of Edward (nicknamed 'the Confessor') saw major developments in the city. Infilling began in the blocks between the original planned streets; lanes subdivided them, providing access to buildings erected in the formerly open backlands. New streets stretched north from the market place known as 'West Cheap', now Cheapside. Excavation has revealed some of the buildings of the late Saxon town and the activities of its people. They were skilled metal-workers; cloth, leather and bonework were being made and decorated. A new bridge was built to replace the lost Roman river crossing. New quays were erected and gradually enlarged along the riverside at Billingsgate and westward. Much of the trade was local; tolls were levied at Billingsgate on boats carrying timber, fish, chickens, eggs and dairy produce, as well as on larger ships from Normandy, France, Flanders and Germany. Yet from the reign of Cnut, if not before, London seems to have had links with the great northern network of trade-routes opened up by enterprising Scandinavian merchants and colonists.

Many of London's multitude of small parish churches had their origin in the late Saxon period, but it was the favour shown by Edward towards a monastery outside the walls which was to dramatically affect London's future geography. A man of great piety, he donated much of his royal income to the building of an abbey dedicated to St. Peter on an island in the marshes of the River Tyburn – Westminster. Next to his magnificent new church, Edward established a royal hall. Future London was to grow around two locations: the mercantile and industrial centre in the 'City', and the centre of royal law and administration at Westminster. It was at Westminster that Edward died and was buried in January 1066; at Westminster too both Harold of Wessex and William of Normandy were crowned in turn as King of England. Faced with 'the restlessness of its large and fierce population' William built castles to control London; within fifteen years of the Norman Conquest work had started on a fortified palace of stone, the White Tower, which was to develop into the Tower of London. Westminster remained the royal centre; William II had a fine new stone palace built, still standing today as Westminster Hall. Many great buildings of stone were new to the Norman city, but it was massive religious buildings which quickly came to dominate the townscape. The Anglo-Saxon minster church of St. Paul was destroyed by fire in 1087; work began on a magnificent replacement in the new Romanesque style, which was to survive, almost intact, until the Great Fire of 1666.

See reconstruction of St. Paul's, page 43.

The building boom reflected a growth in London's population; the largest town in the kingdom, it may have had 10,000 or even 15,000 inhabitants by 1100, and over 30,000 by 1200, many of whom were newcomers to England. Merchants from the Norman towns of Rouen and Caen settled in London to take advantage of the new market that the Conquest opened up. By 1130 a Jewish community, probably originally immigrants from Rouen, was well established in London, centred on 'Jews' Street' (Old Jewry). Providing essential finance for the activities of kings, nobles and merchants, they nevertheless attracted the envy and suspicion of their neighbours; the London Jewry was burnt down and 30 Jews killed in riots following the coronation of Richard I in 1189.

'Old' London Bridge, begun in 1176, stretched across the river on 19 solid stone piers. Vital to London's communications with the rest of the kingdom, it was an object of wonder to foreign visitors. See page 162 for a reconstruction of London Bridge.

Local administration devolved from the general *folkmoot* to a more select group of leading citizens, dignified by the Old English title 'aldermen', responsible for the 'wards' into which the city was divided. In 1191, Richard I formally recognised the commune of London; henceforth London was a corporate entity which could negotiate directly with the king. By the beginning of the 13th century the future pattern of London's development was well-established and most of its basic institutions already existed. The Londoner William Fitz Stephen, writing in 1173, had no doubt that of all the world London was 'the most noble city'.

Map below The extent of Norman London at its peak is contrasted to the size of the city in 1820.

EXTENT OF LONDON 1200

LONDON 1820

+ St. Paul's

+ Westminster Abbey

LONDON 1820

LUNDENWIC

So far archaeology has revealed little trace of Anglo-Saxon settlement in the inner London area in the century or so after the collapse of the Roman province of Britain. Isolated Saxon huts have been identified in the west of London, and in 1990 a number of buildings were revealed in excavations at Hammersmith, only six miles from Roman *Londinium*. Pagan burials are known further afield; a group of wealthy cemeteries in the Croydon area suggests a concentration of settlement there. But the immediate vicinity of the abandoned Roman city seems to have been largely avoided.

By the end of the 6th century the London area was part of the kingdom of the East Saxons. When Christian missionaries arrived in England in 597, they established the first major churches in old Roman towns: Canterbury, Rochester and London, the seat of the bishop for the East Saxons. Later in the 7th century churches were built elsewhere in the kingdom, at Tilbury and Bradwell-on-Sea, while a nunnery at Barking and a monastery at Chertsey in Surrey (*Suthregeona*), were founded by Erkenwald, who was Bishop of London after 675.

Only in 1985 did excavations in Covent Garden confirm that the 8th-century Saxon trading town, *Lundenwic*, did not lie near St. Paul's Cathedral, within the old Roman city walls, but to the west, along the Strand. Further traces of wooden buildings, rubbish pits, gravelled yards and industry have since come to light in this area. Churches like St Martin-in-the-Fields may have been founded at this time, and the name Aldwych, 'the old port', preserves the memory of the settlement.

Lundenwic and other royal 'emporia' were river-ports handling trade among the Anglo-Saxon kingdoms and with Merovingian Europe, where there were major ports at *Quentovic* (a 'lost' site recently rediscovered south of Boulogne) and Dorestad on the Rhine. Fragments of pottery from northern France and pottery and millstones from the Rhineland found in London demonstrate aspects of this trade; excavations near Charing Cross have revealed remains of a reinforced river embankment on which ships could be beached for unloading.

The port of *Lundenwic* was abandoned following Viking raids in the 840s and 850s. After halting the advance of the Danish invaders Alfred, King of Wessex and effectively king of all that remained of 'English' England, re-established London in 886 and encouraged settlement within the Roman city walls. A system of 'burghs', fortified towns, was established to resist further Danish attacks.

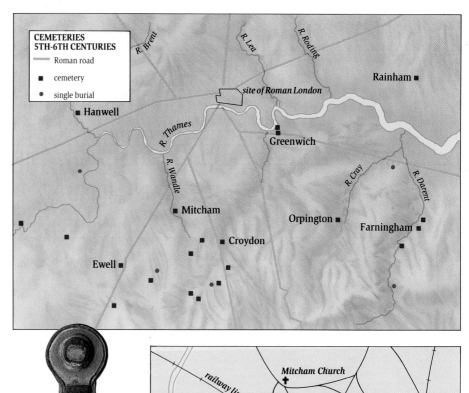

CEMETERIES 5TH-6TH CENTURIES
— Roman road
■ cemetery
● single burial

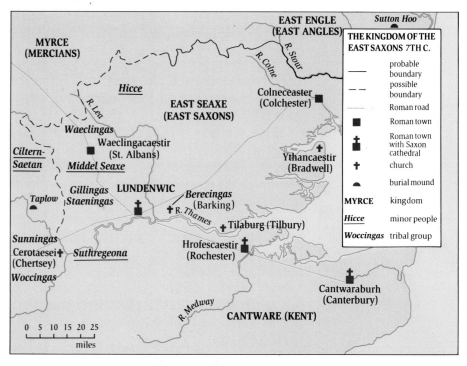

Anglo-Saxon settlement *5th- and 6th-century cemeteries reflect the spread of settlement in the London area (map top), apparently avoiding the vicinity of the old Roman city. The silver-gilt brooch (left) came from burials discovered near Mitcham station, Surrey, between 1888 and 1922 (map above). Few domestic sites from this period have been excavated. At Harmondsworth (lower left), Saxon huts stood in a landscape of Roman and prehistoric fields.*

The East Saxons *The extent of the 7th-century kingdom was largely that of the medieval bishopric of London – Essex, Middlesex and part of Hertfordshire – but included land in the north-west later lost to Mercia (map below). Within the kingdom place-names reveal other smaller groups, perhaps once independent, such as the Gillingas (Ealing), the Berecingas (Barking), the Staeningas (around Staines) and the Waeclingas (around St Albans).*

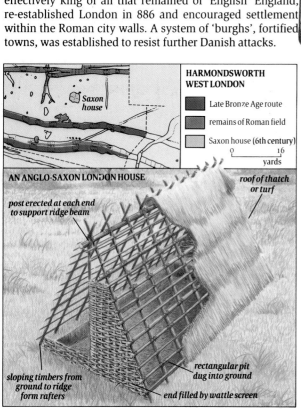

HARMONDSWORTH WEST LONDON
■ Late Bronze Age route
■ remains of Roman field
■ Saxon house (6th century)
0   16
yards

AN ANGLO-SAXON LONDON HOUSE
post erected at each end to support ridge beam
roof of thatch or turf
sloping timbers from ground to ridge form rafters
rectangular pit dug into ground
end filled by wattle screen

THE KINGDOM OF THE EAST SAXONS 7TH C.
— probable boundary
-- possible boundary
— Roman road
■ Roman town
✚ Roman town with Saxon cathedral
✝ church
⌂ burial mound
MYRCE kingdom
Hicce minor people
Woccingas tribal group

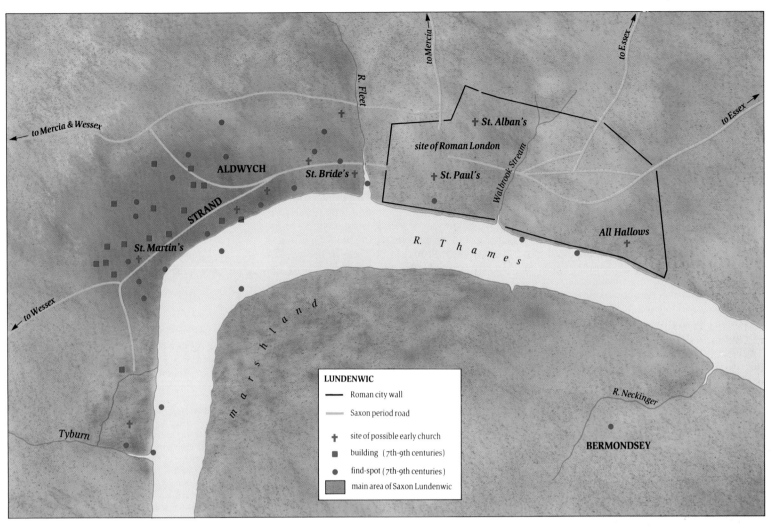

LUNDENWIC

to Mercia

to Essex

to Essex

R. Fleet

† St. Alban's

site of Roman London

ALDWYCH

St. Bride's †

† St. Paul's

Walbrook Stream

to Mercia & Wessex

STRAND

R. Thames

All Hallows †

St. Martin's †

to Wessex

m a r s h l a n d

R. Neckinger

Tyburn †

BERMONDSEY

LUNDENWIC

— Roman city wall

— Saxon period road

† site of possible early church

■ building (7th-9th centuries)

● find-spot (7th-9th centuries)

▨ main area of Saxon Lundenwic

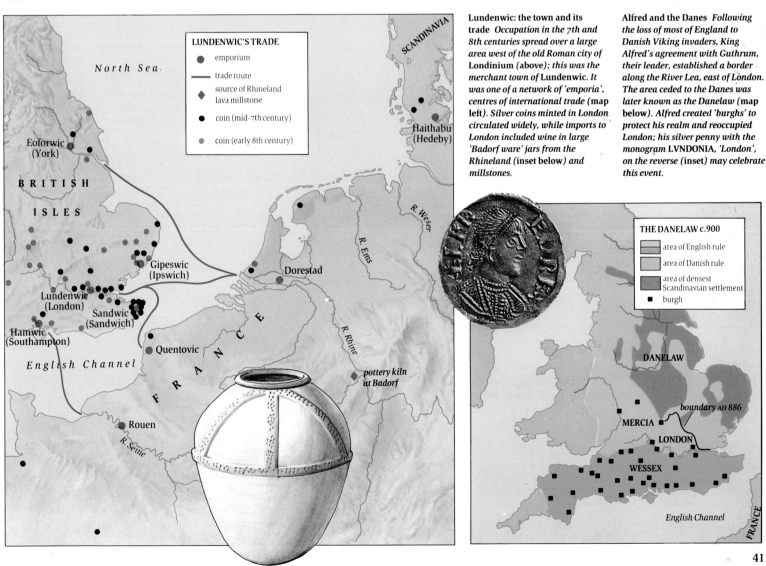

LUNDENWIC'S TRADE

● emporium

— trade route

◆ source of Rhineland lava millstone

● coin (mid-7th century)

● coin (early 8th century)

North Sea

SCANDINAVIA

Haithabu (Hedeby)

Eoforwic (York)

B R I T I S H

I S L E S

R. Weser

R. Ems

Gipeswic (Ipswich)

Dorestad

Lundenwic (London)

Sandwic (Sandwich)

F R A N C E

R. Rhine

Hamwic (Southampton)

Quentovic

English Channel

pottery kiln at Badorf

Rouen

R. Seine

Lundenwic: the town and its trade *Occupation in the 7th and 8th centuries spread over a large area west of the old Roman city of Londinium (above); this was the merchant town of Lundenwic. It was one of a network of 'emporia', centres of international trade (map left). Silver coins minted in London circulated widely, while imports to London included wine in large 'Badorf ware' jars from the Rhineland (inset below) and millstones.*

Alfred and the Danes *Following the loss of most of England to Danish Viking invaders, King Alfred's agreement with Guthrum, their leader, established a border along the River Lea, east of London. The area ceded to the Danes was later known as the Danelaw (map below). Alfred created 'burghs' to protect his realm and reoccupied London; his silver penny with the monogram LVNDONIA, 'London', on the reverse (inset) may celebrate this event.*

THE DANELAW c.900

▨ area of English rule

▨ area of Danish rule

▨ area of densest Scandinavian settlement

■ burgh

DANELAW

boundary AD 886

MERCIA

LONDON

WESSEX

English Channel

FRANCE

SAXON AND NORMAN LONDON

The medieval street-plan of the City of London originated in the Saxon period when the main centre of population of *Lundenwic* lay along the Strand. Long curving streets like Fenchurch and Lombard Street probably developed naturally as routes through the ruins of the Roman city, avoiding obstructions such as the prominent remains of the Roman *basilica*. After 886, when King Alfred initiated settlement within the Roman walls, new streets seem to have been laid out in two areas, some of the blocks being subsequently subdivided by lanes. Roman ruins survived for a long time; a document of 889 relating to a property near Queenhithe refers to 'an old stone building known as *Hwaetmundes stan*' – probably the remains of Roman public baths now known to have stood on the site.

Recent excavations have provided ever-increasing knowledge of the appearance and economy of late Saxon London. Wooden buildings of different types have been discovered, many of them with sunken floors, a feature of much earlier Saxon domestic architecture. Evidence has been recorded of industry – metalworking, boneworking, clothworking – and of trade – new waterfronts at Billingsgate and further west provided berths for cargo vessels. Finds of pottery and millstones from the Rhineland reflect a continuation, or revival, of trade with that area; amber, walrus ivory and whetstones of Norwegian stone demonstrate trade with the north and Scandinavia.

After bearing the brunt of the renewed Danish wars of the late 10th century, London flourished during the 11th century. Developments in the reign of Edward the Confessor, after 1042, were to have a profound long-term effect on its geography. Edward devoted much of his energy, and his finances, to the building of a new monastery dedicated to St. Peter in the west of the city – in the area later known simply as 'the west minster' – with a fine church of Norman style. More important, he established a new royal palace alongside it; an older palace within the city walls, perhaps near Aldersgate, was probably abandoned at this time. Edward's Norman successors continued the development of Westminster as a centre of royal government.

One of the first acts of the new Norman king,

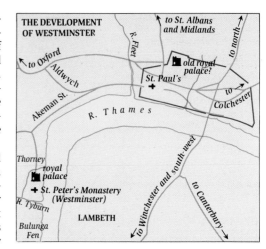

THE DEVELOPMENT OF WESTMINSTER

London and Westminster *Edward the Confessor's new London palace stood to the west of the city, adjacent to St. Peter's monastery (the 'west minster') at Thorney ('thorn island'), where the River Tyburn flowed into the Thames (map top). It may have replaced an older palace situated within the city walls. Traces of an earlier monastic church have been found beneath the floor of Westminster Abbey (right). Built for Edward the Confessor, it was inspired by churches he had seen when in exile in Normandy. His funeral, shown in the Bayeux Tapestry (above), took place just a week after the church was consecrated, in January 1066. His tomb became a shrine and place of pilgrimage.*

LATE SAXON LONDON HOUSES

Late Saxon London *In the late 9th century new streets, flanked by timber houses (above), were laid out in the area of the old Roman city, parts of which were still visible (map above).*
Viking raiders were followed by settlers and traders (map right) whose trade-routes extended to the Baltic and along the Russian rivers. A runic grave-slab (inset) from St. Paul's commemorates a Scandinavian resident of London in the days of King Cnut.

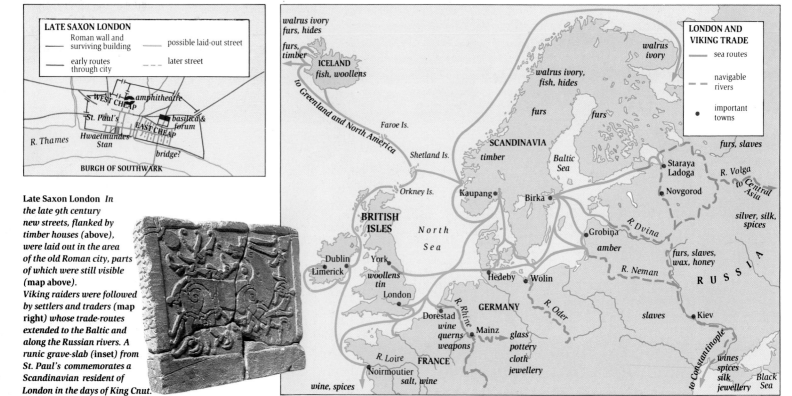

London 1200 *By 1200 London's suburbs stretched to Southwark and Westminster (map right). There were over 100 parish churches. New features of the Norman city were its castles; great monastic houses; hospitals, including those like St. Giles's and St. James's founded as refuges for lepers expelled from the town; the stone-arched London Bridge, begun in 1176 to replace earlier timber structures; and St. Paul's, the Romanesque cathedral which replaced a Saxon church destroyed by fire in 1087. Shown in an artist's impression (below right), its early 13th-century spire rose some 450 feet over the city.*

The London area in 1086 *The Domesday Book compiled for William I in 1085-86 provides a detailed economic survey of the counties around London (map below). Farming settlements bore names which survive today. Fish were netted in the Thames and its tributaries; water mills powered by the same streams ground corn from the openfields.*

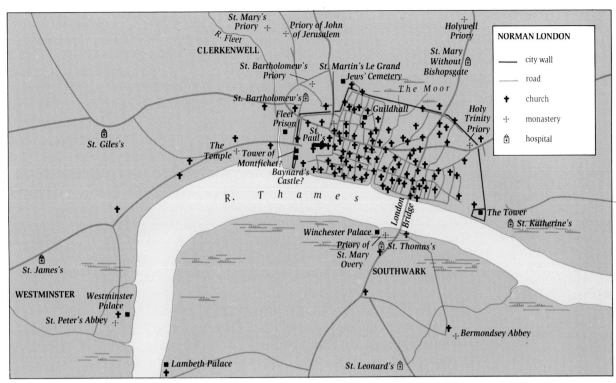

NORMAN LONDON

- city wall
- road
- ✝ church
- ✚ monastery
- 🏠 hospital

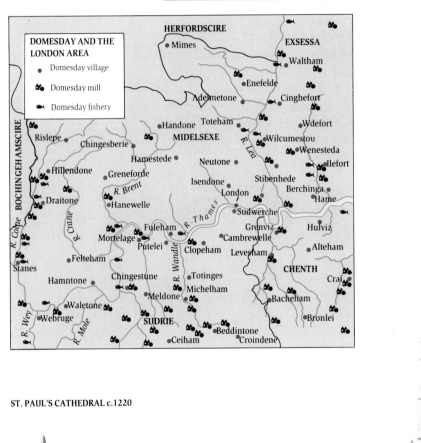

DOMESDAY AND THE LONDON AREA

- • Domesday village
- Domesday mill
- Domesday fishery

ST. PAUL'S CATHEDRAL c.1220

William, in 1066 was to establish castles to control London. One, in the east, was the core of the Tower of London (*page 141*). Others, near St. Paul's, seem to have been those later known as Baynard's Castle and the Tower of Montfichet. Both were demolished in the 13th century, and although traces have been found in excavations, their exact sites and layout are still unclear.

The combined evidence of documentary sources and archaeology allows the reconstruction of a much more detailed picture of London at the end of the 12th century than at any earlier period. This is enhanced by William Fitz Stephen's account of the city in about 1173 in which he describes many features: London's walls and gates; its many parish churches and great monasteries, most of them built in the previous 100 years or so; the royal palace to the west and the suburb stretching towards it (recolonising the site of old *Lundenwic*); the weekly horse-fair at Smithfield; the Moor (Moorfields) outside the wall to the north, where Londoners skated in winter. To William Fitz Stephen London was truly 'the most noble city'.

CHAPTER 4
MEDIEVAL LONDON

BETWEEN THE 12th century and the end of the 15th London expanded and then contracted in physical size and population, while continuing to increase in political and economic importance. Its appearance undoubtedly changed, though perhaps less markedly than in later centuries. At the end of the period, as at the beginning, it was still shaped by Roman walls and Saxon or early Norman street patterns, and the area of building had not spread far beyond the City itself, except in the two distinct settlements of Westminster and Southwark. The materials of which London was built changed only slowly: some stone was used, especially for churches and public buildings, but timber-framing and plaster predominated. Thatch was banned from the City in the 12th century, and brick was appearing more and more by the 15th century. However, there were considerable changes in London's society and government: a complex and, on the whole, effective system of administration developed, and London law, custom and culture came to be both distinctive and deeply influential on the rest of the country.

It is hard to establish the size of London's medieval population with any precision, but it was clearly substantial by the mid-12th century. There is evidence for continued growth through the later 12th and 13th centuries and it seems probable that the City reached its maximum size, with a population of perhaps 80,000, in the early years of the 14th century. Sustained expansion of the kind London experienced in the 12th and 13th centuries was the result of continued population growth in the rest of the country, and high levels of migration from the country to the capital. Many of the incomers are identified by their surnames – Robert of Clacton, William of Lincoln, Nicholas of Reading. Many foreign settlers in London came from northern France (until 1204 Normandy was under English rule), and the Gascon wine trade also brought a number of merchants. The Norman kings encouraged the settlement of Jews in English cities because of their value as sources of loans and capital. In London the area round the street still known as Old Jewry was the physical centre of the Jewish settlement, and remained so until the decline of the community in the later 13th century and its final expulsion in 1290. Foreign merchants, from Flanders and Italy especially, were present in London to profit from the lucrative export trade in wool as well as government finance.

London's exceptional size and wealth in the middle ages were due to a number of factors: its convenient location at the centre of land and water routes, with easy access to the Continent; the City's long-established role as a port; the growth of Westminster as the focus of government, and the growth of government itself. During the 12th and 13th centuries important government offices either shifted from Winchester to London, like the treasury, or ceased to move round the country with the king and took up permanent headquarters at Westminster. As their business increased, they took on staff and began to accumulate records. Though Westminster and the City remained separate in the medieval period, the presence and activity of the royal court and government offices nearby was a powerful stimulus to London's economy, attracting customers for the City's luxury imports and fine craftsmanship, and increasing the demand for services and supplies of all kinds.

The Black Death, a devastating pandemic, swept Europe and arrived in England in 1348. By 1350, it is possible that half the population of the capital died. Contemporary authors wrote of hundreds of burials a day in the new burial grounds

During the 12th and 13th centuries Westminster became increasingly prominent as a focus of royal government. The centralisation of the justice system and the increased importance of the King's courts at Westminster attracted litigants, petitioners and hangers-on. King Henry III invested in massive building projects. The appearance of medieval Westminster is reconstructed on pages 48-49.

The government of the City was clearly an oligarchy, not a democracy: 13th century aldermen were drawn from an elite of mercantile and landowning families with court connections, and held office for long periods. Below This 15th-century manuscript shows Simon Eyre, a City Alderman, in his robes of office. He became Lord Mayor in 1444.

opened to the east of the Tower and north-west of the City, and it appears that even among the richest in society the death rate was ten times normal. The first plague was followed by others, in 1361, 1369 and 1375; although these had a much smaller impact, plague became established in London and other towns and throughout the 15th century. In 1377 the population of London, including Southwark and Westminster, was probably between 35,000 and 45,000. In the immediate aftermath of the epidemics, the flow of migrants to London may have increased to take up the jobs and opportunities made available, but quite soon the increased attractions of the countryside – where land was now cheaper and the rewards of labour higher – began to take effect. From the late 14th century there is evidence of stagnating or gradually falling population levels in London: rents fell, houses stood unlet and fell into disrepair, vacant lots appeared in the over-crowded City streets. The built-up area may have contracted to a narrow fringe round the walls.

Above *Woodcut showing a view of London, from the 'Chronycle of England', published by Wynkyn de Worde in 1497.*

Population decline did not mean economic collapse for London. Living standards for the majority may indeed have improved as the shortage of labour pushed up wages; space and housing were in more generous supply, and the encroachment of the suburbs on the green fields was halted. London in fact increased in importance relative to the rest of England. Though England's involvement in warfare with France, especially in the 1340s and 1350s and from 1415 to 1453, was itself very costly, the financing of warfare and the marshalling of supplies and troops may well have given Londoners some opportunities for profit. Increasingly, from the late 14th century, English overseas trade was channelled through the Port of London, as the London-dominated associations of wool exporters and cloth merchants took control of the country's two main export products. A strong trade axis between London and Antwerp developed in the later 15th century and the City's role as a marketplace for imported goods expanded.

The trading relationship between London and the Continent is mapped on page 52.

London's self-government was achieved in the 12th and 13th centuries as a result of strong pressure from the City on the Crown, forcing or buying concessions in times of weakness. Even before the Conquest, London had its own laws, courts and customs, but the drive now was for Londoners to take control of the City as a whole. In the 12th century they won the right to appoint their own sheriffs – hitherto appointed by the Crown – and collectively took on responsibility for the annual payment to the Crown of £300, in exchange for the grant of royal revenues in the City. In the reign of John (1199-1216) they established a commune, a collectivity to which all Londoners swore allegiance, with a leader or mayor, an achievement that was confirmed in Magna Carta. The City's 24 wards were clearly established by 1127. The divisions may originally have been intended to allocate responsibility for the defence of the City, but they also became units of local government. Each was headed by an alderman, at first possibly a hereditary figure and always a man of wealth and importance. He presided over the local forum or wardmote, and in the 13th century and later the mayor and aldermen together formed the principal adminstrative council of the City. While the overall direction of the City was in the hands of the council of aldermen, local custom and law shaped the day-to-day working of the City. A range of civic courts dealt with civil litigation as well as criminal justice, reflecting some of the most important concerns of the citizens. The long-established court of Husting was concerned mostly with property rights, including the recovery of unpaid rents from tenants; the sheriffs' court dealt with minor assault and trespass, while the mayor's court heard cases of commercial debt and contract.

The demographic upheaval of the Black Death and succeeding plagues had a disruptive effect on wages, prices and the expectations of ordinary Londoners; attempts by employers and landlords to hold wages down were bitterly resented. Foreign wars and changes in patterns of trade caused tensions among the mercantile rulers of the City, some of whom were prepared to exploit popular discontents in their struggle for dominance. London's problems interacted with the wider political conflicts of the Good Parliament of 1376, the Peasants' Revolt of 1381, which paralysed the City for several days, and the crisis of the Lords Appellant in 1387-8. In the last of these a former mayor of London, Nicholas Brembre, was impeached and executed. Richard II quarrelled with London in the 1390s, and this probably ensured the City's acquiescence in the usurpation of Henry Bolingbroke (Henry IV) in 1399. Internally, London was more peaceful in the 15th century; the dominance of the great overseas merchants in London politics was confirmed – hardly anyone who was not a member of the half dozen leading City companies reached the Court of Aldermen, let alone the mayoralty.

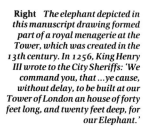

If taxation lists of the period are to be believed, only a tiny proportion of London's inhabitants had sufficient goods to be worth taxing, though some of these few were obviously very wealthy indeed. See map of comparative wealth and poverty, assessed by tax, page 52.

Throughout the Middle Ages, control of London was crucial to political success, either for the monarch or for a claimant or usurper. In the Wars of the Roses, both sides were anxious to hold on to London, but while the rulers of the City preferred to play a conservative part, tending to support the government in power (to which they were committed in loans and financial support) until the scales were already tipping in the other side's favour, there was probably stronger Yorkist feeling among the people. In 1461, when the City government was prepared to send supplies to Margaret of Anjou, whose army was threatening the City, the people of London shut the gates and turned back the carts: 'the commons of the city and the heads were of two opinions and minds', as a contemporary wrote.

While Westminster was undoubtedly the royal capital, in that the offices of government, royal palace and mausoleum were situated there, the City was used by the Crown as a key element in the presentation of the monarchy to the people. Coronation processions and royal 'entries' used the streets of the City, and indeed the people, as setting and audience for magnificent displays of wealth and power. Richard II's quarrel with London was resolved, not only

Right *The elephant depicted in this manuscript drawing formed part of a royal menagerie at the Tower, which was created in the 13th century. In 1256, King Henry III wrote to the City Sheriffs: 'We command you, that ...ye cause, without delay, to be built at our Tower of London an house of forty feet long, and twenty feet deep, for our Elephant.'*

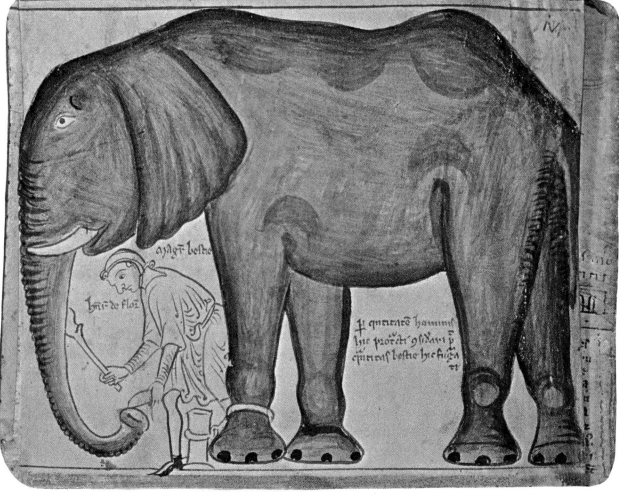

Above *Doorknocker in the form of a grotesque face with a ring in its mouth, late 14th-early 15th century. It was found in Thames Street.*

For a full discussion of London's guilds and livery companies, and a location map, refer to pages 62-63.

by a substantial payment from the City, but with an elaborately-staged procession through the City's streets, emphasising the City's contrition and subordination. Later entries, in which the City was a more equal and perhaps more willing partner, celebrated Henry V's victory at Agincourt, the return of the young Henry VI after his coronation in France, and his marriage to the French princess, Margaret of Anjou. The Tudor dynasty quickly became expert in the construction and presentation of the image of monarchy in this way. Most processions followed the same route, starting with an entry across London Bridge, continuing by way of Gracechurch Street and Cornhill to Cheapside, where pageants were set up at the Conduits and the Eleanor Cross, and terminating at St Paul's.

For a map of medieval markets see page 148.

Ideas of civil rights and responsibilities had been evolving in the 13th century, and with them the definition of citizenship as a specific status, not an automatic right. In return for his civic oath and contribution to civic taxation, the citizen acquired rights to trade retail as well as wholesale, to buy and sell real property and to benefit from the protection of the City's courts. From 1319 new citizens were admitted only with the approval of the men of the trade they intended to practice, effectively giving control of access of citizenship to the emerging associations of traders and craftsmen. It is difficult to estimate the proportion of citizens in London's adult male population in the later Middle Ages, but it seems likely that it was fairly low, probably less than one quarter. Men and women who were not citizens were referred to as 'foreigns', even if London-born; most made their living from unskilled and casual labour, though many illegally practised crafts that were in theory restricted to citizens. One well-recorded feature of medieval London was its guilds and fraternities. We know most about those associated with a particular craft or occupation through their descendants, the Livery Companies, but guilds were formed for many purposes, social, religious, political, as well as economic. In essence they were voluntary associations for mutual support, with a strong emphasis on brotherhood and friendliness; some appear to have been no more than burial clubs and communal chantries, while others evolved into institutions wielding considerable political power. Before 1150 London had guilds of weavers, saddlers and bakers; in 1180 there were at least 19 guilds, including four specifically associated with a trade (goldsmiths, pepperers, clothworkers and butchers). The number of recognised associations seems to have increased continuously, and there were over 100 in the early 15th century. In the 14th and 15th centuries craft groups were obliged to obtain mayoral approval and submit their regulations (often including very detailed descriptions of craft practice) to the City government. The larger companies acquired, through donation and bequest, a considerable amount of real property and with the rent income and members' subscriptions they built halls for their meetings and celebrations, and maintained a lavish communal life. They also devoted money to charitable and pious purposes, especially chantries and commemorations for deceased members.

Map below *The extent of Medieval London at its peak is contrasted to the size of the city in 1820.*

By 1500, the institutions and bureaucracy of City government had established deep roots, and patterns of social behaviour, focused on household, parish and City company, were also well established. It would be many years before the discovery of the New World in 1492 had an impact on London's economy. However, within the next 50 years the religious ideas proliferating on the continent were, through the process of the Reformation, to have a profound effect on both the City's character and appearance.

EXTENT OF LONDON 1500

LONDON 1820

✝St. Paul's

✝Westminster Abbey

LONDON 1820

THE MEDIEVAL CITY

In 1300, the street-pattern of central London was predominantly influenced by the Roman city wall and the skeleton network of streets laid down from late Saxon and early Norman times. There were certainly some slight alterations in street alignments and building lines, and in the arrangements of houses, yards, gardens and private alleys, but to a great extent the pattern within the walls was already set. Most of the population of perhaps 80,000 was tightly packed within the walls and the nearer suburbs, with only a small proportion living in the outer settlements of Southwark and Westminster.

The most important topographical change in the City between 1100 and 1500 was the reclamation of the waterfront, to the south of both Thames Street and the line of the decayed Roman riverside wall. Much of this reclamation proceeded piecemeal: recent excavations on a number of sites have revealed successive timber and stone quayfronts, as the owners of individual riverfront properties built their wharves further and further south out into the river. The process was all but complete by 1500, by which time a strip of land between 120 and 350 feet in width had been reclaimed. By contrast, the two main public landing places at Queenhithe and Billingsgate remained as inlets in the waterfront, as did the smaller haven of Dowgate. London Bridge (rebuilt in stone in the late 12th and early 13th centuries) remained the only road link between the City and the southern counties, but a number of small landing-places and stairs along the waterfront testify to the importance of river transport for both goods and passengers. The river-front was also affected by the development of the Tower as a concentric fortress, with an enlarged moat and barbican gate. The Walbrook stream had disappeared into a culvert that flowed into the Thames west of Dowgate.

By the early 14th century, thanks to the survival of many records of property transactions, we know the names of almost all of the streets that then existed. Some of these names have since altered, Westcheap becoming Cheapside (while the name Eastcheap survived), and Candlewick Street being simplified to Cannon Street, while the important late medieval estate called Leadenhall did not give its name to the street until the 15th century. Many medieval street and area names (still recognisable today) commemorate trades and activities, though the names sometimes stayed while the trades moved on: Ironmonger Lane was no longer the centre of that trade by 1300, and the bellfounders had moved out from Billiter Lane (the lane of the 'bell-yetteres' or bell-founders) to the suburbs well before 1500.

Westminster (reconstructed below, c.1500) was separate from the City, with its own government and a distinct character. For the whole of the Middle Ages the manor of Westminster belonged to the abbey, and the settlement that grew up there, small by comparison with the City but equal in size to some provincial towns, was dominated by the abbey and the royal palace. Henry III (1216-1272) rebuilt the abbey itself and enlarged the palace, adding St Stephen's chapel. His successors were too often absent on foreign wars to make many changes, but Richard II rebuilt the Norman Westminster Hall, and over a long period accommodation for royal courts and government offices was added. The economy of the Westminster area was shaped by these important neighbours, and came to specialise in lodgings, meals and accommodation, luxury retail and services. The first printing press was established by William Caxton at Westminster, within the abbey close, in 1476, only later moving to the City. Though Westminster and the City were linked by the Strand, it was the river which provided direct and rapid communication.

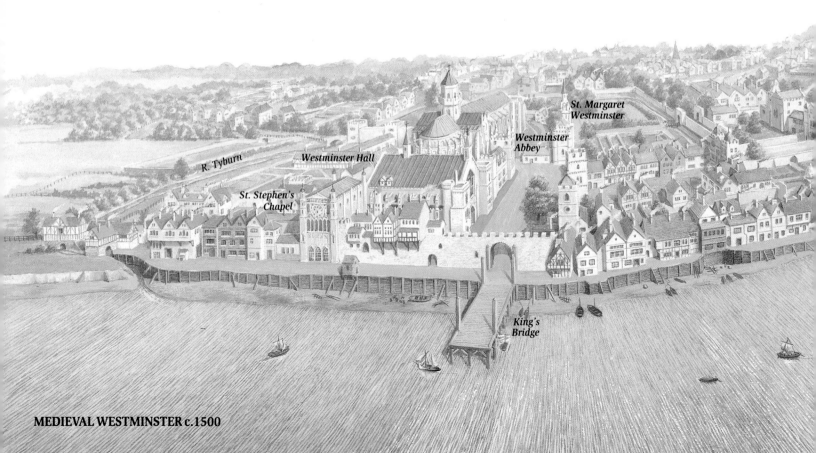

MEDIEVAL WESTMINSTER c.1500

R. Tyburn

Westminster Hall

St. Stephen's Chapel

St. Margaret Westminster

Westminster Abbey

King's Bridge

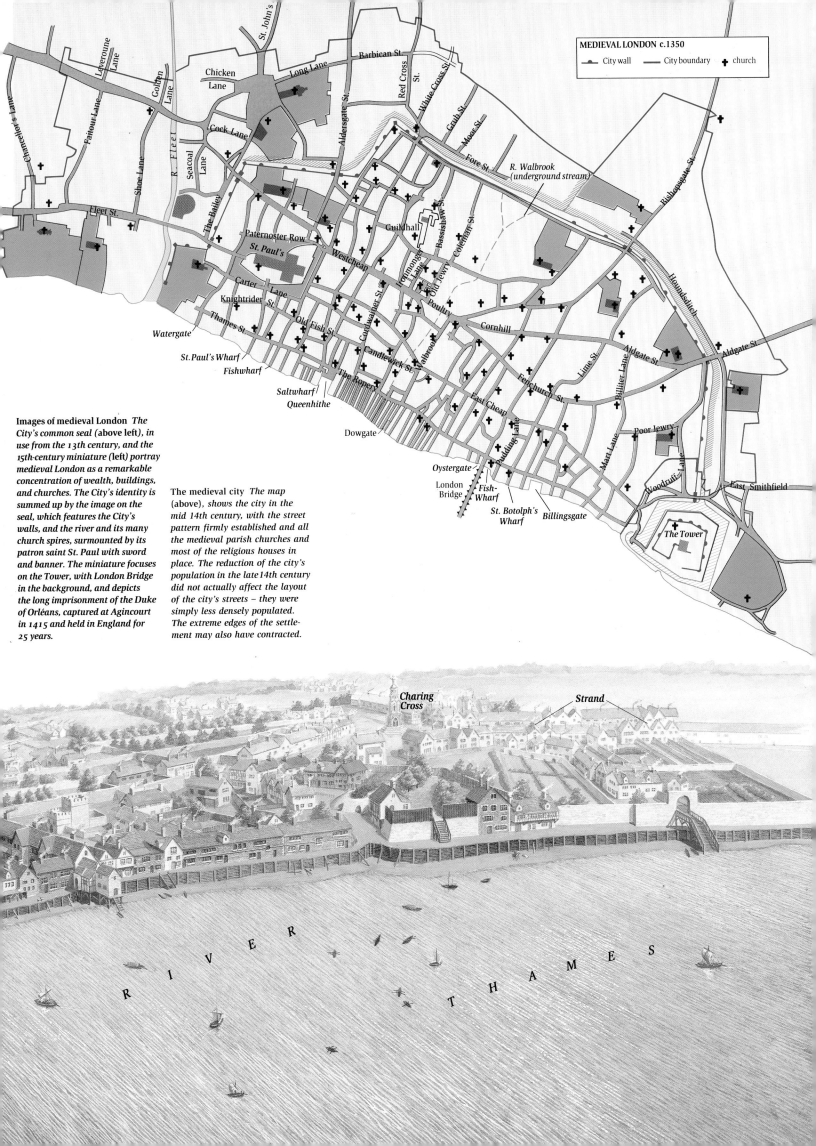

MEDIEVAL LONDON c.1350

City wall   City boundary   church

Images of medieval London *The City's common seal (above left), in use from the 13th century, and the 15th-century miniature (left) portray medieval London as a remarkable concentration of wealth, buildings, and churches. The City's identity is summed up by the image on the seal, which features the City's walls, and the river and its many church spires, surmounted by its patron saint St. Paul with sword and banner. The miniature focuses on the Tower, with London Bridge in the background, and depicts the long imprisonment of the Duke of Orléans, captured at Agincourt in 1415 and held in England for 25 years.*

The medieval city *The map (above), shows the city in the mid 14th century, with the street pattern firmly established and all the medieval parish churches and most of the religious houses in place. The reduction of the city's population in the late 14th century did not actually affect the layout of the city's streets – they were simply less densely populated. The extreme edges of the settlement may also have contracted.*

THE MEDIEVAL CHURCH

The Church played an important role in shaping the appearance and character of medieval London. The City was the seat of an important bishopric and great cathedral; several large and wealthy monasteries, friaries, and nunneries were founded in the City or on its outskirts, and there were over a hundred parish churches in and near the City. The Church's influence permeated everyday life, as people ordered their working day by church bells, made contracts paying 'God's penny', and swore oaths on the gospels.

The Bishop of London was a figure of national importance and close to the top of the ecclesiastical hierarchy. His diocese covered Middlesex, Essex, and part of Hertfordshire, and diocesan courts and officials thus attracted much business from outside the city. Many abbots and bishops from outside London needed to attend Parliament or the monarch, and so found it useful to have a house or inn in the City or nearby in the more spacious western suburbs. The Bishop of

Winchester, whose diocese included Southwark, had a large house there, of which some remains still survive; the locations of others are commemorated by modern place names such as Salisbury Court or Ely Place.

St. Paul's must have been one of the finest and largest of Gothic cathedrals, dating partly from the 12th and partly from the late 13th century. It offered a splendid space both for services and for civic ceremonial. The mayor and aldermen attended mass there on special days, while many civic processions concluded with a service there. St. Paul was regarded as the city's patron saint until the 13th century, when he had to share that position with the murdered Archbishop of Canterbury, Thomas Becket (died 1170), whose cult developed at that time. The citizens founded the Hospital of St. Thomas of the military order of Acre in his honour, on the site of the house in which he had been born in Cheapside.

Most of the big religious houses in London were founded between c.1100 and 1250, and undoubtedly City wealth contributed to their upkeep, even when the first founder was royal or noble. The monasteries tended to form closed communities, repaying the laity's material gener-

Churches, saints and pilgrims *Few medieval churches remain: Austin Friars (right) was destroyed by fire in 1940, while the silver paten (above right) came from St. Michael Crooked Lane, burnt in 1666. The pilgrimage to the shrine of Thomas Becket at Canterbury was immortalised by Geoffrey Chaucer (above) in his Canterbury Tales. Many metal badges (right), showing the saint on horseback, have been found in London, probably pilgrimage souvenirs.*

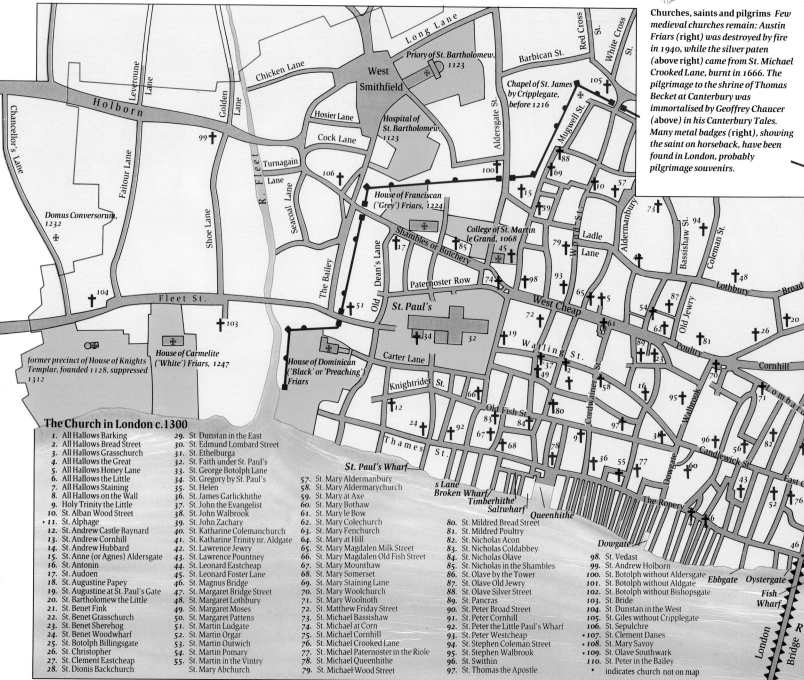

The Church in London c.1300

1. All Hallows Barking
2. All Hallows Bread Street
3. All Hallows Grasschurch
4. All Hallows the Great
5. All Hallows Honey Lane
6. All Hallows the Little
7. All Hallows Staining
8. All Hallows on the Wall
9. Holy Trinity the Little
10. St. Alban Wood Street
• 11. St. Alphage
12. St. Andrew Castle Baynard
13. St. Andrew Cornhill
14. St. Andrew Hubbard
15. St. Anne (or Agnes) Aldersgate
16. St. Antonin
17. St. Audoen
18. St. Augustine Papey
19. St. Augustine at St. Paul's Gate
20. St. Bartholomew the Little
21. St. Benet Fink
22. St. Benet Grasschurch
23. St. Benet Sherehog
24. St. Benet Woodwharf
25. St. Botolph Billingsgate
26. St. Christopher
27. St. Clement Eastcheap
28. St. Dionis Backchurch

29. St. Dunstan in the East
30. St. Edmund Lombard Street
31. St. Ethelburga
32. St. Faith under St. Paul's
33. St. George Botolph Lane
34. St. Gregory by St. Paul's
35. St. Helen
36. St. James Garlickhithe
37. St. John the Evangelist
38. St. John Walbrook
39. St. John Zachary
40. St. Katharine Colemanchurch
41. St. Katharine Trinity nr. Aldgate
42. St. Lawrence Jewry
43. St. Lawrence Pountney
44. St. Leonard Eastcheap
45. St. Leonard Foster Lane
46. St. Magnus Bridge
47. St. Margaret Bridge Street
48. St. Margaret Lothbury
49. St. Margaret Moses
50. St. Margaret Pattens
51. St. Martin Ludgate
52. St. Martin Orgar
53. St. Martin Outwich
54. St. Martin Pomary
55. St. Martin in the Vintry
    St. Mary Abchurch

57. St. Mary Aldermanbury
58. St. Mary Aldermarychurch
59. St. Mary at Axe
60. St. Mary Bothaw
61. St. Mary le Bow
62. St. Mary Colechurch
63. St. Mary Fenchurch
64. St. Mary at Hill
65. St. Mary Magdalen Milk Street
66. St. Mary Magdalen Old Fish Street
67. St. Mary Mounthaw
68. St. Mary Somerset
69. St. Mary Staining Lane
70. St. Mary Woolchurch
71. St. Mary Woolnoth
72. St. Matthew Friday Street
73. St. Michael Bassishaw
74. St. Michael at Corn
75. St. Michael Cornhill
76. St. Michael Crooked Lane
77. St. Michael Paternoster in the Riole
78. St. Michael Queenhithe
79. St. Michael Wood Street

80. St. Mildred Bread Street
81. St. Mildred Poultry
82. St. Nicholas Acon
83. St. Nicholas Coldabbey
84. St. Nicholas Olave
85. St. Nicholas in the Shambles
86. St. Olave by the Tower
87. St. Olave Old Jewry
88. St. Olave Silver Street
89. St. Pancras
90. St. Peter Broad Street
91. St. Peter Cornhill
92. St. Peter the Little Paul's Wharf
93. St. Peter Westcheap
94. St. Stephen Coleman Street
95. St. Stephen Walbrook
96. St. Swithin
97. St. Thomas the Apostle

98. St. Vedast
99. St. Andrew Holborn
100. St. Botolph without Aldersgate
101. St. Botolph without Aldgate
102. St. Botolph without Bishopsgate
103. St. Bride
104. St. Dunstan in the West
105. St. Giles without Cripplegate
106. St. Sepulchre
• 107. St. Clement Danes
• 108. St. Mary Savoy
• 109. St. Olave Southwark
110. St. Peter in the Bailey

• indicates church not on map

Map labels

Long Lane
Chicken Lane
West Smithfield
Priory of St. Bartholomew, 1123
Barbican St.
Red Cross St.
White Cross St.
Chapel of St. James by Cripplegate, before 1216
Holborn
Leveroune Lane
Golden Lane
Hosier Lane
Cock Lane
Hospital of St. Bartholomew, 1123
Aldersgate St.
Mugwell St.
Chancellor's Lane
Faitour Lane
R. Fleet
Turnagain Lane
Seacoal Lane
House of Franciscan ('Grey') Friars, 1224
Shambles or Butchery
College of St. Martin le Grand, 1068
Wood St.
Ladle Lane
Aldermanbury
Bassishaw St.
Coleman St.
Domus Conversorum, 1232
Shoe Lane
The Bailey
Old Dean's Lane
Paternoster Row
West Cheap
Lothbury
Broad St.
Fleet St.
St. Paul's
Watling St.
Old Jewry
Poultry
Cornhill
House of Carmelite ('White') Friars, 1247
House of Dominican ('Black' or 'Preaching') Friars
Carter Lane
Knightrider St.
Cordwainer St.
Walbrook
Lombard St.
former precinct of House of Knights Templar, founded 1128, suppressed 1312
Old Fish St.
Candlewick St.
Thames St.
St. Paul's Wharf
s Lane
Broken Wharf
Timberhithe
Saltwharf
Queenhithe
Dowgate
Ebbgate
Oystergate
Fish Wharf
The Ropery
London Bridge
East C

osity with prayer, though the hospitals offered medical care and shelter. The houses of friars, founded in the 13th century, had a more active role in society, preaching and hearing confessions, although their involvement with the world lowered their reputation with some. Later medieval foundations in and near London included the Cistercian monastery of St. Mary Graces by the Tower, founded by Edward III in 1350, and the Charterhouse to the north west of the city, founded by Sir Walter Manny in 1371: both had their roots in the wealth won from the French wars and in the devastating experience of the Black Death of 1348-9.

It seems to be a characteristic of cities that were large and wealthy in the 11th and 12th centuries, such as Norwich, York and Winchester, that they had a large number of parish churches, which usually survived into the later Middle Ages. London itself had over 100 parish

churches by 1200 and their geographical spread indicates the concentration of population and wealth in the City. While the range of church dedications was quite wide, there were many churches which were dedicated to the same saint, and most thus acquired an identifying 'surname', usually indicating their location but sometimes commemorating the founder.

The large number of parish churches in London ensured that most people could belong to a relatively small and intimate group for religious worship. The churches themselves were an important physical focus for parish life, secular as well as religious; many were rebuilt or substantially extended in the later 14th or 15th centuries, largely through the bequests and gifts of parishioners. The comparative wealth of the London livings, and the other attractions of the capital, meant that the City's churches were staffed by churchmen with higher than normal levels of literacy and academic attainment.

The doctrine of Purgatory, which made prayers for the dead a crucially important part of religious observance, took hold in the 14th century, and led to the proliferation of endowments for chantries (prayers for the dead) and anni-

versary masses in the City churches. Much wealth came to the Church in this way: the churches were adorned with new altars, with lamps and candles, and with subsidiary chapels, and the parish clergy were augmented by an increasing number of chantry priests. Many of the wealthier Londoners set up permanent chantries, usually in their parish churches or St. Paul's, while a very few, including mayor Richard Whittington (died 1423) were able to found colleges of priests to pray for their souls. Poorer individuals, without the resources to pay for prayers in perpetuity, joined small local fraternities which acted as collective chantries, guaranteeing a proper burial and prayers for members when they died. Charity and care for the living went along with this concern for the dead: Whittington's college included an almshouse for thirteen 'poor folk', and many fraternities helped members in sickness and trouble. The City companies played a part in this, administering chantries and bequests for the poor and running almshouses and schools.

Although the evidence of wills and endowments suggests that most Londoners were firmly committed to the rituals and beliefs of the

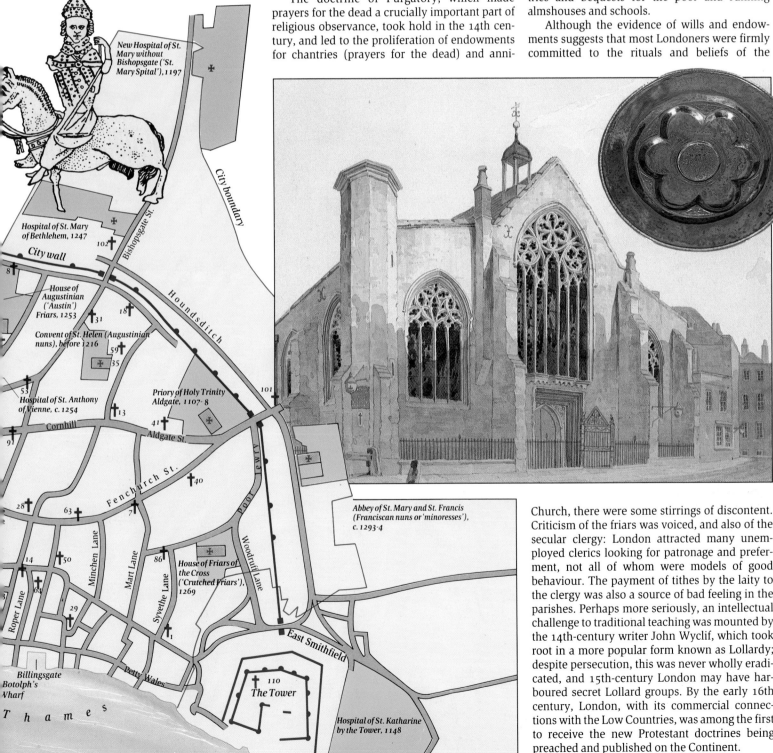

New Hospital of St. Mary without Bishopsgate ('St. Mary Spital'), 1197

City boundary

Hospital of St. Mary of Bethlehem, 1247

City wall

House of Augustinian ('Austin') Friars, 1253

Houndsditch

Convent of St. Helen (Augustinian nuns), before 1216

Bishopsgate St.

Priory of Holy Trinity Aldgate, 1107-8

Hospital of St. Anthony of Vienne, c. 1254

Cornhill

Aldgate St.

Fenchurch St.

Minchen Lane

Mart Lane

Syvethe Lane

House of Friars of the Cross ('Crutched Friars'), 1269

Woodruff Lane

Roper Lane

Billingsgate Botolph's Wharf

Petty Wales

East Smithfield

T h a m e s

The Tower

Abbey of St. Mary and St. Francis (Franciscan nuns or 'minoresses'), c. 1293-4

Hospital of St. Katharine by the Tower, 1148

Church, there were some stirrings of discontent. Criticism of the friars was voiced, and also of the secular clergy: London attracted many unemployed clerics looking for patronage and preferment, not all of whom were models of good behaviour. The payment of tithes by the laity to the clergy was also a source of bad feeling in the parishes. Perhaps more seriously, an intellectual challenge to traditional teaching was mounted by the 14th-century writer John Wyclif, which took root in a more popular form known as Lollardy; despite persecution, this was never wholly eradicated, and 15th-century London may have harboured secret Lollard groups. By the early 16th century, London, with its commercial connections with the Low Countries, was among the first to receive the new Protestant doctrines being preached and published on the Continent.

TRADE AND TRADERS

Early medieval England largely produced raw materials for export. Its principal product was wool; other exports included Cornish tin, hides, sheepskins, and some foodstuffs. Many English ports on the east and south coasts flourished as centres for the export of wool to the clothmaking cities of the Low Countries and Italy. London was certainly a major wool export centre, and in the early 14th century about one third of the national total was shipped from the city's wharves. More than half of this was in the hands of foreign merchants: many Flemings and Italians used their greater capital resources and financial skills to organise the English trade.

Even more important for the city's economy, however, was London's role as a marketplace for imported goods – fine textiles, spices, furs, small manufactured goods, and especially wine, imported from the English territories in Gascony via Bordeaux and La Rochelle, and stored and sold from cellars in the Vintry. As the great fairs of the 13th century declined, nobles, leading ecclesiastics, and the royal household turned to London for all the luxuries their lifestyle demanded, and London merchants and craftsmen profited from supplying these needs. Many of these imports – French wine excepted – came to England not directly from their country of origin, but via the markets of the Low Countries, especially Bruges.

The start of the Hundred Years' War (c.1337-1453) between England and France severely damaged the Bordeaux wine trade and led to trade embargoes with the Low Countries, and very large duties on the export of unprocessed wool. This gave English cloth manufacturing a chance to develop, and by the early 15th century cloth, rather than wool, was becoming the dominant export, and London was

Wealth and Trade London was one of the centres of a complex network of European trade and distribution routes (below left). Overseas trade was concentrated on the waterfront below the Bridge, where the first customs house was built in the late 14th century, but wine was landed upstream at the Vintry, and Baltic trading focused on the Steelyard. Goods were carried up the steep city lanes from the river to the merchants' houses and shops in Westcheap (modern Cheapside), where there were specialist retail areas (map below). Commerce and trade were the source of medieval London's prosperity, as is clearly shown by the map of wealth distribution in the early 14th century (below): the richest areas are the centre and waterfront.

increasing its advantage over other English ports. Political tension increased the risks of trade, while population loss had had an important impact on the economy. In these more difficult times, the larger merchant had the advantage over the smaller, and London merchants individually, and through the associations of the Staple (for wool exports) and the Merchants Adventurers (for cloth export), were able to take a leading position.

After a period of stagnation in the mid-15th century, there seems to have been a general recovery in European trade. The French wine trade grew again with the ending of hostilities, and English cloth exports began to expand rapidly from the 1470s. English trade was increasingly channelled through London: by 1500 about 45 per cent of England's wool exports and 70 per cent of its cloth were passing through the port of London, much of this then going on to Antwerp or Calais. Antwerp and its hinterland also offered high-quality cloth-finishing skills, so London merchants developed an easy and profitable business exporting

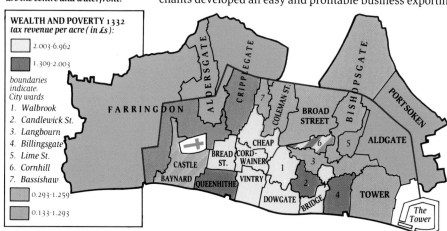

WEALTH AND POVERTY 1332
tax revenue per acre (in £s):

2.003-6.962

1.309-2.003

boundaries indicate
City wards
1. Walbrook
2. Candlewick St.
3. Langbourn
4. Billingsgate
5. Lime St.
6. Cornhill
7. Bassishaw

0.293-1.259

0.133-1.293

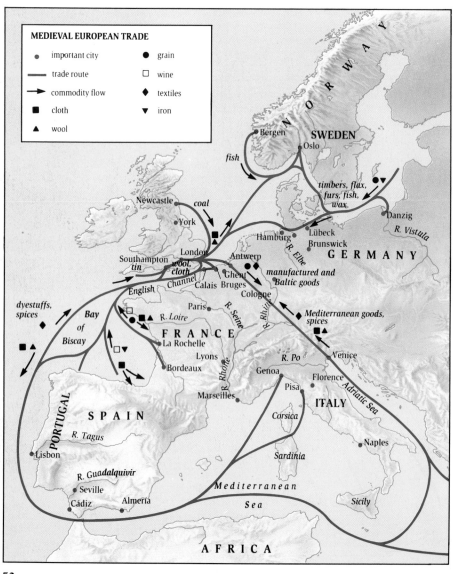

MEDIEVAL EUROPEAN TRADE

• important city
— trade route
→ commodity flow
■ cloth
▲ wool
● grain
□ wine
◆ textiles
▼ iron

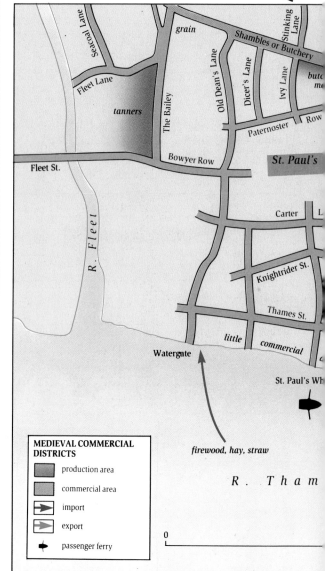

MEDIEVAL COMMERCIAL DISTRICTS

production area

commercial area

→ import

→ export

◆ passenger ferry

undyed, unfinished woollen broadcloths, to the dismay of English clothworkers.

Imports into London in the late 15th century were made up of some raw materials (dyestuffs, spices, fruits, and fish) and a wide range of manufactures, from knives, armour, and weapons, to threads, textiles (especially linen) and haberdashery. Soap, glassware, cheap jewellery, mirrors, spectacles, paper and printed books ('diverse histories') are among the imports listed in a customs account from the 1480s. English craftsmen, including Londoners, objected to this flood of imports which threatened their livelihood, but even Acts of Parliament were powerless to stop it. The evidence suggests that the late 15th-century trade expansion was of greatest benefit to merchants and perhaps also to retailers, but provided little stimulus to London's manufacturing and finishing industries. However, the boom in the import of cheap goods is evidence of high consumer demand in the lower levels of society, which can be attributed to the better standards of living they were then enjoying.

A medieval shop *Westcheap (modern Cheapside), the retailing centre of the city, was lined with shops, some less than 8 or 10 feet square. There were also several larger buildings called selds, in which traders owned or rented space for a bench or table, much like an eastern bazaar or a covered market. There may have been 400 retail units in Westcheap in 1300. The drawing (right) reconstructs a mercer's shop on Soper Lane and Westcheap. Mercers sold mostly linen and imported silk textiles, and perhaps also ribbons, kerchiefs, and trimmings. The goods were displayed outside the shop.*

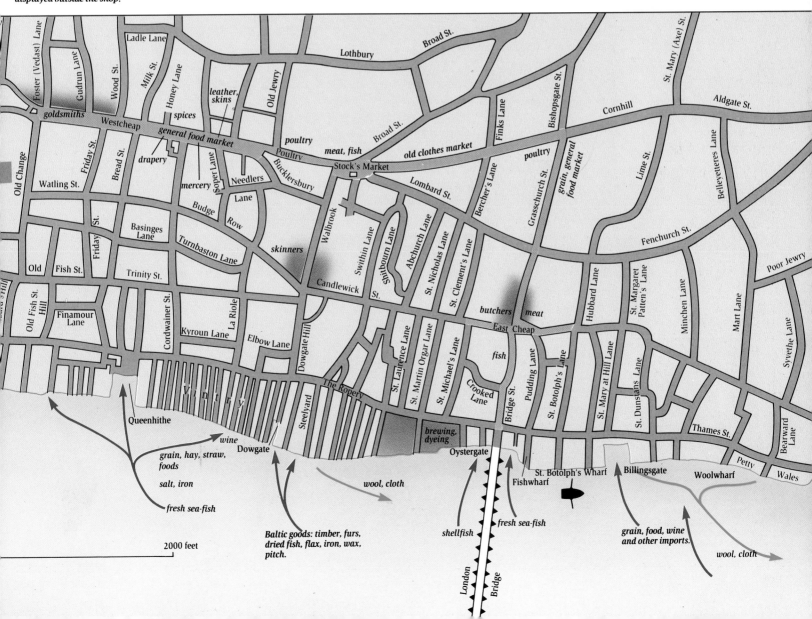

CHAPTER 5
TUDOR AND STUART LONDON

WHILE, IN 1500, the topography of London reflected its history of gradual development over the previous three centuries, it was to undergo rapid and violent changes during the second half of the 16th century, the principal change being the Dissolution of the Monasteries (1536). One amongst many of the causes for the Dissolution was the widely-held view that the monasteries were out-of-date, stifling, degenerate institutions. In addition, this was a highly materialistic age, the aim of ambitious men being the acquisition of land and rank. Shrinking government revenues, Henry VIII's extravagance in building his palaces, and troubles overseas, had all contributed to an urgent need for cash.

Three groups were the principal beneficiaries of the sales or grants of monastic lands resulting from the Dissolution: local landowners who were often tenants of the former houses; courtiers in favour at Court and government agents and officials of the Court of Augmentations which handled the transfers. Thus, Blackfriars passed to Sir Thomas Cawarden, Master of the Revels; Austin Friars to Sir William Paulet, Lord Treasurer. In the countryside outside London a ring of rural mansions sprang up within monastic houses, such as at Bermondsey Abbey which was granted to Sir Thomas Pope, Treasurer of the Court of Augmentations. Urban property owned by the monasteries or by former chantries (endowments of parish churches, confiscated by the Crown at the Reformation, and resold or given away) were released onto the market; in certain areas of the city as much as 60% of the property may have been in ecclesiastical ownership before the Dissolution. Colonisation of the former religious precincts, especially by aliens or foreigners, was common. In about 1586 a survey of the shell of the church of Holy Trinity Priory, Aldgate, found houses sprouting from above the Romanesque arches of the roofless choir. This precinct was inhabited by a number of

Please refer to pages 50-51 for a map of the pre-Dissolution churches and monastic precincts of London.

Right *The painting depicts the coronation procession of King Edward VI in 1547. A traditional route was followed, from the Tower of London to Westminster, and the way was lined with specially-constructed viewing stands from which City Liverymen viewed the procession. Landmarks passed on the way – St. Mary le-Bow, St. Paul's, Ludgate and Charing Cross are all clearly visible. The rural suburb of Southwark on the south bank of the river forms the background.*

Dutchmen, probably refugees, some of whom were making their characteristic Delftware vessels and tiles. Stone monastery buildings were much sought after by craftsmen who used fire in the manufacture of their products, such as glass-makers.

Before 1563, a contemporary historian wrote: '... fair houses in London were plenteous, and very easy to be had at low and small rents, and by reason of the late dissolution of religious houses, many houses in London stood vacant, and not any man desirous to take them.' Between 1550 and 1700 the population of London grew from 80,000 to over half a million. In 1500 it was a national capital inhabited by only four per cent of the English population; by 1700 it had become a major metropolis with almost ten per cent. The resultant pressure on housing in the already overcrowded medieval city necessitated massive building in areas such as the West End – within easy reach of the royal court at Whitehall – and outwards into the suburbs. Buildings and alleys covered a wide area of former fields east of the city, fanning out from the existing ribbon developments on the approach roads and rural hamlets.

This growth is even more remarkable because it was frequently checked by the ravages of plague. Nearly a quarter of the city's inhabitants died of plague in 1563; in 1665 more than 80,000 people died (a figure equivalent to the combined 17th-century populations of Norwich, Bristol, Newcastle, York and Exeter). Plague was concentrated both socially and geographically. Rats and fleas spread the disease throughout overcrowded alleys, shacks and tenements. Graveyards were soon overflowing and new burial grounds had to be established to take the strain. Though the City authorities had attempted to

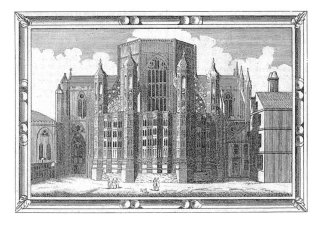

Please refer to pages 64-65 for a map of London's expansion in the 17th century.

Above *An engraving of Henry VII's chapel at Westminster Abbey. The foundation stone was laid in January 1503, and it was still unfinished when Henry died in 1509, although he was buried under the chapel floor. The task of completing the chapel fell to his successor, Henry VIII. The chapel, with its magnificent fan-vaulted ceiling which contrasts with sculptural decorations of extraordinary intricacy, is considered to be one of the great masterpieces of medieval architecture.*

The EXECUTION of KING CHARLES the FIRS
before the Banqueting House Whitehall, January 30.16

See page 63 for a reconstruction of a Tudor house based on a Survey by Ralph Treswell.

introduce special measures during the plague of 1518 (such as the isolation of those infected), their response was usually too little and too late.

John Stow's *Survey of London* (1598) is a major starting-point for any understanding of London's medieval and Tudor topography, social history and customs. Stow saw the city spreading in every direction, and lamented the speed and brutality of the many changes during his lifetime, though he took pride in the increased prosperity of the Elizabethan city and the newer buildings which reflected that wealth. He walked round every ward and looked into every church recording what the Reformation had left in its destructive wake. His view of London was imbued with a love of the past and the already vanished or fast-disappearing city of his youth.

A series of maps based on John Stow's Survey can be seen on pages 60-61.

By the end of the 16th century, many of the City's older noble houses, arranged around a courtyard with a lofty hall, no longer existed, or were barely recognisable. The Inn (town house) of the Earl of Oxford had been let to poulterers for the stabling of horses and housing of poultry. The hall of the Earls of Norfolk at Brokenwharf was a brewery; several town-houses of monasteries outside London had become taverns, even before 1535. Apart from company halls, houses with courtyards were still rare; they had either been built over or encroached upon. Set back from the street behind rows of smaller dwellings, they were sometimes only one room deep but up to five storeys high.

By 1600 a certain zoning of the better residences was becoming apparent. Stow and his near-contemporary, the playwright Dekker, both noticed that certain streets were favoured by the 'worthiest citizens' for their large houses: St Mary Axe, Lime Street, Milk Street and Bread Street for example, all quiet enclaves close to, but sufficiently separated from, Cheapside. The growth of

the royal court at Westminster led to a general drift westwards amongst London's 'worthiest citizens'. Henry VIII built several palaces in the London area, and one, at Bridewell, within the City itself. Whitehall became the principal palace of the Tudor and Stuart monarchy; always a large and rambling complex near the river, it was added to by Henry VIII (the Cockpit, tennis courts and an area for pageants) and by Charles I. Former residences of bishops and priors along the Strand were appropriated by prominent nobles and courtiers. New buildings erected on land lying between the City and Westminster were now largely built in brick. James I was determined that it would be said of him: 'that we had found our Citie and suburbs of London of stickes, and left them of bricke, being a material far more durable, safe from fire and beautiful and magnificent'. In the 1630s Inigo Jones, Surveyor to the King's Works, designed a speculative development at Covent Garden (formerly belonging to Westminster Abbey) for the Duke of Bedford. At the end of the Duke's garden a square of houses arose, modelled on a continental piazza. To the north a similar square was constructed north of Oxford Street, near Southampton House and forming the nucleus of the future Bloomsbury. Jones also laid out a third square at Lincoln's Inn Fields. All these houses were built in fashionable brick.

See Hollar's engraving of Covent Garden, page 65.

London's position as the nation's economic and political centre was based on its rising dominance in the export of cloth which brought wealth to Londoners and made the city an attraction for immigrants. Wages were as much as 50% higher in London than in the provinces; it was rumoured that London's streets were paved with gold. During the first half of the 16th century London was part of a trade network in northern Europe dominated by Antwerp; but as a collection and distribution point for goods in England, London prospered. When Antwerp was damaged by war, London merchants sought wider opportunities. In 1555, Londoners set up the Muscovy Company to expand trade with Russia, and the Levant Company followed. The East India Company was formed in 1600, and within 20 years was responsible for 5% of metropolitan imports. From 1620 the Americas were a further growth area.

In September 1666 the City was largely destroyed by the Great Fire, which started in Pudding Lane near London Bridge. About three quarters of the city within the walls was destroyed, together with St. Paul's cathedral, a multitude of churches and thousands of homes. Following the destruction wrought by the fire, the opportunity was taken to widen many streets, removing obstructions such as markets and churches. Apart from considerable widening of the quay sides, the only new street actually constructed in the city was King St. and its continuation, Queen St., which led from Guildhall to the wharf. The erection of a flamboyant Monument to the Fire (1677), and the laying out of these streets was the only tangible evidence of a short-lived desire to rebuild London in the style of the Paris of Louis XIV. Up to 1666 the city was largely timber-framed, with few stone buildings and a sprinkling of brick. After 1666 the city must have presented a remarkable contrast, with three quarters of the area within the walls rebuilt almost totally in brick, bordered (at least to the north and east) by older timber-framed buildings. The 1667 Act for rebuilding the City classified buildings into four groups with standards for each group, which brought about a uniformity of street frontage which could already be seen in Bloomsbury. In total, Sir Christopher Wren designed 51 post-fire parish churches in the city, of which 23 remain. The variety of Wren's church steeples was soon to become one of the architectural attractions for foreign visitors to London. Undoubtedly his greatest work is the rebuilding of St. Paul's Cathedral (1675-1711). Wren intended to give the London skyline a dome equal to those in Rome and Paris, and the city has been proud of its most important building ever since. Even today, planning legislation requires that uninterrupted views of Wren's masterpiece are retained from several points outside the City.

See page 68 for a map tracing the progress of the Great Fire, and page 69 for a discussion of post-Fire building regulations, and a map of the Sir Christopher Wren's churches.

Map below The extent of Stuart London at its peak is contrasted to the size of the city in 1820.

A full reconstruction of Sir Christopher Wren's vision of the rebuilt city can be seen on pages 168-169.

EXTENT OF LONDON 1680

LONDON 1820

✝ St. Paul's

✝ Westminster Abbey

LONDON 1820

THE TUDOR CITY

Based on the Agas panorama (c. 1559), the reconstruction of Tudor London (*below*) shows a residential city of gabled houses, royal palaces, law courts, markets, walled gardens and innumerable churches. The north and south banks of the city are linked by the single thread of London Bridge, still surviving from the 12th century. London was poised on the brink of change: although it was still essentially a medieval city, the Dissolution of the Monasteries and its aftermath were to have profound consequences.

As in previous centuries, the underlying natural contours of the city continued to be smoothed out and valleys filled in or encroached upon. The part of the River Walbrook which flowed within the City walls had already been embanked by 1500, and was largely covered over by 1600. Concern with the polluted state of the Fleet, and attempts to deepen it and remove wharves and privies, continued throughout the 16th century. The ditch surrounding the City walls was also filled in over much of its length by the encroachment of new buildings and streets.

London still had a noble skyline of church spires, though the greatest of all at St. Paul's had been removed after being hit by lightning in 1561. The city was studded with fine buildings: royal palaces, civic halls and parish churches. Besides the Tower of London, there were two royal

palaces in London in the Tudor and Stuart period. Baynard's Castle, on the waterfront between Queenhithe and the Fleet, was rebuilt in 1501 by Henry VII as a royal residence in the City. The ambitions of Henry VIII went much further, for his apartments at the Tower and at Westminster had both been destroyed by fire early in his reign, and he was therefore looking for a site on which to build a palace fit for a European prince. Bridewell Palace was built between 1515 and about 1523. Though positioned in the middle of a built-up area south of Fleet Street, its site was of a suitable size because the buildings were raised largely on reclaimed land at the confluence of the Fleet and Thames. Like the contemporary palace at Hampton Court in Middlesex, it was meant to be approached principally from the river.

London's civic buildings and amenities were constantly being improved. During the 16th and early 17th centuries several gates were rebuilt: Ludgate (1586), Aldgate (1608) and Aldersgate (1610). The last two were in solid Renaissance style, like many in towns in continental Europe. Buildings attached to Ludgate and Newgate continued to be used as prisons, and during the 16th century ward prisons were established in Bread Street, the Poultry, Wood Street and in Southwark. The Guildhall chapel and college buildings were lost at the Reformation, but the chapel was restored for civic religious services. Several medieval hospitals were dissolved in 1538-47 but others survived: St. Bartholomew's Hospital was refounded in 1544.

THE SURVIVING BUILDINGS OF PRE-FIRE LONDON
(dates indicate earliest building, or fragment of building, on site)

1. Staple Inn (1580)
2. Lincoln's Inn (1490)
3. Middle Temple Hall (1571)
4. Temple (church, 12th century)
5. Inner Temple Gateway (1610-11)
6. St. Dunstan's Porch (1586)
7. Barnard's Hall (early 15th century)
8. Britton's Court (14th century)
9. St. Bride's (Saxon)
10. Apothecaries' Hall (medieval)
11. St. Sepulchre (15th century)
12. Giltspur St. (medieval wall)
13. Gray's Inn (1560)
14. St. Etheldreda's (1290)
15. St. John (12th century)
16. Charterhouse (14th century)
17. St. Bartholomew the Less (15th century)
18. St. Bartholomew the Great (12th century)
19. St. Paul's Chapter-house (1336)
20. Barbican (city wall)
21. St. Giles Cripplegate (1545-50)
22. London Wall (14th-15th century)
23. Guildhall (late 13th century)
24. St. Mary Le Bow (late 11th century)
25. St. Mary Aldermary (tower, 1511)
26. 34 Watling St. (14th-century undercroft)
27. Abchurch Lane (14th-century undercroft)
28. Merchant Taylor's Hall (14th century)
29. St. Ethelburga (14th century)
30. St. Helen's Bishopsgate (12th century)
31. St. Andrew Undershaft (1520-32)
32. All Hallows Staining (15th-century tower)
33. St. Katherine Kree (1628-31)
34. St. Olave (15th century)
35. All Hallows Barking (Saxon)
36. Mitre Street (14th or 15th-century arch)
37. St. Mary Overie (12th century)
38. Clink Street (west gable of Bishop of Winchester's hall, c.1330)

THE TUDOR CITY c.1570

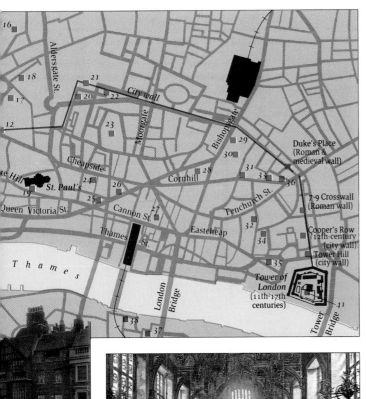

Surviving fragments of the late medieval and Tudor city can still be seen today (map left). They include (below, from left to right) the frontage of Staple Inn, Holborn (1580); Middle Temple Hall (1571). The church of St. Helen Bishopsgate, contains medieval and Tudor monuments to the city's prominent inhabitants.

The reconstructed view (c. 1570, below), is simplified. In reality there were many more buildings packed in the narrow streets. As the following pages show, historians and archaeologists are now filling out the details of this broad impression.

Conduits can be identified from early 16th-century documents as being sited near the main City gates. Water was brought to the City from Hertfordshire and from springs in Islington and Bloomsbury by extensions to the water supply system. The New River was constructed by Hugh Middleton in 1609-13, and augmented by water from the Lea in 1618; it entered the City in wooden pipes, and was distributed to houses.

In 1606-16 the moor on the north side of the City was finally drained and laid out as public gardens, a feature of other major European cities later in the century. Bowling alleys are known within the City walls during the later 16th century, as are tennis courts – both on private properties and in dissolved monastic buildings (such as the former nave of Blackfriars). Cockpits and theatres were mostly situated in the suburbs outside the main City (Shoreditch and Southwark), except for the Fortune Theatre in Golden Lane (1600) and private theatres such as Burbage's, also in Blackfriars.

London was already the national centre for legal education: the Inns of Court and Chancery were all to the west of the City. Some of their surviving buildings date from this period: Middle Temple (Hall of 1571, Middle Temple Gateway 1684, New Court of 1676); Inner Temple (Prince Henry's Room or Inner Temple Gateway, 1611); Lincoln's Inn, Staple Inn and Gray's Inn possess halls of 1492, 1581 and 1560 respectively. Other professions, such as the physicians, also established colleges in London at this time.

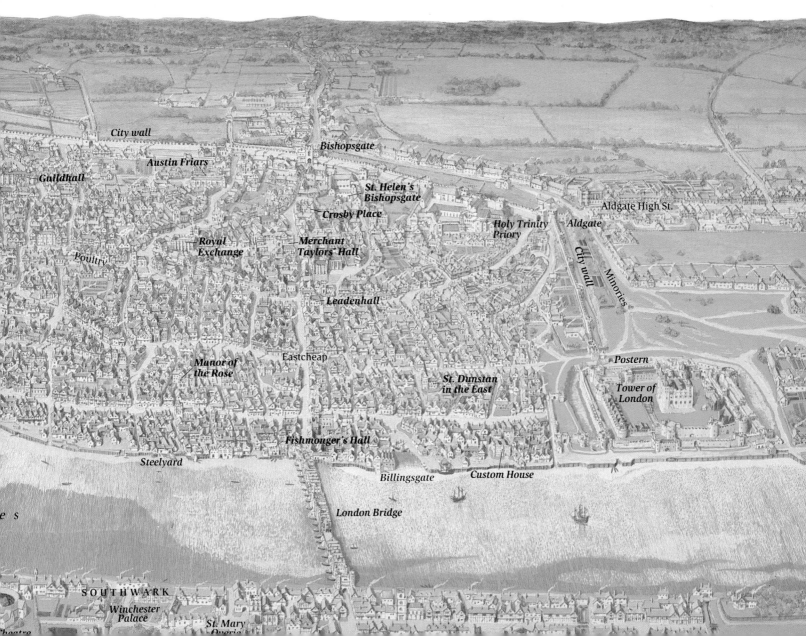

JOHN STOW'S LONDON

John Stow was born in London in 1525, son of a tallow chandler of Throgmorton Street. He became a merchant tailor, and lived for the whole of his life in the City inside Aldgate. A fascination with history and an observant eye encouraged him to write a *Survey of London*. In this he describes the City's churches and institutions, houses and workshops, people and their customs, and the history of each parish in great detail. He alludes often to his own experience of London life, describing the farm just outside Aldgate where he bought milk as a boy, lamenting the spread of 'mean cottages and tenements' out into the surrounding countryside and mentioning the developers of the Tudor city as men he himself knew and had observed. He is, in turn enthusiastic about the vitality of the growing city, and censorious of the commercial ethic of those developers whose wealth was built on the exploitation of poor tenants bewildered by the pace of change. Stow writes as though he were conducting a tour of his native city and from his unique description we can gain a rare insight into its character and development at that time.

London emerges as a tripartite capital. The City at its centre, most of it enclosed by a wall, was its commercial and industrial heart, densely populated, prosperous and busy. It was the historic core of the capital and Stow was fascinated by its complex history and contemporary vitality. Though already built-up, it continued to change in the 16th century, with much rebuilding on old sites, and new building in monastic precincts after the Dissolution of the monasteries. Its many churches and Company halls were augmented by great 16th-century institutions such as the Royal Exchange and it retained a typical medieval mix of inhabitants, from alder-

LONDON OCCUPATIONS

1. Apothecaries	18. Cutlers	35. Heralds	52. Skinners
2. Armourers	19. Curriers	36. Ironmongers	53. Saddlers
3. Butchers	20. Cordwainers	37. Innholders	54. Shipwrights
4. Bakers	21. Clergy	38. Joiners	55. Silversmiths
5. Brickmakers	22. Drapers	39. Leathersellers	56. Salters
6. Brokers	23. Dyers	40. Lawyers	57. Spurriers
7. Blacksmiths	24. Embroiderers	41. Mercers	58. Stationers
8. Basketmakers	25. Fletchers	42. Masons	59. Sailor's Victuallers
9. Brewers	26. Fishmongers	43. Mariners	60. Tenter Stretchers
10. Barber surgeons	27. Founders	44. Merchant Tailors	61. Tallow Chandlers
11. Bowyers	28. Fruiterers	45. Pewterers	62. Upholsterers
12. Clothworkers	29. Grocers	46. Painter Stainers	63. Vintners
13. Carpenters	30. Girdlers	47. Parish Clerks	64. Weavers
14. Clerks	31. Glaziers	48. Physicians	65. Wiredrawers
15. Cheesemongers	32. Glassmakers	49. Plumbers	66. Woodmongers
16. Cooks	33. Goldsmiths	50. Printers	67. Wax Chandlers
17. Corn-millers	34. Haberdashers	51. Plasterers	

● *indicates Company hall*

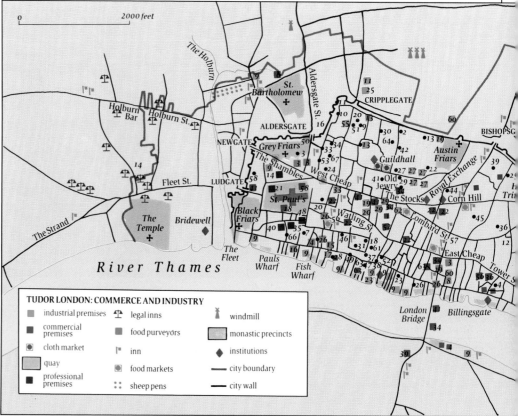

TUDOR LONDON: COMMERCE AND INDUSTRY

- industrial premises
- commercial premises
- cloth market
- quay
- professional premises
- legal inns
- food purveyors
- inn
- food markets
- sheep pens
- windmill
- monastic precincts
- institutions
- city boundary
- city wall

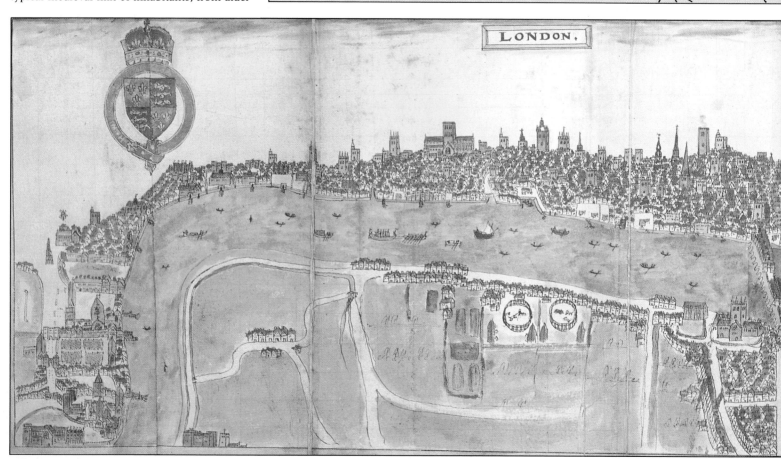

LONDON,

man to pauper, rich and poor living cheek by jowl in a tangle of streets, alleys and courts.

To the west of the City a second area, an exclusive suburb stretching along the Strand towards Westminster, was developing. The landed elite, government servants and lawyers lived in gracious houses along the Thames bank, or in the legal quarter around Fleet Street. It is of particular interest because it heralds a new kind of suburban environment. Unlike the City, with its mixture of rich and poor, craftsmen and merchants, the West End was economically and socially more homogeneous, built for, and inhabited by, political and professional groups.

The third area, a more diffuse suburban growth to the north and east, and in Southwark and its environs, provided a counterbalance to the West End. It housed a growing army of craftsmen and semi-skilled workers who serviced the trade and wealthy population of the capital. Living in an altogether shabbier area dominated by manual work, and with few residential amenities, such workers grew rapidly in numbers and helped increase London's population threefold in the course of the century. It was here that the shipwrights and sailors settled who made London's growth as a centre of overseas trade possible. The social tone of the area was very different to that in the west.

At the time Stow wrote his *Survey*, London was undergoing momentous change. The capital was bursting its medieval bounds and, impelled by the growth of capitalist enterprise and supported by large-scale immigration from the rest of England, the expansion which was to transform the old City into a metropolis had already begun.

Commerce and industry *The map (left) shows that commerce and shops tended to be concentrated in the centre, and industry around the city walls, especially in the east.*
London Housing *The map (top right) shows a clear pattern: an exclusive West End, a prosperous but mixed centre, and a much poorer suburban sprawl, especially towards the East End.*
Residents and Social Institutions (right, centre) *Landed gentry, the governing elite and professionals were based in the West End and Westminster, close to the centre of national government (map centre). Aldermen lived in the city, emphasising its business orientation. Hospitals are near the city wall, within ex-monastic precincts.*
16th-century Tudor development (below right) *The 16th-century witnessed the erection of eight great buildings, 21 new churches, much industrial and commercial development, especially towards the east, and the provision of many amenities, especially within the city.*
Map below *Bird's eye view of London from the* **Particular Description of England**, *William Smith, 1588.*

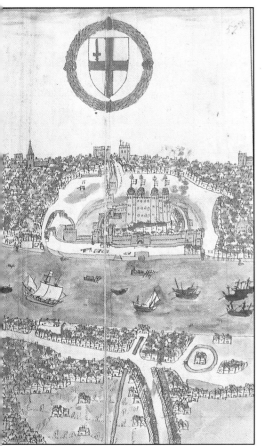

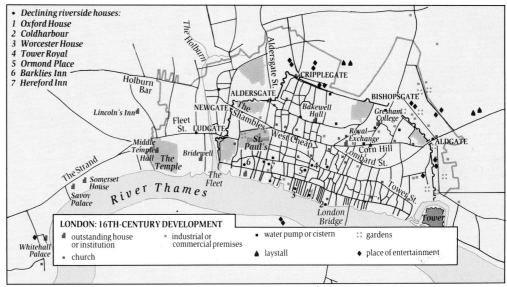

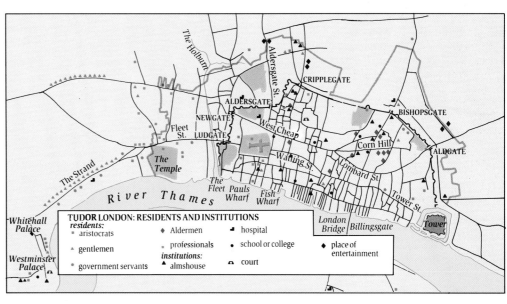

TUDOR TRADE AND COMMERCE

London's status as Britain's largest port was firmly established during the 16th and 17th centuries. In 1559, certain landing places along the quayside, major alleys and watergates of the medieval period, were designated official quays by statute. The chief 16th century quays were still Queenshithe and Billingsgate, but by 1600 ships were setting out on international voyages from the new suburbs downstream of the City: Deptford, Wapping and Ratcliffe. The monopoly of the old-established medieval inlets had been superseded. During this period, companies of merchants were also founded to widen London's international trade. One of the most prominent was the East India Company (founded 1600), although it had no buildings of its own until the 18th century. The Royal Exchange (1571) also helped to stimulate international trade. Duties were levied on merchants at the Custom House, rebuilt in 1559 after being destroyed by fire. It was destroyed again by the Great Fire, and rebuilt by Wren.

Several company halls, the headquarters of the guilds and livery companies, were rebuilt or augmented during the 16th and 17th centuries. Companies took advantage of the Dissolution of the Monasteries and the Reformation: the Leathersellers acquired the large dormitory of the former nunnery of St. Helen Bishopsgate in 1542, and the Butchers acquired the parsonage of the suppressed parish of St. Nicholas Shambles in 1549. Many smaller companies (such as the Embroiderers, Fletchers, or Innholders) now had halls: at least 47 are known by 1600. Taverns were used by some companies as an alternative to their halls. The Weavers, for instance, had a hall but dined in taverns. The Cheapside area was thick with taverns which must have had a regular trade in company meetings and feasts.

The houses and shops of London before the Great Fire can be reconstructed from plans, documents and later engravings. The wealth of the Tudor capital was displayed to the world in large shops on its principal streets. Houses of prominent tradesmen incorporated both cellars and warehouses for merchandise, while well-appointed rooms above the shop provided the living quarters. A typical household would have been crowded; replete with extended family, apprentices and servants.

A mercantile capital The map (bottom) shows the location of the Tudor company halls, now situated in all parts of the City, although only two were outside the walls. By 1600 the guildhalls were no longer at the centre of their specialist retail areas, as they had been in the 13th and 14th centuries. The shields of the 12 greater Livery Companies are depicted (below) in their order of precedence at the Mayor's Feast at Guildhall every year.

The composition of London's imports (above right), showing the relative amounts of manufacturers' foodstuffs and raw materials imported. The number of ships entering London from foreign ports (below right) shows a great increase over the period. By 1686, trade across the Atlantic and with the Far East also played an important role in London's commerce.

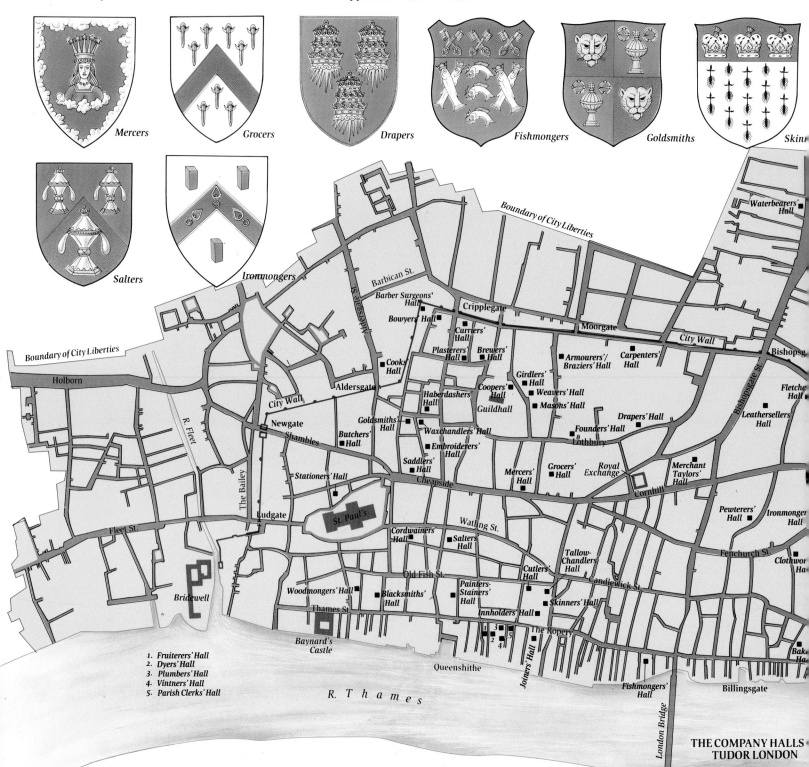

Mercers　　Grocers　　Drapers　　Fishmongers　　Goldsmiths　　Skin...

Salters　　Ironmongers

1. Fruiterers' Hall
2. Dyers' Hall
3. Plumbers' Hall
4. Vintners' Hall
5. Parish Clerks' Hall

**THE COMPANY HALLS
TUDOR LONDON**

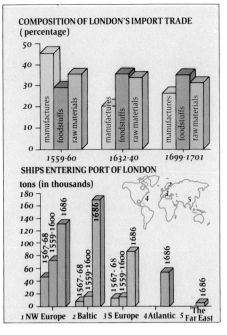

COMPOSITION OF LONDON'S IMPORT TRADE (percentage)

50
40
30
20
10
0

manufactures foodstuffs raw materials — 1559-60
manufactures foodstuffs raw materials — 1632-40
manufactures foodstuffs raw materials — 1699-1701

SHIPS ENTERING PORT OF LONDON

tons (in thousands)
180
160
140
120
100
80
60
40
20
0

1567-68 1559-1600 1686 — 1 NW Europe
1567-68 1559-1600 1686 — 2 Baltic
1567-68 1559-1600 1686 — 3 S Europe
1686 — 4 Atlantic
1686 — 5 The Far East

Byrsa Londinensis vulgo the Royal Exchange.

Merchant Taylors

Haberdashers

Vintners

Clothworkers

rndsditch

Aldgate

'klayers' 'Tilers'

Poor Jewry

Tower of London

The Royal Exchange *(above right)*, *founded by Thomas Gresham, an Elizabethan financier, was opened in 1571. It provided a purpose-built headquarters in which merchants could conduct their business. Modelled on the bourse in Antwerp, it sought to promote London as a commercial capital. Above the great open piazza were a hundred small shops, a thriving centre for milliners, armourers, apothecaries, booksellers and goldsmiths. The niches above the covered walks facing on to the piazza were decorated with statues of English monarchs.*

A Pre-Fire house in Cornhill, *facing the Royal Exchange, was surveyed in 1612 by Ralph Treswell for the Clothworkers' Company which owned it. The details have been reconstructed (right) from engravings of other London houses. The Tudor house was a compact domestic unit; at the ground floor street front there was a shop, with a warehouse behind. Living quarters were on the first floor and above. A fine hall graced the front of the house, while a gallery led to further rooms over the rear kitchen. Merchandise was stored below in large cellars with an entrance to the* street. *This house shared a well with the property next door; in an earlier century the well may have been dug in the courtyard of a larger house, which was later subdivided in this way to make space for smaller houses, closely-packed. There was little, if any, adjacent private open space in the city centre.*

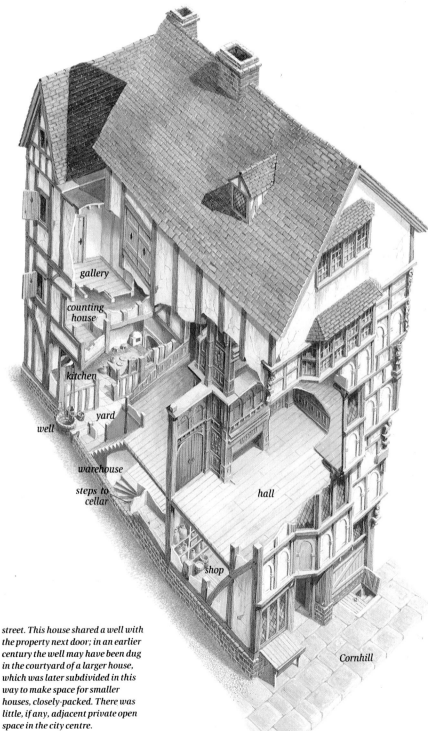

gallery

counting house

kitchen

yard

well

warehouse

steps to cellar

hall

shop

Cornhill

THE GROWTH OF LONDON AND THE NEW SQUARES

London in the 16th century saw very marked population growth in the parishes which fell outside the City walls. The pressure on existing housing which this created resulted in the appearance of a new generation of buildings and alleys which were spread over a wide area of former fields to the east of the City, fanning out from the ribbon-developments which already flanked the approach roads into London. Much of this building took place in previously 'rural' hamlets, especially Ratcliffe, Limehouse, Shoreditch, Whitechapel and around St. Katharine's Hospital. In this piecemeal way, the East End of London was born.

Although open space to the north and west of the City was still being used for animal grazing and cultivation, these activities were also being gradually forced out to Islington and surrounding villages, as development began to fill in the fields and open spaces which lay outside the City walls. The village of Clerkenwell, for instance, was absorbed by the northern spread of the City. By the end of the 16th century, in the wake of the dissolution of the monasteries, bishops and priors had been ejected from their fashionable mansions along the Strand. A string of residences built for nobles and courtiers such as Essex House, Arundel House and Salisbury House, replaced them. By the 1630s, the Crown had abandoned its attempts to prevent expansion and now sought to limit growth by allowing only those developments

17th-century expansion *The painting (left), by John Collet, shows Covent Garden c.1770-1780, many years after it was bulit in the 1630s. By this time it had come to be used as a general market – the centre of fashion had by then moved further west to Westminster and St. James's. Originally, the Third Duke of Bedford's garden bordered the piazza on the south (left of painting) – it was his cousin and successor who secured a licence for building gentlemanly residences to the north of his garden.*

The map (below left) graphically illustrates the spread of London in the first part of the 17th century. Areas of the City, both inside and outside the walls, grew at different rates, and in 1640 certain trades could be identified with certain locations (graphs right). Merchants and officials lived and worked in the old City; but those concerned with distribution, victuals and labouring were found largely in the new suburbs. Hollar's view of west central London in the 1650s (below right) shows a bird's eye view of Covent Garden and many of the surrounding developments. Bedford House is clearly visible to the south of the square.

which reached certain standards. In this way areas such as Covent Garden and Lincoln's Inn – spacious squares flanked by elegant houses for gentlemen and aristocrats – came to be built. The principal architect of this revolutionary new look was Inigo Jones, a Londoner by birth, but heavily influenced by the Italian Renaissance and the architecture of Palladio. Over the next 27 years he not only designed certain specific buildings, such as the Banqueting Hall in Whitehall, but was influential in the layout of many other schemes. Spacious squares, surrounded by well-appointed houses with uniform Italianate facades began to extend westwards toward Royal Westminster. No shops or other signs of commerce were allowed.

The East End, on the other hand, was fast becoming a mixture of houses and small industrial concerns: bell-founding, glass-making, ivory and horn working and, later, silk-weaving and paper-making. These industries flourished outside the City because of several factors: the lower cost of rents; the exclusion of certain trades from practising within the walls; the failure of the City authorities to control the industries springing up in these areas. Within the City itself, traditional small-scale industry and manufacturing continued to thrive: carpenters, cobblers, tailors and printers were all still based within the walls. However, by the 17th century a division between the East and West End was emerging which was to have profound long-term consequences on the geography of London – government and service industries were based to the west of the City, financial services were located in the City itself, and manufacturing spread out to the east.

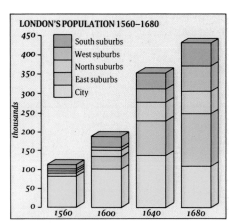

LONDON'S POPULATION 1560–1680

- South suburbs
- West suburbs
- North suburbs
- East suburbs
- City

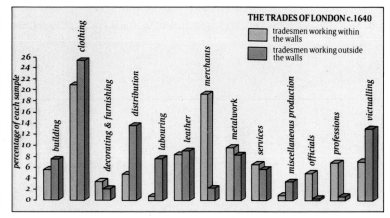

THE TRADES OF LONDON c.1640

- tradesmen working within the walls
- tradesmen working outside the walls

TH OF 17TH CENTURY LONDON
ieval city wall
ndary of City Liberties
f London in:

Parish of St. John, Hackney

to Harwich →

THE HAMLET OF MILE END

gardens

Hangman's Acre

Back Lane

Rose Lane

SHADWELL

Bell Wharf

Ratcliffe Dock

Globe Stairs

Stepney Church

S. Giles Fields

S. Giles

Piazza in Coventgarden

Bedford House

LONDON BEFORE THE GREAT FIRE

The 17th-century City lived through the Civil War in the 1640s, the Great Plague of 1665, the Great Fire of 1666, and the subsequent Great Rebuilding: as a consequence, London in 1700 bore little resemblance to the town in 1600. When the century opened, London lay cramped and crowded within its medieval walls with open fields to north and east, and a rich suburb expanding westwards towards Westminster. Within the City, surrounding some 100 churches, company halls and other public buildings were streets, lanes and alleys cluttered with tall timber-framed buildings. Rich and poor lived side by side, although the wealthier inhabitants often occupied secluded and more spacious sites set back from the street frontage. Houses of

artisans and shopkeepers opened directly onto the road, with little space for a yard behind. The poorest inhabitants lived in single rooms, usually in an upper storey.

Commercial life was in the hands of some 100 companies whose halls were a noteworthy feature of the City. These guilds exercised control over all aspects of trading and manufacture through such measures as their insistence upon a seven-year apprenticeship before granting the freedom to ply a particular trade in London. As a direct result of such restrictions, new businesses often set up outside the walls where the long arm of the guilds could not always reach them. These suburbs developed dramatically as the population rose from 200,000 in 1600 to 375,000 by 1650, an increase due almost entirely to an influx of migrants from the poorer

The Civil War Defences of London 16413-7 (below left) *During the Civil War, London was protected by the largest circuit of town defences in Europe. It incorporated an earthen bank and ditch running between a series of forts and batteries. This map shows the approximate position of the defences, but includes alternative alignments (as between Hyde Park Corner and Millbank) where the actual line is in doubt.*

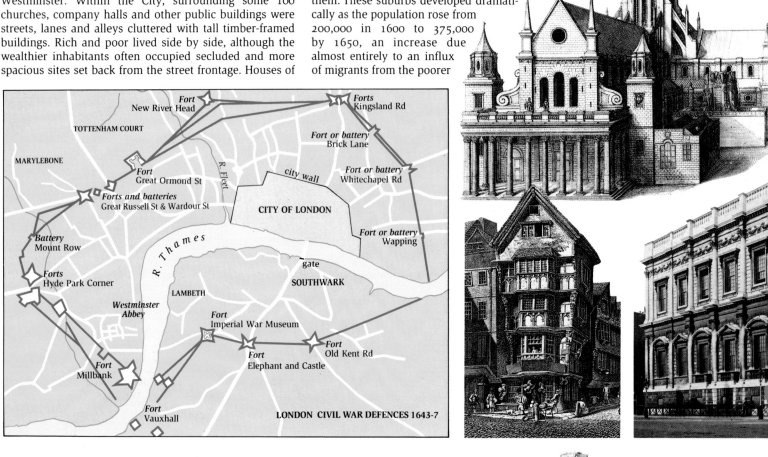

LONDON CIVIL WAR DEFENCES 1643-7

rural areas of England. In the early 17th century the northern and eastern suburbs grew fastest and were filled with artisans and small industries. The western suburb began expanding slightly later, providing services and luxury goods for the capital.

However, this was a century remembered more for its upheavals than for prosperous expansion. In December 1641, following the ousting of the pro-Royalist regime at the election to the Common Council, London became the capital and chief port for the Parliamentarians against the King. The Royalist army was repulsed at Turnham Green in November 1642, after which work began on a massive project to defend the City and its new suburbs. Comparison of the area enclosed by the medieval town wall with that enclosed by the 17th-century work shows how much larger London had become, although extensive areas of open ground were deliberately included within the Civil War circuit. The eleven mile-long defence was a costly

undertaking beset with difficulties, highlighted by the fact that strict Puritans were obliged to countenance Sunday working to get the job done. At the close of the Civil War, London was on the winning side, which did not endear its citizens to later Stuart kings. However, the new defences were never attacked, and were so thoroughly dismantled after 1647 that little trace remains today.

London was rich, powerful and a fortress of Protestant interest but it was also noted for its dirt, overcrowding and squalor. Plague, most unwelcome of all imports brought into this prosperous port, thrived in such conditions, killing over 25,000 Londoners in both 1603 and 1625, and 10,000 in 1636. But the worst and last visitation was in 1665, when over 70,000 citizens died. Large communal graves were dug outside the City, for the churchyards were full: these grim statistics were conscientiously compiled by the parish clerks in the Bills of Mortality, weekly lists of deaths and their cause.

The Great Plague of 1665 *hit the poorest quarters hardest (map left). The Bills of Mortality (below) recorded losses. Migration into the city meant that population still increased (graph).*

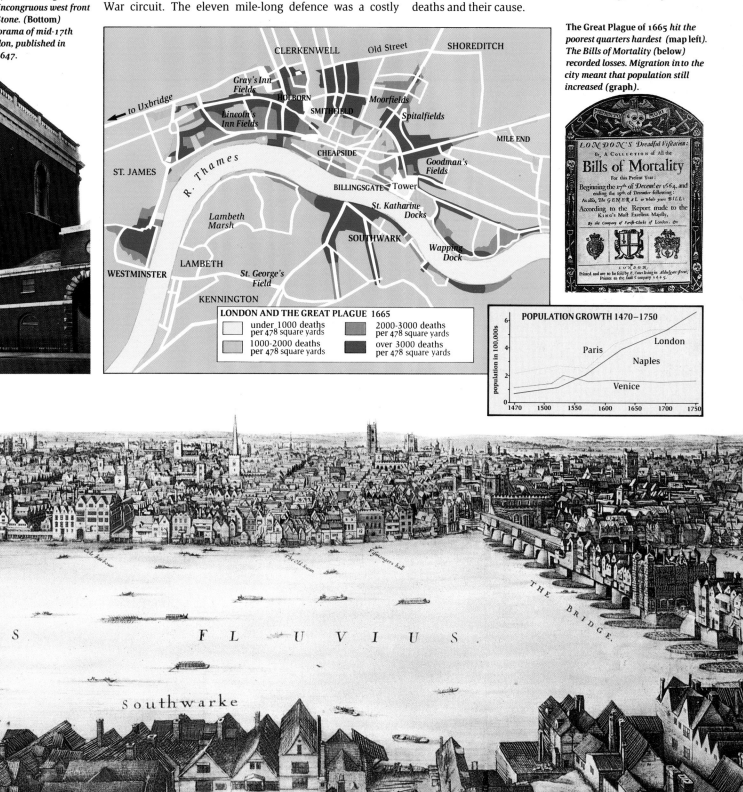

LONDON AND THE GREAT PLAGUE 1665

- under 1000 deaths per 478 square yards
- 1000-2000 deaths per 478 square yards
- 2000-3000 deaths per 478 square yards
- over 3000 deaths per 478 square yards

POPULATION GROWTH 1470–1750

THE GREAT FIRE AND THE GREAT REBUILDING

London, like any ancient town, was prey to fire, and had suffered from many conflagrations. What made the disaster of 1666 different from the rest was its sheer scale. It all began by accident or negligence, in Thomas Faryner's bakehouse in Pudding Lane. The City was tinder-dry after a long hot summer. It was early on Sunday morning, the 2nd September, and by the time Londoners were awake to the danger, the fire was already out of hand. With hindsight it is easy to blame Londoners for not acting faster on that first morning. Many citizens understandably refused to allow their houses to be pulled down to create adequate fire breaks; others, like Samuel Pepys, simply thought the fire a long way off and of little consequence, and went back to bed. They had seen fires before, and this one looked no different. They were soon proved to be wrong. Fanned by a driving east wind, it jumped whatever firebreaks had been created. In spite of the best endeavours of parish and ward officials, and even of the Lord Mayor, Sir Thomas Bludworth, the fire moved faster than the firefighters: the City could not save itself. Charles II, arriving from Whitehall to inspect the damage, could do little more. The fire raced west along the waterfront, which was packed with combustibles such as timber, pitch and oil, reaching Three Cranes (present-day Southwark Bridge) by the afternoon. The Londoners gathered up their possessions and fled by foot, road and river.

The fire burnt all night and gained momentum throughout Monday. By order of the King, the Duke of York (later James II) was placed in control of the city, and his guards tried to prevent disorder and looting. Firebreaks at Queenhithe were ineffective, and the fire advanced west towards the River Fleet, and north beyond Cornhill and the Royal Exchange. Belated arrangements were made to check its progress on the northern and western sides of the City with the establishment of Fire Posts, each manned by 130 men with orders to create firebreaks. Had such a system been used earlier, much might have been saved.

By Tuesday morning, even these measures seemed inadequate, and the militia from Middlesex, Hertfordshire and Kent were ordered into the City to prevent riots and fight the fire. But not even the Fleet River served as an effective firebreak: later that day, having destroyed St. Paul's, the Guildhall, Custom House, and much else, the flames burst out of the City gates, leapt the Fleet, and attacked Fleet Street, threatening, for the first time, Whitehall and the Royal residences. Gunpowder was used to clear firebreaks and save the Tower of London.

On Tuesday night all seemed lost, for the Fire was advancing in a vast arc that stretched from Temple church in the west, to Smithfield and St. Giles Cripplegate in the north, to Leadenhall Market and All-Hallows by the Tower in the east. At this point the wind dropped, allowing the tired fire-fighters to check, control and dowse the flames, which they did throughout Wednesday and Thursday. By Friday, they had succeeded and, exhausted, could stop and count the cost. The devastation which confronted them was horrific. London was a vast, unrecognisable blackened ruin. The destruction within the walled area was worse than that suffered by the City in the Blitz: the Great Fire destroyed over 13,000 houses, 87 churches, 52 company halls, and much more besides. The total loss was estimated at £10,000,000 at a time when the City's annual income was £12,000. Could the City ever be rebuilt? Some contemporaries thought not.

The Great Fire A contemporary painting by an unidentified Dutch artist (below), shows the fire raging uncontrollably. The map (below) plots the daily progress of the fire. The total area devastated in September 1666 was greater than the extent of the city destroyed during 1940-1 (inset map). Wenceslaus Hollar drew St. Paul's after the fire in 1666 (bottom). The burning roof had crashed into the crypt, destroying thousands of pounds worth of books and other goods stored there.

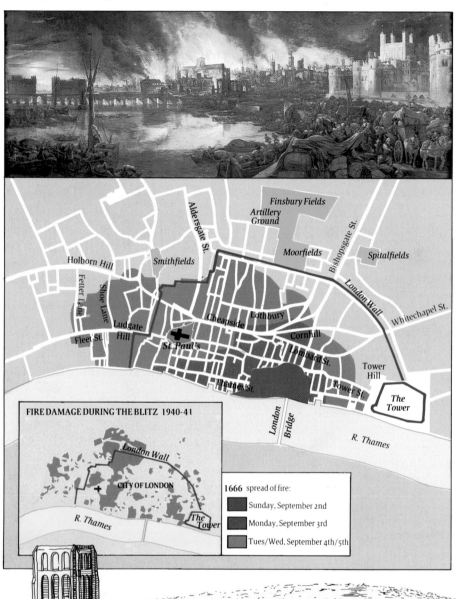

FIRE DAMAGE DURING THE BLITZ 1940-41

1666 spread of fire:
- Sunday, September 2nd
- Monday, September 3rd
- Tues/Wed, September 4th/5th

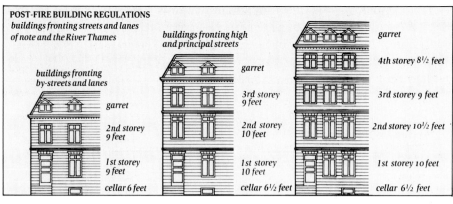

Rebuilding the City The 1667 Rebuilding Act decreed that terraced brick buildings of uniform design would replace timber-framed houses (diagram). New buildings in back streets were two-storeyed; those on other streets, three-storeyed; those on main roads, four-storeyed. (Right) John Evelyn's (rejected) plan for a redesigned City, incorporating a radically new pattern of roads and roundabouts. 51 of the 87 City churches destroyed in the fire were rebuilt (map below), all of them designed by Sir Christopher Wren. About 25 survive today. The rebuilding programme culminated in the building of Wren's masterpiece, St. Paul's Cathedral, which was completed in 1711 (below right).

WREN'S SURVIVING CHURCHES

(*s indicate only tower or shell survives)

1. St. Paul's
2. St. (Anne and St.) Agnes
3. St. Andrew Castle Baynard (by the wardrobe)
4. St. Benet on Thames
5. St. Clement Candlewick Street (Eastcheap)
6. St. Edmund
7. St. James Garlickhithe
8. St. Lawrence Jewry
9. St. Magnus the Martyr
10. St. Margaret Lothbury
11. St. Margaret Pattens
12. St. Martin Ludgate
13. St. Mary Abchurch
14. St. Mary Aldermary
15. St. Mary-le-Bow
16. St. Mary at Hill
17. St. Mary Woolnoth
18. St. Michael Cornhill
19. St. Michael Paternoster
20. St. Nicholas West Fishmarket (Cold-abbey)
21. St. Peter Cornhill
22. St. Sepulchre (Newgate)
23. St. Stephen Walbrook
24. St. Vedast
25. St. Alban Wood Street*
26. St. Augustine by St. Paul*
27. St. Dunstan towards the Tower (in the East)*
28. St. Mary Somerset*
29. Christchurch*
30. St. Olaf Jewry*
31. St. Brides
32. St. Mary Aldermanbury (rebuilt in Fulton, USA)

THE GREAT REBUILDING

There were many who saw London as an inelegant and insanitary city; the aftermath of the fire presented an opportunity to create a radically new town. The rebuilding plan had to reconcile general improvements to the City with the rights of individual property owners. The proposals promptly submitted by men such as John Evelyn and Christopher Wren (*page 168/9*) were superficially attractive, but incorporated street plans at variance with the medieval pattern and complexities of property ownership. Such bold schemes were rejected by the Corporation in favour of a more practical plan. The six commissioners appointed to

redesign the City drew up the most comprehensive town-planning legislation ever seen in England. Over 100 streets and lanes were widened, gradients diminished and two new streets, King Street and Queen Street, laid out. Timber was banned and the majority of the new buildings were of a uniform red brick design. The rebuilding of all private houses and company halls had to be financed from whatever funds individuals had saved from the flames, whereas a tax on coal raised £736,000 for the public works programme. This included the canalisation of the Fleet River, extensions to the Thames Quay, and the rebuilding of the parish churches, St. Paul's and public buildings such as the Guildhall.

The speed of the recovery was as remarkable as the results: by 1671, 9,000 houses and several major public buildings were complete. Work then began on rebuilding some 50 churches, and finally St. Paul's, which was built between 1675 and 1711. By that date, London was a cleaner, safer city with well-ordered streets of red brick and white stone buildings. All this work was achieved in spite of wars against the Dutch which saw an enemy fleet in the Thames in 1667, the Monmouth Rebellion in 1685, the Glorious Revolution in 1688, and some of the hardest winters on record. The Great Fire was a tragedy, but the Great Rebuilding was a triumph: London now presented as elegant a facade to the 18th century as any of Europe's capital cities.

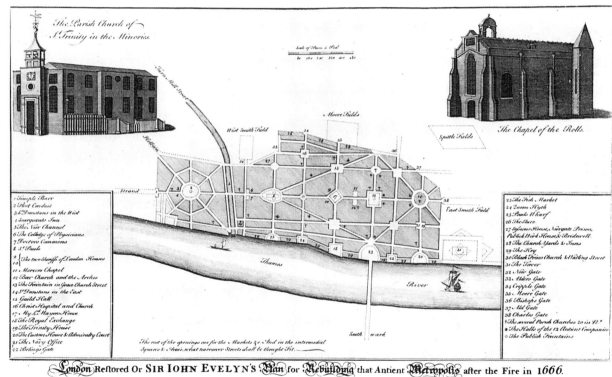

London Restored Or SIR IOHN EVELYN'S Plan for Rebuilding that Antient Metropolis after the Fire in 1666.

POST-FIRE CHURCHES

✝ churches destroyed, not rebuilt

✝ churches rebuilt, tower or shell surviving

✟ churches which survived the Fire

✝ churches rebuilt after Fire, still surviving

✝ churches rebuilt, subsequently destroyed

CHAPTER 6
GEORGIAN LONDON

In 1700 LONDON was not only abominably insanitary, having no main drainage, no publicly provided collections of household refuse and utterly inadequate supplies of clean water, but it lacked many other basic services. It had no public transport, no street lighting, no minimum standards of paving, no paid police force, no fire brigade and no restrictions on the sale of liquor, tobacco or dangerous drugs. During the course of the 18th century, City merchants, shipowners and bankers gained large fortunes from maritime trade and the acquisition of new colonies in North America, the West Indies, the Indian Ocean, the Pacific and Africa. Some of this wealth was spent on fine architecture, some found its way into metropolitan improvements and a great deal was invested in suburban development. By the end of the Napoleonic Wars, steps had been taken to improve sanitation, street lighting, paving, policing and public open spaces. Elegant residential suburbs had arisen in the West End and manufacturing districts had expanded in the east. Outlying villages such as Islington, Blackheath and Twickenham had been transformed into satellite towns. Had a citizen of 1700 returned to London and its environs in 1815, he would not have recognised more than one street in ten, or one building in a hundred.

As the West End was developed, so the rent rolls of the great estates increased (see map page 75). The Duke of Bedford's estate, for example, yielded £3700 in 1732 and over £8000 in 1771.

This period saw the growing prosperity of urban landowners and the rise of the newly enriched middle classes. Many aristocrats enjoyed rich pickings from the fruits of offices of state, as well as income from investments in joint stock companies, banks, insurance companies and overseas commerce. Large fortunes made out of trading ventures and commercial activities established many new families among the residents of fine houses in Mayfair and rapidly expanding estates north of Oxford Street. The expansion of the Bank of England, the opening of the new Stock Exchange, the founding of many private banks, the establishment of almost all the leading insurance companies and a consolidation of the wealth of the livery companies secured the preeminence of the City among financial capitals at the beginning of the 19th century.

The ranks of the newly enriched middle classes were swollen not only by financiers but by a throng of manufacturers and tradesmen. Much industrial activity was situated on the fringes of the built-up area as noxious and dangerous industries were forbidden to locate within the boundaries of the City of London. The insanitary, water-polluting business of dressing and tanning leather was concentrated south of the Thames on the banks of the River Wandle. Foul-smelling glue making, soap boiling, tallow candle-making works were located a little further downstream along the Thames in Rotherhithe and Blackwall. Woodworking industries were kept out of the City because of the dangers of fire; the furniture industry was gradually forced to leave Shoreditch for Camden Town and other places at a safe distance from the City. Sawmills, wheelwrights' shops and cooperages making wooden barrels were situated in Deptford and Greenwich. Small foundries in Poplar, Wapping and Limehouse were connected with trades directly serving shipbuilding and ship repairing yards along the Thames. The docks were beginning to attract their own processing industries, such as sugar refining. Further afield at Stratford in Essex were large flour mills and substantial breweries were established at Greenwich, Hammersmith and Chiswick. The advancing edge of the built-up area closed over a multitude of brick and tile works whose smoking kilns blackened the air in Marylebone, Paddington and Kensington.

For a map of Huguenot churches, please refer to page 136.

The silk weavers of Spitalfields were located closer to the City and nearer to retail shops in the West End. Huguenot refugees had settled there in 1685 and in 1745 no fewer than 133 master weavers employed thousands of journeymen and apprentices. The twisting lanes and alleys of Clerkenwell were crowded with hundreds of watch- and clock-makers, cutlers, surgical instrument-makers, jewellers, diamond-cutters, coach-makers, upholsterers, locksmiths and many other craftsmen. In Westminster in 1749 a directory of trades listed dozens of victuallers, tailors, dressmakers, peruke makers, shoemakers, carpenters, butchers, chandlers, bakers and distillers. It was a royal village, supporting a multitude of tradesmen who catered for the luxurious tastes of the ruling elite.

Georgian London generated a prospering middle range of people, whose incomes were derived from shrewd investments overseas, holdings in government stock, rents from urban property, earnings from professions as varied as acting, portrait painting, publishing, preaching, advocacy, medicine, underwriting, stock-broking, merchanting, shopkeeping, building and catering. Parts of London took on the character of specialised business districts. The Inns of Court provided chambers for lawyers, hospitals were adjoined by consulting rooms for physicians and surgeons, St Paul's Cathedral and Westminster Abbey provided quarters for clergy and also premises for robemakers, printers, publishers, booksellers and journalists. Drury Lane and the Haymarket attracted theatres; St Martin's Lane furnished studios for painters and shops for artists' suppliers. A large number of coffee houses and clubs served as meeting places and centres for gathering and passing on news and views on professional, social and political matters. The places that catered for these transactions spread over much of the West End during the Georgian era. By 1815 the centre of London's clubland had shifted decisively westwards to St James's Street and the most prestigious clubs were housed in elegant architect-designed buildings.

While the rich and the moderately well-to-do gained enormously in numbers, power and wealth during the 18th century, those who earned wages were less favoured. Wages persistently lagged behind the price of bread and household necessities. Poor accommodation and exorbitant rents were causes of far more serious grievance and discontent throughout the century and housing problems grew steadily worse as time passed by. Appalling conditions prevailed in the dilapidated courts and alleys and crumbling tenements that sheltered many wage-earning families. Such was the press of people coming to London in search of employment that modest houses took in more lodgers

A map showing the location of the Georgian clubs of St. James's, and their dates of opening, can be found on pages 76-77.

Below The Thames from Somerset House Terrace towards Westminster, *by Canaletto, 1750-51. His portrayal of the Thames, like the Grand Canal in Venice crowded with brightly coloured barges and ferries, hardly exaggerates the amount of traffic on the waterway, but his depiction of scores of baroque churches and Palladian palaces built in gleaming white Portland stone gives a heightened impression of the grandeur and urbanity of the city.*

than they could decently accommodate, and stables and sheds were converted into makeshift dormitories. But almost as fast as new rooms were occupied old ruinous buildings collapsed. In 1738, Samuel Johnson remarked that London was a place 'where falling houses thunder on your head'. Leaking roofs, damp walls, floors awash with effluent from impeded drains, bad ventilation, infestation with vermin, darkness, cramped, evil-smelling garrets and basements were just some of the problems that had to be contended with. Death rates continued to exceed birth rates and infant mortality was particularly high in all districts where poorer families were concentrated. Consumption and dysentery were endemic. Smallpox was not brought under control until the end of the 18th century. Periodic visitations of typhus and cholera killed off large numbers and cold, wet winters took a heavy toll of the most vulnerable age groups. Only continual replacement by a massive influx of young men and women from the countryside ensured an overall increase in the population.

In 18th-century London, drunkenness, prostitution, pickpocketing and personal assaults were rife, scarcely to be brought under control as long as the number of very poor people continued to increase and the gap between rich and poor widened. In 1797, Patrick Colquhoun, a magistrate in Tower Hamlets, estimated that London's criminal underworld still numbered 115,000, about one in eight of the population, and many of these people were concentrated in the old rookeries of Seven Dials, Chick Lane and Field Lane, Bethnal Green, Petticoat Lane, Houndsditch and Southwark. The containment of crime was assisted by the introduction of street lighting in the City in 1736. A Watching and Lighting Act for Spitalfields in 1738 contributed to an appreciable reduction in street robberies, burglaries and other offences.

In 1700 an observer who climbed to the gallery around the dome of St Paul's Cathedral would have been able to view London in its entirety. Pastures and scattered farms extended from Lamb's Conduit Fields to the distant hilltop hamlets of Hampstead and Highgate; isolated clusters of buildings at Covent Garden, Lincoln's Inn and Leicester Fields stood in the midst of fields; Hyde Park and St James's Park were surrounded by open countryside. An insistent impression of London at this time would have been its rusticity. Within the City itself gardens and orchards lay behind taverns and City company halls, in the Inns of Court and around hospitals, schools and almshouses. Chestnut trees and planes spread their shade over the city squares and fragrant lavender and rose bushes bloomed in private gardens off the Strand. Every Monday morning Londoners were awakened by a cacophony of animal sounds

Coal fires poured soot and sulphur dioxide into the atmosphere, necessitating frequent bathing, washes of clothes and changes of household linen. Even London's vegetables were contaminated with soot and smoke. They were expensive and unpleasant to eat, and a French observer complained in 1765: 'All that grow in the country about London, cabbage, radishes and spinnage, being impregnated with the smoke of sea-coal, which fills the atmosphere of the town, have a very disagreeable taste.' A full discussion of London's history of pollution can be found on pages 134-135.

Left *Vauxhall Gardens, by Thomas Rowlandson, (1756-1827). Originally called Spring Gardens (see map, page 142), they opened just after the Restoration in 1660. Admission was free, and the gardens appealed to all classes of society. In the 18th century the Gardens were greatly elaborated. Supper boxes, Gothic ruins, Chinese pavilions, cascades and many other attractions were added. In the words of a contemporary ballad: 'Here they drink, and there they cram/ Chicken, pasty, beef and ham,/ Women squeak and men drunk fall,/ Sweet enjoyment of Vauxhall.*

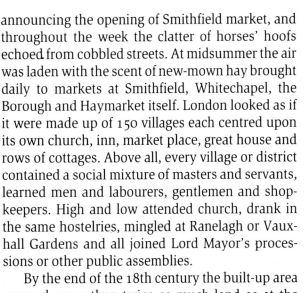

announcing the opening of Smithfield market, and throughout the week the clatter of horses' hoofs echoed from cobbled streets. At midsummer the air was laden with the scent of new-mown hay brought daily to markets at Smithfield, Whitechapel, the Borough and Haymarket itself. London looked as if it were made up of 150 villages each centred upon its own church, inn, market place, great house and rows of cottages. Above all, every village or district contained a social mixture of masters and servants, learned men and labourers, gentlemen and shop-keepers. High and low attended church, drank in the same hostelries, mingled at Ranelagh or Vaux-hall Gardens and all joined Lord Mayor's proces-sions or other public assemblies.

See page 79 for a contemporary illustration of Smithfield market and a map of Georgian markets.

By the end of the 18th century the built-up area covered more than twice as much land as at the beginning, and social segregation had advanced to an even greater extent. Surveying a panorama from the dome of St Paul's in the early 19th century, it would have been difficult to discern the edge of the urban area. Not only had fog and smoke impaired visibility but ribbons of houses and inns stretched along many roads to the far horizon. Everywhere fields were marked out as building plots and new pits were dug for gravel and brick-earth. Daniel Defoe recalled a time when Brick Lane, in Spitalfields, 'had been a deep dirty road, frequented by Carts fetching Bricks that way from Brick-kilns' in fields at Whitechapel. Like hundreds of other country lanes it became a well-paved city street. The New Road, constructed in the middle of the 18th century to by-pass the northern fringes of the built-up area, was reported fifty years later to be 'skirted on both sides with houses' for much of its length. To the South, St George's Fields were rapidly being covered with buildings.

For a discussion on brickmaking see page 133.

The most remarkable changes that came over London during the 18th century were the planning and building of separate residential districts for the rich in the West End and an exodus of middle-class residents from much of the City and from the East End. From the Strand north to Holborn, respectable families were moving out. By the beginning of the 19th century premises in Tottenham Court and mews on the Brewers' Company estate, had begun to be taken over by knife grinders, cab drivers, hawkers of fruit and vegetables, rag and bone collectors and dustmen. Around the Polygon in Somers Town, a shanty town of do-it-yourself houses was springing up. By this time, the middle classes had mostly left the City and its immediate environs east of Aldgate and on the south bank, and their houses were subdivided into tenements; colonies of moderately affluent people had retreated to Blackheath, Dulwich, Brixton, Putney, Kew and Richmond.

The layout of John Nash's scheme in Regent's Park and the building of Regent's Street completed the segregation of upper class neighbourhoods to the west from inferior quarters to the east. The exclusiveness of the West End was sealed by putting up a continuous line of shops along the east side of Regent's Street and by closing entries and alley-ways into Soho (see map page 76).

However deep its social divisions had grown, Londoners still managed to stick together. The royal family continued to live in St James's or at Carlton House. They did not move to a British Versailles. Rioting mobs that rampaged through the streets in 1780 (the anti-Catholic Gordon riots) did not storm the Tower of London or massacre aristocrats. Victories against the French were still occasions for patriotic celebrations by all sections of society. Friedrich Wendeborn, a German visitor at the end of the 18th century, envied even the poorest Londoners for the liberty and independence they possessed. 'A foreigner will at first hardly be pleased with the manner of living in London', he wrote, 'but if he has sense enough to perceive and value that freedom in thinking and acting which is to be enjoyed in England, he will soon wish to conclude his days there'.

Map below The extent of Georgian London at its peak is contrasted to the size of the city today.

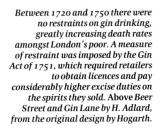

Between 1720 and 1750 there were no restraints on gin drinking, greatly increasing death rates amongst London's poor. A measure of restraint was imposed by the Gin Act of 1751, which required retailers to obtain licences and pay considerably higher excise duties on the spirits they sold. Above Beer Street and Gin Lane by H. Adlard, from the original design by Hogarth.

EXTENT OF LONDON 1820

LONDON 1990

St Paul's

LONDON 1990

GREAT ESTATES AND BUILDING DEVELOPMENT

During the 18th century London expanded over a larger area and at a faster rate than at any previous period. Buildings advanced unevenly in different directions from the centre and building booms alternated with periods of inactivity. Demand for houses was stimulated by rising commercial prosperity and by an increase in population from over 500,000 in 1700 to over 900,000 in 1801 and well over 1,000,000 in 1811. Movement away from the centre of the metropolis was precipitated by a thickening pall of smoke caused by a changeover in domestic fuel from wood to Newcastle coal, by a proliferation of epidemics (no longer plague, but of consumption, dysentery, smallpox, typhus and dropsy) and by high incidences of alcoholism, violence and other social disorders. Opportunities to meet the demand for new accommodation were seized by owners of great estates. The Crown and the church were less active than City livery companies and charitable foundations; and all these public bodies were much less active than private landowners in promoting building development.

A few great estates led the development of the West End as a fashionable residential district. Landowners and their agents designed the layout of streets and squares and drew the boundaries of building plots. Speculative builders were invited to submit plans and elevations of their proposed buildings and, having gained approval for these plans, were granted building leases to carry out the work. The builder sold a leasehold interest in the house he built to an occupier. The occupier was given security of tenure for a long lease of up to 99 years, at the end of which possession of the premises reverted to the ground landlord. The landlord retained a freehold interest in the soil, controlled the use of the land and buildings erected on it and charged a ground rent to the occupier.

While the rich moved west from Soho along Piccadilly and Oxford Street, the poor crowded into districts forming a belt around the City of London: St Giles', Clerkenwell, Spitalfields, and eastward into Bethnal Green, Whitechapel and Wapping. Once the social reputations of the east and west were widely acknowledged the process was cumulative: where the poor moved in the rich moved out. Archenholtz, visiting England in about 1780, remarked: 'the East End, especially along the shores of the Thames consists of old houses; the streets there are narrow and ill-paved; inhabited by sailors and other workmen who are employed in the construction of ships and by a great part of the Jews. The contrast between this and the West End is astonishing.'

The outer edges of the built-up area were scarred with gravel workings, brick pits, smoking kilns and tileries, stinking piles of horse manure, ashes, night soil and indescribable dumps of garbage. The fringe had a 'floating' population of dustmen, carters, rag-pickers, bone-boilers, horse-dealers and washerwomen. It also provided accommodation for pig keepers, dairymen, market gardeners, tanners, candle-makers and it offered space for some rough sports including dog-fighting, bull-baiting, boxing and, on occasion, duelling. It was not an idyllic rural retreat.

The first 18th century building boom began after the Treaty of Utrecht in 1713. Mayfair, west to Hyde Park and north to Oxford Street, was laid out in spacious squares and neat terraces. North of Oxford Street, the Cavendish-Harley family began developing their estate before the boom fizzled out in the 1730s. A fresh burst of activity opened with the Peace of Paris in 1763 and lasted until 1793 when war with France again halted building. During this golden age of Georgian architecture, John and Robert Adam designed the Adelphi, and adorned new developments in Piccadilly, Berkeley Square, Cavendish Square, Portland Place and Fitzroy Square with their gracefully proportioned buildings. Sir William Chambers designed Somerset House and left his mark on Piccadilly, the Albany, Berners Street and Whitehall. Many other architects contributed to the elegance of developments on the Portman, Berners, Portland, Bedford, Southampton–Fitzroy and Foundling Hospital estates. When peace returned in 1815 building activity was slow to recover. By 1820, the Prince Regent had commissioned a grandiose scheme for the development of Regent's Park and the

MAJOR ESTATES IN CENTRAL LONDON

1. Angell	50. Kensington Gore
2. Audley	51. Kilburn Priory
3. Bartholomew's Hospital	52. Ladbroke
4. Battle Bridge	53. Lambeth Wick
5. Bedford, Duke of	54. Lambs Farm
6. Berkeley	55. Leicester
7. Berners	56. Lloyd-Lisson
8. Brett	57. Maddox-Pollen
9. Brewers' Company	58. Maryon-Wilson
10. Cadogan	59. Mawby
11. Calthorpe	60. Mercers' Company
12. Camden Charities	61. Mildmay (Newington Gre
13. Camden, Earl of	62. Minet
14. Charterhouse	63. Morden College
15. Chelsea Hospital	64. New River
16. Choumert	65. Norland
17. Christ Church College, Oxford	66. Norris
18. Christie	67. Northampton
19. Church Commissioners (various estates)	68. Penton
	69. Phillimore
20. Church Commissioners (formerly Bishop of London)	70. Pickering
	71. Portland-Soho
21. Clothworkers' Company (Packington)	72. Portland (Cavendish, Harl
22. Conduit Mead	73. Portman
23. Corporation of London Bridge House	74. Powell
24. Craven	75. Rugby School
25. Crown	76. St. John's College Cambrid
26. Curzon	77. St. Quintin
27. Dartmouth, Lord	78. St. Thomas's Hospital
28. De Beauvoir	79. Salisbury
29. De Crespigny	80. Sir John Cass Charity
30. Duchy of Cornwall	81. Skinner Company
31. Edwardes (Lord Kensington)	82. Slade
32. Eton College	83. Sloane-Stanley
33. Eyre	84. Smith's Charity
34. Foundling Hospital	85. Somers
35. French School	86. Southampton
36. Gascoigne	87. Stonefield (Richard Cloude
37. Graham	88. Sutton
38. Grand Junction Canal Company	89. Talbot
39. Grosvenor, Duke of Westminster	90. Thornhill
40. Gunter	91. Thurloe
41. Hall	92. Tredegar, Lord
42. Harrison	93. Trinity House
43. Harrow School	94. Tyssen-Amhurst
44. Holland (Ilchester)	95. Vallotten
45. Hope	96. Vaughn
46. Hutchins	97. Vauxhall, Manor of
47. Inderwick	98. Walcott
48. Ironmongers'	99. Wenlock
49. Jesus College, Oxford	100. Wright

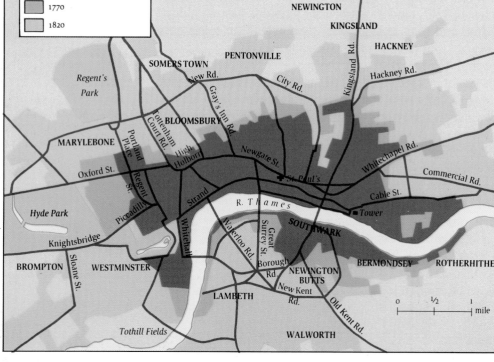

THE GROWTH OF LONDON

- 1720
- 1770
- 1820

HIGHGATE

STAMFORD HILL

NEWINGTON

KINGSLAND

NEWINGTON

HACKNEY

SOMERS TOWN

PENTONVILLE

Hackney Rd.

Regent's Park

New Rd.

City Rd.

Kingsland Rd.

Hackney Rd.

BLOOMSBURY

Gray's Inn Rd.

MARYLEBONE

Portland Place

Tottenham Court Rd.

High Holborn

Newgate St.

Whitechapel Rd.

Oxford St.

Regent St.

St. Paul's

Commercial Rd.

Hyde Park

Piccadilly

Strand

R. Thames

Cable St.

Tower

SOUTHWARK

Knightsbridge

Whitehall

Waterloo Rd.

Great Surrey St.

BERMONDSEY

ROTHERHITHE

BROMPTON

WESTMINSTER

Sloane St.

Borough Rd.

NEWINGTON BUTTS

New Kent Rd.

LAMBETH

Old Kent Rd.

0   ½   1

mile

Tothill Fields

WALWORTH

opening of a broad carriage-way to Carlton House and later to Trafalgar Square. At the same time, the Bedford estate was planning Tavistock Square and Gordon Square, the Grosvenor estate was laying out Belgrave Square and Eaton Square as part of a general design for Belgravia, the Bishop of London was embarking on the development of his lands at Paddington and dozens of small proprietors were calling in surveyors, architects and builders.

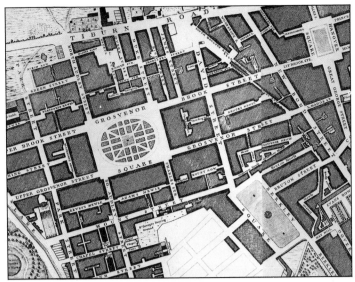

The growth of London (map below left) *Around Westminster the capital had begun to encroach upon Tothill Fields and Chelsea. To the south, Southwark spread over St. George's Fields into Newington. Long fingers of artisan housing and industrial premises stretched out to the north and west. But nowhere was private enterprise more vigorous than on the rural fringes: by the end of the 17th century the villages of St. Marylebone, Pentonville and Camden Town had been absorbed into the metropolis.*

Grosvenor Square (centre left) *This engraving of 1754 shows one of the earliest and grandest developments in Mayfair, the largest of its kind in London, covering six acres.*

The development of Mayfair *John Rocque's map (top right) of 1744 shows the regular pattern of streets in Mayfair beginning to take shape. Grosvenor Square is already built and the Chelsea Water Company's reservoir has been constructed in Hyde Park. South Mayfair remains in pasture around Shepherds Market. Richard Horwood's map of 1799 (centre right) shows the building of southern Mayfair completed by the development of Berkeley Square and Curzon Street.*

Great Estates *The map (below) shows the concentration of the largest estates in the West End. Some estates were owned and managed by public institutions such as the Church, City guilds, hospitals and universities. Others were owned by families, such as the Grosvenors, Portlands, Portmans and Curzons whose large tracts of land in the area left a legacy of orderly Georgian development.*

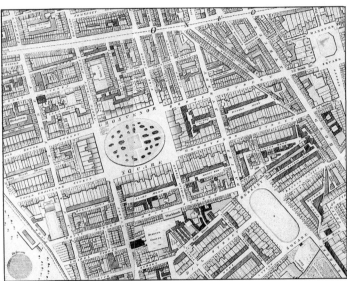

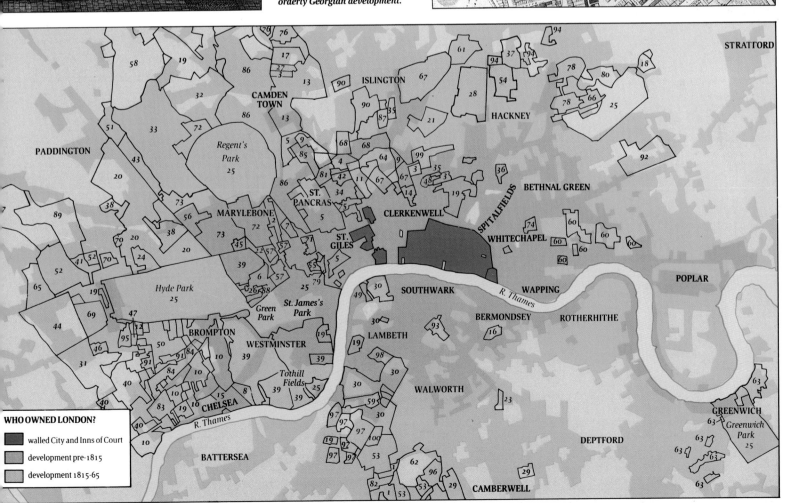

WHO OWNED LONDON?

- walled City and Inns of Court
- development pre-1815
- development 1815-65

PLANNING AND PUBLIC BUILDING

Private developers left London a substantial legacy of 18th-century buildings: noble yet comfortable mansions, orderly terraces of fine-looking commodious houses, elegant rows of shops, coffee houses, clubs, coaching inns, and solid commercial and industrial structures including shipyards and warehouses. A no less valuable legacy of public buildings comes from this period. When private house-building slackened, public bodies remained active. Government offices in Whitehall, Newgate Prison, docks, naval yards and barracks were all built or added to in the war years. At the same time, new bridges were thrown across the Thames and road improvement schemes were started, but not primarily for military purposes – London has no elaborate fortifications or military roads.

Public works replenished and consolidated the social infrastructure in the periods following building booms. Hence the provision of churches, almshouses, schools, museums and hospitals. The hospitals built in the 18th century reflect the rational humanitarian spirit of the era. Old medieval foundations were re-built and new hospitals were constructed – either through private generosity, as with Guy's and the Foundling Hospital, or through annual-subscription schemes as was the case with St. George's, the Middlesex and Westminster Hospitals, all of them built between 1720 and 1760.

Many churches designed to replace those destroyed in the Fire of London were built before 1713. Others designed by Hawksmoor, Archer, Flitcroft, Gibbs and Dance were commissioned by Parliament in 1711 and some of them were not completed until after 1730. A later phase of

Georgian public buildings The Adelphi (below), John and Robert Adams, 1768. This river frontage has now been replaced by Savoy Place and Victoria Embankment Gardens. Right (clockwise) Christ Church, Spitalfields. This baroque design by Nicholas Hawksmoor was completed in 1727. The Bank of England, Threadneedle St. Designed by George Sampson and built 1732-36, it is now entirely encased in later building. The Mansion House Designed by George Dance the elder and built 1739-53. The portico remains unaltered. The Imperial War Museum, Lambeth Designed by J. Lewis in 1812 to rehouse old Bedlam, now Bethlehem Hospital for the insane. A dome was added in 1838.

church building was initiated by the Church Building Act in 1818. One million pounds was spent furnishing new residential districts with Anglican churches. Dissenting congregations also built impressive places of worship during the 18th century.

The period between the Peace of Paris and the beginning of the French wars (1763-1793) was the golden age of Georgian architecture, dominated by two geniuses, William Chambers and Robert Adam. Virtually none of their work survives today – the two contrasting riverside developments of Somerset House (Chambers) and the Adelphi (Adam) were their most outstanding achievements. But it was in the London of George IV that a distinct urban vision is clearly evident. The King wanted to make London a truly magnificent capital, and city improvements and building initiatives, financed by the State, were undertaken on a large scale. John Nash's plans for Regent Street, with its dramatic vistas and the close attention paid to the siting of buildings, epitomises this era. It was also

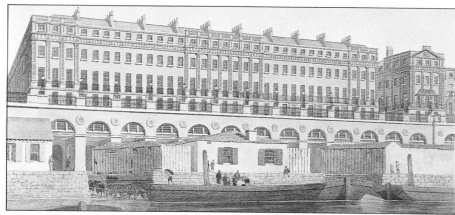

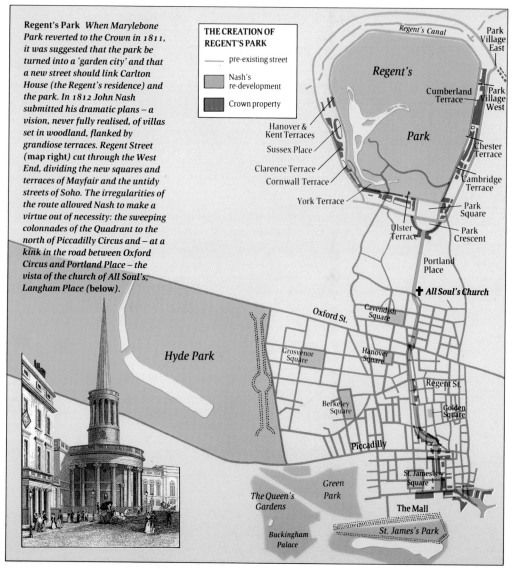

Regent's Park When Marylebone Park reverted to the Crown in 1811, it was suggested that the park be turned into a 'garden city' and that a new street should link Carlton House (the Regent's residence) and the park. In 1812 John Nash submitted his dramatic plans – a vision, never fully realised, of villas set in woodland, flanked by grandiose terraces. Regent Street (map right) cut through the West End, dividing the new squares and terraces of Mayfair and the untidy streets of Soho. The irregularities of the route allowed Nash to make a virtue out of necessity: the sweeping colonnades of the Quadrant to the north of Piccadilly Circus and – at a kink in the road between Oxford Circus and Portland Place – the vista of the church of All Soul's, Langham Place (below).

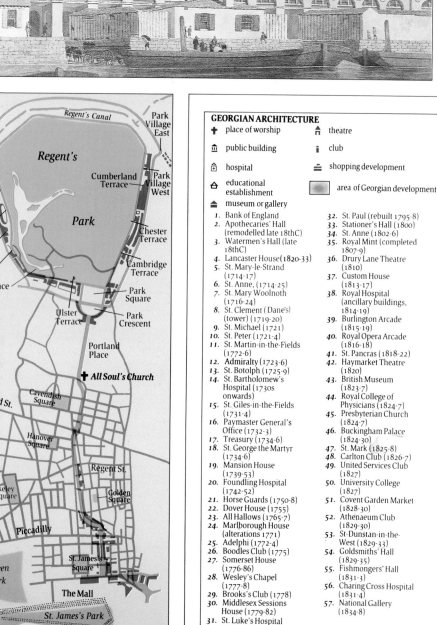

THE CREATION OF REGENT'S PARK

— pre-existing street

Nash's re-development

Crown property

GEORGIAN ARCHITECTURE

✝ place of worship
🏛 public building
🏥 hospital
⌂ educational establishment
🏛 museum or gallery

🏛 theatre
🛈 club
🛒 shopping development
▨ area of Georgian development

1. Bank of England
2. Apothecaries' Hall (remodelled late 18thC)
3. Watermen's Hall (late 18thC)
4. Lancaster House (1820-33)
5. St. Mary-le-Strand (1714-17)
6. St. Anne, (1714-25)
7. St. Mary Woolnoth (1716-24)
8. St. Clement (Dane's) (tower) (1719-20)
9. St. Michael (1721)
10. St. Peter (1721-4)
11. St. Martin-in-the-Fields (1722-6)
12. Admiralty (1723-6)
13. St. Botolph (1725-9)
14. St. Bartholomew's Hospital (1730s onwards)
15. St. Giles-in-the-Fields (1731-4)
16. Paymaster General's Office (1732-3)
17. Treasury (1734-6)
18. St. George the Martyr (1734-6)
19. Mansion House (1739-53)
20. Foundling Hospital (1742-52)
21. Horse Guards (1750-8)
22. Dover House (1755)
23. All Hallows (1765-7)
24. Marlborough House (alterations 1771)
25. Adelphi (1772-4)
26. Boodles Club (1775)
27. Somerset House (1776-86)
28. Wesley's Chapel (1777-8)
29. Brooks's Club (1778)
30. Middlesex Sessions House (1779-82)
31. St. Luke's Hospital (1782-4)
32. St. Paul (rebuilt 1795-8)
33. Stationer's Hall (1800)
34. St. Anne (1802-6)
35. Royal Mint (completed 1807-9)
36. Drury Lane Theatre (1810)
37. Custom House (1813-17)
38. Royal Hospital (ancillary buildings, 1814-19)
39. Burlington Arcade (1815-19)
40. Royal Opera Arcade (1816-18)
41. St. Pancras (1818-22)
42. Haymarket Theatre (1820)
43. British Museum (1823-7)
44. Royal College of Physicians (1824-7)
45. Presbyterian Church (1824-7)
46. Buckingham Palace (1824-30)
47. St. Mark (1825-8)
48. Carlton Club (1826-7)
49. United Services Club (1827)
50. University College (1827)
51. Covent Garden Market (1828-30)
52. Athenaeum Club (1829-30)
53. St-Dunstan-in-the-West (1829-33)
54. Goldsmiths' Hall (1829-35)
55. Fishmongers' Hall (1831-3)
56. Charing Cross Hospital (1831-4)
57. National Gallery (1834-8)

Map labels: Regent's Canal · Park Village East · Regent's Park · Cumberland Terrace · Park Village West · Hanover & Kent Terraces · Sussex Place · Chester Terrace · Clarence Terrace · Cornwall Terrace · Cambridge Terrace · York Terrace · Park Square · Ulster Terrace · Park Crescent · Portland Place · All Soul's Church · Cavendish Square · Oxford St. · Hanover Square · Grosvenor Square · Regent St. · Hyde Park · Berkeley Square · Golden Square · Piccadilly · The Queen's Gardens · Green Park · St. James's Square · Buckingham Palace · The Mall · St. James's Park

during this period that work was begun on two major national institutions: Smirke's British Museum (originating in George IV's gift of his father's library to the nation) and Wilkins' National Gallery.

The return of peace and prosperity was marked by a resurgence of private building as well as the construction of theatres, banks, insurance offices and clubs. These buildings still leave their imprint on London: the clubs of St James's, their grandiose architecture owing much to the pioneering style of Smirke's United Services Club (1816-1817); the first designed shopping streets, such as Woburn Walk, St Pancras (1822) and covered arcades, such as Burlington Arcade (1815-1819); the facade of Nash's Theatre Royal, Haymarket (1820-21).

The map (below) shows the major churches and public buildings of the Georgian period, and indicates (in brown tone) the principal areas of Georgian residential expansion.

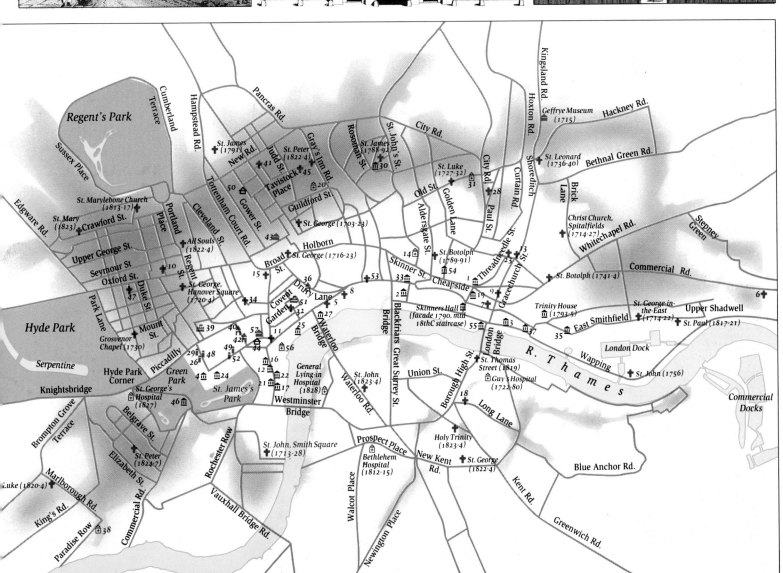

FEEDING LONDON

At the beginning of the 18th century, the country within five or ten miles of London was entirely tributary to the metropolitan market. It supplied fruit, vegetables, meat and milk to the urban population as well as providing grazing land and hay for livestock. With changes in diet throughout the ensuing century, per capita consumption of fresh food increased, and London's agricultural hinterland had to meet that demand. The surface of land in orchards, nurseries and market gardens increased, extending westwards along the Thames as far as Brentford and northwards up the Lea to Homerton.

Not only did urban dairy herds and numbers of pigs increase but numbers of town horses grew rapidly. More and more were employed in ploughing as horses took the place of oxen. Ever greater numbers were used for riding, drawing coaches, carts, wagons and for towing barges on navigable waterways. Vast quantities of hay and oats were shipped and carted into London and vast quantities of manure were hauled out. Agriculture was directed at least as much towards feeding domestic animals as humans.

In 1800, the innermost edge of the built-up area was scarred with gravel workings, clay pits, brick kilns and mountains of metropolitan rubbish. At night, The Rev. Henry Hunter depicted it as forming 'a ring of fire and pungent smoke' around the City. Beyond the pits and tips lay the pastures where London's horses and cattle were put out to graze and butchers fattened their stock. On the Taplow Terrace west of the River Lea, strips of loam soil were intensively cultivated in orchards and market gardens. Further out lay an extensive area of meadows that supplied London with hay. A tract of arable farming persisted on the lighter soils of west Middlesex. A surprisingly large amount of land remained agriculturally unproductive. It was estimated in 1775 that 200,000 acres of waste land lay within 30 miles of the capital. Hounslow Heath and Finchley Common, very close to the built-up area, were described in 1793 as 'fitted only for Cherokees and savages' while Epping Forest was notorious for sheltering robbers. Firewood, gravel and sparse grazing for commoners' cattle were the meagre resources of these wild tracts.

At the beginning of the 18th century London was already the central market place for British agriculture. The prices of the principal grain crops were fixed at Mark Lane by the Tower. Cattle and sheep bred in the uplands of Wales and Scotland were driven to the east Midlands and East Anglia and from there as fatstock they made their last journey to Smithfield Market. Three dozen other markets dealt in different commodities: fruit and vegetables at Covent Garden, hides and leather at Bermondsey, and so on, and as the built-up area spread, new market places such as the Fleet, Oxford and Shepherds Markets, were established.

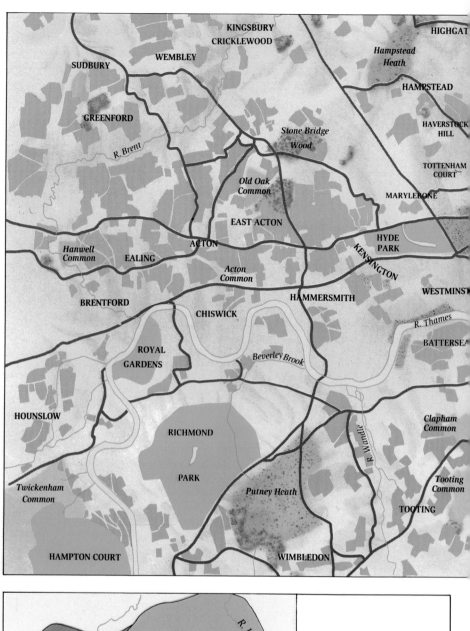

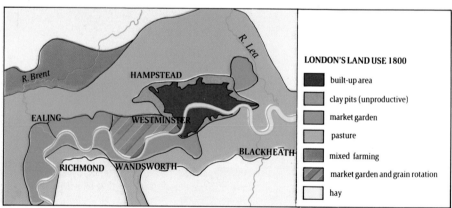

LONDON'S LAND USE 1800
- built-up area
- clay pits (unproductive)
- market garden
- pasture
- mixed farming
- market garden and grain rotation
- hay

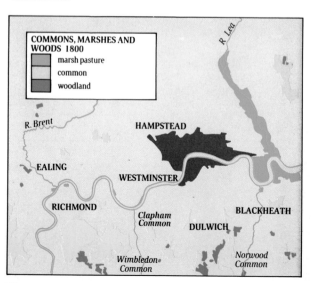

COMMONS, MARSHES AND WOODS 1800
- marsh pasture
- common
- woodland

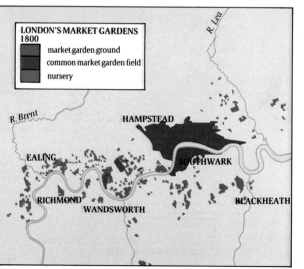

LONDON'S MARKET GARDENS 1800
- market garden ground
- common market garden field
- nursery

Land use, 1746-1800 *The map (top) is based on John Rocque's map of London and its environs (1746) and shows that the countryside beyond the pastoral fringe had a mixture of arable and agriculturally unproductive land. In 1800, Thomas Milne's survey (maps above, left and centre) indicate that land use was much more regularly zoned than it had ever been, the zoning being produced by market forces, not by state regulation. They also show the use of each field or parcel of land around London. A large area is unproductive marsh, wood or heath. The loams of the Thames terraces are intensively cultivated as nurseries and market gardens.*

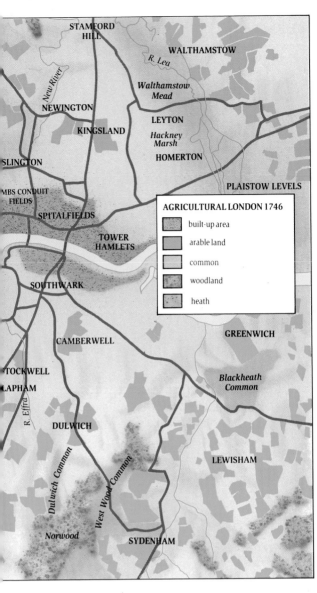

AGRICULTURAL LONDON 1746

STAMFORD HILL

WALTHAMSTOW

R. Lea

Walthamstow Mead

NEWINGTON

LEYTON

KINGSLAND

Hackney Marsh

HOMERTON

SLINGTON

PLAISTOW LEVELS

MBS CONDUIT FIELDS

SPITALFIELDS

TOWER HAMLETS

SOUTHWARK

CAMBERWELL

GREENWICH

TOCKWELL

Blackheath Common

LAPHAM

R. Effra

DULWICH

LEWISHAM

Dulwich Common

West Wood Common

Norwood

SYDENHAM

	built-up area
	arable land
	common
	woodland
	heath

 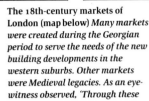 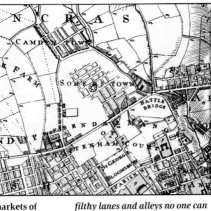

The loss of arable land around 18th century London *is clearly evident when Rocque's map of 1746 (above left) is compared to Milne's survey of 1800 (above right). The brick and gravel pits have gone and Somers Town and Camden Town have been built.*

The 18th-century markets of London *(map below) Many markets were created during the Georgian period to serve the needs of the new building developments in the western suburbs. Other markets were Medieval legacies. As an eye-witness observed, 'Through these*

filthy lanes and alleys no one can pass without being butted with the dripping end of a quarter of beef, or smeared with a greasy carcase of a newly slain sheep.' The Monday market, Smithfield, shown (below) in 1811.

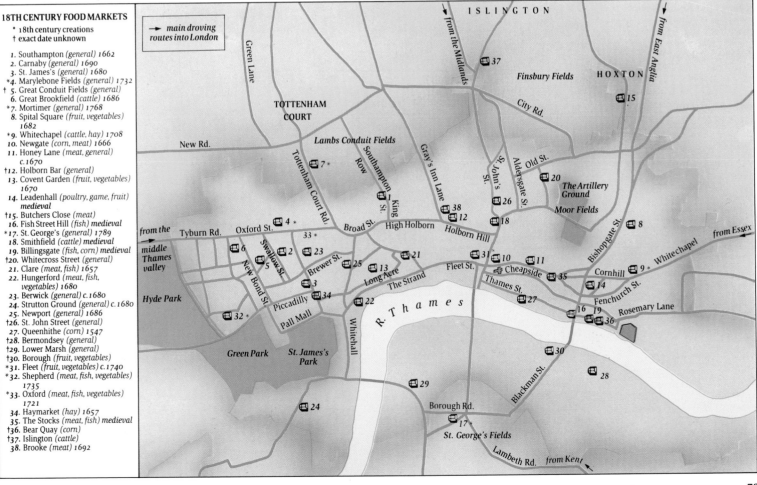

18TH CENTURY FOOD MARKETS

* 18th century creations
† exact date unknown

1. Southampton (general) 1662
2. Carnaby (general) 1690
3. St. James's (general) 1680
*4. Marylebone Fields (general) 1732
†5. Great Conduit Fields (general)
6. Great Brookfield (cattle) 1686
*7. Mortimer (general) 1768
8. Spital Square (fruit, vegetables) 1682
*9. Whitechapel (cattle, hay) 1708
10. Newgate (corn, meat) 1666
11. Honey Lane (meat, general) c.1670
†12. Holborn Bar (general)
13. Covent Garden (fruit, vegetables) 1670
14. Leadenhall (poultry, game, fruit) medieval
†15. Butchers Close (meat)
16. Fish Street Hill (fish) medieval
*17. St. George's (general) 1789
18. Smithfield (cattle) medieval
19. Billingsgate (fish, corn) medieval
†20. Whitecross Street (general)
21. Clare (meat, fish) 1657
22. Hungerford (meat, fish, vegetables) 1680
23. Berwick (general) c.1680
24. Strutton Ground (general) c.1680
25. Newport (general) 1686
†26. St. John Street (general)
27. Queenhithe (corn) 1547
†28. Bermondsey (general)
†29. Lower Marsh (general)
30. Borough (fruit, vegetables)
*31. Fleet (fruit, vegetables) c.1740
*32. Shepherd (meat, fish, vegetables) 1735
*33. Oxford (meat, fish, vegetables) 1721
34. Haymarket (hay) 1657
35. The Stocks (meat, fish) medieval
†36. Bear Quay (corn)
†37. Islington (cattle)
38. Brooke (meat) 1692

→ main droving routes into London

ISLINGTON

from the Midlands

from East Anglia

Green Lane

Finsbury Fields

HOXTON

City Rd.

TOTTENHAM COURT

New Rd.

Lambs Conduit Fields

Old St.

The Artillery Ground

Moor Fields

from the middle Thames valley

Tyburn Rd.

Oxford St.

Broad St.

High Holborn

Holborn Hill

from Essex

Hyde Park

New Bond St.

Brewer St.

Long Acre

The Strand

Fleet St.

Cheapside

Cornhill

Whitechapel

Thames St.

Fenchurch St.

Piccadilly

Pall Mall

Rosemary Lane

Green Park

St. James's Park

R. Thames

Whitehall

Blackman St.

Borough Rd.

St. George's Fields

Lambeth Rd.

from Kent

ROADS AND BRIDGES

As London expanded it depended more and more on road transport to supply the growing metropolis with food, fodder and fuel, to carry people from home to work and to link the provinces with the capital. During the 18th century the volume of traffic on the roads into London increased prodigiously. A new and rapidly growing body of road-users consisted of those making daily journeys to work from the outskirts of London and later from places further afield. In 1725, Daniel Defoe remarked upon the large number of businessmen holding jobs in the City, at the Treasury or at Court, who lived in Epsom, where 'they look as if they had left all their London thoughts behind them'. A century later, William Cobbett observed that 'great parcels of stock jobbers ....skip backward and forward on the coaches, and actually carry on stock jobbing in Change Alley, though they reside at Brighton. There are not less than twenty coaches that leave the Wen every day for this place.'

Roads subjected to increasing wear and tear were less and less able to bear the loads thrust upon them, and in wet winter weather some roads became almost impassable. By transferring the responsibility for road maintenance from parishes to turnpike trusts and granting trusts powers to collect tolls on different classes of road users, parliament hoped improvements would be carried out, especially on heavily used roads. The earliest turnpikes were set up on the Great North Road and by 1750 trusts were established along most of its length from London to the Scottish border. The road to Harwich was turnpiked at an early date and sections of roads to Birmingham, Bristol and Portsmouth were in the hands of turnpikes before mid-century. Remarkable reductions in journey times were achieved in the second half of the 18th century. In 1706 and still in 1754 a coach from London to York took four days. By 1774 the schedule had been reduced to two days. Similar reductions were achieved on journeys to Shrewsbury and other places. Techniques of building and surfacing roads were improved, and by the end of the century investment in turnpikes was booming. Between 1790 and 1835 the number of coach passengers travelling to and from London multiplied sixteen-fold. Although stretches of turnpike roads over the clays of Middlesex, Surrey and Essex

The principal roads from London (map below) *The direct roads supplied the capital with food and fuel, acting as the nerve fibres of the nation's communication system. Marble Arch Turnpike, shown in a watercolour by Thomas Rowlandson, 1750 (right). A horseman, coach and trap are shown speeding away from the toll-gate on the newly laid road surface. Horse-drawn carts and coaches had difficulty climbing the steep ascents to Hampstead, Highgate and Muswell Hill (map right). Many inns offered both travellers and horses rest and refreshment.*

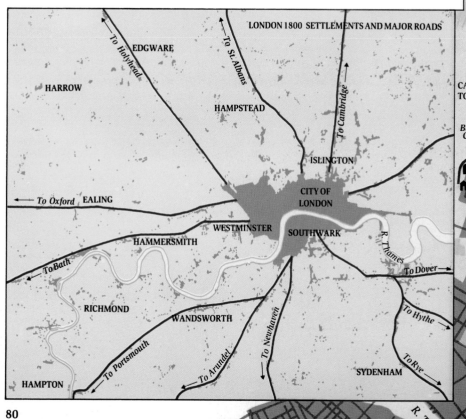

LONDON 1800 SETTLEMENTS AND MAJOR ROADS

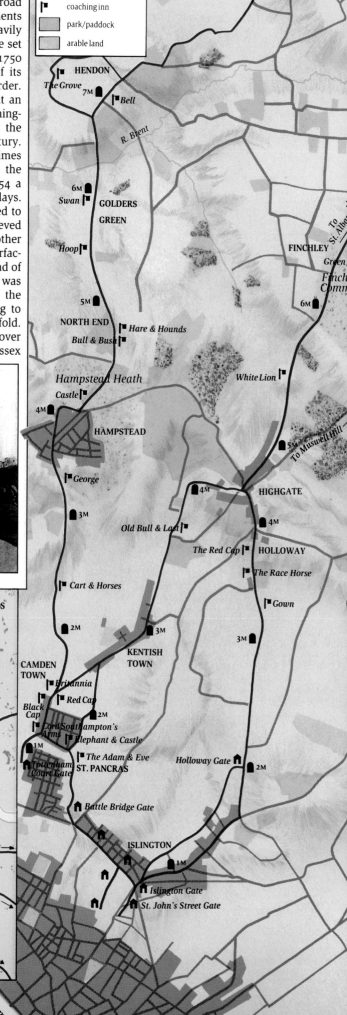

COACHING ROADS 1800
- coaching road
- toll gate
- milestone
- coaching inn
- park/paddock
- arable land

were still in no better condition than parish roads, some of the most serious defects had been remedied.

Within the built-up areas, problems caused by traffic congestion were growing worse rather than bettter. A little relief had been afforded by road-widening schemes and by laying down hard surfaces, but the greatest benefits were obtained from the building of new arterial roads. The most spectacular was the New Road, on the line of the present Marylebone, Euston and City Roads, built in 1756-61 to by-pass the built-up area between Watling Street, present Edgware Road, and Ermine Street, the present Kingsland Road. The building of Westminster Bridge in 1750 and the approach to St George's Circus provided a southern by-pass.

Until 1750 the Thames was a major barrier to road transport. There was only one bridge across the river, London Bridge, and that was encumbered with houses and shops. Upstream, the next bridge was at Putney. To cross the water or go up and down, people and goods were taken by boat. The building of six new bridges between 1750 and 1827 and the removal of buildings from London Bridge in 1759 greatly assisted movement and opened the south bank to development.

Road improvements *In the early 18th century the main roads out of London were legacies from the Roman period (map top right). Medieval road builders had filled the spaces between Roman roads with an intricate network of narrow streets, lanes, alleys and courts. Every kind of traffic from droves of livestock and funeral processions to galloping courtiers exercised their rights of way and many streets also served as open sewers. An engraved view of Cornhill looking westwards (centre right), shows carriages and coaches jostling for space with draymen and wagoners. More was done to improve London's roads between 1750 and 1835 than had been done in the previous thousand years. On the north side of the metropolis, a by-pass was constructed in 1756-61 which was 120 feet wide (Marylebone, Euston and City Rd.). South of the Thames, road improvements – such as St. George's Circus, Westminster Bridge Rd. and Blackfriars Rd. – followed the building of new bridges. The first and most important was Westminster Bridge designed by the Swiss engineer Charles Labelye in 1750. It is illustrated by Canaletto in 1749 before the central arches were completed (below). In the post-Napoleonic war years, three new bridges – Vauxhall (1816), Waterloo (1817) and Southwark (1819) – were opened. Before these bridges were built, however, Londoners had to cross the Thames by boat. The King, Lord Mayor and wealthy citizens owned their own barges. Other Londoners boarded ferries at scores of stairs and landing stages (map below). The illustration of Westminster Bridge (below) shows a number of beached ferries and their passengers. The new bridge meant substantial losses in income for the Thames ferrymen, who protested bitterly, and were eventually awarded £25,000 in compensation.*

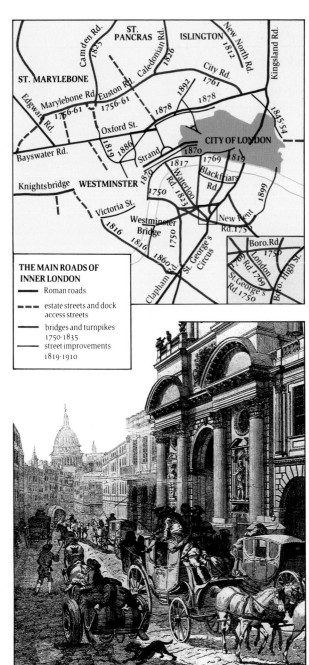

THE MAIN ROADS OF INNER LONDON
— Roman roads
--- estate streets and dock access streets
— bridges and turnpikes 1750-1835
— street improvements 1819-1910

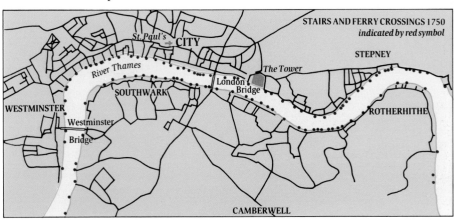

STAIRS AND FERRY CROSSINGS 1750
indicated by red symbol

LONDON'S WATERWAYS

Throughout the 18th century London's maritime trade increased steadily, but the dominance of the port was challenged by an even more rapid expansion of trade at Liverpool, Bristol and other west coast ports. London retained a clear lead in the import of goods from Europe and the east: tea, coffee, pepper, spices, silks, precious metals, gemstones, porcelain, mahogany, teak and raw wool. Western ports were attracting more and more trans-Atlantic trade, importing raw cotton, tobacco, sugar, rum, furs, pine and taking increasing quantities of cocoa, rice and wine. Manufactured goods – cotton cloth from Lancashire, woollens from Yorkshire and Staffordshire pottery – were leaving the country by way of western ports.

The Thames continued to handle an enormous amount of coastal traffic. Portland stone, lime, a great variety of other building materials, shipbuilding timber as well as large quantities of staple foods reached London by water. By far the largest volume of coastal shipping was engaged in carrying sea coal from Newcastle and Sunderland: amounts carried tripled, to 1.4 million tons per annum, over the 18th century. The wherries returned laden with ballast, some of which was dug from pits at Charlton and some dredged from Dagenham Breach.

In 1700 about 435,000 tons of shipping were registered in London; by 1790 the volume had risen to 509,000 tons. In 1800, in addition to 1200 or more sea-going vessels, the waterway was crowded with 1200 coal barges, 500 timber barges, 800 lighters and over 1000 smaller craft handling freight. Ships waited days for favourable winds, high tides and berths at legal quays where customs officers could inspect cargoes. Delays added to the risks of damage and plundering. In 1801 West Indies merchants lost on average 2% of their revenue through theft on the waterway. Partly because of silting in the channel, large ships, drawing 6.7 tons of water could not sail beyond Blackwall.

To relieve crowding in the Pool of London between the Tower and London Bridge, it was decided to construct secure, deep water basins off the busy waterway. The London Docks downstream from the Tower were first opened in 1801. They were accompanied by the building of massive blocks of warehouses in Wapping. The cutting of the West India Docks across the neck of the Isle of Dogs was followed by the opening of Commercial Road in 1803 and the erection of ranges of well-guarded warehouses. Smaller docks for the East India trade were dug at Blackwall, accompanied by an extension of East India Dock Road. The Surrey Commercial Docks were connected to the river by the Grand Surrey Canal and traffic from the midlands was brought by the Grand Junction Canal to Paddington and round the northern fringes of London to the Western Docks at Limehouse.

Above the Pool of London the Thames was a major inland waterway, with a catchment area extending over much of southern and central England. Surrey, parts of Sussex and Hampshire were reached by the Mole and Wey navigations. Reading was connected by the Kennet-Avon Canal with Bath and Bristol and the navigable Thames was linked to Birmingham through the Oxford and Coventry Canals.

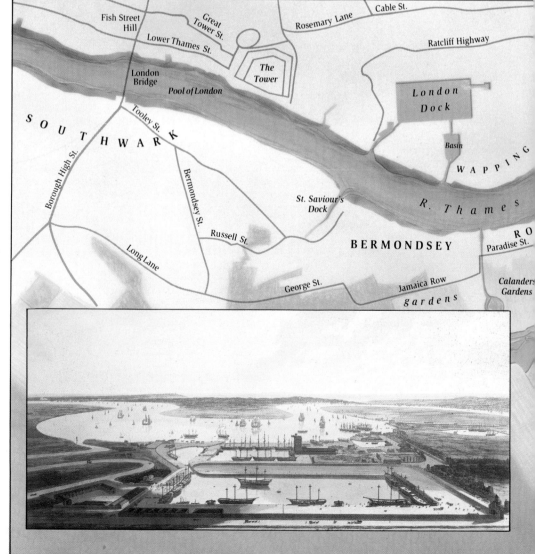

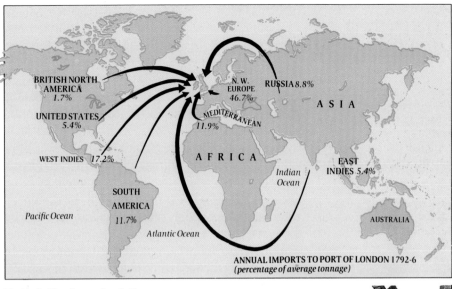

ANNUAL IMPORTS TO PORT OF LONDON 1792-6
(percentage of average tonnage)

BRITISH NORTH AMERICA 1.7%
UNITED STATES 5.4%
WEST INDIES 17.2%
SOUTH AMERICA 11.7%
N. W. EUROPE 46.7%
RUSSIA 8.8%
MEDITERRANEAN 11.9%
EAST INDIES 5.4%

ASIA
AFRICA
AUSTRALIA
Pacific Ocean
Atlantic Ocean
Indian Ocean

The Pool of London, c.1820 (left)
The Rheinbeck Panorama depicts the Thames below London Bridge crowded with the masts of ships moored alongside wharves and warehouses. Above London Bridge lighters, barges and ferries carried passengers and freight upstream.
World trade 1792-96 (map above)
Arrows showing the comparative tonnage of goods imported to London indicate that increasing volumes of trade were with Britain's colonial territories. By far the largest growth in tonnage handled during the 18th century was the coastal trade in coal. The Coal Exchange, in Mark Lane (right), was built in 1747.

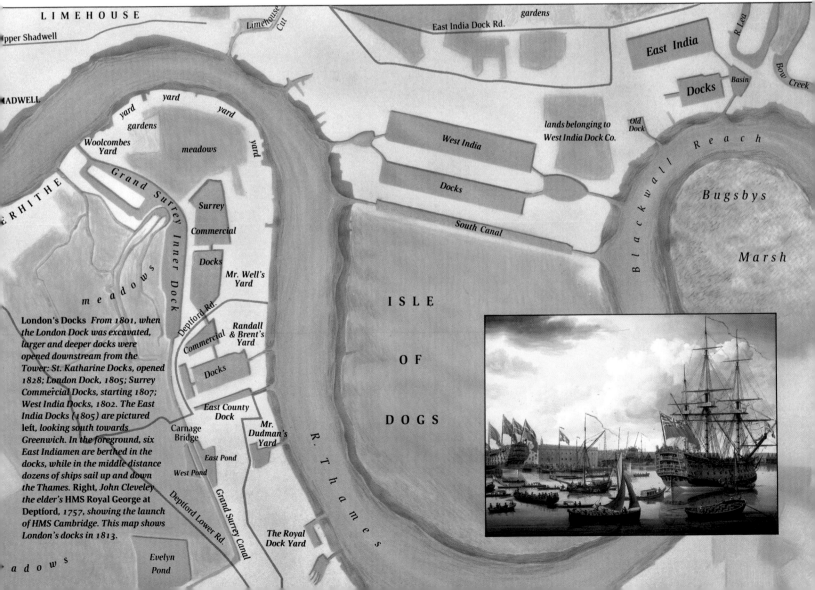

London's Docks *From 1801, when the London Dock was excavated, larger and deeper docks were opened downstream from the Tower: St. Katharine Docks, opened 1828; London Dock, 1805; Surrey Commercial Docks, starting 1807; West India Docks, 1802. The East India Docks (1805) are pictured left, looking south towards Greenwich. In the foreground, six East Indiamen are berthed in the docks, while in the middle distance dozens of ships sail up and down the Thames. Right, John Cleveley the elder's HMS Royal George at Deptford, 1757, showing the launch of HMS Cambridge. This map shows London's docks in 1813.*

CHAPTER 7
VICTORIAN LONDON

THE 19TH CENTURY saw an unprecedented explosion in London's population: in 1801 the population of Greater London was just over one million, by 1911 it had increased to over seven million. The enormous physical expansion that accompanied this population growth meant that new forms of public transport became increasingly necessary: horse buses after 1829; railways in the 1830s; an underground railway system from the 1860s; trams in the 1870s; deep-level electric 'tube' trains in the 1890s. However, access to buses, trams and trains was linked to ability to pay, and this led to an increasing segregation of different classes. The better-off moved to suburbia first, while the poor remained in inner London within walking distance of work. The introduction of 'workmen's fares' in east London in the 1860s and more widely after 1883, and the electrification of tramways during the early 1900s, meant that the regularly employed could also move to new working-class suburbs, but casual labourers on building sites and in docks and markets still needed to live close to places where they might be able to find employment.

Refer to pages 90-91 for a detailed discussion of the evolution of transport in Victorian London.

In Regency London, many of the poor still lived close to the rich, directly dependent on them for their employment. Gradually the class structure changed. New business practices and the expansion of administration called into being a 'lower middle class' of clerks and book-keepers, caricatured by George and Weedon Grossmith in the person of Mr Pooter, in *The Diary of a Nobody* (1892). Mechanisation in industry led to 'de-skilling', as artisans found themselves superseded by machines. This, in turn, led to the creation of a labour aristocracy of skilled workers, especially engineers, who built and maintained the new machines. These new classes provided an additional source of demand for manufactured goods. Instead of providing made-to-order goods for the elite, London's artisans increasingly concentrated on mass production of off-the-peg items for the newly affluent. The links between manufacturer and customer became impersonal and indirect. Factories replaced workshops and domestic industry; and the owners of businesses were less likely to know their employees by name.

Refer to pages 92-93 for a discussion of changes in the industrial structure of London.

In such a rapidly growing city, where for most people status was now earned rather than inherited, families sought to emphasise their position in society by retreating into one-class residential areas, ensuring that they had the best address that they could afford. In the novels of George Gissing, characters

See pages 88-89 for a discussion of the evolution of London housing over the 19th century.

often lived not just in genuine districts of London, but in particular streets; to know their address was to know their status, their morals and their lifestyle. Having become part of London's class of *nouveaux riches*, the Frothinghams (in *The Whirlpool*, 1897) 'after obscure prosperity in a southern suburb ... fluttered to the northern heights' to a house in 'Fitzjohn Avenue', almost certainly Fitzjohn's Avenue, Hampstead. Later, after Bennet Frothingham's suicide, following the collapse of his 'Britannia Loan, Assurance, Investment, and Banking Co.', his widow and daughter moved to a flat in Swiss Cottage, supposedly a new and simple lifestyle: 'Just one servant, who can't make mistakes, because there's next to nothing to do. No wonder people are taking to flats.' Flats were frequently associated with an amoral modernism. They were at first called 'French Flats' to signify both their origin and the cosmopolitan, reputedly scandalous lifestyle of their inhabitants. In *The Whirlpool* Gissing situated the much travelled, dissolute and, significantly, childless Carnabys first in a house in Hamilton Terrace, St John's Wood, but later in Oxford and Cambridge Mansions, a real block of apartments, built in 1882, just four years before the setting of the novel, on the edge of Hyde Park. By contrast, the more conventional, established middle-class Harvey Rolfe moves, after his marriage, to a house in Pinner, outside the built-up area of London, but conveniently situated on the Metropolitan Railway, which had reached Pinner from Baker Street in 1885. Farther down the social hierarchy were suburbs like Crouch End, to which Sidney and Clara Kirkwood moved (in *The Nether World*, 1889) in an effort to escape the oppressive poverty of Clerkenwell. Crouch End in 1885 was 'still able to remind one that it was in the country a very short time ago. The streets have a smell of newness, of dampness; the bricks retain their complexion, the stucco has not rotted more than one expects in a year or two; poverty tries to hide itself with venetian blinds'.

Gissing was the literary counterpart to social researchers like Charles Booth, who worked in conjunction with School Board officers whose job was to ensure that all children of school age attended elementary schools, thereby acquiring an enormous fund of information about families in their area. Charles Booth calculated that more than 30% of Londoners were living below the poverty line in the 1890s. He reckoned that 'questions of employment', especially irregular earnings, accounted for 68% of cases of poverty, 'questions of circumstances', such as illness or large families, for 19%, and 'questions of habit' – drunkenness and thriftlessness – for only 13%. Yet it was the latter causes of poverty, focusing on the inadequacy of the individual, which attracted most attention among Booth's contemporaries. London was described as a 'modern Babylon', a decadent society sure to suffer ultimate destruction. The East End was variously described as an 'inferno', a 'city of dreadful night', 'outcast London', inhabited by 'people of the abyss', trapped in a 'nether world'. These perceptions could be sustained because, of course, most middle-class Londoners never went anywhere near the East End. Increasing residential segregation provided an environment in which ignorance and prejudice could flourish, reinforced by a press which seized every opportunity to spread panic among middle-class readers with tales of incest, crime, riot and the breakdown of social order. Not surprisingly, these rumours made the middle classes even less likely to settle in east London.

The housing problem of Victorian London was attributable to both the city's size and its industrial structure. To live within walking distance of work in, for instance, Covent Garden, Smithfield, London Docks or the major railway termini, was to live

See pages 102-103 for a full discussion of Charles Booth's social research. An examination of the Victorian church on page 97 is also based on research by Charles Booth.

Left A bird's eye view of the City from the west (1832-46). Although London was already beginning to spread, the massive expansion of the 19th century had not yet begun and countryside was still visible at the edges of the city. By 1900, successive building booms had transformed London into a scattered metropolis which continued to absorb, organically, the villages and local communities still visible on this panorama. This scattered city was tied together by the innovation and adaptation of new modes of transport; carriages, cabs, omnibuses, trams, railways. By 1900, railway-based tentacles of suburban growth were beginning to extend away from the city, precursors of an even greater dispersal that would come to dominate the 20th century.

The social researcher, Charles Booth, estimated that 30% of Londoners were living in poverty in 1890. Among his contemporaries William Booth, founder of the Salvation Army, blamed the habits of the poor as much as their economic circumstances. While he acknowledged the lack of sanitation and the inhumanity of sweated labour he stressed: 'Drunkenness and all manner of uncleanness, moral and physical, abound ... As in Africa streams intersect the forest in every direction, so the gin-shop stands at every corner with its River of the Water of Death flowing seventeen hours of the twenty-four for the destruction of the people'.
Below *a typical Victorian slum street, Little Saffron Hill, in the 1890s.*

within an inner ring of pre-Victorian houses that had decayed into overpriced and overcrowded slums by the middle of the 19th century. Moreover, compared to smaller cities, London had proportionally more casual workers who needed to live close to places where they might find work. London was also the country's largest industrial city; in 1861, there were more manufacturing workers in London than the entire population of Manchester, but most worked for very small firms. Unlike industrialists in northern towns, few London employers were either sufficiently wealthy or felt it necessary for the success of their business to provide housing for their employees. Consequently, working-class Londoners depended for their accommodation on speculative builders and private landlords. Under the Cross Act (1875), the Metropolitan Board of Works could engage in slum clearance but was not permitted to undertake new housing construction. The cleared sites were sold to philanthropic or 'five per cent' agencies like the Peabody Trust and the East End Dwellings Company. Rebuilding *in situ*, however sanitary the new buildings, did not relieve problems of congestion so, from the 1880s, more reformers favoured a suburban solution, building cottage estates and garden suburbs, although these were only practicable when cheap fares on public transport became widely available. Suburban living might be healthier, but it had its drawbacks. Food was more expensive in suburban shops than inner-city markets, workers could not go home for lunch, there were few jobs for women. Although housing was better in quality, it was rarely any cheaper. For casual labourers a suburban base was impracticable.

Refer to pages 100-101 for a discussion of slum conditions and philanthropic solutions to the housing problem.

In 1800 most Londoners lived in terraced houses. Middle-class Londoners often shared the use of a communal garden, situated in the centre of a square, as in Bloomsbury, Bayswater and Islington. Working-class Londoners were more likely to live facing an enclosed court, sharing more mundane facilities, like communal privies or water taps. Alternatively, they took rooms in old

Below *The Charing Cross Hotel in the Strand. Designed by E. M. Barry and built in 1863-4, it was one of the first major buildings in London to be faced with artificial stone. Railway hotels were a 19th century phenomenon, attached to all the mainline railway termini. Charing Cross Hotel was built over Charing Cross station, which in turn was built on the site of the old Hungerford Market at the same time as the hotel. Charing Cross was the terminus of the South Eastern Railway.*

middle-class houses that were subdivided among several families. Nobody lived in purpose-built flats. By 1900, detached and semi-detached villas with their own private gardens typified middle-class suburbs; the better-off working classes lived in suburban terraced housing; in inner London an increasing proportion of both rich and poor lived in purpose-built flats – the latter in philanthropic dwellings and, after 1890, in blocks of council flats, the former in luxury apartments that first appeared near Victoria Street in the 1850s and around the Albert Hall in the 1870s.

London was also a city of migrants. Between 1841 and 1871, almost a million migrants came to London; many moving directly from East Anglia or the West Country to suburban Essex, Middlesex and Surrey. Whereas the majority of English migrants were skilled artisans or domestic servants, migrants from Ireland congregated in the poorest districts of inner London and in the worst jobs, as casual labourers or in tailoring or shoemaking sweatshops. By 1861 there were an estimated 178,000 Irish in London, nearly all Catholic, concentrated in Holborn, St Giles (which was nicknamed 'Little Dublin'), Whitechapel and Southwark. There was already a long established Jewish population in London, but London Jewry increased rapidly in the 1880s, as Jews fled from persecution in eastern Europe. By 1914, there were 140,000 Jews in London. They concentrated in the East End, working like the Irish before them in the sweated trades, but also establishing their own businesses, usually employing co-immigrants. Their presence provoked resentment and violence, especially at times of economic depression.

Refer to pages 136-137 for a discussion of both Jewish and Irish immigration in the 19th century.

For all the squalor of slum life, Victorian London was also the 'Heart of the Empire' and the financial capital of the world. New types of building for new business and administrative activities ranged from the office blocks of the City and Whitehall to the department stores of the West End. Mainline railway termini contained spectacular train sheds, such as W. H. Barlow's magnificent 240-feet single span in glass and iron at St Pancras, and Brunel's more cathedral-like pattern of aisles and transepts at Paddington. There were also exhibition buildings, museums, art galleries and concert halls. The Crystal Palace, designed for the 1851 Great Exhibition in Hyde Park, was intended to be temporary, but was subsequently re-erected in south London. The Victoria and Albert Museum opened in 1857, the National Portrait Gallery moved to its present site in 1895, and the Tate Gallery was erected on the site of Millbank Prison in 1897. The non-denominational University College was founded in 1826, while a rival, Anglican-inspired, King's College opened in 1831. More starkly utilitarian buildings included workhouses, prisons, hospitals and infirmaries.

For the story of Harrods see page 149.

For further details of hospitals, refer to page 96, and for a discussion of prisons refer to page 99.

Both the layout and the geographical distribution of pubs, theatres and even churches reflected the Victorian compulsion to classify and segregate their patrons, audiences and congregations. Even in death, social geography was critical. Private cemetery companies divided their cemeteries into separate consecrated and unconsecrated areas for different denominations and graded the size and cost of grave plots. Pubs were divided up into separate compartments, or at least into lounge, saloon and public bar. Huge theatres like the Britannia, Hoxton, which had seats for 3,450 people, contained a variety of tiers with differential pricing. There were also music halls, predominantly suburban in location, although the most prestigious were in the West End, notably the Alhambra Palace in Leicester Square. As early as 1896 the Alhambra began to incorporate film scenes in its performances, including newsreels of events such as the Queen's Diamond Jubilee in 1897, and also short melodramas, initially filmed on an open-air set on the roof of the music hall. By the early 1900s a film industry had been created, with studios concentrated in upwind, fog-free suburbs like Ealing. London was entering the age of mass consumption, mass transit and mass media.

Cemeteries are discussed on page 166.

The Great Exhibition is discussed in full on pages 94-95.

Map below *The extent of Victorian London at its peak is contrasted to the size of the city today.*

EXTENT OF LONDON 1900

LONDON 1990

+St. Paul's

LONDON 1990

THE GROWTH OF 19TH-CENTURY LONDON

In the 1820s William Cobbett described London as 'The Great Wen', a cancer-like growth on the face of Britain, already embracing a population of one and a half million. By the end of Queen Victoria's reign, London's population included four and a half million inhabitants within the administrative limits of the County of London, but another two million in 'Greater London' beyond the jurisdiction of the newly-created London County Council. The map (right) plots this spectacular expansion.

Rates of housebuilding fluctuated much more than rates of population growth. In general, builders over-reacted to economic booms and slumps. In response to boom conditions they would build too many houses too late, so that, by the time the houses were ready for occupancy, demand was on the wane, leaving a glut of unsaleable properties. When prosperity returned, builders would react too slowly, and working-class families would be forced to take in lodgers to help pay increasing rents, or to 'double-up' with other families.

As in Georgian and Regency London, the great landed estates in the West End controlled development by specifying what kinds of buildings could be erected and how they were to be used, forbidding industrial and commercial uses, and sometimes employing gatekeepers to regulate access by non-residents. Some landowners introduced forms of land-use zoning: along the northern edge of Mayfair, between Grosvenor Square and Oxford Street, the Duke of Westminster promoted the construction of blocks of working-class 'model dwellings' as a kind of cordon sanitaire between the luxury houses in the heart of his estate and the commercial disorder of Oxford Street. The houses could also accommodate the army of artisans and tradesmen who relied upon the patronage of the rich.

It is interesting to compare the terraced houses, squares and crescents of Bayswater and Belgravia with the detached villas and individual private gardens on the Clapham Park Estate, developed by Thomas Cubitt in south London. Substantial terraces continued to be built in west London but many quickly became unfashionable; in North Paddington, for instance, large terraced houses were subdivided for occupancy by several working-class or lower middle-class families. There were always more middle-class houses built than there were families to occupy them. Hence the decline into seedy shabbiness of estates in Notting Hill and south of King's Cross, the latter brilliantly depicted in George Gissing's The Nether World (1889) and Arnold Bennett's Riceyman Steps (1923).

The new lower middle class of clerks, bookkeepers and schoolteachers mostly found homes in suburbs like Holloway and Camberwell, linked to the City by horse trams and suburban trains. These suburbs, in turn, were a cut above respectable working-class districts, which were located close to major industrial zones – around the Great Eastern Railway works in Stratford, for example, or following the line of the London and Greenwich Railway through Bermondsey and Deptford. London was not only developing into a 'monster city', but its population was becoming distributed into distinctive social areas by its status and ability to pay.

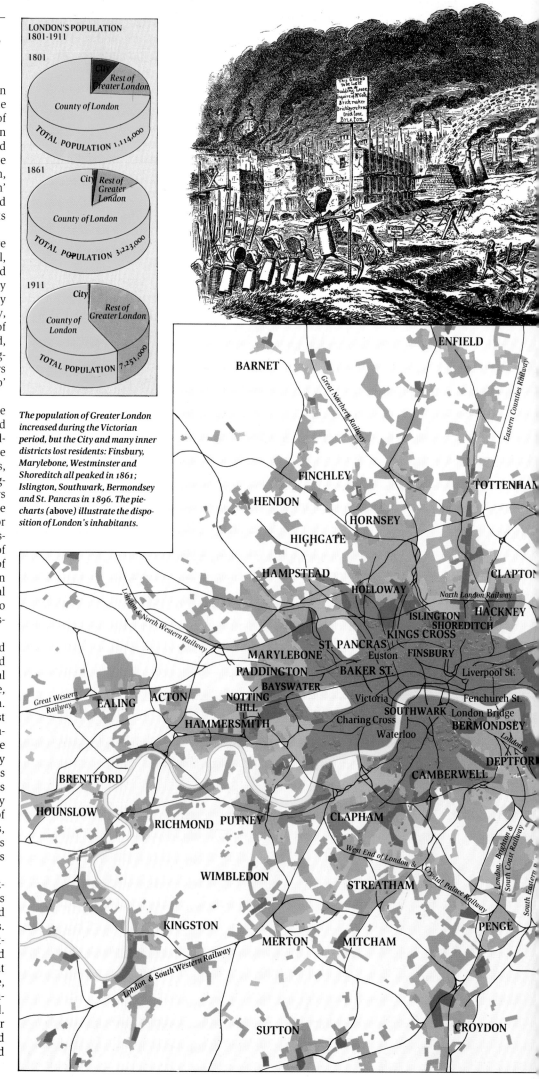

LONDON'S POPULATION 1801-1911

1801
City
Rest of Greater London
County of London
TOTAL POPULATION 1,114,000

1861
City
Rest of Greater London
County of London
TOTAL POPULATION 3,223,000

1911
City
Rest of Greater London
County of London
TOTAL POPULATION 7,251,000

The population of Greater London increased during the Victorian period, but the City and many inner districts lost residents: Finsbury, Marylebone, Westminster and Shoreditch all peaked in 1861; Islington, Southwark, Bermondsey and St. Pancras in 1896. The pie-charts (above) illustrate the disposition of London's inhabitants.

The urban expansion As early as 1829 George Cruikshank
dramatically caricatured the trauma of urban expansion in
his cartoon of 'London Going Out of Town – The March of
Bricks and Mortar' (left). Cruikshank lived in Amwell Street,
close to Islington Fields, which were undergoing just such an
invasion in the 1820s: farmland first became brickfields;
streets were marked out by developers such as Thomas Cubitt;
finally, a host of small speculative builders would acquire
'building leases'.

Until the 1840s Camberwell was still a place of gentility
and even rusticity. Glengall Terrace (top right), built between
1843 and 1845 in Grecian style, was an elegant but modest
expression of the suburb's respectability. Avondale Square
(1875) (right) comprised coarser, mid-Victorian, terraced
houses with attics as well as basements, a middle-class island
in otherwise decaying North Camberwell. Houses on Ivydale
Road (1900) (below right) were smaller, two-storey brick,
slate-roofed; bay windows were de rigueur, their extension
over both floors indicating that Ivydale Road was a
respectable lower middle-class address.

19th-century developments In the
early part of Victoria's reign, the
residential areas of London were
still laid out on Georgian lines, in
squares, terraces and crescents.
Bayswater, for example (bottom
left), was laid out in the 1830s and
1840s. Substantial mansions were
often still terraced – such as the
example (below), 'Albert Houses',
Queen's Gate, South Kensington,
which had 20 rooms. In more distant
suburbs, where land was cheaper,
detached villas were preferred, as at
Clapham Park (1862) (bottom
right). Houses would have looked
something like this villa from
Lansdowne Road, Holland Park,
Bayswater 1844 (right).

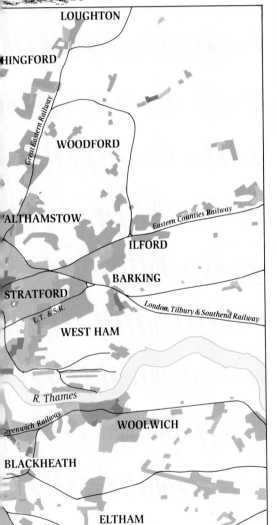

LONDON'S GROWTH 1800-1914

1800	1900
1845	1914
1860	— main railways, 1914
1880	

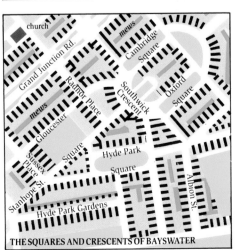

THE SQUARES AND CRESCENTS OF BAYSWATER

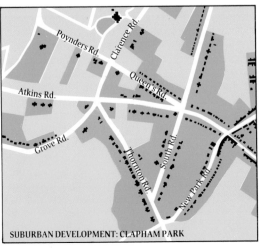

SUBURBAN DEVELOPMENT: CLAPHAM PARK

THE TRANSPORT REVOLUTION

At the beginning of the 19th century, 'public transport' in London comprised short-stage coaches, hackney carriages, and ferry-boats, all affordable only by the better-off. So, when George Shillibeer introduced his 20-seater coaches in 1829, which operated along the New Road from Paddington to the City with fares as low as sixpence, he followed the Parisian practice of calling them 'omnibuses', for *all* the people. Even sixpence was beyond the means of most Londoners, and horse-buses remained a middle-class form of transport. Nevertheless, by 1850 there were more than 1300 buses on the streets of London; fierce competition between rival private operators, larger vehicles with rooftop seats, and tax cuts all allowed cheaper fares. Reduced profit margins led inevitably to rationalisation later in the 1850s, and most operators were taken over by the French-backed London General Omnibus Company. By 1875, the company could boast almost 50 million passengers per annum.

Many omnibus routes linked the City and West End to the earliest railway termini, sited on the edge of the built-up area, often some way short of the termini we know today. The London & Southampton Railway ended at Nine Elms (1838), extending to Waterloo ten years later; the first 'West End' terminus of the London & Brighton and South Eastern Railways was at Bricklayers Arms, off the Old Kent Road (1844). On these peripheral sites, land was relatively cheap and there were less likely to be delays negotiating the compulsory purchase of property that was already in profitable use. Even so, some demolition was usually necessary: the cutting of the path of the London & Birmingham Railway through Camden Town is dramatically described by Dickens in *Dombey and Son*.

The earliest passenger railway in London was the London & Greenwich (1836), running south-east from London Bridge on a four-mile viaduct. The company planned a 'promenade' alongside the viaduct, and hoped to utilise the arches for shops and houses, but in fact, urban railway viaducts everywhere brought blight, not prestige, to their surroundings. While the London & Greenwich was a commuter railway from the beginning, companies to the north and west of London regarded suburban trains as an awkward inconvenience, getting in the way of more lucrative mainline traffic. When the Great Western opened in 1838, the first station out of Paddington was West Drayton, thirteen miles away. Suburban traffic did become more important; but for the moment it remained primarily middle-class.

The Railway Mania of the 1840s generated so many competing schemes, all vying for access to the heart of the city, that a Royal Commission recommended in 1846 that no further railway lines should be built in central London. Where extensions were subsequently authorised, for example from London Bridge to Charing Cross (opened in 1864) and Cannon Street (1866), large-scale demolition proved necessary, displacing at least 76,000 of the poorest Londoners between 1855 and 1900, and intensifying overcrowding in the slum districts that survived. From the 1860s some companies were obliged by government to run cheap workmen's trains, to compensate for the displacements they had caused: this was the price paid by the Great Eastern Railway for its extension from Bishopsgate to Liverpool Street (1874). More generally, the Cheap Trains Act (1883) required all companies to offer workmen's fares on early morning and evening services, as directed by the Board of Trade. By October 1911, out of 390,000 passengers carried each weekday by twelve leading companies from stations 4-30 miles distant from central London, more than 105,000 travelled on workmen's tickets. There were also workmen's fares on new tube railways, such as the City & South London (1890), and on trams. Nevertheless there were still few working-class commuters on the Great Western or London

& North Western lines; the East End–West End dichotomy in London's social geography was re-emphasised and extended.

Railways left their mark in other ways. Mainline termini attracted grand hotels. Regular shopping trips to the West End became possible, stimulating the growth of department stores. Offices, shops and newly fashionable luxury flats lined Victoria Street, linking Victoria Station to Westminster. Farther out, tangles of junctions and marshalling yards isolated communities on 'the wrong side of the tracks'. Engine sheds and railway works became foci for suburban employment, for example in Stratford and Battersea.

Apart from three short-lived demonstration lines laid by an American promoter, George Francis Train, in 1861, the earliest horse tramways in London date from the 1870s. Horse trams were cheaper than buses, carried more passengers and operated earlier in the morning and later at night. They were the making of modest but respectable suburbs like Holloway and Camberwell. But the electrification of tramways in the early 1900s heralded a new era. The electric tramcar became the 'gondola of the people', used for weekend trips to parks, countryside and football matches, as well as for journeys to work. Neither horse-drawn nor electric trams were allowed in the City or West End; the only connection between north and south London systems was through the Kingsway Tunnel, opened in 1908.

The railway era *Two paintings depict the romance of the railway: John O'Connor's 'St Pancras Hotel and Station from Pentonville Road: Sunset' (1884) (above) and Camille Pissarro's 'Lordship Lane Station, Upper Norwood' (1871) (bottom). The clock-tower at King's Cross (1852) can be seen to the right of St Pancras Hotel, begun in 1867. Lordship Lane, on the Crystal Palace & South London Junction Railway, opened in 1865. A few substantial villas had already been built in the area (below left). By 1888 (below right) much more development had occurred to the north of the station and around the railway station at Forest Hill. The map (top right) shows the evolution of London's railways, and includes shallow underground lines (such as the Metropolitan and District), but omits the first deep tubes which date from the 1890s. Two graphs (right) show patterns of suburban passenger traffic in October 1911, by company and by distance travelled.*

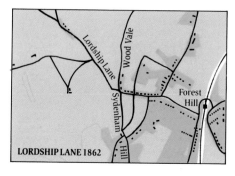

LORDSHIP LANE 1862

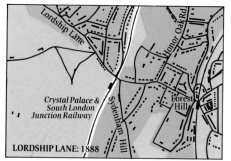

Crystal Palace & South London Junction Railway

LORDSHIP LANE: 1888

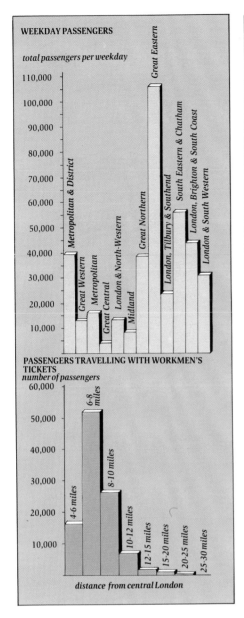

WEEKDAY PASSENGERS

total passengers per weekday

Bar chart values on y-axis: 110,000 / 100,000 / 90,000 / 80,000 / 70,000 / 60,000 / 50,000 / 40,000 / 30,000 / 20,000 / 10,000

Bars labelled: Metropolitan & District, Great Western, Metropolitan, Great Central, London & North-Western, Midland, Great Northern, London, Tilbury & Southend, South Eastern & Chatham, London, Brighton & South Coast, Great Eastern, London & South Western

PASSENGERS TRAVELLING WITH WORKMEN'S TICKETS

number of passengers

60,000 / 50,000 / 40,000 / 30,000 / 20,000 / 10,000

Bars: 4-6 miles, 6-8 miles, 8-10 miles, 10-12 miles, 12-15 miles, 15-20 miles, 20-25 miles, 25-30 miles

distance from central London

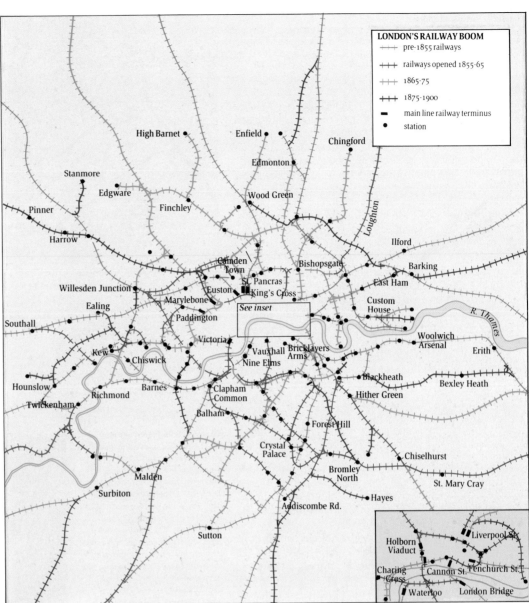

LONDON'S RAILWAY BOOM
- pre-1855 railways
- railways opened 1855-65
- 1865-75
- 1875-1900
- main line railway terminus
- station

London's trams The map (right) shows the chronology of tramway construction prior to 1895. Services were run by private companies, of which the largest was the North Metropolitan, one of whose cars is shown outside the West Ham Union workhouse in 1890 (far right). The City remained the preserve of cabs and horse buses (bottom right); the bridge across Ludgate Hill carried the only mainline railway to cross central London.

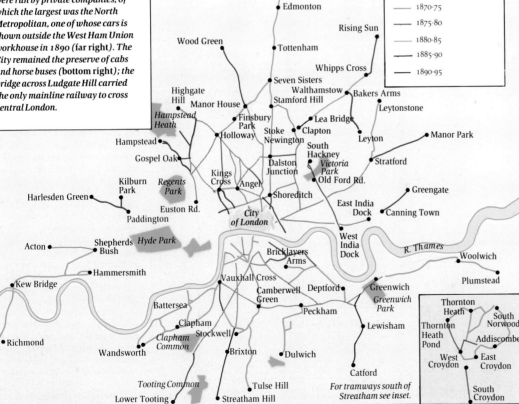

THE GROWTH OF LONDON'S TRAMWAYS
- 1870-75
- 1875-80
- 1880-85
- 1885-90
- 1890-95

For tramways south of Streatham see inset.

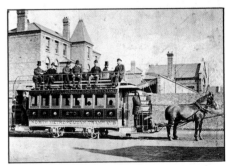

VICTORIAN INDUSTRY

19th-century London was a city of small workshops. In 1851, 86 per cent of industrial employers in London had less than ten workers; and only 17 establishments employed more than 250 persons. During the century, some major London industries declined as a consequence of technological change and the extension of free trade. London's shipbuilding industry, for example, collapsed in the 1860s; timber hulls had been replaced by iron, and the more spacious sites on the Clyde and Tyneside provided the room to build and launch much larger vessels.

18th-century trades, such as clothing and shoe-making, had concentrated in small workshops, mostly in the West End, close to the homes of wealthy customers. During the 19th century, they were replaced by a mass-produced industry, making lower quality goods for purchase 'off the peg' by the new proletariat and lower-middle class. New forms of production were much larger in scale, but involved less skill and a more intricate division of labour: workers were now more often female, paid by the piece, employed either in sweat-shops or, as outworkers, in their own homes.

By the early 20th century, however, new and larger-scale industries in suburban London – Greenwich, Woolwich, West Ham, Enfield – were increasing in importance. In 1907 there was an average of 20 employees per factory or workshop in Greater London; but the figure in Woolwich was 69. The Edwardian era also saw the beginning of new electrical industries, vehicle manufacture (such as the Matchless motorcycle works, established in Plumstead in 1899), and an entertainment industry, prompting the location of film studios in relatively unpolluted areas of west London such as Ealing. Printing and photographic industries also moved to the pure air of western suburbs; Kodak located in Harrow as early as 1892.

The fortunes of Thameside industrial districts such as West Ham and Poplar were closely associated with the 19th-century expansion of the docks, intended to rid the river of growing congestion. By 1830, the docks dealt with most overseas trade, much of it with the British Empire, but riverside wharves still accommodated coastal traffic (from the British Isles), almost three-quarters of the total. Labour relations in the docks were often strained, reflecting a system of mainly casual labour which denied most dockers any guarantee of regular work, and culminating in a bitter Dock Strike in 1889. On average, 55 per cent of labourers seeking dock work were turned away. It is not surprising that dockers' wives and children were obliged to seek paid work themselves. Boys left school at the earliest opportunity to become newsvendors and errand boys, work which appeared well-paid in the short term, but which condemned them for life to the ranks of the uneducated and unskilled, thereby perpetuating a vicious circle of poverty.

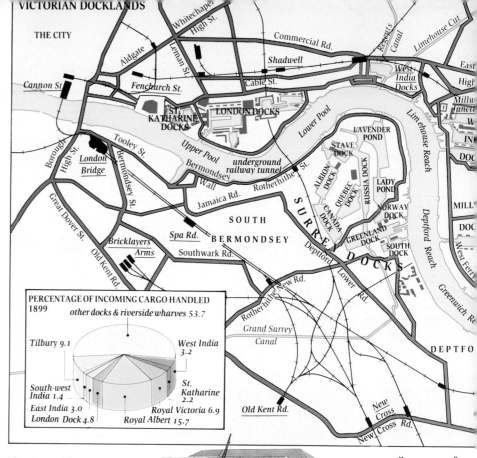

VICTORIAN DOCKLANDS

PERCENTAGE OF INCOMING CARGO HANDLED 1899

other docks & riverside wharves 53.7

Tilbury 9.1

West India 3.2

South-west India 1.4

St. Katharine 2.2

East India 3.0

London Dock 4.8

Royal Victoria 6.9

Royal Albert 15.7

Victorian Docklands *The map (top) and table (far right) show the evolution of London's docks. The Port of London Authority took control of the entire system in 1909. By the 1880s, between 50,000 and 100,000 men depended on the docks and riverside wharves for their employment. The graph (above right) shows the number of vessels engaged in foreign trade entering and leaving the Port of London. The average size of ships increased dramatically between 1861 and 1899. Net registered tonnage rose from 3.1 million tons in 1861 to 9.2 million in 1899 (imports), and from 1.6 million to 6.0 million tons (exports). The photograph (right) shows the busy waterfront between Yardley's Wharf and Braithwaite & Dean's Wharf (on the south bank, backing onto Rotherhithe Street), as it existed in 1937. The labour-intensive nature of cargo-handling in the docks in pre-container days is clearly evident (far right).*

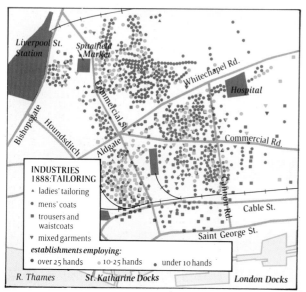

INDUSTRIES 1888: TAILORING

▴ ladies' tailoring
• mens' coats
■ trousers and waistcoats
▾ mixed garments

establishments employing:
● over 25 hands   ● 10-25 hands   • under 10 hands

R. Thames   St. Katharine Docks   London Docks

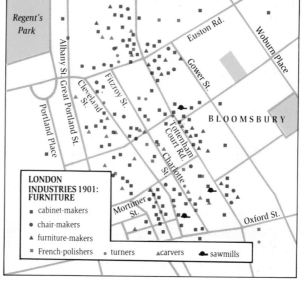

LONDON INDUSTRIES 1901: FURNITURE

■ cabinet-makers
● chair-makers
▴ furniture-makers
■ French-polishers   • turners   ▴ carvers   ⚓ sawmills

London's Industry *The map (far left) plots the vast number of tiny tailoring workshops in the East End, each with its own speciality. By the 1900s small workshops were giving way to much larger clothing factories, such as Schneider's in Whitechapel (above), which employed mainly young immigrants. Similar patterns and degrees of specialisation characterised other industries, such as furniture manufacturing around Tottenham Court Road (map left) and Curtain Road. The map (right) shows large factories in the County of London in 1898. West Ham, just outside the LCC area, had a wide range of industry (far right): from the Royal Docks, sugar refineries, chemical works and distilleries in the lower Lea Valley, to the Stratford railway works in the north.*

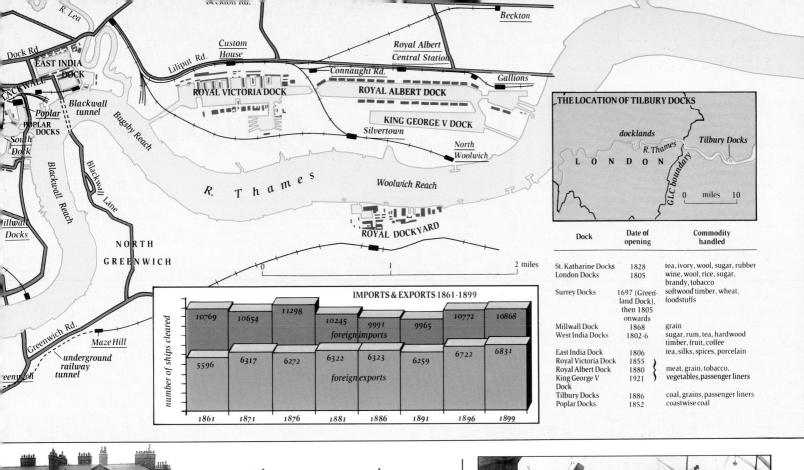

R. Lea
Beckton Rd.
Beckton
Custom House
Liliput Rd.
Royal Albert Central Station
Connaught Rd.
Gallions
EAST INDIA DOCK
ROYAL VICTORIA DOCK
ROYAL ALBERT DOCK
Blackwall tunnel
POPLAR DOCKS
KING GEORGE V DOCK
Poplar
South Dock
Bugsby Reach
Silvertown
North Woolwich
Millwall Docks
Blackwall Lane
Blackwall Reach
R. Thames
Woolwich Reach
NORTH GREENWICH
ROYAL DOCKYARD
Greenwich Rd.
Maze Hill
underground railway tunnel
Greenwich

0   1   2 miles

THE LOCATION OF TILBURY DOCKS

docklands
R. Thames
Tilbury Docks
LONDON
GLC boundary
0   miles   10

IMPORTS & EXPORTS 1861-1899

number of ships cleared

foreign imports

10769	10654	11298	10245	9991	9965	10772	10868

foreign exports

5596	6317	6272	6322	6323	6259	6722	6831

| 1861 | 1871 | 1876 | 1881 | 1886 | 1891 | 1896 | 1899 |

Dock	Date of opening	Commodity handled
St. Katharine Docks	1828	tea, ivory, wool, sugar, rubber
London Docks	1805	wine, wool, rice, sugar, brandy, tobacco
Surrey Docks	1697 (Greenland Dock), then 1805 onwards	softwood timber, wheat, foodstuffs
Millwall Dock	1868	grain
West India Docks	1802-6	sugar, rum, tea, hardwood timber, fruit, coffee
East India Dock	1806	tea, silks, spices, porcelain
Royal Victoria Dock	1855	meat, grain, tobacco, vegetables, passenger liners
Royal Albert Dock	1880	
King George V Dock	1921	
Tilbury Docks	1886	coal, grains, passenger liners
Poplar Docks	1852	coastwise coal

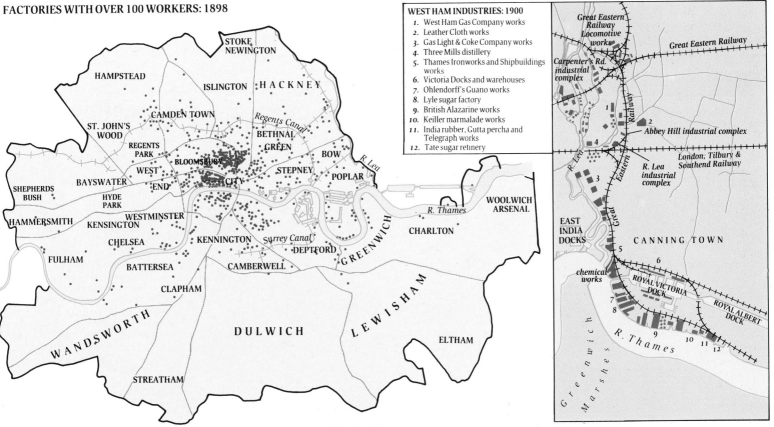

FACTORIES WITH OVER 100 WORKERS: 1898

STOKE NEWINGTON
HAMPSTEAD
ISLINGTON
HACKNEY
CAMDEN TOWN
Regents Canal
ST. JOHN'S WOOD
BETHNAL GREEN
REGENTS PARK
BLOOMSBURY
BOW
BAYSWATER
WEST END
CITY
STEPNEY
POPLAR
R. Lea
SHEPHERDS BUSH
HYDE PARK
WESTMINSTER
HAMMERSMITH
KENSINGTON
R. Thames
WOOLWICH ARSENAL
CHELSEA
KENNINGTON
Surrey Canal
GREENWICH
CHARLTON
FULHAM
DEPTFORD
BATTERSEA
CAMBERWELL
CLAPHAM
DULWICH
LEWISHAM
ELTHAM
WANDSWORTH
STREATHAM

WEST HAM INDUSTRIES: 1900

1. West Ham Gas Company works
2. Leather Cloth works
3. Gas Light & Coke Company works
4. Three Mills distillery
5. Thames Ironworks and Shipbuildings works
6. Victoria Docks and warehouses
7. Ohlendorff's Guano works
8. Lyle sugar factory
9. British Alazarine works
10. Keiller marmalade works
11. India rubber, Gutta percha and Telegraph works
12. Tate sugar refinery

Great Eastern Railway Locomotive works
Great Eastern Railway
Carpenter's Rd. industrial complex
Abbey Hill industrial complex
London, Tilbury & Southend Railway
R. Lea industrial complex
EAST INDIA DOCKS
CANNING TOWN
chemical works
ROYAL VICTORIA DOCK
ROYAL ALBERT DOCK
Greenwich Marshes
R. Thames

HEART OF EMPIRE

The 19th century saw the transformation of the City of London from a bustling trading and mercantile centre to the commercial heart of the world's largest and wealthiest city and the financial centre of a vast Empire. The Napoleonic Wars had dealt a blow to rival European financial centres and increasingly London was seen as a safe haven – the merchant bankers of London became world financiers, investing vast sums in overseas development and industry.

This explosion in financial and business activity was accompanied by the extensive redevelopment of the City itself. The Bank of England had lost its monopoly on joint-stock banking in 1824, opening the way for the appearance of a large number of new joint-stock banks such as Barclays and the Midland. New headquarters were built for the new joint-stock banks and large insurance companies, especially in the area near

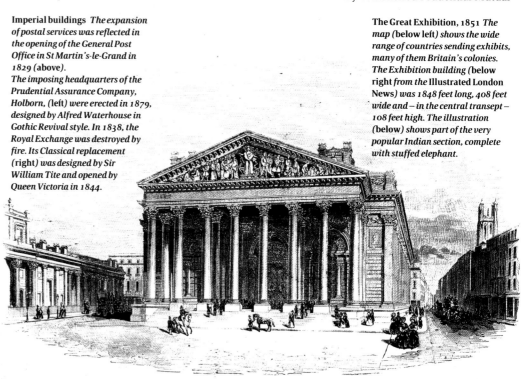

the Bank of England. The model of these buildings was the Italian palazzo, with public rooms on the ground floor, the board room on the first floor, other offices (some for letting) on the second floor, and a caretaker's flat in the attic. Although these buildings often had more space than their owners required, they promoted an impressive corporate image and allowed for future expansion. As the number of small firms

proliferated, and as banks and other major institutions began to occupy all the space in their own buildings, so speculatively-built office blocks became popular. In the 1860s, as schemes increased in size and therefore in capital requirements, limited companies took over from individual speculators. Many of these buildings were now occupied by several firms: in 1881 a City Corporation daytime census counted 1,320 lettings in only 26 buildings.

London directories recorded about 570 stock and share brokers in 1861, but over 5000 in 1901. 86 bankers were listed in 1861, 224 in 1901, of whom 85 were classified as 'foreign and colonial'. Insurance became big business: marine insurance revived from a slump after the Napoleonic Wars and the number of Lloyd's underwriters increased from 189 in 1849 to over 400 in 1870. London fire insurance companies established agencies and branch offices all over the world. By 1839 there were also 72 life assurance offices in London. In 1852, a deputation of working men asked the recently established Prudential Mutual

Imperial buildings *The expansion of postal services was reflected in the opening of the General Post Office in St Martin's-le-Grand in 1829 (above).*
The imposing headquarters of the Prudential Assurance Company, Holborn, (left) were erected in 1879, designed by Alfred Waterhouse in Gothic Revival style. In 1838, the Royal Exchange was destroyed by fire. Its Classical replacement (right) was designed by Sir William Tite and opened by Queen Victoria in 1844.

The Great Exhibition, 1851 *The map (below left) shows the wide range of countries sending exhibits, many of them Britain's colonies. The Exhibition building (below right from the Illustrated London News) was 1848 feet long, 408 feet wide and – in the central transept – 108 feet high. The illustration (below) shows part of the very popular Indian section, complete with stuffed elephant.*

NEW ROYAL EXCHANGE.—(FROM THE ARCHITECT'S DRAWING.)

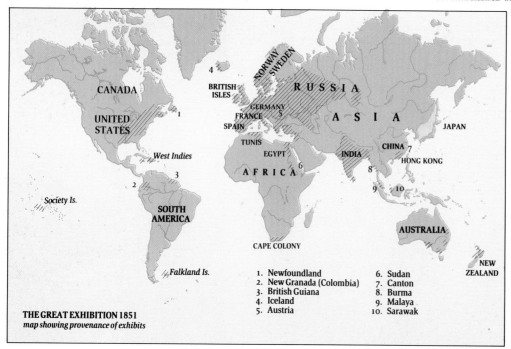

THE GREAT EXHIBITION 1851
map showing provenance of exhibits

1. Newfoundland
2. New Granada (Colombia)
3. British Guiana
4. Iceland
5. Austria
6. Sudan
7. Canton
8. Burma
9. Malaya
10. Sarawak

Assurance, Investment and Loan Association whether it would entertain assurances as small as £20, payable by weekly instalments. By 1875 the Prudential alone had more than 2 million 'industrial' assurance policies.

Britain's role in the world was also celebrated in 'The Great Exhibition of the Works of Industry of all Nations', staged in Joseph Paxton's 'Crystal Palace' in Hyde Park in 1851. Exhibits were requested from all over the world, including 'natural' wonders such as the Koh-i-Noor diamond, machinery, manufactures and crafts and fine arts. In all, over 100,000 exhibits were provided by 13,937 exhibitors: 7,381 from the United Kingdom and its Dependencies, 6,556 from the rest of the world. Exhibits from the former were grouped westward of the central transept, while those of other nations were placed together eastward. The position of each country's display was determined by its latitude, Mediterranean and tropical states nearer the transept than temperate nations. In all, about two million people visited the exhibition between 1st May, when it was formally opened by Queen Victoria, and 15th October, most paying several visits. In an effort to attract all social classes, entrance fees were varied: on two days when admission cost £1, only a few wealthy visitors came; on 80 days when the charge was a shilling, nearly 4.5 million tickets were sold. The British Museum was one of many attractions where tourist attendance increased. But there was disappointment that only 58,427 foreigners arrived in England between April and September 1851. Even then, it was reported, 'the proverbial expense of a London season must, in many instances, be a serious impediment against frequent visits'.

Administering the Empire involved a major expansion of government offices in Whitehall (map below right). Less than 20 years after the Board of Trade offices were built in 1827, they were remodelled by Sir Charles Barry, incorporating existing buildings, such as Dorset House, behind a new facade. The whole block now constituted 'New Treasury Offices' (top right). More spectacular, especially when viewed from St. James's Park, were the New Government Offices designed by Sir George Gilbert Scott to house the Foreign, Colonial, India and Home Offices (top left), and erected south of Downing Street (1868-73). The march of Government southwards continued with the 'New Public Offices' (1899-1915); other new buildings included the War Office (1899-1906) and the Admiralty extension (1895).

North of Whitehall lay Trafalgar Square, completed in 1843 with the construction of Nelson's Column. In the popular imagination, Trafalgar Square truly was the 'heart of Empire'; the setting for mass meetings, demonstrations (sometimes violent), victory celebrations and New Year revels.

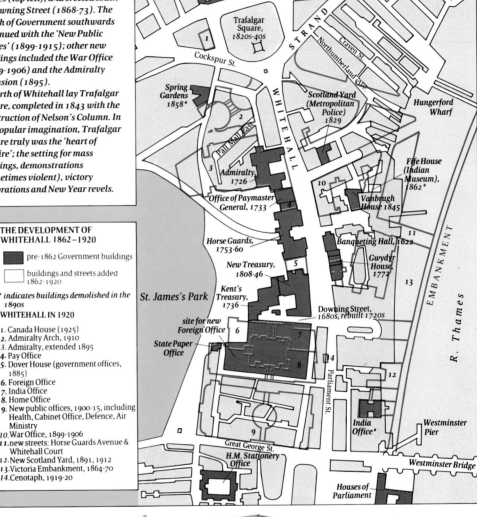

THE DEVELOPMENT OF WHITEHALL 1862–1920

■ pre-1862 Government buildings

□ buildings and streets added 1862-1920

* indicates buildings demolished in the 1890s

WHITEHALL IN 1920

1. Canada House (1925)
2. Admiralty Arch, 1910
3. Admiralty, extended 1895
4. Pay Office
5. Dover House (government offices, 1885)
6. Foreign Office
7. India Office
8. Home Office
9. New public offices, 1900-15, including Health, Cabinet Office, Defence, Air Ministry
10. War Office, 1899-1906
11. new streets: Horse Guards Avenue & Whitehall Court
12. New Scotland Yard, 1891, 1912
13. Victoria Embankment, 1864-70
14. Cenotaph, 1919-20

VICTORIAN LONDON: HOSPITALS AND CHURCHES

DISEASE AND HOSPITALS

Although by 1861, there were 80 hospitals in London, there was still little provision for the poor, apart from the sick wards of workhouses, until a public hospital system was established in 1867, funded out of Poor Law rates. The first Metropolitan Asylums Board hospital, opened in 1870, was only for paupers suffering from smallpox or scarlet fever,

Victorian Hospitals The map (below) shows the main hospitals of central London in 1902. Many of London's specialist hospitals (see table, right) had opened since 1800. They included eye hospitals, orthopaedic hospitals, children's hospitals and several which catered only for women. But apart from isolation and mental hospitals, and a few Poor Law infirmaries, there were still very few hospitals in South London.

HOSPITALS NOT NAMED ON MAP (CENTRAL LONDON)

1. Belgrave Hospital for Children (1866)
2. Royal Dental Hospital (1901)
3. British Lying-In Hospital (1749)
4. Cancer Hospital (1851)
5. Chelsea Hospital for Women (1871)
6. City Orthopaedic Hospital (1851)
7. Central London Throat & Ear Hospital (1875)
8. Evelina Hospital for Sick Children (1869)
9. French Hospital (1867)
10. General Lying-in Hospital (1765)
11. Hospital for Consumption & Diseases of the Chest (1841)
12. Central London Ophthalmic Hospital (1843)
13. Hospital for Diseases of the Throat (1863)
14. West End Hospital for Nervous Diseases (1878)
15. Hospital for Sick Children (1852)
16. Hospital for Women (1843)
17. Hospital for Women & Children (1866)
18. London Homeopathic Hospital (1849)
19. National Hospital for Diseases of the Heart (1857)
20. National Hospital for the Paralysed & Epileptic (1859)
21. National Orthopaedic Hospital (1836)
22. Queen Charlotte's Lying-in Hospital (1739)
23. Royal Hospital for Children & Women (1816)
24. Royal Orthopaedic Hospital (1838)
25. Royal South London Ophthalmic (1857)
26. Royal Westminster Ophthalmic (1816)
27. London Temperance Hospital (1873)
28. St. Peter's Hospital for Stone and other Urinary Diseases (1860)
29. Victoria Hospital for Children (1866)
30. Cheyne Hospital for Women (1874)
31. St. Saviour's Cancer Hospital (1875-7)
32. Italian Hospital (1884)
33. New Hospital for Women (Elizabeth Garrett Anderson) (1872)
34. Paddington Green Children's Hospital (1883)
35. Western Ophthalmic Hospital (1856)
36. Lock Hospital (for prostitutes) (1862)

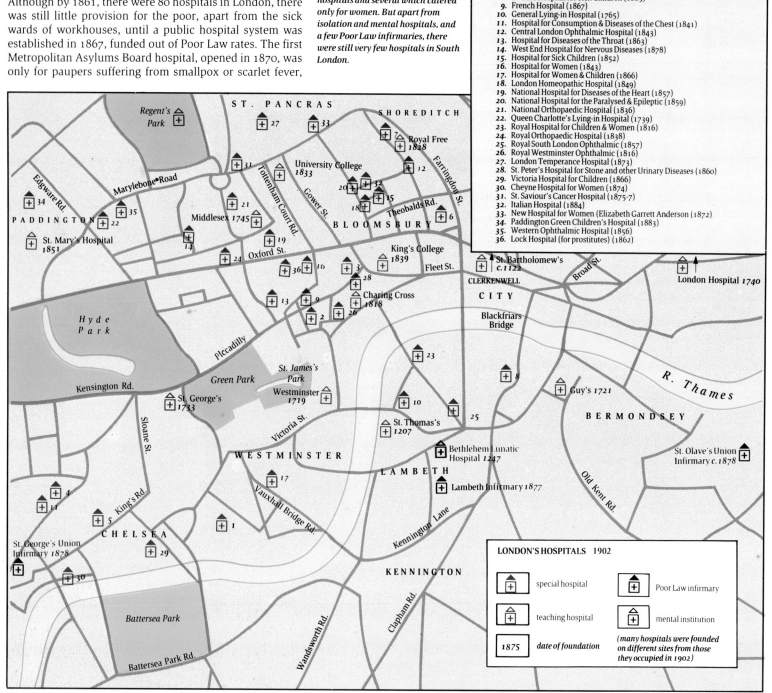

LONDON'S HOSPITALS 1902

special hospital

Poor Law infirmary

teaching hospital

mental institution

1875 *date of foundation*

(many hospitals were founded on different sites from those they occupied in 1902)

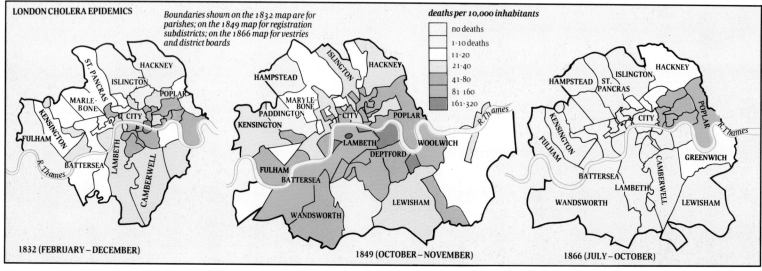

LONDON CHOLERA EPIDEMICS

Boundaries shown on the 1832 map are for parishes; on the 1849 map for registration subdistricts; on the 1866 map for vestries and district boards

deaths per 10,000 inhabitants

no deaths
1-10 deaths
11-20
21-40
41-80
81-160
161-320

1832 (FEBRUARY – DECEMBER)

1849 (OCTOBER – NOVEMBER)

1866 (JULY – OCTOBER)

but free admission was later extended to people who were not paupers, and to those suffering a wider range of illnesses. By 1877, five fever hospitals, on the edges of the built-up area, in Hampstead, Homerton, Stockwell, Fulham and Deptford provided 1,450 beds. By 1890 there were a further 26 Poor Law Infirmaries with a total of 13,203 beds.

London suffered major epidemics of cholera in 1832, 1848-49, 1853-54 and 1866. Thinking that the disease was spread by the inhalation of miasmas, invisible noxious gases emitted by excrement and rotting waste, sanitary reformers tried to prevent the spread of cholera by flushing sewers and watering streets to wash away the waste. In fact, cholera is primarily water-borne, and the first effect of this enthusiastic cleansing was to make the 1848-49 epidemic, with a death rate of 6.6 per 1,000 persons, even worse than that of 1832, where the death rate was 3.4 per 1,000. The distribution of deaths in successive cholera epidemics clearly reflects the comparative cleanliness of the water supply in different parts of London. In 1832, for example, cholera was worst in south London, where the Southwark Waterworks Company supplied appallingly polluted water pumped direct from the Thames, opposite a sewer outfall near London Bridge. In North London, water supplied by the New River Company was relatively pure, and death rates were correspondingly lower.

VICTORIAN CHURCH-BUILDING

Despite massive church-building campaigns in the first half of the century, by 1851 the Church of England could provide sittings for only 17% of Londoners. Nonetheless, the lack of adequate buildings could not really be blamed for the finding that fewer than 25% of Londoners attended church on Census Sunday in 1851. Fifty years later the *Daily News* found that attendances still amounted to only 22% of the population. Allowing for 'twicers' who attended morning and evening services, it seemed that 19% of Londoners went to church, but the proportion varied from 40% in wealthy suburbs to less than 10% in parts of the East End. Anglicans were most numerous in areas of high social status, especially in west London, while non-conformists predominated in upper working-class and lower middle-class districts. There was also a regional factor; church-going was more common in north-east than in south-west London, reflecting the strength of nonconformity in East Anglia where many north-east Londoners had originated. Charles Booth's survey in the 1890s shows a proliferation of Baptist chapels south of the Thames, perhaps attributable to the charismatic impact of Charles Spurgeon, who became minister of a chapel in Southwark when aged only 19, and subsequently based his ministry at the Metropolitan Tabernacle, Elephant & Castle. In contrast, Catholicism was at its strongest in inner west London, where a poor Irish population combined with wealthier 'old Catholics' and with high Anglicans who had converted to Catholicism, following the example of the theologian, author and cardinal, John Henry Newman.

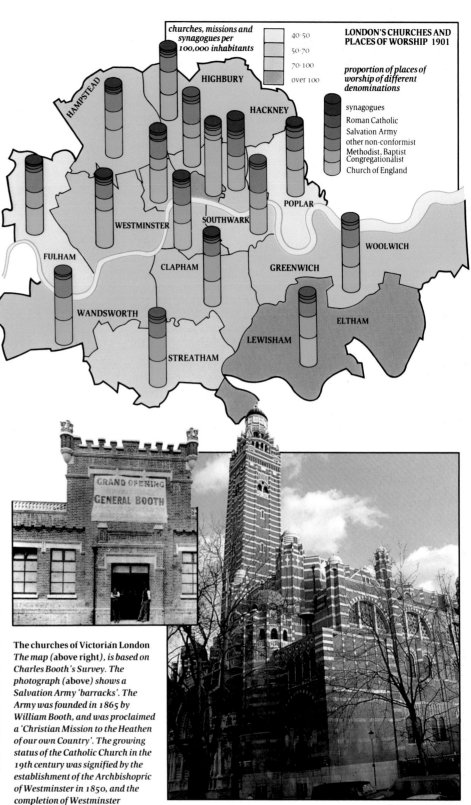

The churches of Victorian London *The map (above right), is based on Charles Booth's Survey. The photograph (above) shows a Salvation Army 'barracks'. The Army was founded in 1865 by William Booth, and was proclaimed a 'Christian Mission to the Heathen of our own Country'. The growing status of the Catholic Church in the 19th century was signified by the establishment of the Archbishopric of Westminster in 1850, and the completion of Westminster Cathedral (above right) in 1903 on a site near Victoria Street.*

Disease and health reform *The maps (left) show the comparative mortality rates in three cholera epidemics. Distribution of deaths closely reflects contaminated water supplies. Drinking water frequently came untreated from the polluted River Thames, a common target in Punch. The cartoon (near left) of 1858, shows 'Father Thames Introducing His Offspring to the Fair City of London'. The offspring are labelled Diptheria, Scrofula and Cholera. John Snow's map of cholera deaths in Soho (right) shows the concentration of cases around a contaminated water-pump in Broad Street, vital evidence for his argument that cholera was water-borne.*

DIPHTHERIA.   SCROFULA.   CHOLERA.

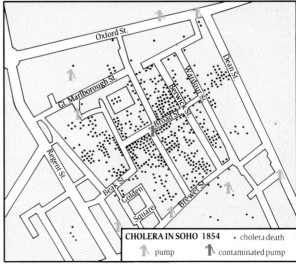

CHOLERA IN SOHO 1854   • cholera death
  pump   contaminated pump

CRIME AND PUNISHMENT

London in the early 1800s was a violent and unsafe place to live. There was no London-wide police force until Robert Peel established the Metropolitan Police in 1829; and parish constables and night watchmen were no match for hordes of pickpockets, 'footpads' and garotters (equivalent to today's muggers). Districts with the most reported crimes included the City, Soho, Covent Garden and Wapping, where ragged mudlarks (who scavenged in the Thames mud) pilfered rope or coal from barges, while lightermen carried off bales of silk worth hundreds of pounds.

Yet, by the 1880s, London was considered 'the safest capital for life and property in the world'. In 1831, 378 persons per 100,000 had been taken into custody for crimes of violence; sixty years later the rate was 216 per 100,000. Improvement was not continuous; crime increased when real wages fell and the price of bread rose, and most petty larcenies occurred in winter, reflecting links between crime and poverty. But overall, the image created by dramatic outrages such as the Whitechapel Murders detracts from the reality that, for most people most of the time, London was getting safer.

Many Victorians believed that there was a hereditary 'criminal class', reproducing itself in London's 'rookeries', such as Saffron Hill (where Fagin lived in Oliver Twist). Thieves' lodging houses, 'dolly shops' (unlicensed pawn-brokers), whose proprietors acted as fences for stolen property, and 'nests of brothels' abounded in rookeries in Whitechapel, Shadwell and Spitalfields in the East End, and Drury Lane and Seven Dials in the West End. Reformers hoped that destruction of such rookeries would mean the end of criminal culture. In practice, slum clearance merely displaced criminal quarters to more suburban districts, such as Hoxton and Poplar, Vauxhall and Deptford, where the 'honest poor' might be led astray by their new, dishonest neighbours. With the ending of transportation to the colonies and a reduction in death sentences, more criminals were eventually released from prison and returned to the slums.

With the end of transportation more prisons had to be built. 'Convict prisons', for prisoners who previously would have been transported, included Millbank (1816-1903),

THE WHITECHAPEL MURDERS 1888

1. *Martha Turner – found at George Yard Buildings (now Gunthorpe St.) 7 August.*
2. *Mary Ann Nichols (also called Polly) – found at Bucks Row (now Durward St.) – 31 August.*
3. *Annie Chapman – found at the rear of 29 Hanbury St. – 8 September.*
4. *Elizabeth Stride – found at the rear of the International Working Men's Club, 40 Berner St. (now Henriques St.) – 30 September.*
5. *Catherine Eddowes – found at 30 (s-w corner) Mitre Square (off Mitre St.), just north of Aldgate – 30 September.*
6. *Mary Jane Kelly – found at 9 Miller's Court, Dorset St. (off Commercial St.) Spitalfields – 9 November.*

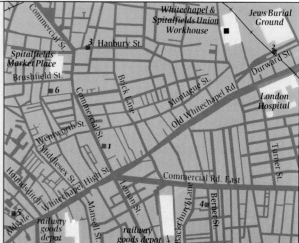

THE FIFTH VICTIM OF THE WHITECHAPEL FIEND.

FINDING THE MUTILATED BODY IN MITRE SQUARE

The Whitechapel murders *Six brutal murders of prostitutes in Whitechapel between August and November 1888 (map above) have been attributed to the mysterious 'Jack the Ripper', variously identified as a failed barrister, a minor aristocrat, an East End Jew (pure anti-semitism), or a doctor (because of his evident skill at dissection). Panic was whipped up by sensational reports in newspapers such as* Illustrated Police News *(left).*

Legal London *The City House of Correction, Holloway (above right), opened in 1882. The map (right) shows the complex array of London's courts and prisons in 1862, centred on the Inns of Court around Chancery Lane. There were over 1500 resident barristers, 3500 solicitors, and more than 5000 law clerks and court officers employed in London's legal district in 1851. The superior courts for civil law were brought under one roof with the opening of the Royal Courts of Justice (or Law Courts) in the Strand in 1882 (from the* Illustrated London News, *top). Designed by G. E. Street, the building contains over 1000 rooms.*

Inns of Court:
1. Lincoln's Inn
2. Temple
3. Gray's Inn
4. Furnival's Inn
5. Staple Inn
6. Sergeant's Inn
7. Clifford's Inn
8. Clement's Inn
9. New Inn
10. Lyon's Inn
11. Symond's Inn
12. Barnard's Inn
13. Thavies' Inn

LONDON'S PENAL SYSTEM

- ■ Inns of Court
- ⚖ Law courts
- 🏛 County courts
- ▲ Police courts
- ●3 prisons

PENAL INSTITUTIONS NOT SHOWN
County Courts:
  Lambeth
  Bow
Police Courts:
  Hammersmith
  Wandsworth
  Greenwich
  Woolwich
Prisons:
  Female Convict Prison, Brixton
  Hulks, Woolwich
  City House of Correction, Holloway
  Surrey House of Correction

Brixton (1819) and Pentonville (1842). When transportation to America ended, and before Australia became an alternative destination, prisoners were incarcerated in old warships called 'hulks', anchored in the Thames at Woolwich. Even in the 1850s two dilapidated hulks continued to accommodate convicts. There were also 'correctional prisons' for sentences of less than two years, for example at Wandsworth (1849) and Holloway (1852), and 'detentional prisons', such as Newgate and Clerkenwell, for prisoners awaiting trial following committal by a magistrate. Nearly 40,000 prisoners passed through these London prisons annually during the 1850s.

A variety of theories influenced prison layout. 'Classification' involved separating young from old, and novices from 'old lags'. The 'silent associated system' prevented prisoners from speaking to one another. Prisoners were placed in radiating wings, supervised by a centrally-positioned warder. Under the 'separate system' prisoners were prevented from mixing or talking by being confined to separate cells for long periods. All three ideas were evident at Holloway (*right*). There were six separate wings, each with its own exercise yard. Front left was for juvenile males, up to the age of 17; front right was for females; the rear, radiating wings were for adult males whose days were spent working a treadwheel, picking oakum (unravelling old rope), shoemaking, mat-making or tailoring.

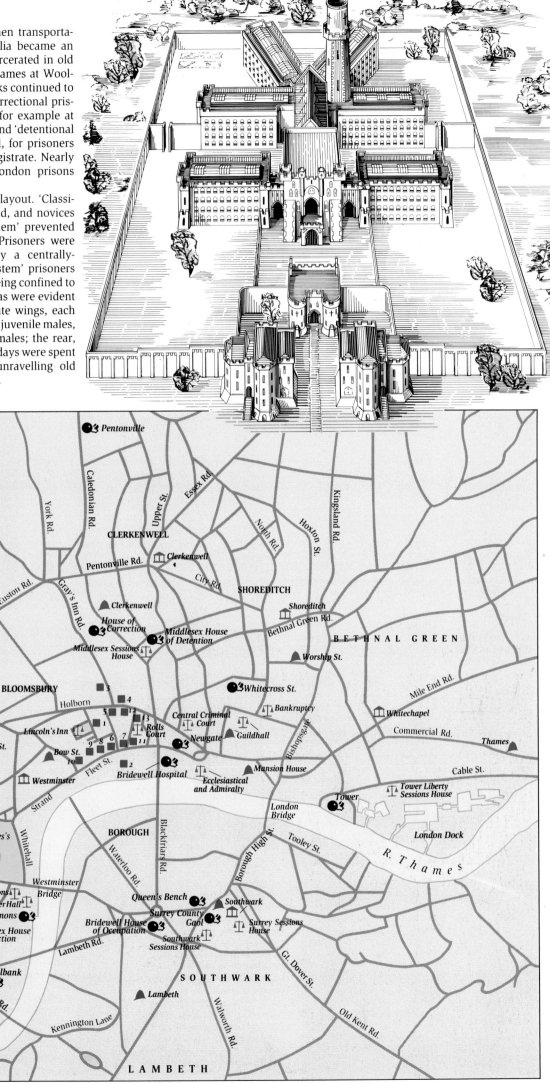

VICTORIAN HOUSING IMPROVEMENTS

The first reaction of many well-to-do Victorians was to blame London's poor for the squalor and degradation in which they lived. But in the early 1840s a series of government inquiries prompted by Edwin Chadwick, a leading sanitary reformer, began a change in attitudes. Two pioneer agencies – the Metropolitan Association for Improving the Dwellings of the Industrious Classes and the Society for Improving the Condition of the Labouring Classes – started to erect 'model dwellings' for the poor. Their example was followed on a larger scale in the 1860s by the Peabody Trust, funded by the benevolence of George Peabody, a London-based American banker, and the Improved Industrial Dwellings Company, founded by Sydney Waterlow, a prominent stationer and printer, later Lord Mayor of London and a Liberal MP. The IIDCo. became the most important example of 'five per cent philanthropy', so called because dividends to shareholders were restricted to a modest five per cent. Another leading reformer, Octavia Hill, preferred to improve existing houses, which she acquired from slum landlords. She also employed 'lady visitors' who would offer advice on home economics at the same time as they collected the weekly rent. Along with many other philanthropists, she feared that cheap housing would discourage self-help and prove literally demoralising.

With the publication in 1883 of Andrew Mearns' tract, *The Bitter Cry of Outcast London*, which linked overcrowding to irreligion, immorality and incest among the labouring classes,

and the setting up of a Royal Commission on the Housing of the Working Classes (1884-85), several more model dwellings agencies were established. The East End Dwellings Company (1884) resolved to provide for the very poorest families, who had hitherto been neglected while the Four Per Cent Industrial Dwellings Company (1885), founded by Lord Rothschild, was mainly intended for poor Jewish families. Yet by 1900, seven of the largest trusts and companies still housed fewer than 80,000 people – at a time when London's population was increasing by more than this every year.

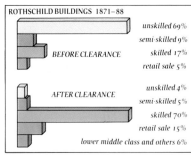

ROTHSCHILD BUILDINGS 1871–88

BEFORE CLEARANCE	
unskilled	69%
semi-skilled	9%
skilled	17%
retail sale	5%

AFTER CLEARANCE	
unskilled	4%
semi-skilled	5%
skilled	70%
retail sale	15%
lower middle class and others	6%

Victorian slums *The worst housing conditions were in shanty towns like 'The Potteries' in North Kensington (top right), inhabited by pigkeepers, brickmakers and day labourers, and in central London 'rookeries', where former middle-class houses were converted to tenements (above). Many more Londoners lived in drab terraced houses, like those in the shadow of railway viaducts near London Bridge, depicted by Gustav Doré in 1871 (top left).*

Philanthropic schemes *The Peabody Trust typically erected 'associated dwellings' (where tenants shared sculleries and toilets), as at Blackfriars Road (1871, above). Both the Trust and the Improved Industrial Dwellings Co. preferred sites near the West End (right), often on land provided by aristocratic landlords or the Metropolitan Board of Works. The Four Per Cent Industrial Dwellings Company did build in the East End, as at Rothschild Buildings (1887), pictured (left) in 1902; the graph shows the different class structure of the new residents.*

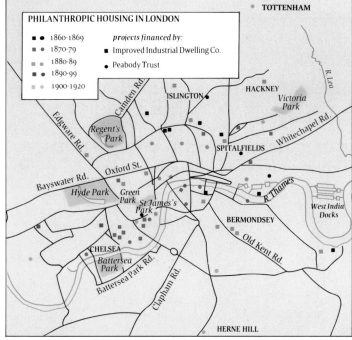

PHILANTHROPIC HOUSING IN LONDON

- 1860-1869
- 1870-79
- 1880-89
- 1890-99
- 1900-1920

projects financed by:
- Improved Industrial Dwelling Co.
- Peabody Trust

TOTTENHAM

Edgware Rd.

Camden Rd.

Regent's Park

HACKNEY

Victoria Park

ISLINGTON

Whitechapel Rd.

SPITALFIELDS

R. Lea

Bayswater Rd.

Oxford St.

Hyde Park

Green Park

St James's Park

R. Thames

West India Docks

CHELSEA

Battersea Park

Battersea Park Rd.

Clapham Rd.

BERMONDSEY

Old Kent Rd.

HERNE HILL

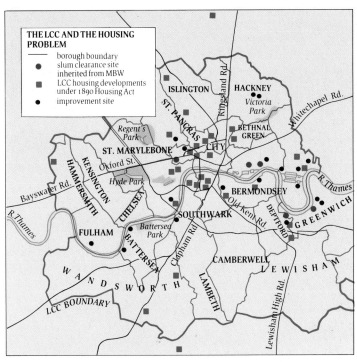

THE LCC AND THE HOUSING PROBLEM

— borough boundary
● slum clearance site inherited from MBW
■ LCC housing developments under 1890 Housing Act
● improvement site

BOUNDARY STREET ESTATE BEFORE RE-DEVELOPMENT (below)

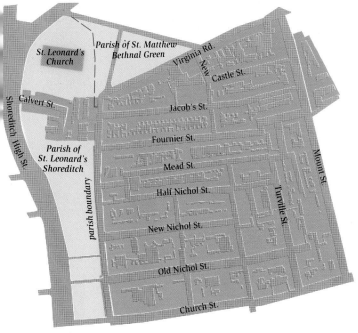

AFTER RE-DEVELOPMENT

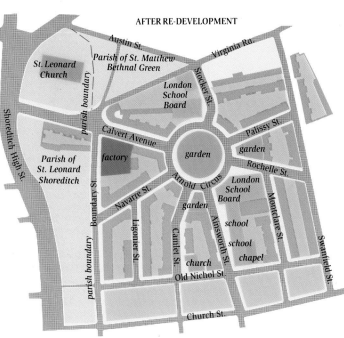

Slum clearance *The London County Council (LCC) inherited several slum clearance sites which its predecessor, the Metropolitan Board of Works (MBW), had hoped to sell to model dwellings companies. Under the 1890 Housing Act, the LCC obtained permission to rebuild cleared slums and to build on greenfield sites. The LCC was also obliged to provide housing where existing dwellings were demolished to make way for urban improvements, such as new streets and schools. The map (above left) shows the sites of LCC housing erected between 1890 and 1914. The most extensive slum clearance scheme in the 1890s was in Bethnal Green, where 'Old Nichol', a maze of narrow streets (left), similar in character to Little Britain, in Southwark (photograph right), was replaced by the Boundary Street Estate completed in 1900. This was distinguished by wide streets and impressive blocks of flats focused on a central garden (below left and right). The LCC then turned to a suburban solution, erecting 'cottage estates' at Totterdown in Tooting, south London (1903, photograph top right) and Old Oak in Acton (1912), just inside the county boundary, and at White Hart Lane (1904) and Norbury (1906), just 'out-county'.*

While many 'model dwellings' were erected on slum sites, the slum-dwellers who were the previous occupants, could not afford the rents demanded even for the plainest one- or two-room flats. New residents in Rothschild buildings (1887), for example, were predominantly skilled workers and their families. This experience was repeated when the London County Council began building council flats in the 1890s: in much of the East End and Docklands, their dwellings proved 'hard to let', primarily because they were too expensive for casually employed dockers and sweat-shop workers.

THE SOCIAL FABRIC

Social surveys of London had been undertaken as early as the 1840s, laying bare the extent of poverty and overcrowding in slums such as St Giles'. Investigative journalists, such as Henry Mayhew, reported on their interviews with the poor, while a few intrepid social explorers dressed as tramps and experienced at first-hand a night in the casual ward of a workhouse. But it was Charles Booth, a successful businessman, who produced the most detailed and comprehensive survey of the living conditions and culture of the London poor. His *Life and Labour of the People of London* ran to seventeen volumes, including series on 'Industry', 'Religious Influences' and 'Poverty'.

The population of London was divided into eight categories; classes A-D constituted 'the poor', on incomes of less than 21s. per week 'for a moderate family'; classes E and F were also 'working class', but better-paid and in regular employment; and classes G and H comprised the middle class and above. Booth was shocked to find that over 30 per cent of Londoners were living in poverty, and that a much larger proportion could expect to pass through periods of real poverty at some stage in their life. While he recognised a residuum of 'loafers and semi-criminals', he paid much more attention to the majority of the poor, whose poverty was no fault of their own. He identified the structural causes of poverty: seasonal and casual labour markets, which required a

pool of excess, underemployed labour; a life cycle in which most working-class families would experience poverty during child-rearing and old age. Unsurprisingly, Booth was prominent in calling for old age pensions.

Booth produced two kinds of 'poverty map' of London. The first showed degrees of poverty in 1889-90 in 134 areas, each of about 30,000 inhabitants. In many districts around the City, from Clerkenwell through Hoxton, Whitechapel and Wapping, to Bermondsey and Southwark, more than 40 per cent of families were classed as poor. More surprising outliers of poverty included parts of Paddington, Notting Hill, Pimlico and Battersea. In a second map, Booth assigned each street to one of seven colours 'according to the general condition of the inhabitants', ranging from black and dark blue, corresponding to classes A and B, to red and yellow, streets inhabited by middle-class, servant-keeping families. This map clearly shows the very fine grain of poverty.

Booth saw beyond the details to generalise about the social geography of London. He noted the tendency for the colours on his maps to lighten with increasing distance from the river, and identified a pattern of concentric rings 'with the most uniform poverty at the centre'. He saw an obvious link between poverty and overcrowding; while one of his assistants (Llewellyn Smith, later a prominent social researcher in his own right) argued that London was breeding an unskilled, unhealthy and morally degenerate race: districts with most poverty were those with the largest proportion of their inhabitants born in the metropolis.

PERCENTAGE IN POVERTY

The lowest class – occasional loafers and semi-criminals (class A).

The very poor – casual labour, hand-to-mouth existence, chronic want (class B).

The poor – including alike those whose earnings are small because of irregularity of employment, and those whose work, though regular, is ill-paid (classes C & D).

The regularly employed and fairly paid working class of all grades (classes E & F).

Lower and upper-middle class and all above this level (classes G & H).

0 5   15   25   35   45   55 %

Wealth and poverty *The graph (left) shows Booth's definition of poverty classes and the percentage of families found in each class. The map (right) is based on Booth's 1889-90 map, showing the degree of poverty in 134 subdivisions of London. The facsimile (top right) from Booth's 'Descriptive Map of London Poverty 1889' shows wealth and poverty on a detailed street-by-street basis, clearly indicating the slums east of Woburn Place and south of King's Cross (dark blue), only a few hundred yards from exclusive Bloomsbury squares and terraces (red and yellow).*

LONDON: COMPARATIVE POVERTY 1890
percentage of people living in poverty (*each area mapped represents c.30,000 people*)

- 60-70%
- 50-59%
- 40-49%
- 30-39%
- 20-29%
- 10-19%
- under 10%

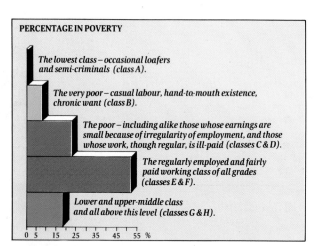

The 'habits of the people' *Booth also produced detailed tabulations and maps of licensed premises. The map (near right) shows the distribution of licensed premises in the East End. Pubs with full licences were concentrated along main streets like Whitechapel Road. There were also numerous beerhouses on side streets, vying with grocers' shops for street-corner sites, and established in the wake of the Beershop Act of 1830, which allowed ratepayers easy access to licences (called 'on' licences) to brew and sell their own beer, but not spirits. After 1834, licences for off-sales (including spirits) were also granted, but there were few off-licences in the East End.*

The geography of pubs throughout London in 1900 is summarised in a second map (far right) based on Booth's statistics. In the City there were large numbers of all kinds of licensed premises – one for every 44 residents (most served visitors and workers). In inner east London there was one pub for every 400 residents, but few off-licences – they proliferated in middle-class suburbia where drinking in pubs was less 'respectable' than drinking at home.

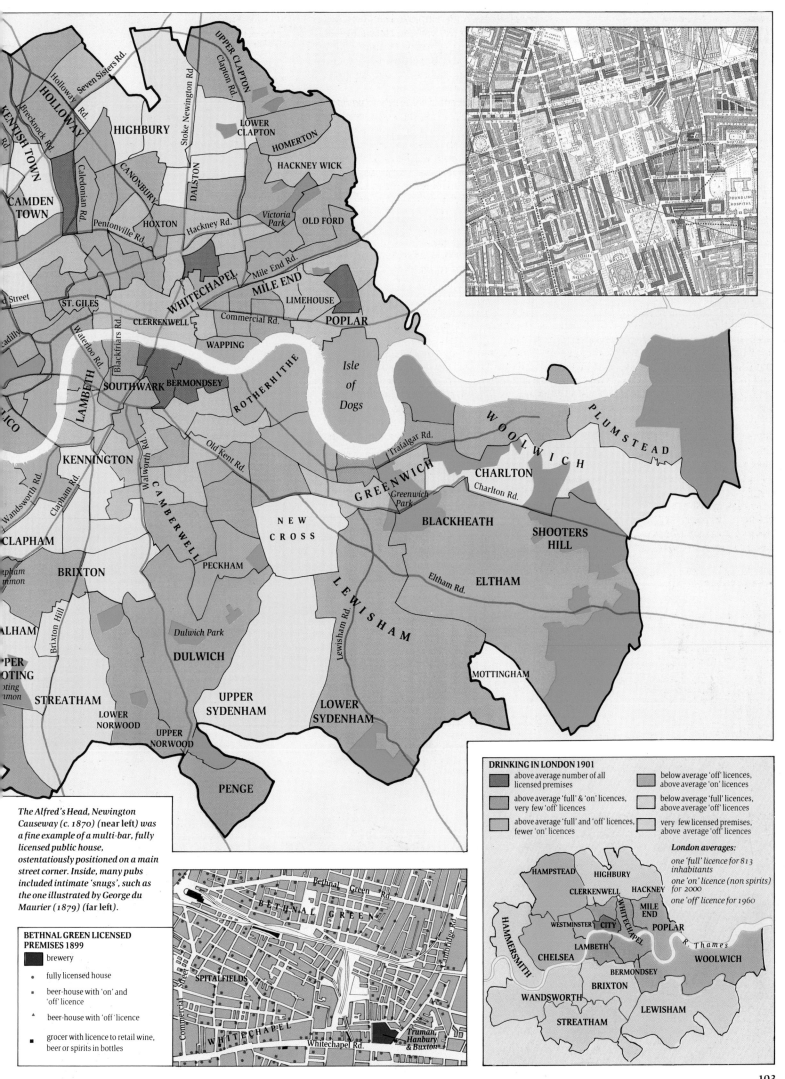

HOLLOWAY

Holloway Rd.

Seven Sisters Rd.

Brecknock Rd.

Caledonian Rd.

KENTISH TOWN

CAMDEN TOWN

HIGHBURY

CANONBURY

Pentonville Rd.

HOXTON

UPPER CLAPTON

Clapton Rd.

Stoke Newington Rd.

DALSTON

LOWER CLAPTON

HOMERTON

HACKNEY WICK

Hackney Rd.

Victoria Park

OLD FORD

ST. GILES

CLERKENWELL

WHITECHAPEL

MILE END

Mile End Rd.

LIMEHOUSE

POPLAR

Commercial Rd.

WAPPING

LAMBETH

Waterloo Rd.

Blackfriars Rd.

SOUTHWARK

BERMONDSEY

ROTHERHITHE

Isle of Dogs

KENNINGTON

Walworth Rd.

Clapham Rd.

Wandsworth Rd.

CAMBERWELL

Old Kent Rd.

NEW CROSS

GREENWICH

Greenwich Park

Trafalgar Rd.

WOOLWICH

PLUMSTEAD

CHARLTON

Charlton Rd.

BLACKHEATH

SHOOTERS HILL

CLAPHAM

BRIXTON

Brixton Hill

PECKHAM

LEWISHAM

Lewisham Rd.

Eltham Rd.

ELTHAM

Dulwich Park

DULWICH

STREATHAM

LOWER NORWOOD

UPPER NORWOOD

UPPER SYDENHAM

LOWER SYDENHAM

MOTTINGHAM

PENGE

The Alfred's Head, Newington Causeway (c. 1870) (near left) was a fine example of a multi-bar, fully licensed public house, ostentatiously positioned on a main street corner. Inside, many pubs included intimate 'snugs', such as the one illustrated by George du Maurier (1879) (far left).

BETHNAL GREEN LICENSED PREMISES 1899

■ brewery

• fully licensed house

• beer-house with 'on' and 'off' licence

▲ beer-house with 'off' licence

■ grocer with licence to retail wine, beer or spirits in bottles

Bethnal Green Rd.

BETHNAL GREEN

SPITALFIELDS

Commercial Street

Cambridge Rd.

WHITECHAPEL

Whitechapel Rd.

Truman, Hanbury & Buxton

DRINKING IN LONDON 1901

■ above average number of all licensed premises

■ above average 'full' & 'on' licences, very few 'off' licences

■ above average 'full' and 'off' licences, fewer 'on' licences

□ below average 'off' licences, above average 'on' licences

□ below average 'full' licences, above average 'off' licences

□ very few licensed premises, above average 'off' licences

London averages:

one 'full' licence for 813 inhabitants

one 'on' licence (non spirits) for 2000

one 'off' licence for 1960

HAMPSTEAD

HIGHBURY

CLERKENWELL

HACKNEY

HAMMERSMITH

WESTMINSTER

CITY

WHITECHAPEL

MILE END

POPLAR

CHELSEA

LAMBETH

R. Thames

WOOLWICH

WANDSWORTH

BERMONDSEY

BRIXTON

LEWISHAM

STREATHAM

METROPOLITAN IMPROVEMENTS

By the mid-19th century, London's fragmented urban infrastructure could not support its growing population. Outside the City of London, there was a patchwork of parish vestries, some relatively democratic and efficient, others inactive or corrupt. There were also special trusts for the provision of street lighting and maintenance of turnpike roads, and no fewer than eight sets of commissioners of sewers. Nobody took responsibility for London as a whole until, in 1855, the Metropolis Management Act established the Metropolitan Board of Works (MBW), which was empowered to levy a rate for improvements.

The most urgent problem for the MBW was sewage. Up until then, sewers had dealt only with surface water and discharged directly into the Thames, from which water companies were still extracting drinking water. Household waste went

Local government *The map (right) shows the vestries and district boards responsible for local government, 1855-99.*
The reform of the sewers *An engraving of the 1840s (below left) shows an open sewer running under the floor of a lodging house in Fish Lane, Holborn. The map (below) shows how the sewage system was improved by the MBW. By the mid-1860s, existing main-line sewers had been diverted into newly-built intercepting sewers. New storm relief sewers, carrying surface run-off, were constructed in the 1880s. To improve flow, pumping stations raised 'low-level sewers' (those closest to the Thames) to provide a gradient for the continuing flow downstream. Abbey Mills Pumping Station (bottom) was opened in 1868, powered by 8-feet Cornish beam engines. It was Gothic in style, with wrought-iron staircases and decorative cast-iron columns.*

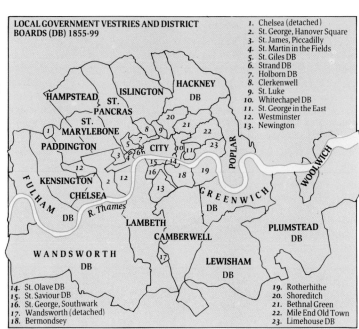

LOCAL GOVERNMENT VESTRIES AND DISTRICT BOARDS (DB) 1855-99

1. Chelsea (detached)
2. St. George, Hanover Square
3. St. James, Piccadilly
4. St. Martin in the Fields
5. St. Giles DB
6. Strand DB
7. Holborn DB
8. Clerkenwell
9. St. Luke
10. Whitechapel DB
11. St. George in the East
12. Westminster
13. Newington

14. St. Olave DB
15. St. Saviour DB
16. St. George, Southwark
17. Wandsworth (detached)
18. Bermondsey

19. Rotherhithe
20. Shoreditch
21. Bethnal Green
22. Mile End Old Town
23. Limehouse DB

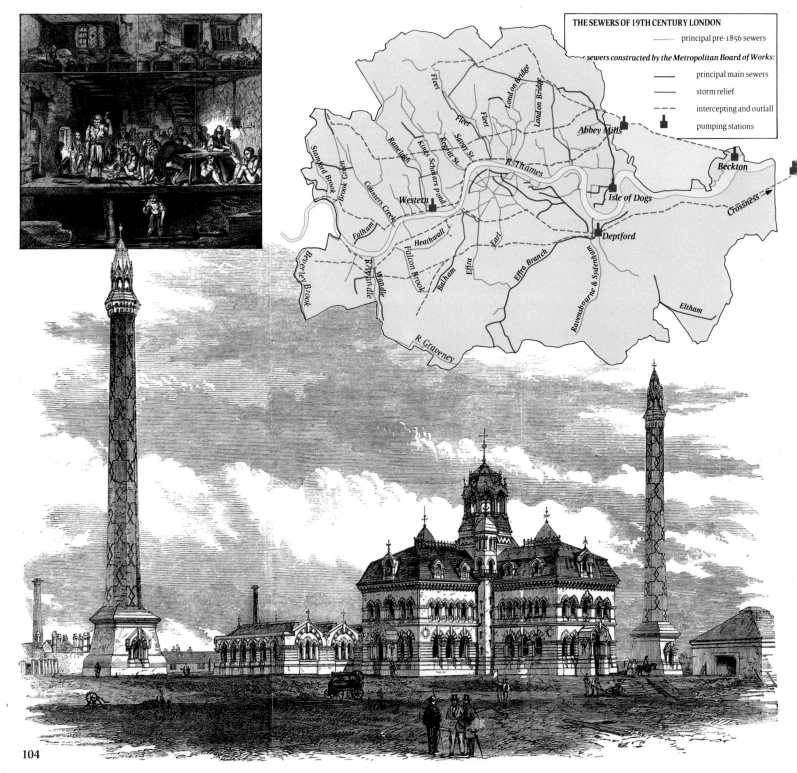

THE SEWERS OF 19TH CENTURY LONDON

— principal pre-1856 sewers

sewers constructed by the Metropolitan Board of Works:

— principal main sewers
— storm relief
--- intercepting and outfall
■ pumping stations

into cesspits, which leaked their contents into adjacent wells, thereby ensuring that water supplies were polluted. Successive cholera epidemics highlighted the severity of the problem, which was finally brought home to parliament by the 'Great Stink' of 1858, when the stench from the Thames became overpowering. Under its chief engineer, Sir Joseph Bazalgette, the MBW constructed 82 miles of 'intercepting sewers'; smaller, pre-existing sewers draining into the Thames were intercepted, so that their effluent could be discharged much farther downstream, at Beckton and Crossness, where outfall sewage works were completed in 1864 and 1865. The MBW also embarked on an ambitious programme of road-building and improvement, combining traffic engineering with slum clearance.

Despite these improvements, the MBW was still regarded as undemocratic and corrupt. The demands of Liberal and East End politicians for a directly-elected, London-wide council, able to transfer resources from rich West End areas to poorer East End districts, led to the creation of the London County Council (LCC) in 1888. In 1899, to counter the power of the LCC, the vestries and District Boards were replaced by 28 new metropolitan borough councils. This new administrative structure lasted until the 1960s, yet, even in the 1880s, the built-up area of London extended beyond LCC boundaries.

Street improvements Holborn Viaduct (under construction near right, and complete far right, below) was built between 1863 and 1869, designed to ease traffic flows between the City and West End, avoiding the steep hills into and out of the Fleet Valley. The only visible section is the cast-iron bridge across Farringdon Street, but the whole viaduct is a quarter of a mile in length, hidden between offices and warehouses. Even more impressive was the Victoria Embankment, completed in 1870. A new river wall was built up to 500 feet out into the Thames, providing 37 acres of new land. A cross-section through the Embankment beneath Charing Cross station (far right, above) shows an underground conduit carrying water and gas mains and telegraph cables; beneath that was the low-level intercepting sewer; the Metropolitan District Railway occupied a tunnel under the newly laid-out Embankment Gardens.

THE THAMES TUNNEL 1843

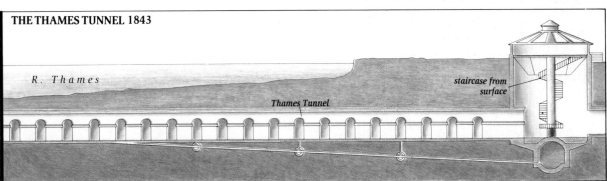

R. Thames

Thames Tunnel

staircase from surface

Street improvements by the MBW (map right) culminated in the construction of Shaftesbury Avenue (1877-86) and Charing Cross Road (1887), and the enlargement of Piccadilly Circus around the statue of Eros. Further improvements were undertaken by the LCC, including the clearance of 28 acres to create Kingsway and Aldwych, opened in 1905, and the building of road tunnels under the Thames, at Blackwall (1897) and Rotherhithe (1908).

The first Thames Tunnel (cross-section above right, entrance above), between Wapping and Rotherhithe, was a private initiative, promoted by the engineer, Marc Brunel. It was started in 1825 but not completed until 1843. Within 15 weeks of its opening, one million pedestrians had paid 1d each to walk through, but once the novelty wore off, it became the haunt of thieves and prostitutes. The tunnel had cost over £600,000, but in 1865 it was sold to the East London Railway Company for only £200,000 and converted into a railway tunnel.

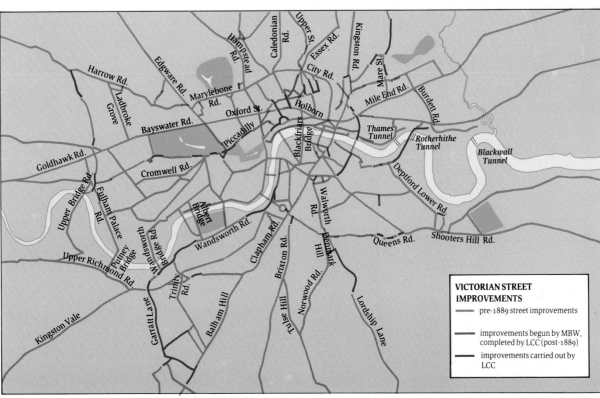

VICTORIAN STREET
IMPROVEMENTS
──── pre-1889 street improvements

──── improvements begun by MBW, completed by LCC (post-1889)

──── improvements carried out by LCC

VICTORIAN COMMERCE

The 19th century saw dramatic changes in the geography and character of London's shops. From the beginning of Victoria's reign there was a movement westwards, along the lines of Holborn, Oxford Street, Fleet Street and the Strand, following the continuing shift of the better-off into West End estates. Later in the century a 'new West End' developed in Knightsbridge, Kensington and Bayswater. Major suburban shopping centres, for example in Clapham, Brixton and Holloway, also acquired their own department stores and, by 1914, chain stores such as Marks & Spencer. In many suburbs single-storey, flat-roofed extensions were built to accommodate shops on what had previously been the front gardens of private terraced houses. By the end of the century new, purpose-built department stores were replacing such makeshift extensions.

There were also changes to London's markets. In 1855, the live cattle market moved from Smithfield to Copenhagen Fields, north of King's Cross. A new meat market opened at Smithfield in 1866, and the insanitary Newgate Shambles was closed. Many central London markets were displaced by the construction of new streets and railways, while others, such as Covent Garden, Spitalfields, Billingsgate and Leadenhall, were rebuilt. These, and other new markets with direct rail access to the rest of the country – at King's Cross, Somers Town and Stratford – increasingly restricted their activities to the wholesale trade. Most markets were owned by the City Corporation, railway companies or local authorities, but a few philanthropic ventures were promoted, including Angela Burdett-Coutts' Columbia Market in Bethnal Green, a spectacular failure from the day it opened in 1869. Street markets were increasingly subject to regulation under public health and nuisance laws, and regarded as sources of public disorder, especially as they did most of their business late on Friday and Saturday nights, and on Sundays. Yet the number of street traders increased, as underemployed casual labourers, in a last attempt to avoid the workhouse, tried to earn a living as streetsellers.

The growth of a new class of shop and office workers, aspiring to middle-class standards, and the building of thousands of modest but comfortable suburban homes that needed furnishing and decorating, increased the demand for consumer goods. As public transport improved, West End stores attracted customers from all over London. William Whiteley commenced business in Westbourne Grove in 1863, the same year as the Metropolitan Railway was opened to nearby Paddington. Selfridge's, opened in 1909, was located immediately above the 'Twopenny Tube', opened in 1900 (now part of the Central Line). Some multiple retailers, such as W. H. Smith, were directly associated with public transport, with their chain of station bookstalls; others, such as J. J. Sainsbury, used improved road and rail transport to distribute standardised, mass-produced goods. J. J. Sainsbury started out with a dairy in Drury Lane in 1869, followed by branches in working-class market streets, such as Chapel Market in Islington. In the 1880s, the firm expanded into more distant middle-class suburbs such as Croydon and Lewisham. By 1914, there were 115 branches of Sainsbury.

Most department stores began as drapers' shops, only gradually increasing the range of goods they sold. By the 1840s there was a cluster of stores on Tottenham Court Road (Shoolbred's, Maples, Heal's), reflecting the concentration of furniture manufacturers in adjacent streets. On Oxford Street, Peter Robinson, linen draper, expanded from one shop in 1833 to take over five adjacent shops by 1860. But the most spectacular 19th-century department store was Whiteley's, described in 1887 as 'an immense symposium of the arts and industries of the nation and of the world'. By 1906, it boasted 159 separate departments, including services such as refreshment rooms, estate and ticket agencies, laundry and dry cleaners. The new department stores

London's Victorian Markets *The distribution of wholesale and retail markets in central London, including the numbers of costermongers at each market, as recorded by the journalist, Henry Mayhew, in 1849* (**map below left**). *The Hay Market moved in 1830 to Cumberland Market, east of Regent's Park. Robert Bevan's painting* (**below**) *depicts hay carts at Cumberland Market in 1915.*
West End stores *The map* (**above right**) *shows the distribution of London's department stores in 1910, and indicates whether they were purpose-built, or originated as bazaars. By 1850, there were at least ten bazaars in central London, usually arcaded structures where individuals could rent stalls, mostly selling items of female dress, and luxury goods such as jewellery and toys.*

LONDON'S MARKETS 1849

street market in 1849 — wholesale market (named on map below)

street markets
(number of costermongers given in brackets)

1. New Cut, Lambeth (300)
2. Lambeth Walk (104)
3. Walworth Road (22)
4.* Camberwell (15)
5. Newington (45)
6. Kent St., Borough (38)
7. Bermondsey (107)
8. Union St., Borough (29)
9. Great Suffolk St. (46)
10. Blackfriars Rd. (58)
11. Brill & Chapel St. (300)
12. Camden Town (50)
13. Hampstead/Tottenham Court Rds. (333)
14. St. George's Market, Oxford St. (177)
15. Marylebone (37)
16. Edgware Rd. (78)
17. Crawford St. (145)
18. Knightsbridge (46)
19. Pimlico (32)
20. Tothill St. (119)
21. Drury Lane (22)
22. Clare St. (139)
23. Exmouth St. (142)
24. Leather Lane (150)
25. St. John's St. (47)
26. Old St. (46)
27. Whitecross St. (150)
28. Islington (79)
29. City Road (49)
30. Shoreditch (100)
31. Bethnal Green (100)
32. Whitechapel (258)
33. Mile End (105)
34. Commercial Rd. (114)
35. Limehouse (88)
36. Ratcliffe Highway (122)
37. Rosemary Lane (119)
38. Columbia

* not shown on map
† wholesale markets opened 1850-1893
wholesale markets not shown on map
† Stratford (fruit, veg.)
† Brentford (fruit, veg.)

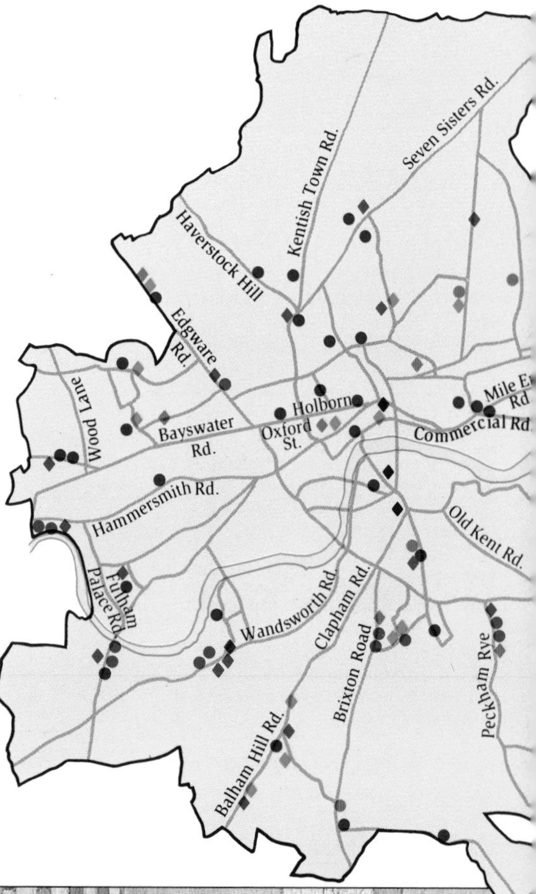

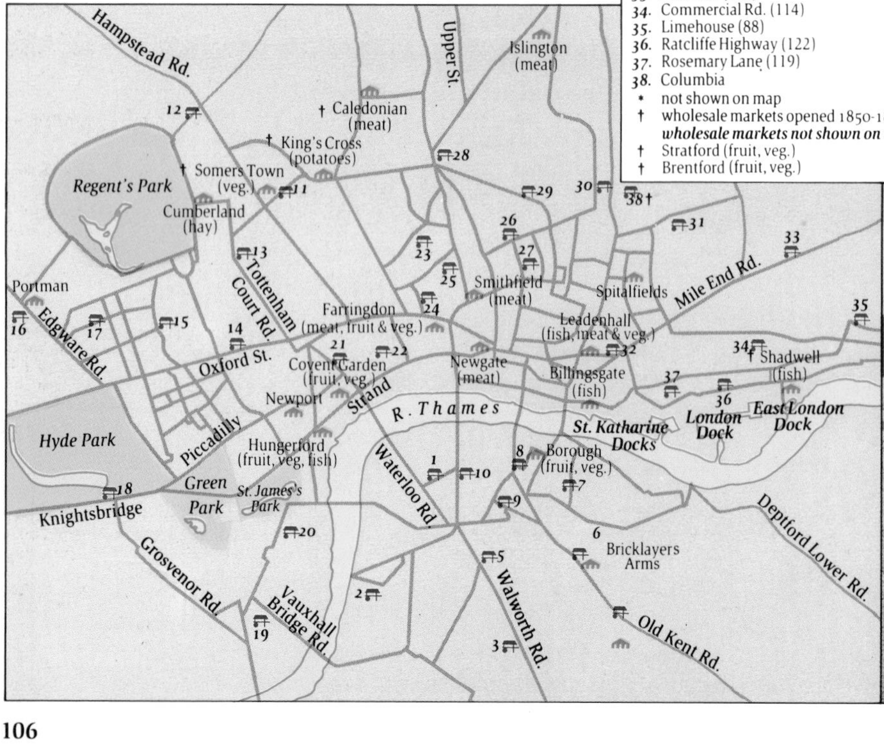

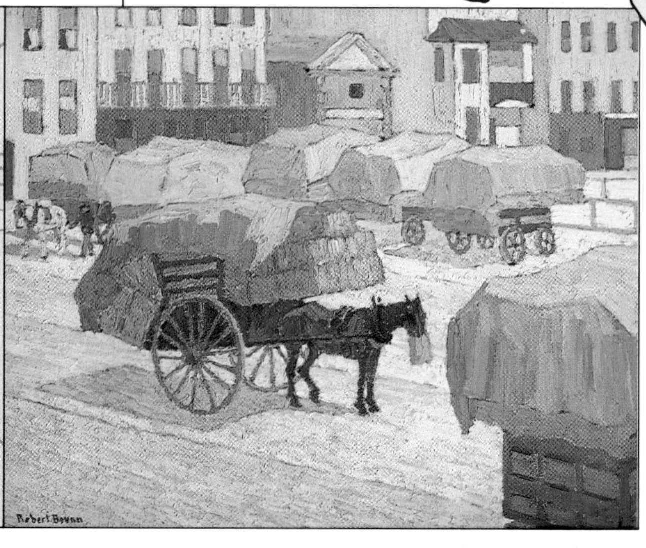

advertised fixed prices, required payment in cash, and encouraged the general public to enter with no obligation to buy. Up until the 1870s they had remained collections of existing small shops knocked together and extended piecemeal. The first purpose-built department store was the Bon Marché in Brixton, opened in 1877, named in imitation of the Parisian store. As West End land values increased, so store-owners preferred to build upwards: multi-storey premises extended from bargain basements to roof gardens. By the 1890s lifts were common, and in 1898 Harrods installed the first moving staircase. An earlier technological innovation had been plate glass, allowing much more impressive window displays, while from the 1880s electric lighting was introduced, increasing the attractions of night-time shopping. To raise capital for these changes, stores ceased to be private family businesses and became limited companies.

The most impressive brand-new department store was built in 1909 by Harry G. Selfridge, who had previously worked in the Chicago store of Marshall Field. Selfridge's was almost a town in its own right; by 1914, 950 men and 2,550 women were employed in its 160 departments. With restaurants, hairdressers, writing, reading and rest rooms, Selfridge's was the equivalent of a gentleman's club for women. Indeed, Selfridge's proclaimed itself as being 'dedicated to the service of women'.

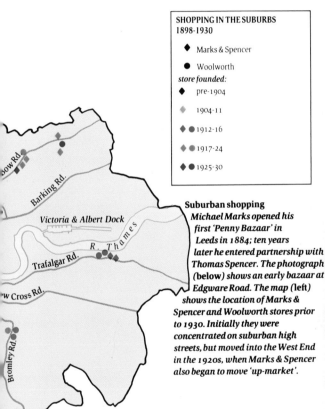

Selfridge's *An advertisement (above) proclaims the opening of London's grandest purpose-built department store in 1909. The cross-section (right) shows the layout of Selfridge's, from the mail order department in the basement to the elegant tea-rooms on the top floor.*

LONDON'S DEPARTMENT STORES 1850–1910

■ bazaar established pre-1850
○ shop established pre-1850
● shop developed 1850-80
▪ department store established after 1880 (named)

SHOPPING IN THE SUBURBS 1898-1930

◆ Marks & Spencer
● Woolworth

store founded:
◆ pre-1904
◆ 1904-11
◆● 1912-16
◆● 1917-24
◆● 1925-30

Suburban shopping
Michael Marks opened his first 'Penny Bazaar' in Leeds in 1884; ten years later he entered partnership with Thomas Spencer. The photograph (below) shows an early bazaar at Edgware Road. The map (left) shows the location of Marks & Spencer and Woolworth stores prior to 1930. Initially they were concentrated on suburban high streets, but moved into the West End in the 1920s, when Marks & Spencer also began to move 'up-market'.

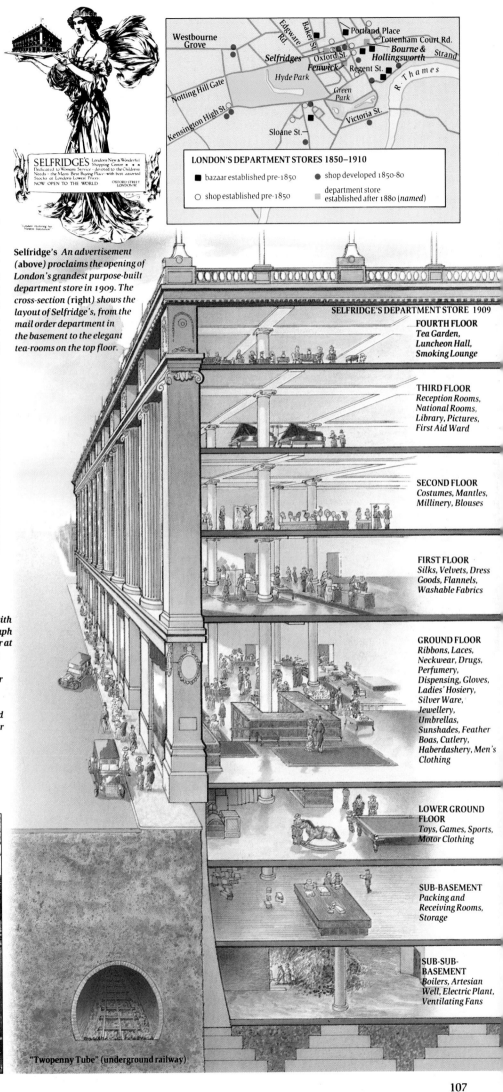

SELFRIDGE'S DEPARTMENT STORE 1909

FOURTH FLOOR
Tea Garden, Luncheon Hall, Smoking Lounge

THIRD FLOOR
Reception Rooms, National Rooms, Library, Pictures, First Aid Ward

SECOND FLOOR
Costumes, Mantles, Millinery, Blouses

FIRST FLOOR
Silks, Velvets, Dress Goods, Flannels, Washable Fabrics

GROUND FLOOR
Ribbons, Laces, Neckwear, Drugs, Perfumery, Dispensing, Gloves, Ladies' Hosiery, Silver Ware, Jewellery, Umbrellas, Sunshades, Feather Boas, Cutlery, Haberdashery, Men's Clothing

LOWER GROUND FLOOR
Toys, Games, Sports, Motor Clothing

SUB-BASEMENT
Packing and Receiving Rooms, Storage

SUB-SUB-BASEMENT
Boilers, Artesian Well, Electric Plant, Ventilating Fans

"Twopenny Tube" (underground railway)

CHAPTER 8
LONDON BETWEEN THE WARS

THE FIRST HALF of the 20th century saw a more radical transformation of London's appearance and physical extent, and of its inhabitants' lifestyles, than any comparable period of history. Passing through an era of spectacular urban expansion in the 1930s, by 1950 London had begun a marked population decline. Dominated at the beginning by Victorian values and social inequalities, by the middle of the century welfare-state idealism offered the prospect of a new beginning. These transformations were partly brought about by continuing dissatisfaction with the city's legacy of social division and physical decay. National life at the time was dominated by these problems, and the political means of resolving them, but the status and size of London focused the issues with special intensity.

London retained its historic status in international affairs through the administration of the Empire and its share of British seaborne trade. Throughout the period it also dominated national, political, administrative, economic and cultural life. In some ways, for example through the growth of manufacturing and office activities and the developing role of central government, the city's role within Britain was even greater by the end of the period than at the beginning. Its most pressing problems in the early years, however, were associated with the Victorian legacy of industry and overcrowded housing. In the wake of the First World War there was a severe housing shortage. Initially, the official response was to placate post-war expectations of 'homes fit for heroes' by subsidising publicly-built, rented ('council') housing. The London County Council (LCC) became by far the largest public sector developer, both within London and on land acquired elsewhere in south-east England. This surge of building nevertheless petered out in the early 1920s, and subsequent rates of council house building fluctuated with the economic and political conditions of the period.

For an examination of the post-war housing boom, refer to pages 114-115.

These pioneering schemes set higher standards, both for amenities and space, than ever before for London working-class housing. The 'Garden City' ideas of Ebenezer Howard, writing in the 1890s, were already being put into practice at Hampstead Garden Suburb, Letchworth and, later, at Welwyn Garden City. These were privately-sponsored examples of what could be done to improve living conditions. They influenced standards of council housing in the 1920s, as well as the programme of new town building which was begun later, in the 1940s. Although supported by Liberal views on state intervention to alleviate the plight of the poor, the move from *laissez faire* solutions towards public intervention was also promoted by the growth at local, and later national levels, of the Labour Party. Within a few years the same models were adapted by private companies, building for middle-class owner-occupiers, to create an even more radical transformation of London's housing. This explosion of house-building would be on a scale that could scarcely have been imagined before the First World War, and which the public sector could never match.

In the early years of the century, plans were already being prepared to improve road and public transport in London. At this time many people worked relatively near to their homes within the city, but longer distance commuting was growing, using surface railways and the few privately-built underground lines. Even before motor lorries, and to a lesser extent private cars, began to multiply in the 1920s, congestion on the streets was a major issue. The First World War itself radically affected transport in London.

Below *The years after the Great War saw a spectacular growth and consolidation of the London Underground system. Under the aegis of Frank Pick, the commercial manager of the Underground Group, this expansion was marked by a series of promotional posters, their unique style owing much to the special typeface called 'Underground', which had been designed during the Great War and was used throughout the system. Posters, such as 'Brightest London' (Horace Taylor, 1924), were not simply design classics. The Underground was well-used at peak travel times, but it also made commercial sense to encourage Londoners to use the Underground for day excursions, shopping trips and evenings in the West End.*

BRIGHTEST LONDON
IS BEST REACHED BY
UNDERGROUND

The expansion of public transport is mapped on page 113.

The rise of 'Metroland' is examined on page 114, and the impact of the mass movement out of the inner city during the 1930s is mapped in detail on page 112.

Many of the munition factories were located around the edge of the city, on sites which were often to become major factory estates in the 1920s and 30s. Movement therefore became less tightly confined to the area of the Victorian city. As people and jobs migrated outwards, so did the diversity of transport; electric railways, the expanding underground, trams and trolley buses, omnibuses and bicycles were each employed in an increasing variety of combinations to get around the city. This 'mobility revolution', based on public transport, owed little to private car ownership.

As well as these transport developments, other forms of new technology wrought significant changes in London's life. These included the spread of electrical power and the telephone, and the introduction of new industrial products and methods of mass manufacture. Large private companies, which could support the growing scale of manufacturing and marketing, became increasingly important. These various opportunities came together most spectacularly in the expansion of the inter-war suburbs, which served a growing demand from small, lower middle-class families who could afford to buy houses for the first time, instead of renting them in the older parts of the city. Stable, if often modest incomes, derived from the growing numbers of office and skilled manual jobs, enabled many to obtain mortgages with affordable deposits

Above The Odeon, Leicester Square, with black granite facade, shortly after it opened in 1937.
Perhaps the most far-reaching development in this period was the expansion of the 'media': mass-circulation newspapers, radio broadcasting, and cinema. American culture – styles of dress, music, dancing and, above all, films – permeated British life between the wars, and had an extraordinary impact on a society which only twenty years before had been largely Victorian in cultural values and attitudes.

A full discussion of London's industry in the first half of the 20th century can be found on pages 116-117.

and repayment rates. Entirely new domestic lifestyles could also be evolved, based on the latest household gadgetry, including gas and electric cookers, vacuum cleaners, refrigerators and the increasingly available 'wireless'. The design of houses, furniture, domestic appliances and vehicles came under the influence of mass production techniques and modernist ideals.

The attractions of the suburbs were not just confined to private citizens. Manufacturing companies, until then concentrated by the availability of transportation links and labour into the crowded areas of inner and east London, were also able to move out. This was encouraged by improved communications, and by road transport for both freight and workers. Expansive industrial estates, mainly of single-storey factories, dotted the suburbs along the new trunk roads and on other readily available vacant land. These estates not only attracted traditional industries, they also became the sites for many new firms, sometimes American-owned, serving the London consumer boom. Typical products included domestic electrical goods, radios, new types of building materials, furniture, processed foods and pharmaceuticals. Whole new industries also developed around the production of aeroplanes, motor vehicles, machine tools and production control apparatus. To some extent the economic depression of the early 1930s favoured the production and building industries of London by keeping down labour, materials and land costs. The city, with its prosperous suburban-dweller, was also the main national focus of demand.

All of these innovations inevitably changed ways of life, especially for women. Although the war had expanded the range of acceptable jobs for single women, few middle-class wives worked; their task was to run the new home in which machines had replaced the servants of the Victorian era. Communities were less 'communal' than in the old city areas, and problems arose, especially for women, in adjusting to the solitary suburban life. This was especially true for working-class women moving to the LCC's 'out county' estates, which were situated outside London. A new type of 'commuter settlement' began to dominate London life, based upon the ideal of separate, semi-detached family homes, with large gardens on extensive estates. Of course, some of the areas of

Below *During the Second World War, London was subjected to massive aerial bombardment (the Blitz), which started in September 1940 and lasted six months. Vast areas of the City were destroyed and 20,000 Londoners lost their lives. Many people took shelter in Underground stations, but six out of ten Londoners preferred to stay in their own homes as the relentless bombing attacks continued. The Germans had hoped that this massive onslaught on the civilian population would quickly undermine morale but Londoners proved resilient. St. Paul's Cathedral was hit by a bomb on 29th December, and the photograph of its mighty dome wreathed in flames came to symbolise for many the city's proud resistance.*

earlier development in inner London continued to support large, older and often still prosperous populations, including upper-class families with servants. With time, however, their residents moved to the owner-occupied life of the outer suburbs, and many 'respectable' areas declined physically and socially, as poorer people moved in to rent part of the large subdivided properties.

Wembley Stadium symbolises much about London at this time. First used for the Cup Final of 1923, it was the centrepiece of the 1924-25 British Empire Exhibition, designed to assert the continuing significance of Britain's imperial power. In all, 27 million people attended the Exhibition over the two years. Like the stadium itself, with its 120,000 capacity, the Exhibition thus heralded the era of mass entertainment. Over the next 15 years a wave of characteristic 'Metroland' suburbia surrounded the site and engulfed much of the rest of Middlesex. After the Exhibition, Wembley became a major focus for suburban industrial development in the 1930s. It was not unique in this; Park Royal, less

than two miles to the south, had been the site of the Royal Agricultural Society showground, abandoned in 1905, and pressed into service for war purposes. By the Second World War it had become the largest industrial area in London.

Between the World Wars, two major problems associated with the 'liberation' of people from the pressures of overcrowded urban life emerged as time progressed. The first was the increasing separation, on a large scale, of areas where the 'haves' and 'have-nots' lived. The latter remained concentrated in the deprived industrial slums around the city centre and in the East End. Secondly, the rapid physical encroachment of the city into the surrounding countryside, often through tentacles of 'ribbon development', attracted increasing concern. By the late 1930s these two consequences of post-war development had created a rare consensus across a broad spectrum of opinion over the need to control the sprawl of London. Two ideals, one for social unity which would be brought about by improving health and social conditions of the poorer people, and the other for the conservation of the countryside, were engaged to support the first efforts to define 'green belts' which would contain the growth of London.

Social conditions in many of the old parts of London remained desperately poor. Punitive Poor Law regulations remained the principal way of relieving poverty until the late 1920s, when national public assistance was first introduced. The alleviation of poverty attempted by the London County Council and the boroughs had created political and social upheavals in the 1920s. In spite of council slum clearance and campaigns to improve child health, education and recreation, many of London's poor were still trapped in slum housing close to the factories and docks of inner London. The poor quality and insecurity of such employment meant that incomes were low, with the ever-present threat of lay-offs, or redundancies. The depression of the early 1930s, although less acute and shorter-lived in London than elsewhere in Britain, created special hardships in the heavy industrial and sweatshop areas of the East End. These overcrowded areas were particularly difficult and expensive to redevelop. A good deal of the worst housing was finally demolished by German bombing during the Second World War, thus establishing conditions in which massive urban renewal schemes, including demolition of the remaining slums, were undertaken after 1950.

Bethnal Green, an area of inner-city deprivation between the wars, is mapped and discussed in detail on page 115.

The movement to halt the suburban sprawl of London, which originated in the late 19th century, gathered pace during the 1930s. After 1935, local authorities in rural areas began to purchase open land around London to preserve it from private development. Initially this was undertaken with the financial backing of the LCC, but later the rural authorities increasingly undertook land purchasing schemes on their own initiative. The motives for this expenditure were mixed; the LCC wished to preserve accessible open spaces for Londoners, while the rural councils wanted to stop established communities being swallowed up by London's growth. Others pointed with alarm to the reduction of farmland, and its effects on national food production. These various interests came together after the Second World War, when it was seen that huge resources would be needed to purchase all the land needed around London. Local authorities were empowered to refuse planning permission for non-conforming developments in the statutory green belt. With the New Towns Act of 1946, this move established the regional framework for planning London's post-war development.

Map below The extent of London on the eve of the Second World War is contrasted to the size of the city today.

Of the many radical changes affecting London during this period, it was the Second World War and its devastating physical impact on the city that was the most important culminating event. It is remarkable that, even in the darkest days of the war, plans were already being laid, not only for reconstruction, but also for government policy in the city. Post-war priorities were not merely to repair the damage of the Blitz, but also to tackle the long-term problems of housing, transportation and social conditions.

EXTENT OF LONDON 1938

LONDON 1990

+ St. Paul's

LONDON 1990

SUBURBANISATION AND SPRAWL

The suburban development of London had been progressing for at least 300 years before the beginning of the 20th century. Nevertheless, suburbs are most commonly associated in the public mind with the vast unplanned housing developments of the 20s and 30s and the rise of 'semidetached' London. This suburban expansion was the result of both high birth rates and increased levels of migration into the capital. However, while Greater London's population grew from 7.5 million in 1921 to 8.7 in 1939, the numbers living in the inner city fell by almost 450,000 as people moved out to the suburbs.

It was the rising numbers of middle class administrative and clerical workers in commercial life and the civil service who enjoyed the new lifestyle, living in the suburbs and commuting to work. But by far the most dynamic changes in lifestyle between the wars were being created by American-style consumerism. A growing range of goods was available for mass sale; cookers, fridges, furniture, radios, processed foods – all the accoutrements of home decoration, gardening and increased leisure time. These commercial changes created employment for a growing number of managers, clerks, production and maintenance workers, drivers, design and sales staff.

THE CARRERAS FACTORY: JOURNEY TO WORK 1936
— London area postal boundary
— postal district boundary
Starting point of journey
up to 100 workers
over 100 workers

Commuting to work increased between the wars, as people left the city for the outer London suburbs (maps below). The new Carreras cigarette factory at Mornington Crescent (below), employed 2,600 workers and the map (above) shows where they lived and how far they had to travel to work.

Far-sighted plans to accommodate the growth of goods and freight traffic were reflected in the construction of new trunk roads around and out of London, including Western Avenue and the North Circular Road. Improved road networks also meant that industry could move out of the crowded inner city. Estates of factories grew up around the periphery of London, taking advantage of the new road transport, both for freight and workers.

As people and jobs migrated outwards, more varied

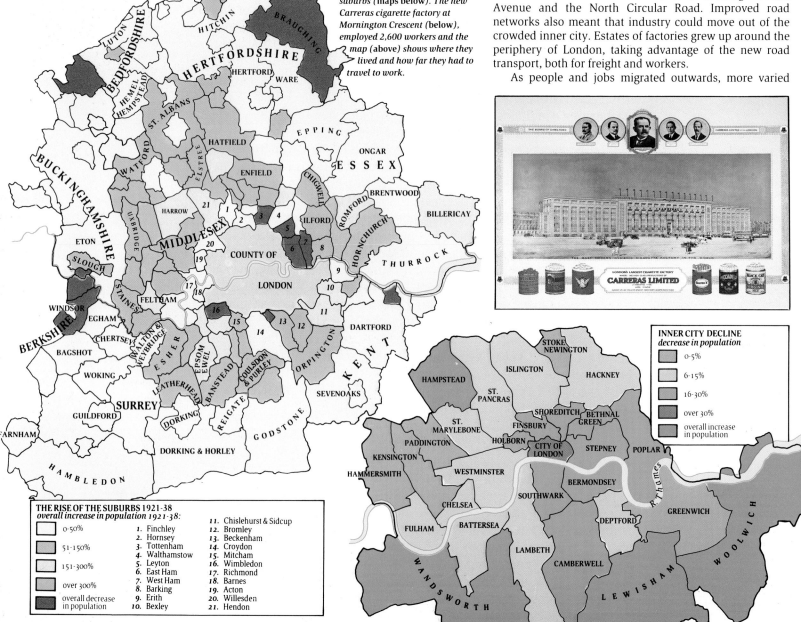

THE RISE OF THE SUBURBS 1921-38
overall increase in population 1921-38:
- 0-50%
- 51-150%
- 151-300%
- over 300%
- overall decrease in population

1. Finchley
2. Hornsey
3. Tottenham
4. Walthamstow
5. Leyton
6. East Ham
7. West Ham
8. Barking
9. Erith
10. Bexley
11. Chislehurst & Sidcup
12. Bromley
13. Beckenham
14. Croydon
15. Mitcham
16. Wimbledon
17. Richmond
18. Barnes
19. Acton
20. Willesden
21. Hendon

INNER CITY DECLINE
decrease in population
- 0-5%
- 6-15%
- 16-30%
- over 30%
- overall increase in population

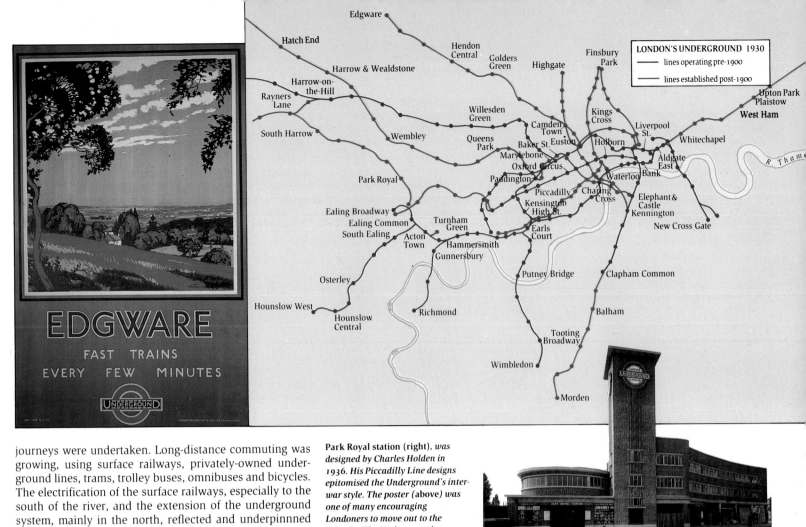

journeys were undertaken. Long-distance commuting was growing, using surface railways, privately-owned underground lines, trams, trolley buses, omnibuses and bicycles. The electrification of the surface railways, especially to the south of the river, and the extension of the underground system, mainly in the north, reflected and underpinned the spread of the city. Indeed, in some cases lines were built in advance of new houses to encourage land development.

Park Royal station (right), was designed by Charles Holden in 1936. His Piccadilly Line designs epitomised the Underground's inter-war style. The poster (above) was one of many encouraging Londoners to move out to the suburbs. The Underground was crucial in laying the foundations for Edgware's popularity.

The public transport revolution
This period was the Golden Age of public transport in London. Its rapid growth was already under way at the beginning of the century, as numerous companies introduced electric and petrol-driven services to displace the steam trains and horse trams of the last century. The expansion and electrification of both underground (map above) and surface railways (map right) were to continue for the next 40 years. The increasing scope of the system was dictated by a few dominant private companies, especially the Underground Group, the Metropolitan Railway and Southern Railways.

LONDON'S RAILWAYS 1930
stations closed post-1930 are named in italics

— main railway lines pre-1900
+++ pre-1900 lines electrified 1900-1930
— steam lines built 1900-1930
+++ electric lines built 1900-1930

The London County Council and some boroughs had also become involved in the electrification of tramways before the Great War. London Transport was established in 1933, finally ending the freewheeling chaos of London's public transport development, with the underground system, buses and trams all placed under the co-ordinating control of a single public corporation.

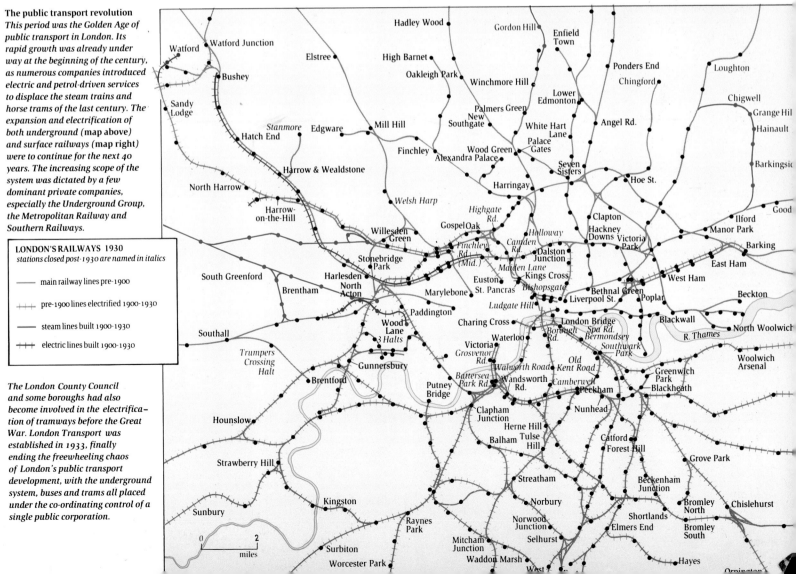

THE SUBURBAN DREAM

The lives of Londoners in the first fifty years of this century were dominated by two world wars and the Depression of the 1930s. One of the immediate effects of the First World War was the loss of able-bodied men and consequently the employment for the first time of women in many factory, transport and clerical jobs. These war-time upheavals, and the new social attitudes that went with them, radically altered the life prospects of many ordinary Londoners, sweeping aside Victorian and Edwardian formality, raising aspirations and offering fundamental and lasting changes in lifestyle.

Unlike the rest of the country, London appeared to benefit from the Depression. It was not dependent on heavy industry, but instead had developed the new manufacturing technologies of the time, and so attracted cheap labour from other parts of the country. At the same time, employment in offices – both government and commercial – was expanding. The result of London's resilience in the depressed job market of the post-war years was a huge housing and consumer boom. The economic and social trauma of the post-war years made home-ownership a stable and attractive prospect to the rapidly-growing middle classes, and from the early 1920s private building contractors began to buy cheap land on the outskirts of the city for building purposes – this was the birth of 'semi-detached' London. Throughout the 20s and 30s this ring of suburban development expanded, and new houses were to spring up especially quickly in areas which were served by new railway and Underground extensions – a trend which was enthusiastically endorsed by London Underground advertising campaigns which celebrated the lure of 'Metroland'. Houses were also built along new arterial roads and by-passes. The more expensive up-market houses were concentrated in the north and north-west of London in areas such as Finchley, Hendon and Barnet, and around the southern fringes, at Coulsdon, Purley and Epsom. Architecturally eclectic, the favoured styles included 'mock Tudor', a reflection of aspirations towards an idealised rural past. Cheaper, more down-market 'semis' sprung up all round London, and – with small down-payments and low mortgage repayments – were well within the reach of many Londoners. Although mass-produced to a few basic designs, styles were subtly varied to give individuality, including mock-Tudor and Gothic details, with variations in porches, bay windows and tiling.

Surrounding the private, family-centred world of semi-detached London, the most dynamic changes in lifestyle between the wars were being created by commercial pressures towards American-style mass consumerism, and its promotion through both advertising and the media. A growing range of goods was being manufactured using new production techniques for mass sale: cookers, washing machines, fridges, furniture, electrical heaters, radios, processed food, lighting accessories, bathroom fittings, popular newspapers, magazines and books – all the accoutrements of home ownership and improvement, gardening and increased leisure time.

At the same time, new styles of mass retailing were opening up, initially in the West End, which could be reached quickly and easily by improved

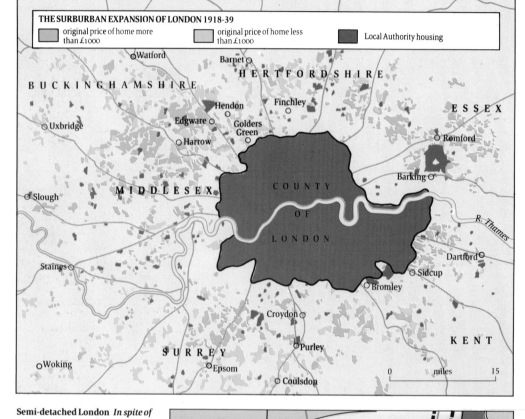

THE SURBURBAN EXPANSION OF LONDON 1918-39

- original price of home more than £1000
- original price of home less than £1000
- Local Authority housing

Watford · Barnet · HERTFORDSHIRE · BUCKINGHAMSHIRE · Hendon · Finchley · ESSEX · Uxbridge · Edgware · Golders Green · Romford · Harrow · Slough · MIDDLESEX · Barking · COUNTY OF LONDON · R. Thames · Staines · Dartford · Sidcup · Bromley · Croydon · KENT · SURREY · Purley · Woking · Epsom · Coulsdon · 0 miles 15

Semi-detached London In spite of the suburbs' reputation for dull uniformity, there were considerable variations in house size, style and cost (map above). The majority of semi-detached houses sold at less than £1,000 (photograph right), and encircled London (map above). Larger, often detached houses were typical of areas away from the growing industrial zones (below right). The map (right) shows a suburban estate at Finchley, with its gardens, shops, tennis court, woodland fringe and 'rustic' street names. The LCC's council estates, built after the First World War, set high standards (below).

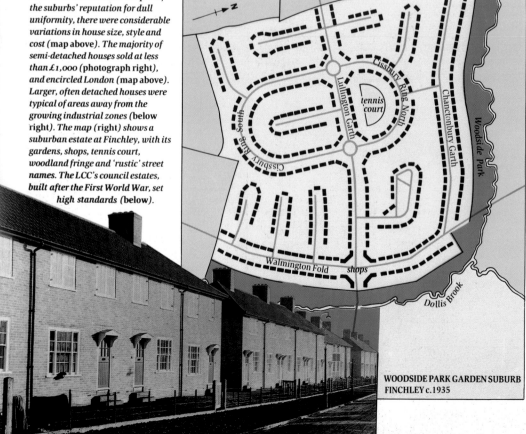

Cissbury Ring North · Lullington Garth · Chanctonbury Garth · tennis court · Cissbury Ring South · Woodside Park · Walmington Fold · shops · Dollis Brook

WOODSIDE PARK GARDEN SUBURB FINCHLEY c.1935

public transport, and then in the expanding suburban centres. The siting of chain stores, such as Woolworths, became the best indicator of the status of local shopping centres. New means of mass communication, such as the popular press and the 'wireless', which had spread to most homes by 1939, also created a common basis of information. But it was the cinema which was to have the greatest impact on the lives of ordinary Londoners. Picture palaces, lavishly decorated and hugely popular, were sited away from the traditional theatrical heartland of the West End, their locations mirroring the new prominence of suburbia. By 1929, there were 266 cinemas in the LCC area alone.

Nevertheless, inherited inequalities between rich and poor, especially in incomes and housing, remained etched on the map of the city. Many of the slums of the 1930s showed little improvement from the beginning of the century. Initially, the LCC and inner city boroughs had embarked on an ambitious policy, building large quantities of high-quality council houses and estates on the semi-rural fringes of London. Typical houses had three or more bedrooms, running water, inside toilets and bathrooms, electricity and spacious gardens. However, the Depression and the economic difficulties it caused meant that these standards were not always maintained, and public and political concern was slow to be translated

into effective large-scale action. The Second World War devastated large parts of London, including many of the areas of worst social deprivation. Whereas the effort of building between the wars had resulted in almost a million suburban, largely private dwellings, the drive after the War was towards rebuilding the inner areas, or alternatively offering people a new life outside London altogether, in the first new towns, established after 1947. In both cases, public intervention, under the new 'Welfare State', was essential – the Victorian combination of market forces and philanthropy deployed to solve social inequality was finally superseded by concerted State action.

METRO-LAND
PRICE TWO-PENCE

UNDERGROUND

'I never had any other desire so strong and so like to covetousness as that one which I have had always, that I might be Master of a small House and a Large Garden, with moderate conveniences joined to them.'
Abraham Cowley

LEAVE THIS
AND

MOVE TO EDGWARE

Inner City deprivation *Poverty dominated many areas of inner and east London throughout the inter-war period. This area of Bethnal Green (right) shows the close inter-mingling of poor housing with industry (in an area dominated by small furniture and sweatshop clothing firms), warehousing, public buildings, and shops in the early '30s. Employment was uncertain in these areas, often casual and in trades with marked seasonal and cyclical fluctuations. Housing lacked most basic amenities, including electricity and water supply; toilet, bathing and cooking facilities had to be shared. Many houses, built over 100 years ago, were now overcrowded, damp and dilapidated (photograph below right). In spite of these deprivations, such areas enjoyed an intense, street-based communal life, often missed by those who moved to new LCC estates (bottom).*

Selling the suburbs *London Underground was instrumental in promoting the suburbs – often new Underground lines preceded development. The booklet (left), Metroland (1928-9), was one of a series put out by the Metropolitan Railway to encourage travel to the suburbs. It contained advertisements from estate agents and building contractors. The line from Golders Green to Edgware was extended 1923-4, and London Underground posters (below left, by William Kermode 1924), promoted the suburban dream – 'that I might be Master of a small House and a Large Garden'.*

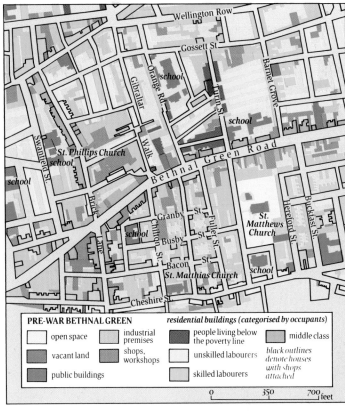

PRE-WAR BETHNAL GREEN   *residential buildings (categorised by occupants)*

- open space
- industrial premises
- people living below the poverty line
- middle class
- vacant land
- shops, workshops
- unskilled labourers
- *black outlines denote houses with shops attached*
- public buildings
- skilled labourers

0   350   700 feet

INDUSTRIAL LONDON

The industrial inheritance of London in 1900 was twofold: the port and the Victorian manufacturing belt. The latter was a densely packed arc of small-scale specialist industries in the inner suburbs to the north, east and south of central London, frequently the successors of long-established crafts which originated in the medieval city. With some small exceptions, these specialised areas were still clearly identifiable fifty years later. They survived by specialising in labour-intensive, skilled production, such as tailored clothing or furniture in the East End and precision engineering in Clerkenwell and Tottenham Court Road, and depended

on a skilled, local labour supply. Many larger firms, however, had already moved out to areas such as Hackney, Camden Town, or even farther afield, where modern production methods could be more easily applied. The migration of London's skilled labour force to the suburbs also continued, especially during the immediate aftermath of the Second World War.

The other old-established industrial zone in London was associated with the port. The diverse shipping activities connected with the port itself were augmented by heavy industries based on large sites, such as marine engineering and repairs, construction materials, timber, food processing, and gas and electricity production. Many of these industries were extensions of port activities, servicing shipping or processing imported materials, which were then

The Victorian Legacy The inheritance of Victorian manufacturing was spread throughout the London County Council area in 1947 (map below), often intermingled with residential areas, and it was the aim of the planning authorities to improve the living environment by disentangling this intermixture. However, the old-established industrial quarters (map below left) – from the tailoring districts of the East End to furniture-making in Shoreditch – had always depended on skilled workers living nearby. Heavier industries were found farther east (map below right), near the port and along the River Lea and the canals. These industries supported shipping, through engineering and repair yards, and processed imported raw materials, such as timber, food and chemicals. In time, however, many industries moved out to London's suburbs, where skilled workers increasingly preferred to live. The wartime bombing of central London and the devastation it caused to the city's traditional industrial heartland, accelerated this trend.

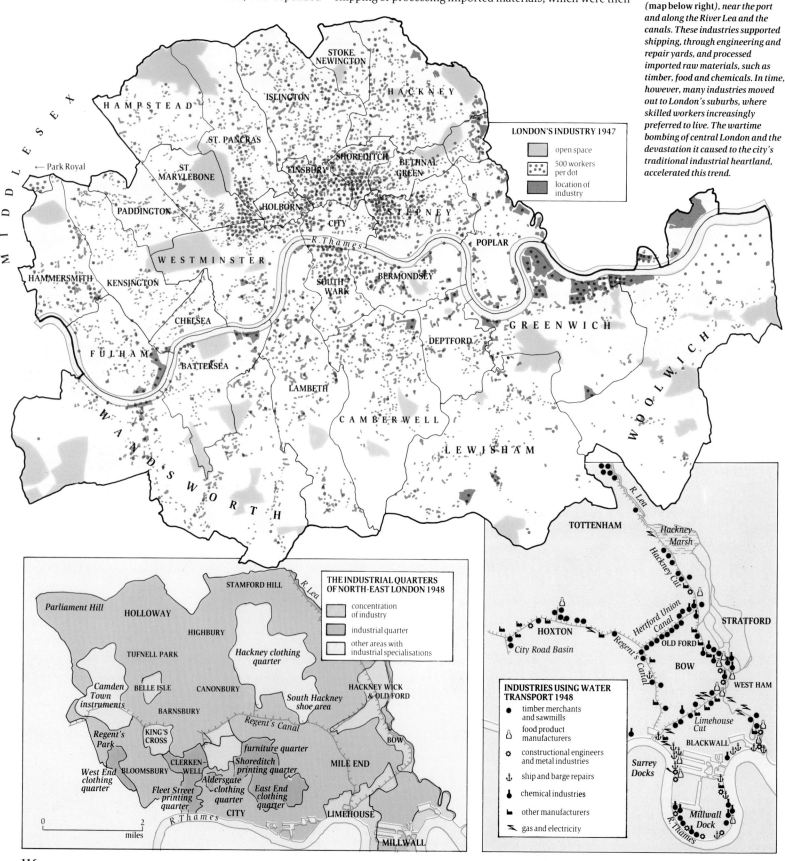

LONDON'S INDUSTRY 1947

- open space
- 500 workers per dot
- location of industry

THE INDUSTRIAL QUARTERS OF NORTH-EAST LONDON 1948

- concentration of industry
- industrial quarter
- other areas with industrial specialisations

INDUSTRIES USING WATER TRANSPORT 1948

- ● timber merchants and sawmills
- ⌂ food product manufacturers
- ✿ constructional engineers and metal industries
- ⚓ ship and barge repairs
- ♟ chemical industries
- ⚒ other manufacturers
- ⚡ gas and electricity

transported by rail to other parts of the British Isles, or re-exported abroad. Whilst they provided some diversity for the economy of the East End, they attracted only low wages, and were particularly susceptible to the depression conditions of the 1930s.

The main change after 1918 was the development of factories outside these old-established areas, especially in the new suburban industrial estates which were located around the outer periphery of the city and accessible to the new trunk road networks. These estates provided an ideal environment for spacious factories, which used electric power and automated processes, and were therefore less dependent on the old craft skills. One such industrial zone in the north-east of London spread along the Lea valley from the port area as far north as Enfield. Inner-city firms, such as clothing and furniture-makers, moved out here, and were joined by newer factories specialising in activities such as electrical engineering.

The same pattern was repeated on an even larger scale in west London, as the spread of suburban housing accelerated. The largest area was in and around Park Royal, built on the site of a failed agricultural show and First World War munitions factory. Although a few large companies moved in later, most firms here were small, occupying ready-made factories. Wembley Stadium (*page 143*), the site of the British Empire Exhibition in 1924, also subsequently offered extensive areas for industrial development. Among other areas that became prominent were Hayes, Southall and Acton, and locations along the new arterial roads and the North Circular, such as Colindale and Cricklewood. The main industries that grew up at this time were general and electrical engineering, vehicle manufacture and food processing. By 1933, the Lea valley industrial zone employed almost 38,000 people, and west Middlesex 75,000.

Inter-war industrial expansion
Many factories were built along London's new arterial roads between the wars, exploiting the new potential of road transport, which freed industry from the old ties of rail and water transport. One of the new industrial areas was the Great West Road, opened in 1925. Many now-familiar factories moved into new premises: Smith's crisps (poster below); Curry's cycles and radios (below); Gillette razors (bottom). These spacious, show-case factories depended on the spread of the electricity supply, from new river-based power stations, such as at Battersea (right).

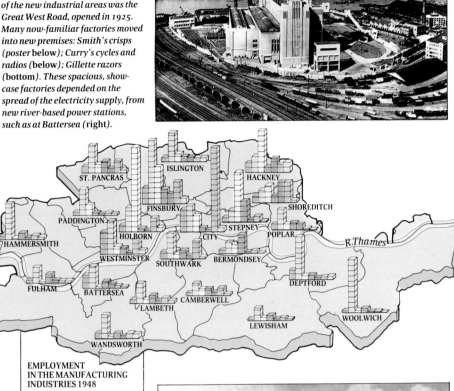

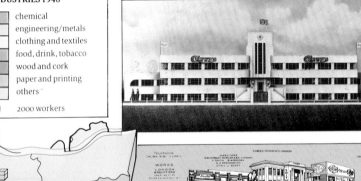

EMPLOYMENT IN THE MANUFACTURING INDUSTRIES 1948

- chemical
- engineering/metals
- clothing and textiles
- food, drink, tobacco
- wood and cork
- paper and printing
- others

☐ 2000 workers

EMPLOYMENT IN THE NON-MANUFACTURING INDUSTRIES 1948

- building
- transport & commercial
- distribution
- insurance, banking, finance
- administration
- professional
- miscellaneous

☐ 2000 workers

Distribution of industry, 1948
Different districts within the city came to specialise in distinct manufacturing trades (map top right), often because of local concentrations of skilled workers, as well as long-established traditions – for example printing in Fleet Street and the City. London was, however, a predominantly service-based city (map above), and by the late '40s its financial, administrative and professional functions had come to dominate the City and the West End.

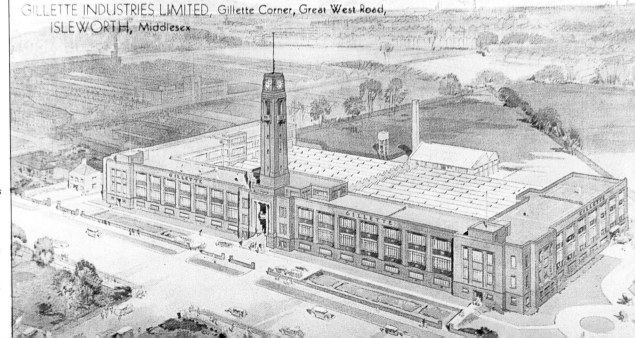

LONDON AT WAR

The outbreak of war in 1914 coincided with the early development of military aviation. German strategists argued that bombing their city would undermine morale so much that Londoners would force their government to sue for peace, and in July 1915 the Kaiser finally gave approval to attacks on the enemy capital. From then until May 1918 London became a limited field of conflict; German attacks were few and far between and were relatively ineffective. At first they were confined to Zeppelin raids; airships dropped 196 tons of bombs on London and the surrounding counties. In 1917 the German air force developed a long-range bomber, the Gotha IV, capable of reaching London. From June 1917 a series of raids was carried out in fine weather; these attacks killed 835 Londoners and injured 1,437. London turned into an armed camp: an outer circle of airfields provided fighter protection; an inner circle of searchlights and anti-aircraft guns covered the capital itself.

The experience of the first war prepared London for the next. In the later 1930s the government developed plans for the air defence of London: air raid shelters were built; civil defence training and advice were established; the evacuation of the city was planned in great detail. The first bombs were not dropped until September 1940, which was the start of the Blitz, a sustained six-month bombardment of London which produced 71 major raids during which 18,000 tons of high explosive bombs were dropped and 20,000 Londoners killed. At first, German aircraft attacked military and port facilities, but in alleged retaliation for British attacks on German cities the bombing became more indiscriminate. Londoners were forced to spend sleepless nights in air raid shelters and Underground stations. The bombing cut gas, water and electricity supplies, and brought transport chaos after heavy raids. Yet the Blitz failed to destroy London's economy or to undermine London's morale. When Londoners were asked by Mass Observation what made them most depressed in the winter of 1940-1, they rated weather above air-raids. Over half of Londoners polled were against vengeance attacks on Germany.

In May 1941, the bombing abruptly came to an end with the great raid of May 10th. Over the next three years

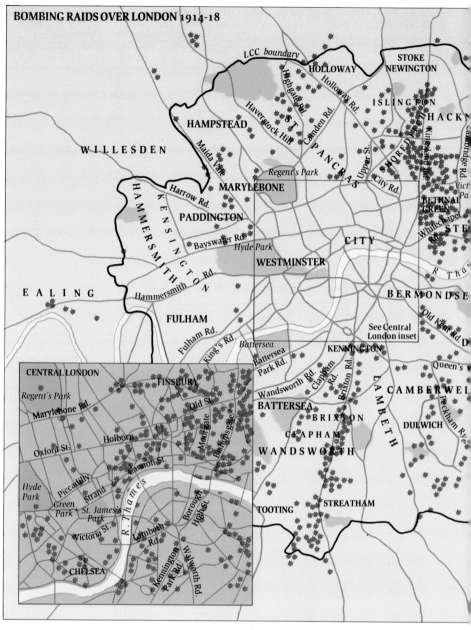

BOMBING RAIDS OVER LONDON 1914-18

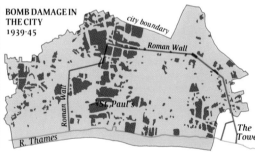

BOMB DAMAGE IN THE CITY 1939-45

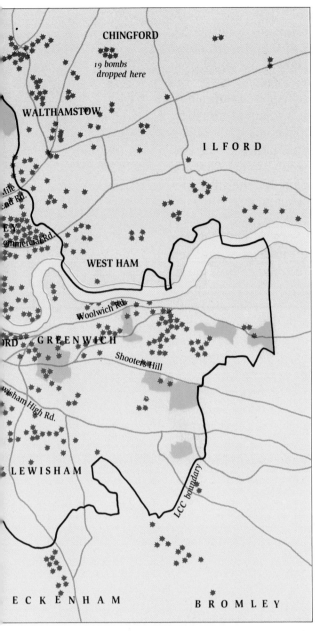

CHINGFORD

*19 bombs
dropped here*

WALTHAMSTOW

ILFORD

WEST HAM

Woolwich Rd.

GREENWICH

Shooter's Hill

wisham High Rd.

LEWISHAM

LCC boundary

ECKENHAM

BROMLEY

Aerial bombardment *On the morning of 13 June, 1917, fourteen German bombers launched the first aircraft attack on London. They dropped four tons of bombs, most of them falling within a mile of Liverpool Street Station. From then until May 1918, a series of raids brought destruction to many parts of the capital (map left). Barrage balloons, anti-aircraft artillery and searchlights were mobilised to combat the raiders. The same system was used during the Blitz, when German air forces attacked London repeatedly between September 1940 and May 1941. The most famous raid took place on 29 December, 1940 when large parts of the city were destroyed (map below).*

London's defences were strengthened and it became a major centre of wartime production, despite its proximity to German air bases. The long lull made the renewed assault in the summer of 1944 harder to take. The Allies were mounting ever-heavier attacks on German cities. Hitler demanded retaliation and launched the new wonder-weapons, the V1 flying-bomb and the V2 rocket, against London. The first bombs landed in June 1944 and continued almost to the end of the war. The damage they did was slight compared with the Blitz, but the attacks were indiscriminate and loss of life was high. The V-weapons killed 9,200 and injured 22,000 others. In total, 29,890 Londoners were killed by enemy action in the capital, and 50,507 were injured. Large areas of central London were reduced to rubble. It was this destruction which helped to accelerate the trend towards centralised town-planning, as well as hastening the re-location of London's population, from city to suburbs, which had begun in the 1930s.

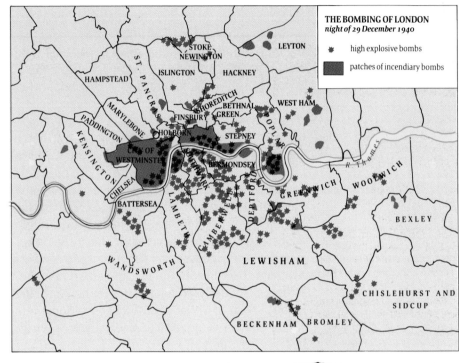

THE BOMBING OF LONDON
night of 29 December 1940

✳ high explosive bombs

▪ patches of incendiary bombs

STOKE NEWINGTON
LEYTON
HAMPSTEAD
ST. PANCRAS
ISLINGTON
HACKNEY
MARYLEBONE
PADDINGTON
SHOREDITCH
BETHNAL GREEN
WEST HAM
FINSBURY
HOLBORN
STEPNEY
POPLAR
KENSINGTON
CITY OF WESTMINSTER
BERMONDSEY
R.Thames
CHELSEA
WOOLWICH
BATTERSEA
LAMBETH
CAMBERWELL
DEPTFORD
GREENWICH
WANDSWORTH
BEXLEY
LEWISHAM
CHISLEHURST AND SIDCUP
BECKENHAM
BROMLEY

London at war *When the Second World War broke out, London was prepared for extensive bombing. In September 1939, 690,000 children were evacuated from London to safer provincial areas and London's main stations were thronged with diminutive travellers (photograph far left). By the time of the Blitz many had returned to the city. Two maps (centre left and below right) give a stark picture of the damage sustained by Greater London and, in particular, the City during the Second World War. An artist's drawing of the City (left), made from a low-flying hot air balloon, gives a vivid impression of the destruction. An exhortation to 'Dig for Victory' is emblazoned across the Royal Exchange, ironically surrounded by vast craters (left), while bowler-hatted office workes pick their way across flattened streets in Moorgate in 1941 (above left). London Underground posters celebrated the City's survival – Walter Bradberry's 'The Proud City' (above right) shows the Chelsea power station lit up by searchlights. Although the worst of the Blitz was over by May 1941, German attacks returned again in 1944, with Hitler's 'Vengeance' weapons, the V1 flying bomb and V2 rocket (map far right).*

THE PROUD CITY

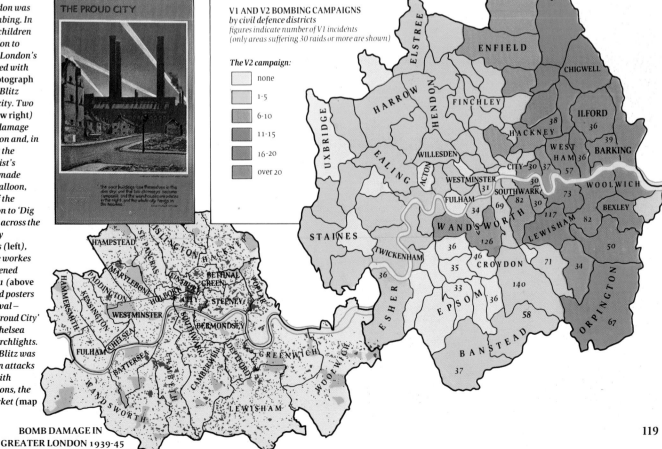

V1 AND V2 BOMBING CAMPAIGNS
by civil defence districts
figures indicate number of V1 incidents
(only areas suffering 30 raids or more are shown)

The V2 campaign:

	none
	1-5
	6-10
	11-15
	16-20
	over 20

ELSTREE
ENFIELD
CHIGWELL
HARROW
HENDON
FINCHLEY
ILFORD
UXBRIDGE
HACKNEY
38
36
WEST HAM
BARKING
WILLESDEN
30
37
36
39
57
EALING
ACTON
CITY
WESTMINSTER
31
SOUTHWARK
WOOLWICH
FULHAM
82
30
73
34
69
117
BEXLEY
82
WANDSWORTH
LEWISHAM
STAINES
126
50
TWICKENHAM
36
46
ESHER
35
CROYDON
71
34
ORPINGTON
33
140
67
EPSOM
36
BANSTEAD
58
37

HAMPSTEAD
ST. PANCRAS
ISLINGTON
HACKNEY
HAMMERSMITH
MARYLEBONE
PADDINGTON
FINSBURY
BETHNAL GREEN
POPLAR
KENSINGTON
HOLBORN
CITY
STEPNEY
WESTMINSTER
SOUTHWARK
BERMONDSEY
DEPTFORD
GREENWICH
WOOLWICH
CHELSEA
FULHAM
BATTERSEA
LAMBETH
CAMBERWELL
WANDSWORTH
LEWISHAM

BOMB DAMAGE IN GREATER LONDON 1939-45

119

CHAPTER 9
POST-WAR LONDON

LONDON HAS CHANGED radically since the Second World War. During the first 20 years of the period, the city's status as the hub of the British Empire progressively diminished, followed within 15 years by the final decline of the old Port of London. In the 1980s, however, the expansion of global financial markets enabled the City to renew and sustain its leading world position. Other areas of the city were also transformed by changes in the international situation; as declining seaborne trade undermined the economy of east London, so the rise of mass air transport focused expansion in the west around Heathrow, and in the south around Gatwick. Perhaps the most startling consequence of this expansion was the growth of London as a major centre of international tourism, a development certainly not anticipated by planners at the end of the war.

London's manufacturing industries and the port – the traditional employers of the working-class – continued to thrive until the mid-1960s, but then began to close down or move out of the city with increasing speed. Unemployment rose, and the quality of the remaining types of employment declined. Jobs tended to be less secure, poorly paid and often associated with the tourist trade, consumer services or construction work. The public sector, always a vital part of the London economy, took up some of the slack in the 1970s, but this was followed by a reduction of employment in national and local government, the health service and public transport.

Changes in city life, however, reflect not only economic and social realities, but also involve attitudes to the image of the city and the buoyancy of its culture. In popular mythology, post-war London's success reached its first peak with the emergence of the pop music and fashion cultures of the 'swinging sixties'. The coming-of-age of the post-war baby-boom, the development of the modern 'media' industries – led by film, television, radio and advertising – all combined to focus world attention on London in the late 1960s. Two decades later, with the burgeoning of business and financial services, the 1980s equivalent of the 1960s was the 'yuppy' generation. Although based more upon commercial opportunism than artistic flair, this offered a similarly dominant, and ephemeral, image of London, reflected in many aspects of life from property development to fashion.

London also enjoyed a post-war artistic revival. During the 1960s and 70s it became an international centre of music, drama and the visual arts. This achievement had been stimulated by public and private sector investment in orchestras, theatre and opera companies and in such developments as the South Bank and Barbican Arts Centres. It also depended on expanding opportunities in higher education and rising middle-class cultural aspirations. The artistic dynamism of the capital was further enhanced by the growing number of immigrants from Ireland, Europe and the Commonwealth. They, in turn, generated new, and exotic, sources of music, fashion and lifestyle.

Another key change in post-war London was a greater awareness of the need to preserve the city's dwindling cultural and architectural heritage. In the view of many, such sentiments emerged much too late; the 20th-century transformation of London's appearance had been radical, piecemeal and generally of mediocre quality. As investment in manufacturing industries was withdrawn, successive booms in private property development satisfied an apparently insatiable demand for modern office accommodation. The West End and the City were transformed by periods of virulent land speculation and soaring land prices. This pattern has been frequently repeated in London's history;

Below In popular mythology, post-war London's success reached its peak with the emergence of the pop music and fashion cultures of the 'swinging sixties'. During this period, London became the focus of world attention as a leading 'style capital', and nowhere epitomised this new-found status as a trend-setting city more vividly than Carnaby Street, with its fashionable boutiques and lively street culture.

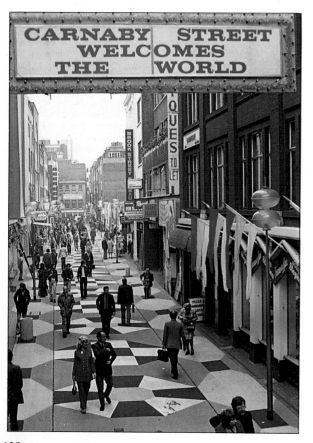

A discussion of London's history of immigration can be found on pages 136-137.

A discussion of the unrealised plans to redevelop Piccadilly Circus in the 1960s, and the contrasting plans to redevelop, and at the same time conserve, Spitalfields in the 1990s can be found on page 129.

the needs of private capital have always taken precedence over regulation or planning. Ironically, less than 25 years on, some of the 1960s developments were being demolished under the influence of the 'heritage' movement, to be replaced by buildings which revived historical styles.

By the 1970s London's 'inner city' problems were multiplying. To the problems of unemployment were added ethnic conflicts, public sector under-funding, unsuccessful education reforms, overstrained welfare and health services, and increasingly inadequate housing provisions. Although most of the houses built in London immediately after the war were intended for council renting, there were long waiting lists which increased as rates of building diminished after the 1960s. Much of the post-war housing built by public authorities was criticised for its alienating design, poor construction and inadequate maintenance. For people moving into the city, whether from northern England, Ireland, Barbados or Bangladesh, neither private ownership nor council renting was readily available. The provision of privately rented accommodation – best suited to the shifting population of the city – declined, and what remained rose in price. The early 1960s were marked by the scandal of 'Rachmanism' whereby unscrupulous developers forced out tenants who were paying low, controlled rents, and redeveloped the properties either to sell them or charge higher rents. The optimistic attitude of the 1950s to house building and urban redevelopment was therefore the prelude to a long-term intensification of London's housing problems. In its most extreme form, this was demonstrated by the growth of the homeless population during the 1980s.

Nowhere were the difficulties of reconciling conflicting interests in post-war London more evident than in formulating a cohesive transport policy.

The gentrification of London's Georgian and Victorian housing is discussed on page 127.

Above *The 1980s saw a period when London's business and financial services burgeoned. London's industrial legacy had been supplanted in the post-war years by the rise of the service sector. By the 1980s, London was exploiting its status as a leading world financial centre, and the square mile of the City was being transformed by a speculative building boom. This view of the Stock Exchange at night clearly demonstrates the mixture of old and new which now characterises the architecture of the City, with the measured classicism of the Mansion House (centre) and the Bank of England (left) contrasted against the looming height of the NatWest Tower in the distance.*

Right *London's status as an international capital in the post-war years was reinforced by the opening of a new London airport, at Heathrow, in 1946. Development proceeded there over the ensuing forty years, the most recent addition being Terminal Four (opened in 1986). In 1977, the Piccadilly Line was extended to Heathrow, improving transport links with the capital. Heathrow is one of the world's busiest airports; Terminal Four can handle up to 4000 passengers an hour.*

A map of the Greater London Development Plan can be found on page 128.

Far-sighted plans for road improvement had been laid down in the 1930s, but they were soon overwhelmed by the scale of traffic growth in the 1960s. The Greater London Development Plan of 1967 proposed a 'motorway box' around the centre of London, orbital motorways in the suburbs, and routes to connect the city centre to the national motorway system. Some argued that the building of motorways in cities simply encouraged the greater use of vehicles, swamping other aspects of city life wherever they decanted traffic into local areas. The most telling opposition, however, came from localities directly threatened by motorway building, and the strength of opposition to the GLDP scheme led to its eventual abandonment. Transport policy has long been the

subject of heated political debate. Questions have been asked about appropriate levels of subsidy required to maintain British Rail and London Transport services, road-building schemes, the introduction of bus lanes, the possibility of taxing road users, and the high proportion of cars in rush-hour queues that were company-subsidised. The rise of tourism and the revival of commuting into central London in the 1980s added to the problems. Meanwhile, it was universally agreed that the quality of public transport continued to decline, dragging the quality of the environment of inner London down with it. In 1986 the completion of the M25 orbital motorway around London vividly demonstrated the problems of preserving the city in an age of escalating road transport. It connected many areas around London for the first time, but generated as much as six times the traffic anticipated, and quickly became a byword for congestion and motorists' frustration.

The construction of the M25, the increase in traffic congestion and the rising pressures on public transport within London are all discussed on pages 126-127.

In 1945 the dominant administrative and political authority in London was the London County Council. This elected authority governed inner London, but its powers to represent the whole of London became increasingly limited by the expansion of the city beyond the LCC boundaries. In spite of opposition, the Greater London Council was instituted in 1965, covering 610 square miles and including most of the built-up area of the city. The whole of Middlesex and parts of the other surrounding counties were added to the former LCC area. A new two-tier metropolitan government was established, with 32 London Boroughs and the City of London. The GLC was given a significant range of powers which included strategic planning, the administration of the LCC's housing and parks, main sewers, drainage and flood prevention, main roads and traffic planning, the fire and ambulance services, and refuse disposal. The old LCC educational service was separately preserved as the Inner London Education Authority, while the outer boroughs administered their own education.

The GLC was able to draw on the rating resources of outer London to help inner London; as the wealth gap between these two areas increased however, political resistance from outer London to this redistribution of finances grew. The GLC's administrations were often out of step with central government, which frequently led to political conflict. One trend which in part reflected this situation was the erosion of the GLC's powers, sometimes as part of wider administrative reforms. In 1974, for example, the establishment of Regional Water and Health Authorities removed the GLC's drainage and flood prevention functions, and the ambulance service. The GLC's housing, both within the city and in the overspill communities outside it, was transferred in 1980-82 to local boroughs. The Council's supervisory control of the London Transport Executive granted in 1970, was removed once more in 1984. By the 1980s, the net effect had been to diminish the GLC's role in London to activities such as transport studies, arts and recreation, and grants to voluntary bodies.

The GLC was finally abolished in 1986 and for the first time for 97 years, there was no unified representative government for the city, a situation unique among major world cities. In 1990, the Inner London Education Authority, the last remnant of the old LCC, was also abolished.

In the post-war decades, the major contrast in development was between the declining east of the city, and the prosperous, 'overheating' west. This was, if anything, encouraged by planned investment in new towns, motorways, Heathrow Airport, research and development, and military establishments. Only by the mid-1980s did the completion of the M25, the final designation of the third London airport at Stansted, the imminent completion of the Channel Tunnel, and the prospect of major investment in eastern Docklands and lower Thamesside, seem to shift the emphasis eastwards. Nevertheless, the huge established pressures for development west of the capital remained the single most powerful force shaping the geography of the London region as the last decade of the 20th century approached.

Map below London as it is today.

EXTENT OF LONDON 1990

+ St. Paul's

LONDON 1990

THE POST-WAR CITY

In the twenty-year period after the Second World War the reconstruction of London was often localised and piecemeal. As economic recovery occurred the initiative for rebuilding in both the City and the West End was taken largely by landowners and commercial developers. Public agencies, however, channelled their efforts into neighbourhoods that had suffered serious war-

The Lansbury Estate in Poplar was built as the Festival of Britain's Exhibition of Living Architecture. This pioneering neighbourhood scheme (left) was intended to act as a blueprint for the redevelopment of the deprived areas of the East End. Unfortunately, its intimate scale, low-rise design and generous provision of open space, was not widely adopted because of the pressure for mass-production building techniques which grew in subsequent years.

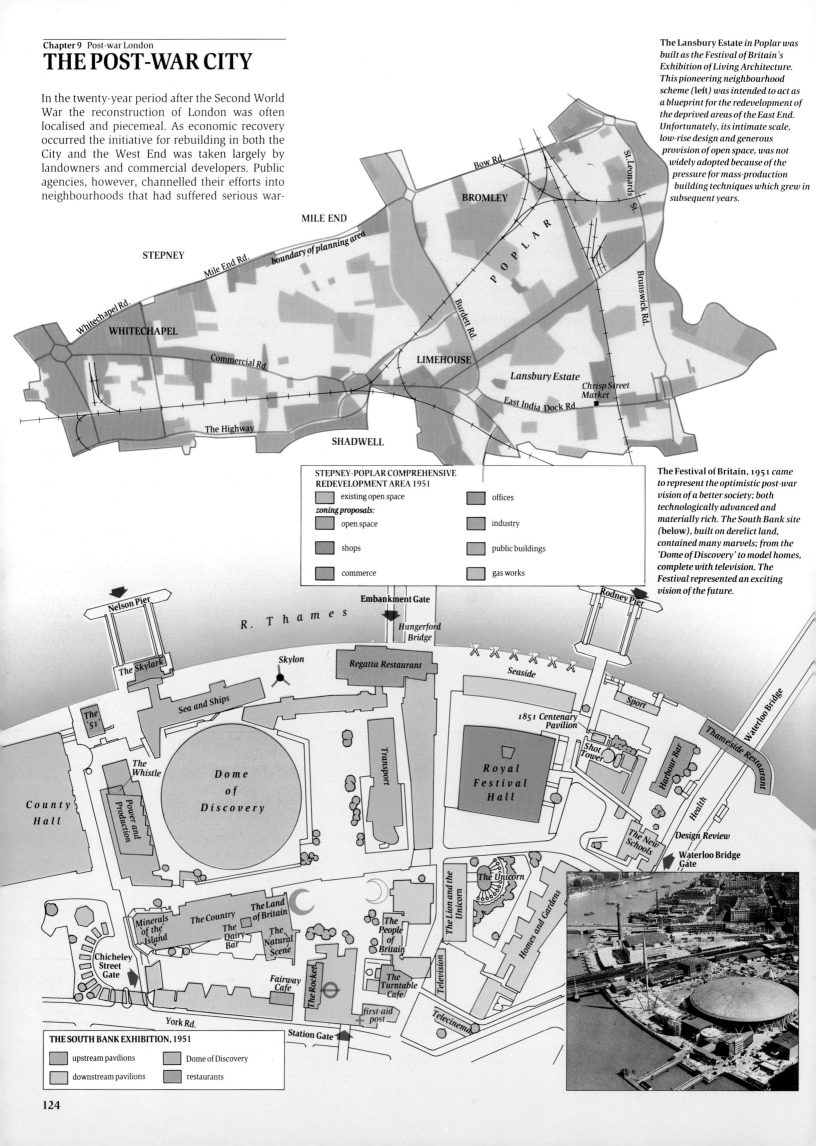

STEPNEY
MILE END
Mile End Rd.
boundary of planning area
Whitechapel Rd.
WHITECHAPEL
Commercial Rd.
The Highway
SHADWELL
BROMLEY
Bow Rd.
St. Leonards St.
P O P L A R
Brunswick Rd.
Burdett Rd.
LIMEHOUSE
Lansbury Estate
Chrisp Street Market
East India Dock Rd.

STEPNEY-POPLAR COMPREHENSIVE REDEVELOPMENT AREA 1951

- existing open space
- *zoning proposals:*
- open space
- shops
- commerce
- offices
- industry
- public buildings
- gas works

The Festival of Britain, 1951 came to represent the optimistic post-war vision of a better society; both technologically advanced and materially rich. The South Bank site (below), built on derelict land, contained many marvels; from the 'Dome of Discovery' to model homes, complete with television. The Festival represented an exciting vision of the future.

Nelson Pier
Embankment Gate
R. Thames
Hungerford Bridge
Rodney Pier
Waterloo Bridge
The Skylark
Skylon
Regatta Restaurant
Seaside
Thameside Restaurant
The '51
Sea and Ships
Sport
1851 Centenary Pavilion
Shot Tower
Harbour Bar
The Whistle
Power and Production
Transport
Dome of Discovery
Royal Festival Hall
Health
Design Review
County Hall
The New Schools
Waterloo Bridge Gate
The Unicorn
The Lion and the Unicorn
Homes and Gardens
Minerals of the Island
The Country
The Land of Britain
The Dairy Bar
The Natural Scene
The People of Britain
Chicheley Street Gate
Television
Fairway Cafe
The Rocket
The Turntable Cafe
first-aid post
Telecinema
York Rd.
Station Gate

THE SOUTH BANK EXHIBITION, 1951

- upstream pavilions
- downstream pavilions
- Dome of Discovery
- restaurants

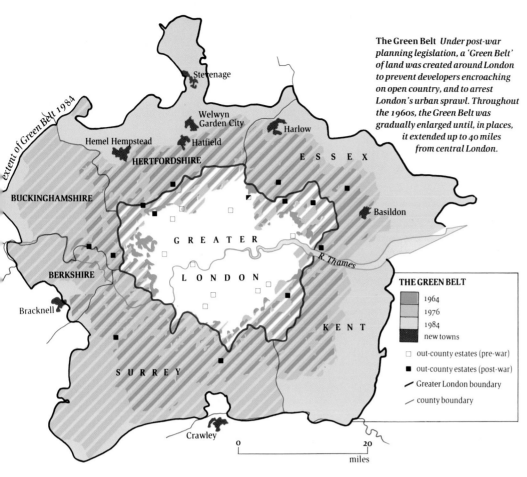

The Green Belt *Under post-war planning legislation, a 'Green Belt' of land was created around London to prevent developers encroaching on open country, and to arrest London's urban sprawl. Throughout the 1960s, the Green Belt was gradually enlarged until, in places, it extended up to 40 miles from central London.*

THE GREEN BELT

1964
1976
1984
new towns

□ out-county estates (pre-war)
■ out-county estates (post-war)
⎯ Greater London boundary
⎯ county boundary

Perhaps the real embodiment of the Festival spirit can be found in the new towns ideology. Although sometimes attached to small existing settlements, they were intended to create a totally new living environment. They were to be medium-sized communities (around 60,000, although most are now well above this), with a preponderance of council housing for poor families from London, set in rural surroundings. They would have enough employment in new industries, often relocated from London, to prevent the need for commuting back to the city. Housing areas were grouped into neighbourhoods of 10-12,000 people, with local shopping and schools clearly segregated from the industrial areas. Green belts – areas of farmland and open country protected by law from urban infringement – were instituted in 1946. The new towns thus presented a planned outlet for the growth of London's population, protecting the countryside of the south-east from indiscriminate sprawl.

The planned approach to London's many long-standing problems, adopted after the Second World War, can claim several successes. The Green Belt has been regularly extended and remains one of the most secure features of British planning policy. In their early years the new towns did suffer teething problems – many residents from crowded, intimate inner-city areas found the planned vision of a better future antiseptic, and suffered greatly from the towns' lack of community. However, over the years many new towns have matured into pleasant, well-established settlements and have been among the major growth centres of the 1980s.

The idealism of the Festival of Britain could not, however, anticipate the 'baby boom' of the 1960s, and the resulting pressures placed on space in the London region. Nor did it foresee the effects of growing affluence on people's housing expectations, or the huge impact of general car-ownership, with its resultant effects on transportation in the city. Furthermore, post-war optimism did not anticipate the collapse of London's manufacturing and port economy some twenty years later. The new towns have been able to adapt to, and even benefit from, these changes. The redevelopment of the East End in the 1950s and '60s appeared to eradicate the legacy of poverty, but many high-rise estates built then developed serious problems in the following years. Characterised by high levels of crime and socio-economic deprivation, they form a major challenge for the future planning of London.

time devastation and into major redevelopment schemes, designed to solve the problems of London's poor housing, social amenities and infrastructure. The 1947 Town and Country Planning Act legislated for the comprehensive redevelopment of the slum areas of the old Victorian city. The largest schemes were in the East End, where the old patterns of roads, services, housing and land-use were almost completely replaced over a twenty-five year period on the basis of a master plan drawn up in the 1940s.

Another approach to redevelopment involved moving people out of the city altogether, initially to the eight 'first generation' new towns, established in 1947-8: Basildon, Bracknell, Crawley, Harlow, Hatfield, Hemel Hempstead, Stevenage and Welwyn Garden City. The two strategies were linked, since the rebuilt inner city areas could house barely half the population that was living there before. The new centres, along with established towns where expansion had been planned, such as Thetford in Norfolk or Swindon

in Wiltshire, were intended to receive 'overspill' from the reconstruction of London.

The idealism of the time was perhaps best symbolised by the South Bank Exhibition, the centrepiece of the 1951 nationwide Festival of Britain. One hundred years after the Great Exhibition in Hyde Park, this was intended to 'demonstrate the contributions to civilisation made by British advances in 'Science, Technology and Industrial Design' against a background representing the living and working world. Built on derelict land on the south bank of the river, the most striking features were the 'Dome of Discovery', containing exhibits on modern exploration, and the cigar-shaped 'Skylon', balanced on wires high above the spectators. After wartime and post-war deprivations, exhibits on 'The Land' of Britain, and 'The People', presented the morale-boosting prospect of a planned, technologically-based Welfare State. The Festival's permanent legacy to London was a section of river wall bounding the site, and the Festival Hall.

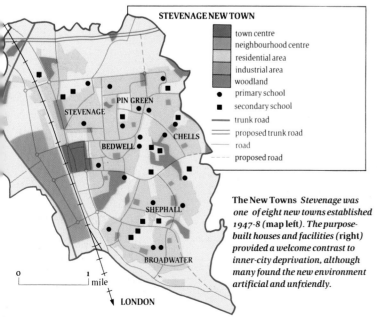

STEVENAGE NEW TOWN

town centre
neighbourhood centre
residential area
industrial area
woodland
● primary school
■ secondary school
⎯ trunk road
⎯ proposed trunk road
⎯ road
--- proposed road

The New Towns *Stevenage was one of eight new towns established 1947-8 (map left). The purpose-built houses and facilities (right) provided a welcome contrast to inner-city deprivation, although many found the new environment artificial and unfriendly.*

LIVING AND WORKING IN LONDON

After the Second World War, the population of inner London, especially of skilled workers, was in decline. Londoners were moving out to the suburbs, satellite settlements and the new towns which were a major component of post-war planning. In the 1960s, inner London was also affected by the collapse of manufacturing. Many national firms closed their London factories, while small firms in the traditional textiles, clothing, furniture and metal-working quarters were going out of business, unable to compete with firms outside the city which were better placed to introduce new technology and compete internationally. Suburbs and new towns offered far better living conditions for the skilled workforce than deprived inner-city areas had done.

A further blow came with the equally rapid collapse of the port, caused by increased competition from continental and smaller east coast ports which used new container-handling and 'roll-on roll-off' ferry technologies. Hampered by poor labour relations and management, London's docks were progressively closed between 1969 and 1981, with the loss of about 25,000 jobs. A workforce of only 2-3,000 was left at the Tilbury and other downstream docks. Firms dependent on the port closed, adding to the general decline of inner city manufacturing.

Meanwhile, London's service role was being reasserted. London was becoming increasingly dependent on financial business in the 1970s and 80s, competing with New York and Tokyo. Most of the new jobs were filled by white collar workers from the suburbs and outside London, leading in turn to an increase in commuting during the 1980s as more people travelled into the city for work. This added to the strains on transportation that the decentralisation of manufacturing should have relieved.

Although offices were encouraged to move out of central London in the 1960s and 70s, this had only a marginal impact on the overall pattern of commuting, especially as demand for office space grew. Massive new schemes for road improvement had also been proposed to relieve some of the strain on public transport, but met with fierce opposition from local and environmental groups during the 1980s. The capital's commuting hinterland now extended to the south coast and into East Anglia, the Midlands and the West Country. In 1939, 8,600,000 people lived within Greater London; the total has now fallen to 6,500,000 as many Londoners have moved to new homes beyond the Green Belt.

Despite this movement away from the city, London's most pressing problem is still a sub- stantial shortfall in satisfactory housing – 32,000 families in London are now recognised as home- less. Local Authorities are confronted with gen- erations of London housing in a serious state of disrepair: solidly-built Victorian houses, inter- war semis and shoddily-built 1960s apartment blocks are all in need of massive investment.

London's residents are highly varied in wealth, age and household size and composition. In 1900, 90 per cent of Londoners rented their

Developing the Green Belt Many Londoners have opted to move away from the problems of inner-city dilapidation into the surrounding countryside. The illustration (top right) shows a proposed shopping centre at Colnbrook, one mile west of Heathrow airport. Its neo-classical style is very different from the architecture of other new shopping centres and may make the project more acceptable to the substantial lobby who oppose the building of homes and other facilities within the Green Belt.

The whole of the M25 orbital motorway cuts through Green Belt land and sites close to motorway junctions offer prime locations for new development of all kinds (map below).

MOVING OUT OF LONDON 1988

- Green Belt 1976
- Green Belt 1984
- proposed new communities
- proposed shopping centres

Gentrification The map (right) shows areas in Islington where significant influxes of professional and managerial residents occurred in the 1960s. The photographs (far right) show the run-down state of Canonbury housing before this trend (above), and the refurbished houses (below). The map (centre right) shows other districts gentrified in the 1960s. Renovation has occurred more widely in recent years.

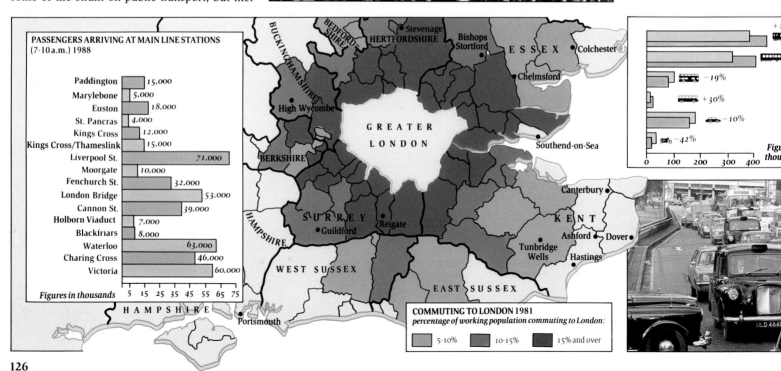

PASSENGERS ARRIVING AT MAIN LINE STATIONS
(7-10 a.m.) 1988

Station	Figures (thousands)
Paddington	15,000
Marylebone	5,000
Euston	18,000
St. Pancras	4,000
Kings Cross	12,000
Kings Cross/Thameslink	15,000
Liverpool St.	71,000
Moorgate	10,000
Fenchurch St.	32,000
London Bridge	53,000
Cannon St.	39,000
Holborn Viaduct	7,000
Blackfriars	8,000
Waterloo	63,000
Charing Cross	46,000
Victoria	60,000

Figures in thousands

−19%
+30%
−10%
−42%

COMMUTING TO LONDON 1981
percentage of working population commuting to London:

- 5-10%
- 10-15%
- 15% and over

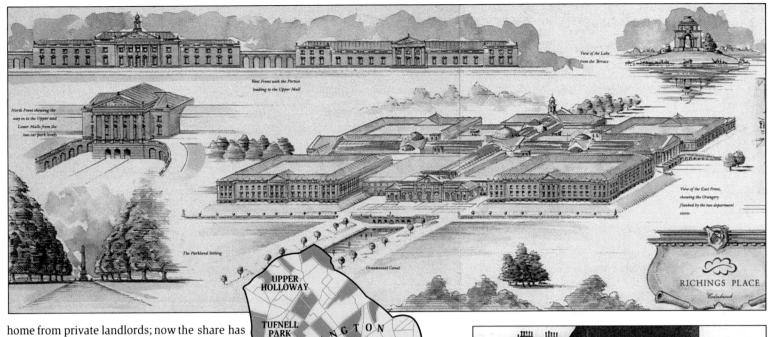

View of the Lake from the Terrace

North Front showing the way in to the Upper and Lower Malls from the two car park levels

West Front with the Portico leading to the Upper Mall

The Parkland Setting

Ornamental Canal

View of the East Front, showing the Orangery flanked by the two department stores

RICHINGS PLACE
Colnbrook

home from private landlords; now the share has dropped to below one-fifth. Just over half of London's homes are now owned by their occupants; the remaining 30 per cent are rented from local authorities or housing associations. The number of new homes built in Greater London declined sharply from 32,000 in 1972 to 8,000 in 1980. Until the building recession of the later 1980s, the figure ran at approximately 10,000 each year. Private rehabilitation of housing stock is widespread, especially in accessible inner suburbs with attractive but often run-down housing dating from Victorian and Edwardian times. Such areas have been discovered by managerial and professional households in recent years and a process of 'gentrification' has taken place – not only the houses themselves, but entire neighbourhoods are transformed by influxes of relatively affluent newcomers, who replace working-class residents. House prices subsequently rise, home-owners replace tenants, and local shops and restaurants cater for a wealthier clientele. In the 1990s London housing is a mosaic of great complexity with respect to age, appearance, comfort and quality. Like New York and Paris, some of the best accommodation in this cosmopolitan city is owned by members of London's large, diverse foreign community.

UPPER HOLLOWAY

TUFNELL PARK

ISLINGTON

Holloway Rd.

HIGHBURY

CANONBURY

Canonbury Rd.

LOWER HOLLOWAY

BARNSBURY

PENTONVILLE
ANGEL

0 ¼
miles

LONDON GENTRIFICATION 1961–1971

CAMDEN
ISLINGTON
HACKNEY
WESTMINSTER
KENSINGTON & CHELSEA
HAMMERSMITH
CITY
TOWER HAMLETS
SOUTHWARK
GREENWICH
WANDSWORTH
LAMBETH
LEWISHAM

LONDON GENTRIFICATION 1961–1971

- 'gentrified' areas 1961
- 1971

COMMUTING TO CENTRAL LONDON (7-10 a.m.)

- 1983
- 1987
- British Rail
- Underground
- Bus
- commuter coach
- private car
- motor/pedal cycle

A city of commuters
London commuters use a wide range of transport (table left). During the 1980s the number of commuters travelling by train, Underground or coach increased, while journeys by bus, private car or cycle declined. The map (far left) shows how London's commuting area has grown since mid-century. In 1951 it hugged the continuously built-up area; by 1981 it extended far into surrounding counties, and long-distance commuters travelled from East Anglia, the Midlands and the West Country. The table (inset) shows that Liverpool St. Station (serving East Anglia and Essex) receives the largest number of commuters, followed by Waterloo and Victoria. Roads, notably the M25, have been built to take some of the strain (map right), but under-investment in transport generally has made road congestion (left) and railway over-crowding (above left) an all too frequent occurrence.

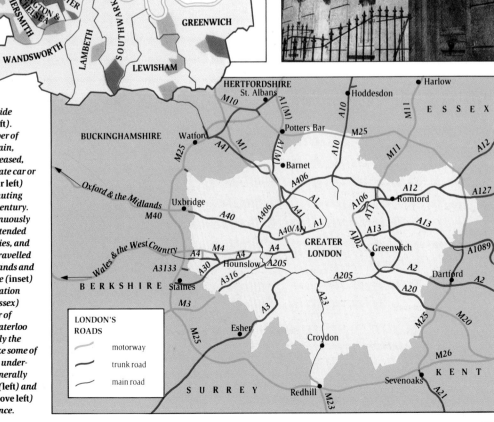

HERTFORDSHIRE
St. Albans
Harlow
Hoddesdon
M10
A1(M)
A10
M11
ESSEX
BUCKINGHAMSHIRE
Watford
Potters Bar
M25
M25
A41
M1
A1(M)
A10
M11
A12
Barnet
A406
A1
A106
A12
Romford
A127
Oxford & the Midlands
M40
Uxbridge
A40
A41
A1
A11
A13
A13
A406
A40(M)
A1
GREATER LONDON
A102
Greenwich
A1089
Wales & the West Country
A4
M4
A4
A4
A2
Hounslow
A205
A205
Dartford
A2
A3133
A30
A316
A20
BERKSHIRE
Staines
M3
A23
M25
M20
A3
Esher
Croydon
M26
KENT
SURREY
Redhill
Sevenoaks
M23
A21

LONDON'S ROADS

- motorway
- trunk road
- main road

PLANS FOR LONDON

During the Second World War it was widely accepted that public authorities would have to take the initiative in post-war reconstruction. The devastation wrought by the war gave London a unique opportunity to tackle the city's long-standing social, housing and environmental problems. The planning pioneer, Sir Patrick Abercrombie, advocated that development should be co-ordinated across the whole city and, as early as 1944, he anticipated reconstruction up to 30 miles from the centre in his 'Greater London Plan'. After 1947, when local authorities and a new Ministry of Town and Country Planning were given powers to control the change of land use, Abercrombie's vision became the basis of London's planning for the next thirty years. These years proved to be a period when market forces increasingly encouraged the decentralisation not only of people, but also of an increasing number of economic and social activities.

The late 1960s were the high-point for urban planners. As a result of his innovative proposals for the Green Belt and the building of the new towns, Abercrombie's influence was still strong, but these schemes were becoming increasingly inadequate to serve fast-developing needs. Population and employment in the London region were growing more rapidly than expected. Affluence created extra housing demands, while ever-increasing car ownership – the number of private cars quadrupled in the London area between 1945 and 1960 – necessitated a massive investment in road building schemes. There were constant pressures to make Green Belt land available for housing, but powerful lobbies still argued for countryside conservation.

A succession of plans was presented to reconcile these regional conflicts. In 1964 the Ministry of Housing and Local Government produced the 'South-East Study'. On top of existing 'overspill' policies for Londoners, it predicted that a further 3.5 million people would need houses by 1981. About one third of these were to be in new growth centres outside London. As in each of the subsequent plans, the Green Belt was to be sustained, or even strengthened. The 1964 Labour Government established Planning Councils for each region, and tightened controls on industrial and office

Planning Greater London In 1965 London gained a new planning body which replaced the London County Council. Like the LCC at its inception, the Greater London Council (GLC) effectively encompassed the whole city, combining the administration of 32 metropolitan boroughs as well as the City of London. One of its tasks was to produce a strategic plan for the capital to act as a framework within which the boroughs could carry out more detailed development planning. This was the Greater London Development Plan (GLDP). The 'Strategy for the South East' had already emerged in 1967 (map below), placing London's development in its wider regional context. The GLDP sought to integrate policies for transportation, housing, land use, retail and commercial activity, historical conservation and recreation and open space.

The Plan was widely criticised. By far the most contentious issue was the imposition of a comprehensive motorway network. But with London's employment contracting rapidly and inner city housing problems escalating, the realism of many of the GLC's planning ambitions was also questioned. Political changes ensured that road proposals were never adopted. A modified plan was eventually approved in 1976 (map right), but even this had only limited relevance in the rapidly changing circumstances of the late 1970s and '80s. Throughout the 1980s there was no coherent regional planning policy for London; development was often held up or radically altered by a conservationist lobby dedicated to ensuring that it was modified to enhance the existing buildings and landscape.

growth in the South-East. 'A Strategy for the South-East' followed in 1967. This proposed promoting major new city regions as counter-magnets to London, linking them to the metropolis by transportation 'corridors'. In the event, only Northampton and Milton Keynes were eventually developed.

In 1970 local authorities co-operated to produce the 'Strategic Plan for the South East'. This acknowledged the uncertainties which had bedevilled earlier plans, and proposed a flexible approach based on five major and seven smaller growth areas. The anticipated population boom predicted in the 1960s had proved to be greatly exaggerated; by the mid-1970s new towns and regional growth centres could only expand at the expense of London itself. A review of the Plan was published in 1976 but, with the decline in economic prospects, regional planning in the South East was effectively abolished in 1979. Throughout the 1980s, regional strategy was seldom considered by the Government, unless under pressure from local authorities and private developers requiring guidance for their activities.

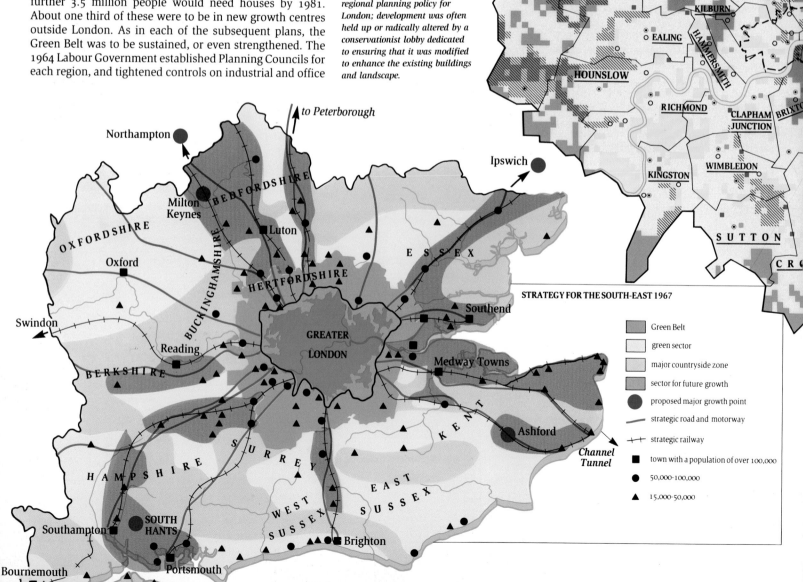

GREATER LONDON DEVELOPMENT PLAN 1976

- Green Belt
- metropolitan open land
- housing problem area
- area of opportunity
- ILFORD strategic centre
- ○ action area
- ■ preferred office location
- ■ preferred industrial location
- ◉ transport interchange
- --- central London
- — borough boundary

ENFIELD · HARROW · WOOD · WEMBLEY · UXBRIDGE · HOLLOW · KILBURN · EALING · HAMMERSMITH · HOUNSLOW · RICHMOND · CLAPHAM JUNCTION · BRIXTON · WIMBLEDON · KINGSTON · SUTTON · CRO

STRATEGY FOR THE SOUTH-EAST 1967

- Green Belt
- green sector
- major countryside zone
- sector for future growth
- ● proposed major growth point
- — strategic road and motorway
- +++ strategic railway
- ■ town with a population of over 100,000
- ● 50,000-100,000
- ▲ 15,000-50,000

to Peterborough · Northampton · Milton Keynes · BEDFORDSHIRE · Luton · OXFORDSHIRE · Oxford · BUCKINGHAMSHIRE · HERTFORDSHIRE · ESSEX · Swindon · Reading · BERKSHIRE · GREATER LONDON · Southend · Medway Towns · Ipswich · HAMPSHIRE · SURREY · WEST SUSSEX · KENT · Ashford · Channel Tunnel · EAST SUSSEX · SOUTH HANTS · Southampton · Brighton · Bournemouth · Portsmouth

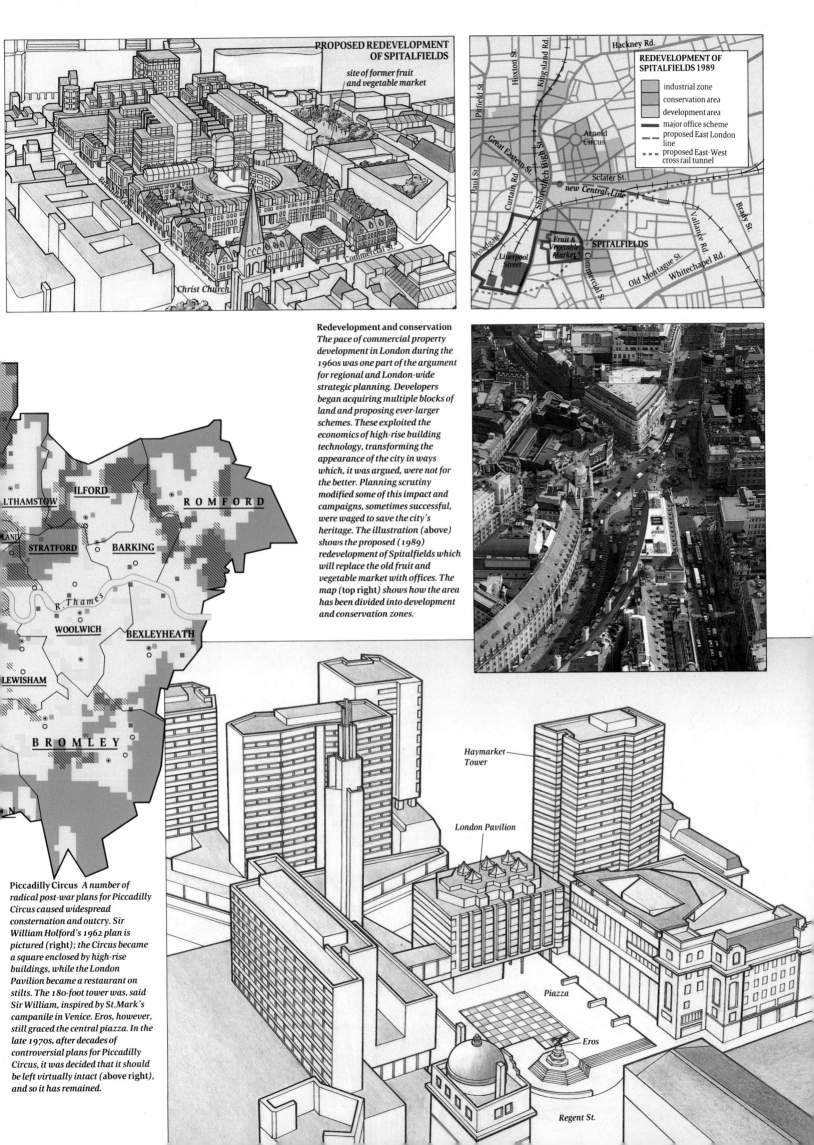

PROPOSED REDEVELOPMENT OF SPITALFIELDS

site of former fruit and vegetable market

Brushfield St.

Commercial St.

Christ Church

REDEVELOPMENT OF SPITALFIELDS 1989

- industrial zone
- conservation area
- development area
- — major office scheme
- --- proposed East London line
- ···· proposed East-West cross rail tunnel

Hackney Rd.

Hoxton St.

Kingsland Rd.

Pitfield St.

Great Eastern St.

Shoreditch High St.

Paul St.

Curtain Rd.

Arnold Circus

Sclater St.

new Central Line

Broadgate

Liverpool Street

Fruit & Vegetable Market

SPITALFIELDS

Commercial St.

Vallance Rd.

Brady St.

Old Montague St.

Whitechapel Rd.

Redevelopment and conservation
The pace of commercial property development in London during the 1960s was one part of the argument for regional and London-wide strategic planning. Developers began acquiring multiple blocks of land and proposing ever-larger schemes. These exploited the economics of high-rise building technology, transforming the appearance of the city in ways which, it was argued, were not for the better. Planning scrutiny modified some of this impact and campaigns, sometimes successful, were waged to save the city's heritage. The illustration (above) shows the proposed (1989) redevelopment of Spitalfields which will replace the old fruit and vegetable market with offices. The map (top right) shows how the area has been divided into development and conservation zones.

LTHAMSTOW ILFORD ROMFORD

LAND STRATFORD BARKING

R. Thames

WOOLWICH BEXLEYHEATH

LEWISHAM

BROMLEY

Piccadilly Circus *A number of radical post-war plans for Piccadilly Circus caused widespread consternation and outcry. Sir William Holford's 1962 plan is pictured (right); the Circus became a square enclosed by high-rise buildings, while the London Pavilion became a restaurant on stilts. The 180-foot tower was, said Sir William, inspired by St.Mark's campanile in Venice. Eros, however, still graced the central piazza. In the late 1970s, after decades of controversial plans for Piccadilly Circus, it was decided that it should be left virtually intact (above right), and so it has remained.*

Haymarket Tower

London Pavilion

Piazza

Eros

Regent St.

CHAPTER 10
LONDON THEMES

FOR VIRTUALLY A thousand years the might of London has been inextricably bound up with its role as the nation's capital, seat of its rulers and home of its Parliament, administration and courts of law. Coupled to that vital power base, London flourished as a centre of commerce and manufacture, with a port that served not only the trade routes of western Europe but of the British Empire and the entire world. London's role as the nation's capital has shaped large parts of the city centre. The Tower of London encapsulates William the Conqueror's royal residence and stronghold, defending the city from water-borne invasion along the Thames Estuary. Royal palaces grace the river banks from Greenwich in the east to Hampton Court in the west, and royal parks provide expanses of public open space in many parts of the metropolis. But it is the palaces and parks, streets and statues of Westminster which most powerfully evoke the presence of 'royal London'. Following the initiative of the Saxon kings, who established a monastery and residence on Thorney Island upstream of *Londinium*, successive monarchs fashioned and refashioned Royal Westminster, making it the official setting for the most solemn occasions of national life, such as coronations, royal weddings and state funerals. The City of London, on the other hand, displays a different kind of grandeur; its dense network of streets, the cathedral of St. Paul's, its historic churches and great buildings all reflect the dynamism generated by trade and international finance. During the working week the City is vibrant and busy, but at night and during weekends it is virtually deserted. The City never used to be like this. For almost all of its history it housed merchants and bankers, craftsmen and apprentices, reaching a peak population of more than 200,000 in the early 18th century. But as the suburbs grew, the old City became a specialised 'central business district' of offices and banks and its residential population declined. Railway building in the second half of the 19th century eradicated numerous workshops and homes, as did the bombing raids and fires of World War II. The City's population dwindled to 4,000 in 1971 but climbed again to 5,300 as apartments were built in the fire-bombed warehouse district of the Barbican.

Just as the City and Westminster acquired increasingly specialised functions during the past two centuries, so did other sections of the expanding capital. Downstream from the City, especially on the north bank, enclosed docks of ever increasing dimensions were dug from the soft alluvium of the Thames floodplain in the late 18th and throughout the 19th centuries. The new enclosed docks captured a large share of Thames trade, but cargo ships continued to be loaded and unloaded at wharves and jetties along the river. Many docks spawned a range of industries, processing goods as diverse as edible oils, timber, chemicals and metals. Each complex of docks was fringed by housing for its employees. Docklands formed a distinctive but oddly isolated part of London, well connected to the rest of the world but with very few public transport links to the heart of the city. During the 1960s and 1970s the docks closed, and many traditional industries died. Imaginative redevelopment schemes have transformed the upstream docks and the Isle of Dogs, but the Royal Docks downstream present a major challenge not only for the 1990s but for the 21st century.

Victorian London witnessed the powerful emergence of a second 'central business district', with a different set of businesses and offices from those in the City. High quality shops had long served the fasionable clientele of Mayfair, and towards the end of the 19th century were joined by new department stores. The West End was London's 'theatreland' and between the wars the most prestigious cinemas were opened there. Unlike the City, where life was geared almost exclusively to the pursuit of commerce, the West End was home to a wide range of cultural activities which included the great museums and galleries as well as many colleges of the University.

Beyond the City, Westminster, the West End and Docklands there developed the largest and in some ways the most diverse component of London – its suburbia. Despised by many Londoners but home to most, the suburbs grew in complexity over the past two centuries. Successive innova-

For a full discussion of the evolution of Westminster and Whitehall, see pages 154-155.

For an examination of the evolution of London's docks see pages 82-83 and 92-93, and for a review of present-day Docklands, see pages 158-159.

See reconstruction of Tower of London and discussion of London's royal parks and palaces, pages 138-139.

The present-day City is examined in detail on pages 152-153.

See discussion of Selfridge's (page 107) and an examination of the evolution of West End shopping districts and the growth of Harrod's on page 149. London's 'theatreland' is examined on page 147.

tions in private and public transport enabled the tide of bricks and mortar to wash further into the London Basin, incorporating villages, hamlets and market towns and engulfing commons, heaths and parks. The term 'urban sprawl' does not do justice to the dramatic rise of Greater London. Every omnibus route, railway line or tramway had to be planned. Each portion of land had to be purchased or leased and provided with an appropriate selection of housing, shops and local facilities. The net result is an intricate fabric of old and new. Historic parish churches, Georgian and Victorian houses fringe ancient village greens. Main roads twist and turn in obedience to long-lost field patterns. Shop fronts stand along building lines defined by tram lines that have long since disappeared, or push forward where suburban front gardens have been built over to provide more retail space.

London's urban sprawl and the gradual absorption of villages into the city boundaries are mapped on pages 156-157.

The growth of London from a population of 15,000 in the reign of William the Conqueror to 8,193,000 in 1951 required massive supplies of building material. In early times most dwellings in London were made of local materials, such as timber and thatch. Only ceremonial buildings or monuments of great importance merited costly imports of stone or the manufacture of bricks and tiles. The Great Fire of 1666 brought this tradition to an end. The new London became a tidy, brick-built town whose houses were roofed with tiles or slates to withstand fire. Brick continued to be the most widespread medium of construction until it was overtaken by concrete in the present century. The opening of railways enabled bricks to be hauled from the great clayfields of the East Midlands, and London no longer had to rely on local brickfields. Prestigious buildings still demand materials from other parts of Britain or from distant corners of the globe.

Sources of stone and brickearth are mapped on pages 132-133.

The growing city needed to be equipped with much more than housing and business premises. The bulk of historic London had developed north of the Thames and only modest expansion occurred on the south bank. Old London Bridge had been rebuilt several times, but many more bridges were needed if London was to become a less 'lop-sided' capital. New bridges were built in the middle of the 18th century and these permitted a burst of suburban growth south of the river. During the subsequent two and a half centuries many more bridges have been added and the old ones replaced; two more will be completed soon. Bridges are far more than functional links in the communication system since they also provide a fascinating complement to the sweep of London's river.

For a reconstruction of old London Bridge, see pages 162-163.

Beneath the city streets there exists another London, vital but little known. The brightly coloured Tube map is familiar to Londoners and visitors alike, but few have a clear idea of where the tunnels run. Even less is known about the 'lost rivers', intercepting sewers and local drains which carry waste and prevent flooding. With so many people concentrated in such a small area the pollution of London's environment has been notorious since medieval times. Remarkable progress in waste disposal and water supply was achieved between 1850 and 1950 but not until recent years was the Thames cleaned up and London's air purged of its smoke and dust. Unfortunately, substances emanating from vehicle emissions now pollute London's air and degrade its environment, posing yet another challenge for scientists and legislators.

Sewers are plotted on pages 160-161.

The pollution of London's environment is mapped on pages 134-135, while the underground city is examined on pages 160-161.

Throughout history monarchs, architects and planners devised schemes for 'improving' London with new thoroughfares, triumphal buildings, docks, cemeteries and a host of other projects. Some of their visions became reality but other ideas remained in draft, forming the London that never was. Even now, bold plans are being announced to revitalise several parts of the capital. Their massive cost, social implications and innovative styles make them highly controversial. Time alone will tell whether they will contribute to the London of the future or will join so many yellowing pages of architectural dreams and lost opportunities.

Unrealised visions of London are illustrated on pages 168-169, while contemporary visions of a future London are examined on pages 170-171.

London's present condition can only be fully understood by appreciating its long and complex history. All the issues mentioned above are explored in the final pages of this Atlas. The present chapter illustrates a selection of themes that characterised London in the past and still continue to do so, while the final chapter concentrates on places within the capital.

THE FABRIC OF LONDON'S BUILDINGS

Good building stone is absent from the London area. For much of the city's history, therefore, many ordinary buildings were made from timber and roofed with thatch. Brick fashioned from local clay was used for more solid buildings; only churches and great civic buildings were constructed from stone, which had to be brought into the city from other regions. Disastrous fires were a frequent occurrence, and after the Great Fire (*page* 68) legislation stipulated that brick and stone, tiles and slates should be used in new buildings. The story of London's building stone reflects, primarily, the evolution over the centuries of transport which enabled materials to be brought from ever greater distances. A similar background characterises the history of the use of brick – when local clays became exhausted, increasingly distant clay deposits had to be used.

Rather than try to trace the history of stone use throughout the whole City of London, it is easier to concentrate on the buildings in one particular area such as Trafalgar Square. The oldest is the church of St. Martin in the Fields which was designed by James Gibbs in 1726, when he encased an older structure of Kentish Ragstone and Reigate Stone, brought by boat from the Medway Valley or North Downs, in a classical exterior of Portland Stone. Portland is the best English building stone, and had already been used extensively by Sir Christopher Wren for rebuilding the City after 1666. This pure white limestone is quarried exclusively on Portland Island off the south coast of Dorset, whence it was shipped to London. For the National Gallery (1835), William Wilkins also chose to use Portland Stone. There, however, loss of the original smooth surface of the limestone through the effects of wind and rain has caused fossil shell fragments to stand out. This demonstrates that the limestone was once a sea-bed sediment, rich in broken debris of oysters.

The National Westminster Bank in the south-west corner of the square was designed in 1871

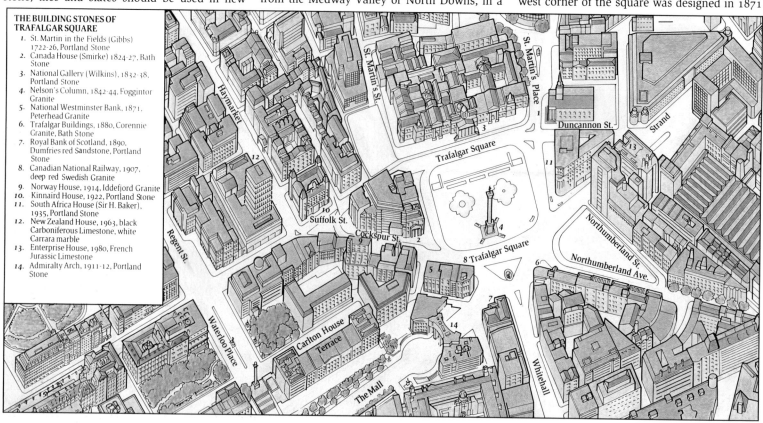

THE BUILDING STONES OF TRAFALGAR SQUARE

1. St. Martin in the Fields (Gibbs) 1722-26, Portland Stone
2. Canada House (Smirke) 1824-27, Bath Stone
3. National Gallery (Wilkins), 1832-38, Portland Stone
4. Nelson's Column, 1842-44, Foggintor Granite
5. National Westminster Bank, 1871, Peterhead Granite
6. Trafalgar Buildings, 1880, Corennie Granite, Bath Stone
7. Royal Bank of Scotland, 1890, Dumfries red Sandstone, Portland Stone
8. Canadian National Railway, 1907, deep red Swedish Granite
9. Norway House, 1914, Iddefjord Granite
10. Kinnaird House, 1922, Portland Stone
11. South Africa House (Sir H. Baker), 1935, Portland Stone
12. New Zealand House, 1963, black Carboniferous Limestone, white Carrara marble
13. Enterprise House, 1980, French Jurassic Limestone
14. Admiralty Arch, 1911-12, Portland Stone

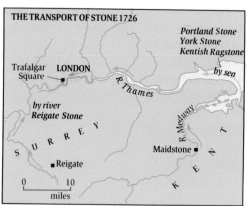

THE TRANSPORT OF STONE 1726

Portland Stone
York Stone
Kentish Ragstone

Trafalgar Square **LONDON**

by sea

by river
Reigate Stone

R. Thames

S U R R E Y

R. Medway

K E N T

Maidstone

Reigate

0   10
miles

The transport of stone *The three small maps (left and below) show the routes that were used to bring various stones to Trafalgar Square: in the early 18th century, before a variety of transport was available; in the 1780s, when the building of canals meant that more distant sources of stone could be used; and in the mid-19th century, when the railway carried stone from far more distant sources to London. Nelson's Column, for example, erected in 1842, was built of several different granites from as far afield as Cornwall and Scotland.*

The building stone of Trafalgar Square *The map (above) shows the location of selected buildings around Trafalgar Square, and the varied stones in which they are built. The modern block of New Zealand House (left), built of Carboniferous Limestone, contrasts with the National Westminster Bank (1877, right), built of Portland Stone, with Peterhead Granite columns. The fine facade of the National Gallery (1835, below) is constructed of Portland Stone, favoured by Sir Christopher Wren.*

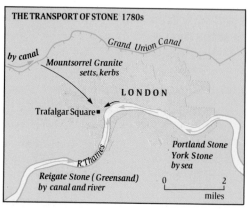

THE TRANSPORT OF STONE 1780s

by canal

Grand Union Canal

Mountsorrel Granite
setts, kerbs

LONDON

Trafalgar Square

R. Thames

Portland Stone
York Stone
by sea

Reigate Stone (Greensand)
by canal and river

0   2
miles

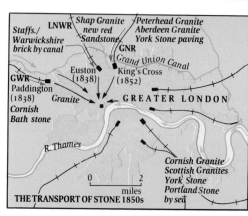

THE TRANSPORT OF STONE 1850s

Staffs./ Warwickshire brick by canal

LNWR Shap Granite new red Sandstone

Peterhead Granite Aberdeen Granite York Stone paving

GNR

Grand Union Canal

Euston (1838)

King's Cross (1852)

GWR Paddington (1838)

Granite

GREATER LONDON

Cornish Bath stone

R. Thames

Cornish Granite Scottish Granites York Stone Portland Stone by sea

0   2
miles

by F.W. Porter, the architect of branches of the original National Provincial Bank. White Portland Stone walling is set off by distinctive pink columns of polished Peterhead Granite from coastal Aberdeenshire. The steep-pitched roofing carries green Lake District Slate from quarries in the Borrowdale Volcanic Series, which is an ancient accumulation of volcanic ash. The Edwardian building, bearing the initials 'C.N.' (Canadian National Railway), is built of deep red Swedish Granite and has interiors of Italian Marble. In the background, rises New Zealand House (1963), a tower mainly of steel and glass. The basal stalk to the building, however, is faced with dense black natural stone. Rare fossils (white against the black surface) prove that this Carboniferous Limestone is from either Tournai in Belgium or from southern Ireland. It is evident from this limited survey that the range of stone being used in this prestigious London square became much greater and more exotic as new and more effective modes of transport – roads, canals, railways, shipping – became available.

Brick had been used for buildings in London since Roman times but after the Great Fire it entered into renewed favour. Sir Christopher Wren claimed: '... the earth around London, rightly managed, will yield as good brick as were the Roman bricks ... and will endure, in our air, beyond any stone our island affords'. Throughout the 17th and 18th centuries, brickearth and London Clay were used to produce 'stock bricks' which varied in colour from bright yellow to deep earthy purple, and were used as good facing bricks. They contrasted with cheaper 'place bricks' which were not intended to be left exposed to the elements. Some stock bricks were carried up the Thames from Kent but most were made in the immediate environs of London.

Natural brickearth contains lime and can yield a yellow or whitish brick, unlike other clays whose iron oxide gives rise to red bricks unless lime or chalk is added. By the early 19th century, local supplies of pure brickearth were almost expended. Builders relied increasingly on a mix of chalk, London Clay and a little brickearth.

As well as suburban brickfields, which extended ever-further into the countryside, more distant sources of clay were used in Kent, Essex and the Upper Thames Valley. The bricks they yielded were carried to London along the Thames or along the new canal system. The opening of the Grand Union Canal through Middlesex in 1794 was soon followed by a proliferation of brickfields along its banks. In the 1830s and 1840s, brickmaking by hand started to be replaced by machinery which meant that it was possible to use the harder clays of the Midlands. The building of the Great Northern Railway in 1852 enabled the clays of Bedfordshire (some 35

Building in Brick Gray's Inn (1676, right) was built using 'stock bricks', a mixture of brickearth and London Clay, ranging in colour from dark red to the deep purple of Bedford Square (below right). The Victorian houses (1880s, below) were built of brick made up of a mixture of chalk, clay and a little brickearth, a practice known as 'soiling'.

miles north of the capital) to be used in steam-powered mechanised brickworks. In 1881, a grey-brown clay deposit, very suitable for brickmaking, was discovered below the brownish top layer of the Oxford Clay at Fletton near Peterborough. This became the new focus of mechanised brickworks. Even more were opened in the 1890s alongside railway lines in Bedfordshire and Buckinghamshire. Despite the great reliance in recent decades on concrete, brick is still popular with many of London's architects and builders at the end of the 20th century.

London Brick The maps (right) show sources of Brickearth and other clays that were used for making London's bricks from medieval times to the 18th century. In 1811, Henry Hunter described London as being surrounded by 'a ring of fire' from brick kilns. For him '... the face of the land is deformed by the multitude of claypits from which is being dug the brickearth fused in the kilns which smoke all around London'. By the early 19th

LONDON BRICK 14th–19th CENTURY

brickearth	terraces	—— extent of city in 1800
general brickearth cover	alluvium	— extent of city in 1900
London clay	— Medieval walled area	■ stock brickmaking centre
		→ transport of brick

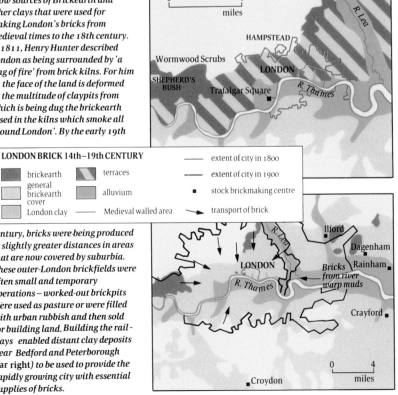

century, bricks were being produced at slightly greater distances in areas that are now covered by suburbia. These outer-London brickfields were often small and temporary operations – worked-out brickpits were used as pasture or were filled with urban rubbish and then sold for building land. Building the railways enabled distant clay deposits near Bedford and Peterborough (far right) to be used to provide the rapidly growing city with essential supplies of bricks.

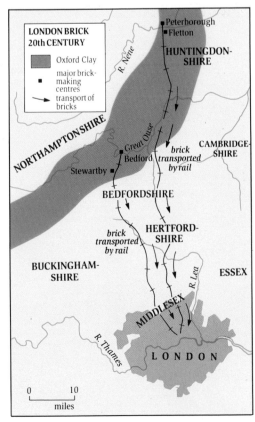

LONDON BRICK 20th CENTURY

- Oxford Clay
- ■ major brick-making centres
- → transport of bricks

LONDON AND POLLUTION

The demographic growth and economic success of London was at the cost of profound damage to the environment. Concern for high concentrations of smoke can be traced back to the 13th century when a commission was established to investigate the problem and recommended that burning of sea coal be prohibited. This was not implemented and in the 17th and 18th centuries there was growing awareness of the adverse effects of smoke on the health of the citizens of London. King James I complained about the soiling of St. Paul's Cathedral in 1620; John Evelyn in 'Fumifugium' (1661) noted decreased visibility; in 1784 the naturalist Gilbert White recorded dark plumes rising from London.

In late Victorian times the word 'smog' was coined to describe the combination of smoke and fog evoked so graphically in the novels of Dickens. The Smoke Abatement Act (1853-56) and the Sanitation Act (1866) had attempted to curb smoke emissions but there was little improvement in conditions until the Clean Air Acts of 1956 and 1968. There were also important changes in energy sources. Industry and commerce switched largely from coal to oil and gas; the railways were powered by electricity rather than steam; domestic central heating using oil, gas and electricity replaced open coal fires. Finally, slum clearance and urban renewal

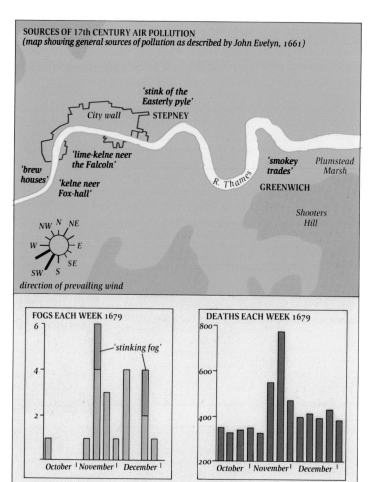

SOURCES OF 17th CENTURY AIR POLLUTION
(map showing general sources of pollution as described by John Evelyn, 1661)

'stink of the Easterly pyle'
City wall   STEPNEY
'lime-kelne neer the Falcoln'
'brew houses'
'kelne neer Fox-hall'
'smokey trades'   Plumstead Marsh
R. Thames   GREENWICH
Shooters Hill

NW N NE
W   E
SW S SE
direction of prevailing wind

FOGS EACH WEEK 1679
'stinking fog'
October   November   December

DEATHS EACH WEEK 1679
October   November   December

London's air pollution, 17th-19th centuries *The sources of air pollution described by John Evelyn in 'Fumifugium' (1661) are shown in the map (left). The city was covered with a 'hellish and dismal cloud of sea-coal' emanating from the premises of brewers, dyers, lime burners, salt boilers and soap makers. Pollution from industries to the west of the city was blown across London by the prevailing south-westerly winds. Smoke damaged plants, buildings, clothes, furnishings and paintings, and human health (graphs left) was impaired. Evelyn advocated the removal of noxious industries 5 or 6 miles east of the city to the Greenwich peninsula, where winds would sweep pollution far downstream. The map (below left) shows the amount of sulphate deposited in rain around London in 1869-70. Highest levels of pollution were found in densely-populated poorer residential districts fringing the City and in the East End, with concentrations extending east of London and into the lower Lea Valley. Some of the atmosphere of Victorian smog is captured in H. Medleycott's painting, 'The Pool of London', c.1880 (bottom). A Punch cartoon of 1905, (below left) illustrated the difficulty of finding one's way round streets where atmospheric conditions led to negligible visibility.*

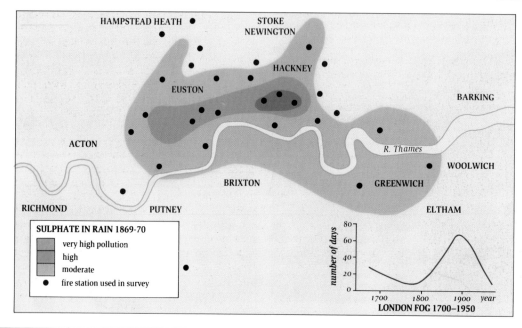

HAMPSTEAD HEATH   STOKE NEWINGTON
HACKNEY
EUSTON   BARKING
ACTON
R. Thames   WOOLWICH
BRIXTON   GREENWICH
RICHMOND   PUTNEY   ELTHAM

SULPHATE IN RAIN 1869-70
- very high pollution
- high
- moderate
• fire station used in survey

number of days
80
60
40
20
0
1700   1800   1900   year
LONDON FOG 1700–1950

A QUALIFIED GUIDE.—*Befogged Pedestrian.* "Could you direct me to the river, please?" *Hatless and Dripping Stranger.* "Straight ahead. I've just come from it."

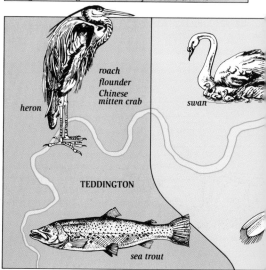

heron
roach
flounder
Chinese
mitten crab
swan
TEDDINGTON
sea trout

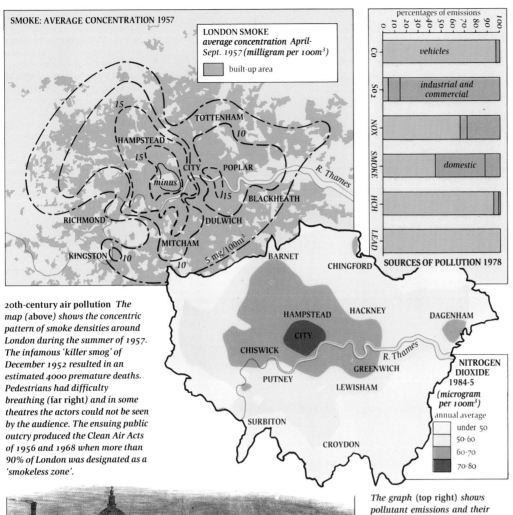

SMOKE: AVERAGE CONCENTRATION 1957

LONDON SMOKE
average concentration April-Sept. 1957 (milligram per 100m³)

built-up area

TOTTENHAM
HAMPSTEAD
CITY   POPLAR
minus
R. Thames
BLACKHEATH
RICHMOND
DULWICH
MITCHAM
KINGSTON   10   10   5 mg/100m³
BARNET

15   10   15   15

SOURCES OF POLLUTION 1978

percentages of emissions
0 10 20 30 40 50 60 70 80 90 100

Co	*vehicles*
So₂	*industrial and commercial*
NOX	
SMOKE	*domestic*
HCH	
LEAD	

20th-century air pollution *The map (above) shows the concentric pattern of smoke densities around London during the summer of 1957. The infamous 'killer smog' of December 1952 resulted in an estimated 4000 premature deaths. Pedestrians had difficulty breathing (far right) and in some theatres the actors could not be seen by the audience. The ensuing public outcry produced the Clean Air Acts of 1956 and 1968 when more than 90% of London was designated as a 'smokeless zone'.*

CHINGFORD
HAMPSTEAD   HACKNEY   DAGENHAM
CITY
CHISWICK   R. Thames
GREENWICH
PUTNEY   LEWISHAM
SURBITON
CROYDON

NITROGEN DIOXIDE 1984-5
(microgram per 100m³)
annual average

under 50
50-60
60-70
70-80

The graph (top right) shows pollutant emissions and their sources in London in 1978. The map (above) shows the pattern of concentration of nitrogen dioxide, an atmospheric gas derived mainly from vehicle exhausts, which contributes to problems experienced by asthma sufferers.

Cleaning the Thames *The cartoon (left) appeared in Punch on 10th July 1858, graphically expressing widespread disquiet about the pollution of the Thames in the mid-1850s. The Thames has undergone an extensive clean-up over the past 30 years. The map (below) shows some of the birds and fish that have since returned to various sections of the Thames: after a century's absence the brown shrimp has returned; numbers of salmon, heron, swan and pochard have risen and shelduck now over-winter in their hundreds in Barking Creek.*

did away with thousands of sources of pollution as densely-packed houses were replaced by centrally-heated flats and spacious suburbs.

London has resolved the problem of smoke pollution but is now faced with a less visible but potentially more dangerous mix of pollutants resulting from vehicular emissions. Carbon monoxide is now the most common air pollutant in London. Vehicles also emit hydrocarbons and the action of sunlight on this mix of emissions can produce toxic secondary pollutants (for example, ozone, aldehydes and various aerosols) known as 'photochemical' smog. Nitrogen dioxide contributes to this smog and acid rain.

The Thames and its tributaries were also fouled by the growth of the city and its industries. Conditions became so bad, and the stench from the polluted river so great, that in 1858 Members of Parliament resolved to strengthen the powers of the newly created Metropolitan Board of Works (*page 104*). By 1865 80 miles of intercepting sewers had been installed, carrying still untreated sewage to outfalls on the Thames, well downstream of the built-up area. During the 20th century many more sewers were constructed and treatment works opened, but the condition of the Thames continued to decline, reaching its worst point in the 1950s. Since then there has been major investment in sophisticated treatment processes. A good indication of recent improvements in water quality is in the fish life in the Thames. In 1957 only eels could survive in London's river. Seventeen years later no fewer than 82 species of fish were living in the London stretches of the Thames, including the first salmon since 1833.

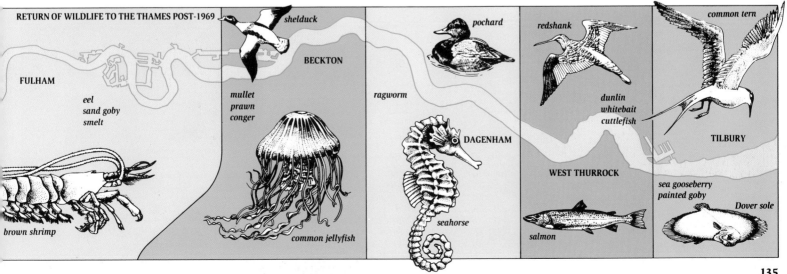

RETURN OF WILDLIFE TO THE THAMES POST-1969

shelduck
BECKTON
pochard
redshank
common tern

FULHAM
eel
sand goby
smelt

mullet
prawn
conger

ragworm

DAGENHAM

dunlin
whitebait
cuttlefish

TILBURY

WEST THURROCK

sea gooseberry
painted goby

Dover sole

brown shrimp

common jellyfish

seahorse

salmon

LONDON AND IMMIGRATION

London has played a dominant role in Britain's long history of immigration, serving as a major port of entry, a place of economic opportunity and a hoped-for refuge from persecution. In medieval times, some foreigners lived in separate neighbourhoods, notably Jewish traders (who had come to England with the Normans but were expelled to the Continent in 1290) and Hanseatic merchants from northern Germany, who occupied the Steelyard alongside the Thames until they too were expelled in 1598. By contrast, Flemings, Dutch, French and Walloons were assimilated far more easily into the indigenous population. In the 1650s, Cromwell allowed persecuted Jews to return to England and Sephardic Jews settled in Whitechapel. They were followed by members of Ashkenazi Jewish communities from central Europe who brought their trades and skills to districts beyond the eastern limits of the City, where they could operate without interference from the guilds and City companies. Huguenots (Protestants) fled from France when religious toleration was ended by the Revocation of the Edict of Nantes in 1685. They moved to two main districts, setting up silk-working establishments in Spitalfields and opening craft workshops in Soho, which became a 'little France'.

Many immigrants came to London because their native countries formed part of the rapidly-expanding British Empire. By the late 18th century small numbers of Chinese had settled near the docks, having arrived on East India Company

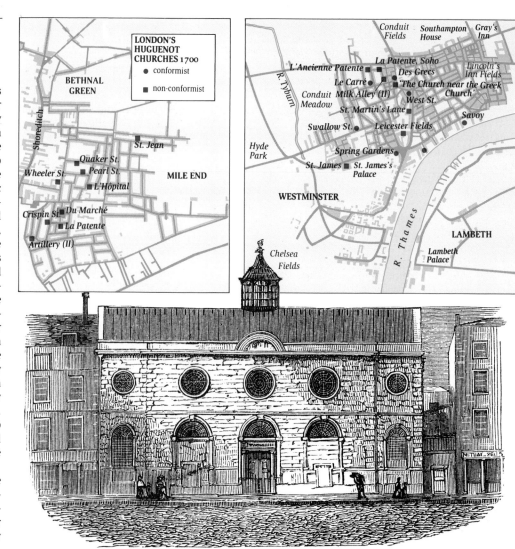

London and the Huguenots *The two maps (above) show the distribution of Huguenot churches in Spitalfields and the West End in the early 18th century. A Huguenot church at Threadneedle Street (1669) is depicted (above) from the* **London Illustrated News.**

Jewish London *The map (left) shows the Jewish East End in 1900. Some streets had more than 90% Jewish inhabitants. The map (below left) shows Jewish dispersal from the East End. As they grew more prosperous during the 19th century, the Jewish population began to move to suburban areas, especially to the north and north-west. The photograph (below) shows Jewish immigrants arriving at London Docks in the late 1930s.*

ships. Indians, many of whom were brought to London as servants of colonial families, also began to arrive in London at this time. Similarly, returning traders, plantation owners, army officers and government officials brought Afro-Caribbean migrants to London as slave-servants prior to the abolition of slavery in Britain in 1807, and in the British Empire in 1834.

Immigration in the 19th century was due to a number of new factors. The Great Irish Famine of the 1840s unleashed a new flood of immigrants, and by 1851 over 100,000 Irish were living in the capital, making up one in twenty of all Londoners. The rookeries of St. Giles, Whitechapel and Southwark contained particularly large numbers, many of whom lived in appalling poverty. Many worked in the docks, or in the workshops and factories of the East End, where sweated labour

THE JEWISH EAST END 1900
Jews as proportion of total population:

| less than 5% | 5-24% | 25-49% | 50-74% | 75-94% | 95-100% |

AT THE DOCKS: ARRIVALS BY A GERMAN STEAMER.

was firmly entrenched.

In 1881, the assassination of Tsar Alexander II gave rise to anti-semitic pogroms in Russia and Poland. Vast numbers of Jews fled westward, hoping to move to North America. Many settled instead in the East End and other parts of London, where they played a major role in the clothing and tailoring trades and in many branches of commerce and finance. 30,000 Germans lived in London before the First World War – the German businesses and restaurants in Charlotte Street earned it the nickname 'Charlottenstrasse'. When war broke out, many returned to the fatherland, while remaining Germans of military age were interned (a fate which also befell the German population at the outbreak of the Second World War). Numerous German families in London adopted English-sounding surnames.

The receptive role of the East End has continued in recent years with many thousands of migrants from the West Indies and the Indian subcontinent settling there in the 1950s and 1960s. Other recent migrant groups have contributed to the complex mosaic of modern London; the West End has its Chinatown, while Greek and Turkish Cypriots settled in North London in sizeable numbers during the 1960s and 1970s. The capital's ethnic and racial diversity contributes a richness and vitality to its restaurants, entertainment, religion, music, cultural and commercial life. Nevertheless, every major immigrant group has faced hostility and discrimination on arrival in London. Jewish and Irish migrants experienced persistent religious intolerance, while Asians have faced a barrage of racist taunts and physical violence. Racial harassment is widespread, but at its most severe in deprived East End boroughs such as Tower Hamlets.

Irish London *The map (above right) shows the distribution of Irish-born residents in London in 1851 – very soon after the Great Potato Famine, which had driven large numbers of Irish people to seek a better life elsewhere. High concentrations of Irish residents can be seen in areas such as St Giles and Whitechapel. Here 'rookeries' (overcrowded tenements) provided meagre accommodation for the new arrivals.*

Commonwealth immigration *The post-war economic boom created an enormous demand for labour, and from the early 1950s Commonwealth immigrants began to flood into the city seeking work. This immigration was to have a profound impact on the social geography of the capital (maps right). Once 'pioneer' immigrants had become established in certain areas, friends and families arriving later tended to cluster together, forming clearly demarcated communities, such as Jamaican Brixton and Punjabi Southall. The photos show: the Whitechapel Mosque; the recently completed Hindu temple at Neasden; new arrivals to London in the 1950s; the Fournier Street Mosque (formerly a Huguenot church and a Jewish synagogue).*

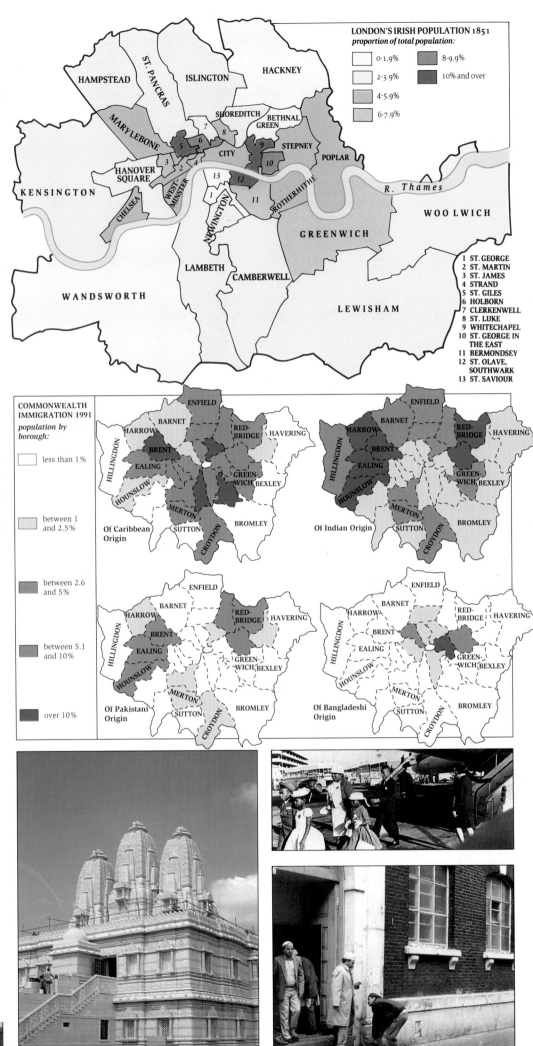

LONDON'S IRISH POPULATION 1851 *proportion of total population:*

- 0-1.9%
- 2-3.9%
- 4-5.9%
- 6-7.9%
- 8-9.9%
- 10% and over

1 ST. GEORGE
2 ST. MARTIN
3 ST. JAMES
4 STRAND
5 ST. GILES
6 HOLBORN
7 CLERKENWELL
8 ST. LUKE
9 WHITECHAPEL
10 ST. GEORGE IN THE EAST
11 BERMONDSEY
12 ST. OLAVE, SOUTHWARK
13 ST. SAVIOUR

COMMONWEALTH IMMIGRATION 1991 *population by borough:*

- less than 1%
- between 1 and 2.5%
- between 2.6 and 5%
- between 5.1 and 10%
- over 10%

Of Caribbean Origin

Of Indian Origin

Of Pakistani Origin

Of Bangladeshi Origin

137

ROYAL LONDON

For most people 'Royal London' conjures up thoughts of palaces, parks and pageantry. It is, of course, much more than that. The unique structure of London owes its origin to the decision of the Saxon king, Edward the Confessor, to locate his main residence upstream of the ancient city, selecting the site where the church of St. Peter – soon to be known as Westminster Abbey – was being built (*page 42*). Edward died before his projects were complete but his choice of Westminster was accepted by many later sovereigns. Courtiers and servants settled nearby, ensuring that Westminster and its environs would be endowed with palaces, parks and the highest functions of government and adminstration in the land. For two centuries following 1650 the Crown and members of the nobility developed their estates on the western side of London, gracing them with splendid streets, fine houses and elegant squares (*page 75*). For virtually a thousand years successive kings and queens have lived in London, often favouring Westminster but sometimes preferring palaces located some distance from the heart of London. Many were built adjacent to the Thames which provided a beautiful setting and guaranteed the most rapid and efficient means of transport during many centuries when road conditions were poor.

These royal homes vary in size and sophistication, from the austere stone of the Tower of London, through the fine Tudor brickwork of St. James's and Hampton Court, to the elegant splendour of Buckingham Palace. Several originated as mansions built for courtiers, passing into royal hands by purchase, gift or seizure at a later stage. All have undergone extensive remodelling and enlargement as fashions changed and standards of space and comfort were transformed. Much the same is true of the royal parks and the private gardens which surround the capital's royal houses. Many began as ancient hunting grounds or were acquired as Church lands or were confiscated by Henry VIII in the 1540s. The imagination of landscape designers, combined with the craft of generations of gardeners, has endowed London with a unique legacy of 5,700 acres of lawns, trees and lakes which act as a haven for many species of birds.

Eleven miles south-west of Westminster – but much further along the meandering Thames – Hampton Court epitomises all that is best about Royal London. Its pure air, woodlands and proximity to the river made it the favourite country home for successive generations of English monarchs. Work on the palace, which was designed to be the grandest in Europe, was started in 1514 by Cardinal Wolsey, Lord Chancellor to Henry VIII. Fifteen years later Wolsey presented it to his King in a futile attempt to remain in favour. Henry greatly loved the palace and had it enlarged. At the request of Charles II and William and Mary its gardens were redesigned, with three broad avenues leading away from the palace and the Long Water extending for three quarters of a mile through the deer park almost to the Thames. Sir Christopher Wren rebuilt the eastern section of the palace in neo-classical style but its brick-built western part remains largely unchanged.

Buckingham Palace has been the London home of the monarchy since Victoria became Queen in 1837. It originated in 1703 as a brick mansion for the Duke of Buckingham, (depicted left in an aquatint of 1819 by W. Westall) and was purchased by George III in 1762. In 1825 George IV engaged John Nash to make substantial alterations and over the next 12 years a new palace of Bath Stone was constructed around the old house. The east wing was built in 1847 and was heightened and faced in Portland Stone in 1912-13 to harmonise with the Victoria Memorial (1910). The present building has some 600 rooms (photograph below); the royal apartments are in the north wing. When the monarch is in residence the royal standard flies over the building and the Changing of the Guard takes place in the forecourt at 11.30 each morning.

London's royal parks Varying greatly in size and appearance, royal palaces and parks grace central London and some of its more fortunate suburbs (map right). Carefully manicured lawns at Green Park (53 acres) and flowerbeds in St. James (93 acres) contrast with plantations and vast expanses of greensward in Richmond Park (2,358 acres). This originated as a hunting preserve enclosed by Charles I in 1637 and is still the home for herds of deer. The oldest royal park in London is at Greenwich and was enclosed in 1433. Almost five centuries later, leases on farmland at Marylebone Fields reverted to the Crown in 1811 and the combined genius of John Nash and the Prince Regent created Regent's Park (472 acres) (page 76).

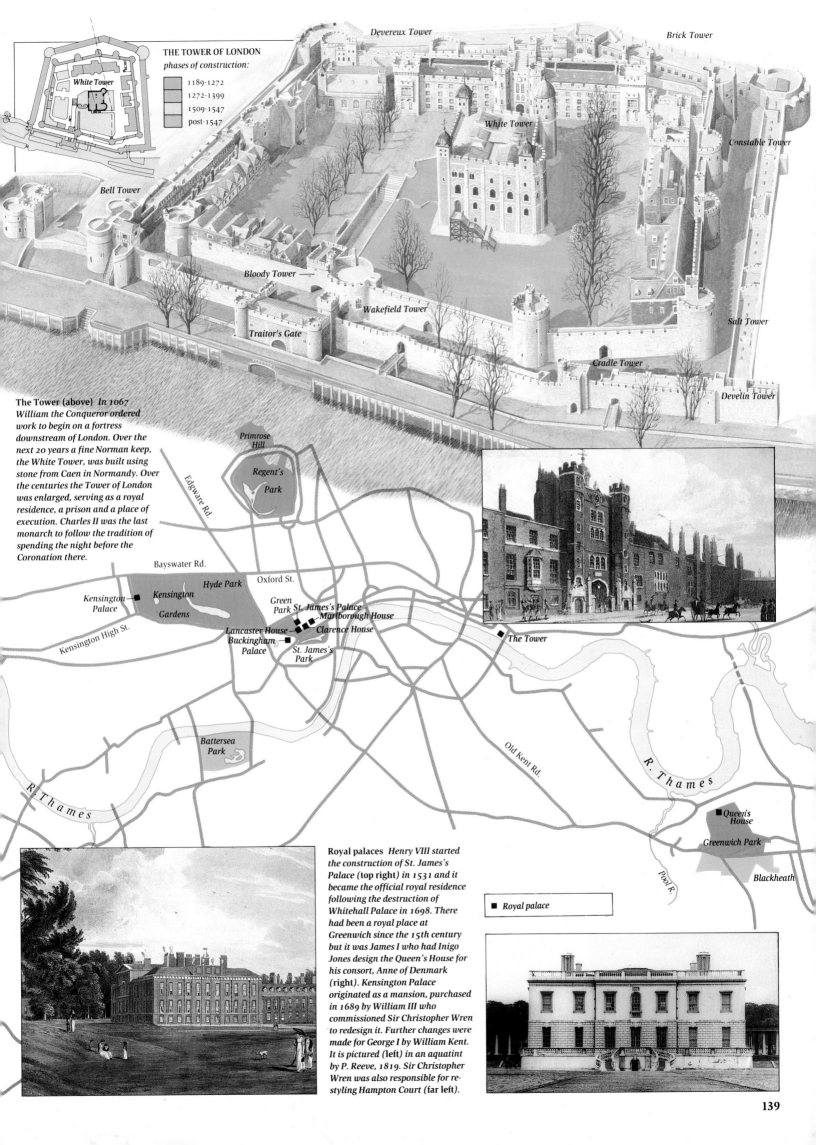

THE TOWER OF LONDON

phases of construction:

- 1189-1272
- 1272-1399
- 1509-1547
- post-1547

White Tower

Devereux Tower

Brick Tower

White Tower

Constable Tower

Bell Tower

Bloody Tower

Wakefield Tower

Salt Tower

Traitor's Gate

Cradle Tower

Develin Tower

The Tower (above) *In 1067 William the Conqueror ordered work to begin on a fortress downstream of London. Over the next 20 years a fine Norman keep, the White Tower, was built using stone from Caen in Normandy. Over the centuries the Tower of London was enlarged, serving as a royal residence, a prison and a place of execution. Charles II was the last monarch to follow the tradition of spending the night before the Coronation there.*

Primrose Hill

Regent's Park

Edgware Rd.

Bayswater Rd.

Oxford St.

Hyde Park

Kensington Palace

Kensington Gardens

Kensington High St.

Green Park

St. James's Palace

Marlborough House

Clarence House

Lancaster House

Buckingham Palace

St. James's Park

The Tower

Battersea Park

R. Thames

R. Thames

Old Kent Rd.

Pool R.

Queen's House

Greenwich Park

Blackheath

■ Royal palace

Royal palaces *Henry VIII started the construction of St. James's Palace (top right) in 1531 and it became the official royal residence following the destruction of Whitehall Palace in 1698. There had been a royal place at Greenwich since the 15th century but it was James I who had Inigo Jones design the Queen's House for his consort, Anne of Denmark (right). Kensington Palace originated as a mansion, purchased in 1689 by William III who commissioned Sir Christopher Wren to redesign it. Further changes were made for George I by William Kent. It is pictured (left) in an aquatint by P. Reeve, 1819. Sir Christopher Wren was also responsible for re-styling Hampton Court (far left).*

139

GREEN LONDON

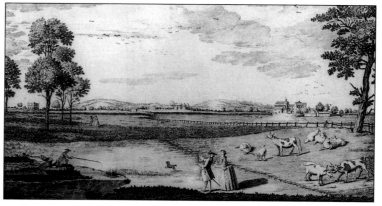

Londoners enjoy a great wealth of public open spaces, ranging from historic royal parks, ancient commonlands and the grounds of great mansions, to municipal parks created in Victorian times or in the present century. Many of London's squares are blessed with public gardens and numerous urban churchyards have been converted into green havens. Greater London has no fewer than 1,700 public open spaces greater than one acre in size, covering in all 67 square miles. Many are much smaller: even the densely built-up City of London has 400 green spaces.

Much of the land now occupied by the great royal parks was acquired from the Church in the 1540s after the Dissolution of the Monasteries. Tudor monarchs kept these areas on the outskirts of London for hunting and other private pleasures, but under the Stuarts and the Prince Regent they were laid out as parks and opened for public enjoyment – central London's largest green space, Hyde Park, originally one of Henry VIII's deer parks, was opened to the public in the early 17th century. The large flocks of deer still in evidence at Richmond Park reveal its antecedents as a royal hunting ground. As the capital grew, so did the need for parks in other districts. Despite various campaigns, it was not until 1842 that funds were granted from the Crown for London's first real public park to be created. This was Victoria Park in Hackney, soon to be followed by Battersea Park (1858) and, later in the century, by Finsbury and Southwark Parks (both opened in 1867). These parks were laid out replete with sweeping drives, ornamental lakes, bandstands and pavilions. Many of them were densely planted – reflecting the Victorian passion for horticulture and botany – places of education and instruction as well as pleasure.

Many areas of commonland around London had been enclosed for building or farming during the 18th and 19th centuries, but those that remained were protected under the Metropolitan Commons Act (1866). In this way, Hampstead Heath and Wimbledon Common were secured for public access. The Corporation of the City of London was also empowered to acquire and conserve land up to 25 miles away from the city for public recreation and enjoyment; Epping Forest (1878) and Highgate Woods (1885) are notable examples.

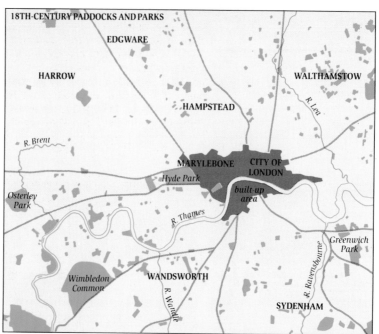

18TH-CENTURY PADDOCKS AND PARKS

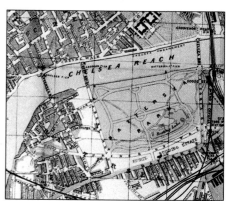

London's parks *The fields at Marylebone (pictured by Chatelin, left), just beyond the 18th-century built-up area, provided opportunities for Londoners to enjoy country air. In 1800 Thomas Milne mapped and recorded numerous stretches of private parkland, many of which were to be replaced by suburban housing (map left).*

By the early 16th century, the royal park of St. James was surrounded by bricks and mortar. Henry VIII had the area drained and formal gardens were laid out for James I. The park was redesigned after the fashion of Le Nôtre, with two avenues and a great linear lake. It is pictured (below) in an engraving of 1794 by Canaleti Delin. In the 1820s the park was improved by John Nash and the lake remodelled.

Battersea Park formed one of the splendours of Victorian urban improvement. Until the mid 16th century this low-lying land had been flooded by the Thames, but parts were later farmed, as John Rocque showed in 1745 (bottom left). Following legislation in 1846 an embankment was built and the ground raised by earth excavated from the Victoria Dock. Reynold's map of 1883 (bottom right) shows the Park's plantations, avenues and lake which proved fashionable for skating. (The 'Ladies' Mile', 1867, is depicted centre.*) The poster (top right), designed by Edward Bawden in 1936, was one of many encouraging Londoners to visit their city's parks.*

Many landscaped parks on private estates were converted into municipal parks and thereby saved from the suburban tide. Even now new parks are being laid out, such as the ambitious Burgess Park along the line of the old Surrey Canal in Southwark. However, public open spaces as shown on map (*below*) can only form part of the picture of green London, since the city also boasts hundreds of thousands of private gardens which yield flowers, fruit and vegetables.

As well as Greater London's impressive legacy of public open spaces and private gardens, there are also numerous sportsfields, reservoirs, sewage works, railway embankments and areas of derelict land. All of these areas support wildlife and most are valuable land for nature conservation. Abandoned wharves in Docklands, for example, have been colonised by plants and animals and now form important wildlife habitats in parts of the city that otherwise lack open spaces. Over 100 sites which contain especially varied or rare plants and animals have been identified, and three dozen of these have been designated as Sites of Special Scientific Interest. There are also five Local Nature Reserves.

The city centre, not surprisingly, is poor in wildlife compared with outer locations, but as a result of increased development and aggressive mechanised farming techniques in the countryside, birds of prey, grass snakes and foxes are being driven into the suburbs, attracted by the plentiful supplies of food. Aquatic habitats are important for many birds, with reservoirs, gravel pits and ornamental lakes supporting large populations of resident and migratory birds. Good examples are found in the Colne and Lea valleys, the royal parks and alongside the Thames in west London (*page 135*). Habitat management affects the number of plants and animals found in various parts of the capital. Sports pitches and some amenity grasslands are managed intensively and do not support much wildlife.

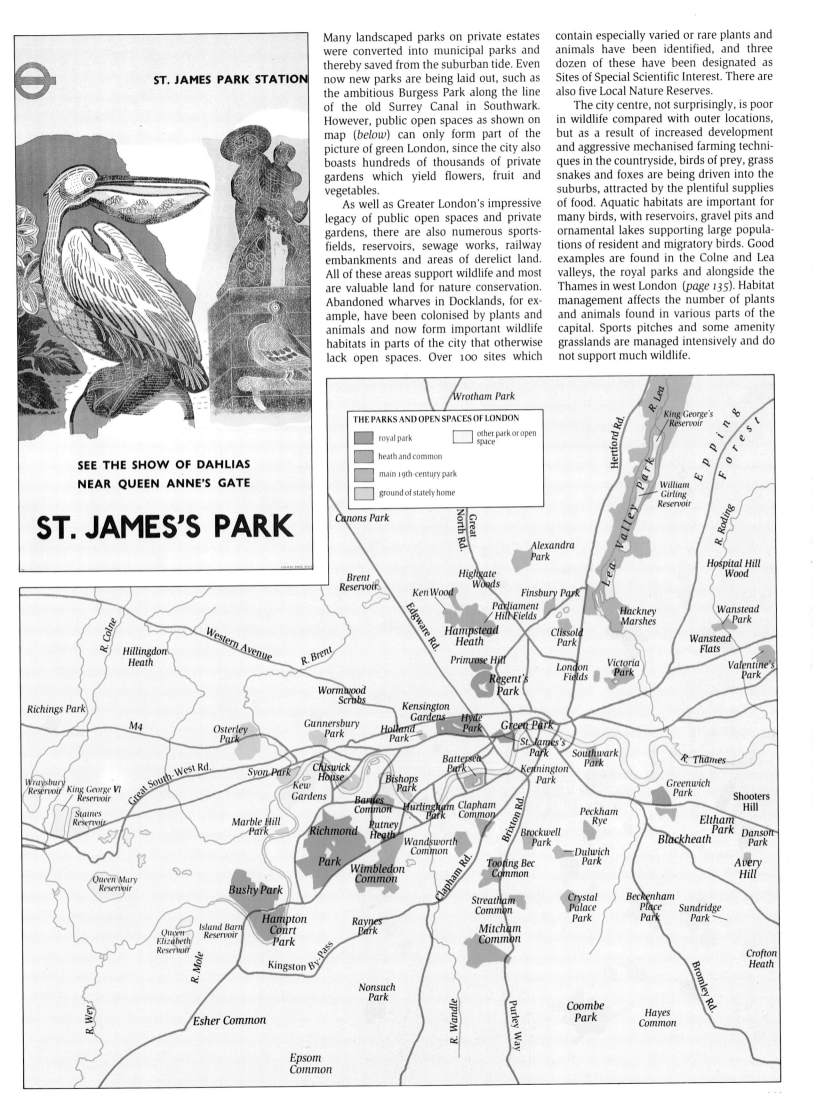

THE PARKS AND OPEN SPACES OF LONDON

- royal park
- heath and common
- main 19th-century park
- ground of stately home
- other park or open space

LONDON AT LEISURE

Medieval Londoners enjoyed varied entertainment: ball games; wrestling; shooting with bow and arrow; swimming; and skating on bone skates on the Thames in cold winters. Mystery plays, pageants and fairs provided more organised entertainment. Cock-fighting and bull- and bear-baiting were certainly known in Tudor times and may have been introduced in the 13th century. They flourished beyond the city limits, especially on the south bank. Animal fighting was not made illegal until 1835.

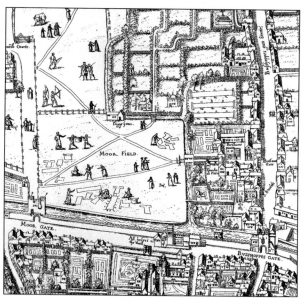

London's Elizabethan theatres *(page 146)* were sited east of the City and in Southwark, which was notorious for its brothels. Ordinary Londoners, not admitted to the Royal Parks until the 17th century, enjoyed visiting the pleasure grounds fringing the built-up areas. Spring Gardens, on the eastern edge of St James's Park, was an early example, but by 1630 scandalous activities were reported along its shady paths. Puritan zeal curtailed

Archery in Moorfields is depicted on the 'Copperplate Map' of 1559 (left), the oldest surviving map of London. Archery was encouraged by law – it supplied trained bowmen in times of war.

Frost fairs and pleasure gardens During exceptionally harsh winters the Thames froze over and frost fairs were held on the ice, with booths, dancing and ox-roasting. Freezing no longer occurred after 1831, when old London Bridge was demolished and river flow accelerated. A frost fair of 1683-4 is depicted in a painting by Hondius (above left). Alongside the Thames, Vauxhall Gardens in Lambeth formed one of the best known pleasure gardens to fringe London in the 18th and 19th centuries (map below). For over a hundred years, until the gardens closed in 1859, Vauxhall supplied music, suppers and entertainment. The Grand Walk at Vauxhall is shown (centre left) in an engraving by Wade and Muller, 1759.

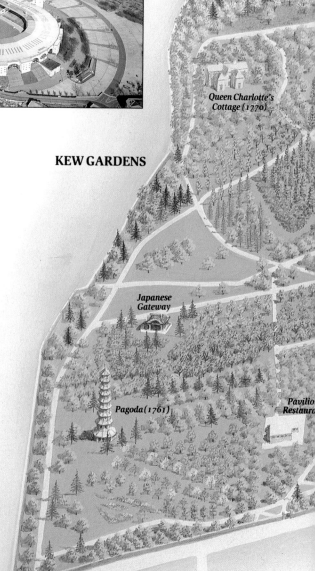

KEW GARDENS

Queen Charlotte's Cottage (1770)

Japanese Gateway

Pagoda (1761)

Lion Gate

Pavilio Restaura

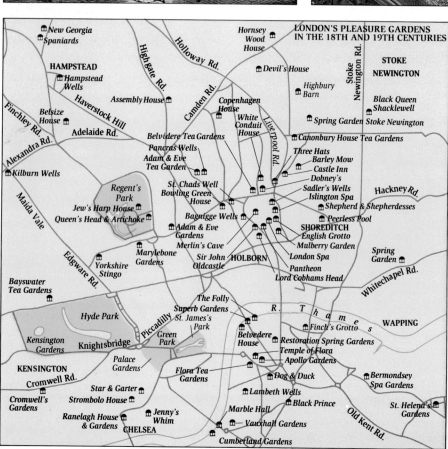

LONDON'S PLEASURE GARDENS IN THE 18TH AND 19TH CENTURIES

New Georgia
Spaniards
HAMPSTEAD
Hampstead Wells
Belsize House
Assembly House
Haverstock Hill
Finchley Rd.
Alexandra Rd.
Adelaide Rd.
Kilburn Wells
Maida Vale
Regent's Park
Jew's Harp House
Queen's Head & Artichoke
Edgware Rd.
Yorkshire Stingo
Bayswater Tea Gardens
Hyde Park
Kensington Gardens
Knightsbridge
KENSINGTON
Cromwell Rd.
Cromwell's Gardens
Star & Garter
Strombolo House
Ranelagh House & Gardens
CHELSEA
Jenny's Whim
Marble Hall
Vauxhall Gardens
Cumberland Gardens
Palace Gardens
Flora Tea Gardens
The Folly
Superb Gardens
St. James's Park
Green Park
Piccadilly
Belvedere House
Temple of Flora
Apollo Gardens
Dog & Duck
Lambeth Wells
Black Prince
Highgate Rd.
Holloway Rd.
Camden Rd.
Belvidere Tea Gardens
Pancras Wells
Adam & Eve Tea Garden
St. Chads Well
Bowling Green House
Bagnigge Wells
Adam & Eve Gardens
Merlin's Cave
Sir John Oldcastle
Marylebone Gardens
Copenhagen House
White Conduit House
Hornsey Wood House
Devil's House
Highbury Barn
Liverpool Rd.
Spring Garden
Canonbury House Tea Gardens
Three Hats
Barley Mow
Castle Inn
Dobney's
Sadler's Wells
Islington Spa
Shepherd & Shepherdesses
Peerless Pool
SHOREDITCH
English Grotto
Mulberry Garden
London Spa
Pantheon
Lord Cobhams Head
HOLBORN
Stoke Newington Rd.
STOKE NEWINGTON
Black Queen
Shacklewell
Stoke Newington
Hackney Rd.
Spring Garden
Whitechapel Rd.
WAPPING
Finch's Grotto
Restoration Spring Gardens
Bermondsey Spa Gardens
St. Helena's Gardens
Old Kent Rd.
R. Thames

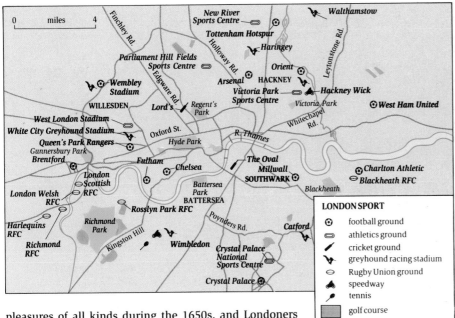

springs, such as Sadler's Wells and Hampstead Wells, became fashionable spas.

During Victorian times Londoners were provided with a great range of leisure facilities, ranging from parks and sports clubs to theatres and music halls. Regent's Park Zoo was opened in 1847, while Joseph Paxton's Crystal Palace was rebuilt at Sydenham after the Great Exhibition of 1851 (page 95), where it offered an exotic setting for all kinds of entertainment until its destruction by fire in 1936. Many of the sporting teams that are supported today by dedicated fans were created or formalised during the 19th century. Rugby clubs were founded at Twickenham and Richmond in the 1860s and several of London's leading Football Association teams date from the 1880s. The All England Lawn Tennis Club was created in 1882, evolving from an earlier croquet association. Cricket was being played in Tudor London, but its rules were not formalised until 1744. Henley was the first site of the Oxford and Cambridge Boat Race (1829), but it was moved to its present Putney to Mortlake course in 1845. Perhaps London's most popular 20th-century sporting event is the London Marathon, run between Blackheath and Westminster Bridge in late April.

London hosts major events throughout the year. Thousands of visitors flock to Soho for the Chinese New Year and to the Notting Hill Street Carnival in late August. The Chelsea Flower Show displays the latest ideas in floriculture and garden design each May and, in November, following a tradition dating back to 1215, the Lord Mayor drives through the City of London in a gilded coach.

pleasures of all kinds during the 1650s, and Londoners rejoiced when the monarchy was restored in 1660. The 18th century was the era of pleasure gardens; Chelsea's Ranelagh Gardens, for instance, opened in 1742 and were graced with an ornamental lake, a Chinese pavilion and a great rococo rotunda. In more modest style, Londoners were attracted to inns and springs within easy access of their homes, and some of London's natural

Sporting London The city boasts a large number of sports facilities, (map above). One of London's most famous sports venues is Wembley Stadium (below left). With a capacity of 100,000, it hosts a variety of events.

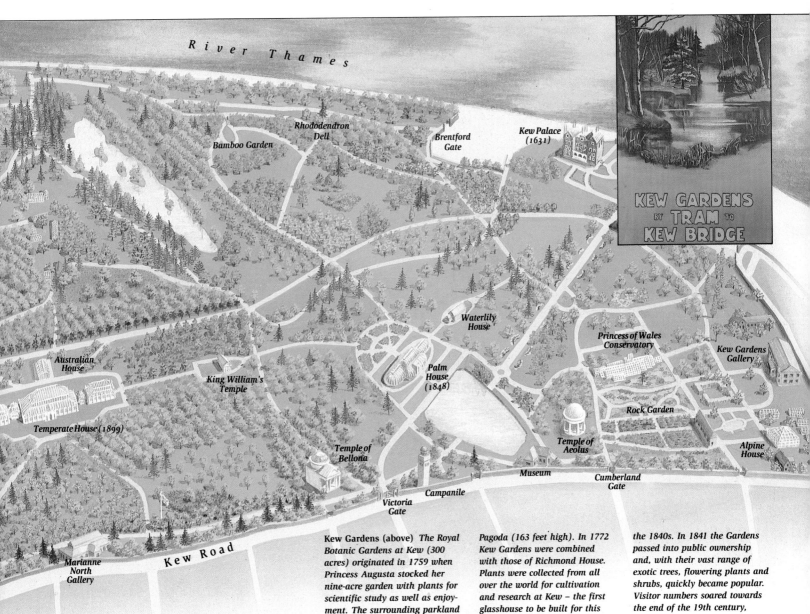

Kew Gardens (above) *The Royal Botanic Gardens at Kew (300 acres) originated in 1759 when Princess Augusta stocked her nine-acre garden with plants for scientific study as well as enjoyment. The surrounding parkland was subsequently adorned with classical temples and the Great*

Pagoda (163 feet high). In 1772 Kew Gardens were combined with those of Richmond House. Plants were collected from all over the world for cultivation and research at Kew – the first glasshouse to be built for this purpose was the Palm House, designed by Decimus Burton in

the 1840s. In 1841 the Gardens passed into public ownership and, with their vast range of exotic trees, flowering plants and shrubs, quickly became popular. Visitor numbers soared towards the end of the 19th century, when public transport made Kew more accessible to Londoners.

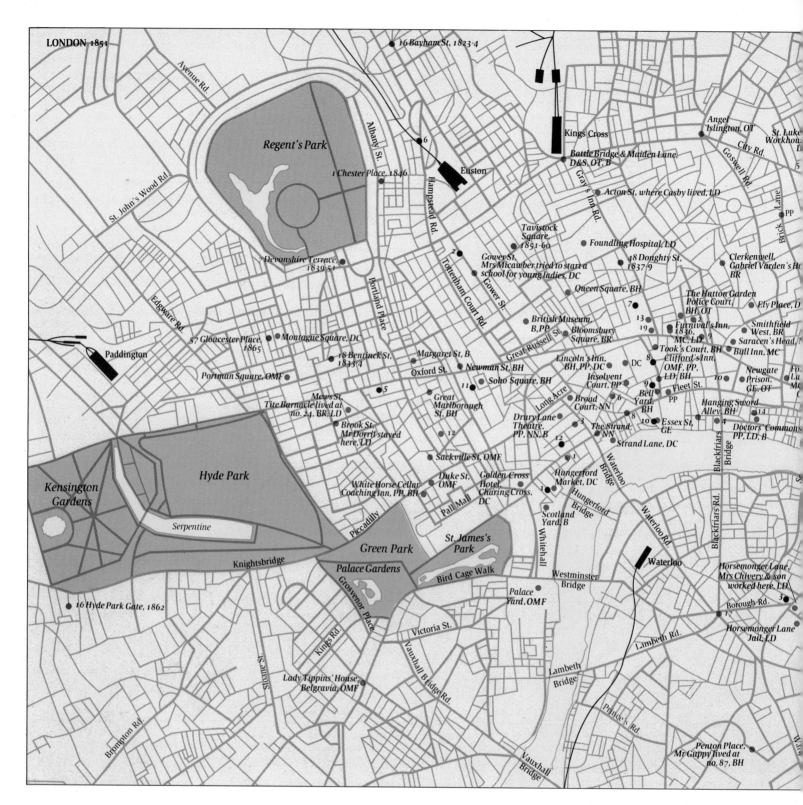

16 Bayham St, 1823-4

Kings Cross

Angel Islington, OT

St. Luke Workhou. D

City Rd.

Goswell Rd.

Brick La.

Battle Bridge & Maiden Lane, D&S, OT, B

Regent's Park

Albany St.

6

Euston

Hampstead Rd.

Acton St, where Casby lived, LD

Gray's Inn Rd.

1 Chester Place, 1846

PP

St. John's Wood Rd.

Tavistock Square, 1851-60

Foundling Hospital, LD

Clerkenwell, Gabriel Varden's H. BR

Portland Place

2

Gower St, Mrs Micawber tried to start a school for young ladies, DC

48 Doughty St, 1837-9

7 Devonshire Terrace, 1839-51

Gower St.

Queen Square, BH

The Hatton Garden Police Court, BH, OT

Ely Place, D

Edgware Rd.

Tottenham Court Rd.

British Museum, B, PP

7

13

19

Furnival's Inn, 1836, MC, LD

Smithfield West, BR

Saracen's Head, BR

57 Gloucester Place, 1865

Montague Square, DC

Bloomsbury Square, BR

Took's Court, BH

Bull Inn, MC

Great Russell St.

Clifford's Inn, OMF, PP, LD, BH

Paddington

18 Bentinck St, 1833-4

Margaret St, B

Oxford St.

Newman St, BH

Lincoln's Inn, BH, PP, DC

DC

9

Newgate Prison, GE, OT

Fo. Lu

MC

Portman Square, OMF

5

Soho Square, BH

Insolvent Court, PP

6

Bell Yard, BH

Fleet St.

PP

10

Mews St, Tite Barnacle lived at no. 24, BR, LD

11

Great Marlborough St, BH

Long Acre

Broad Court, NN

8

3

10

Essex St, GE

Hanging Sword Alley, BH

14

Brook St, Mr Dorrit stayed here, LD

12

Drury Lane Theatre, PP, NN, B

12

The Strand, NN

Strand Lane, DC

Doctors' Commons PP, LD, B

Sackville St, OMF

1

Waterloo Bridge

Blackfriars Rd.

Kensington Gardens

Hyde Park

Serpentine

White Horse Cellar Coaching Inn, PP, BH

Duke St, OMF

Golden Cross Hotel, Charing Cross, DC

1

Hungerford Market, DC

Hungerford Bridge

Pall Mall

Scotland Yard, B

Blackfriars Bridge

Piccadilly

Green Park

St. James's Park

Whitehall

Waterloo Rd.

Knightsbridge

Palace Gardens

Bird Cage Walk

Westminster Bridge

Waterloo

Horsemonger Lane, Mrs Chivery & son worked here, LD

16 Hyde Park Gate, 1862

Grosvenor Place

Palace Yard, OMF

Borough Rd.

3

Kings Rd.

Victoria St.

Lambeth Rd.

Horsemonger Lane Jail, LD

Sloane St.

Vauxhall Bridge Rd.

Lambeth Bridge

17

Brompton Rd.

Lady Tippins' House, Belgravia, OMF

Prince's Rd.

Penton Place, Mr Guppy lived at no. 87, BH

Vauxhall Bridge

THE LONDON OF CHARLES DICKENS

Throughout its history London has inspired works of literary genius but none can be compared with the powerfully evocative writings of Charles Dickens (1812–1870). Born in Portsea, Dickens came from Chatham to London when he was ten to join his family in their new home in Camden Town. It was not to be an easy or settled existence in the metropolis; just two days after his twelfth birthday the talented and ambitious boy was sent to pack boot-blacking in Warren's warehouse near Hungerford Stairs, reached after a 3-mile walk from Camden Town. A few days after the job started his father, John Dickens, was imprisoned as an insolvent debtor in Marshalsea Prison, Southwark. In the fashion

of the time (evocatively described in *Little Dorrit*) his wife and children moved into the Marshalsea with him, and young Dickens was forced to find new lodgings close to the prison. Although his father was released after a few months and Charles was able to attend school again, the squalor, smell and despair of Marshalsea remained imprinted in his mind. From an early age he was exposed to the hardship he was to portray so powerfully in his novels.

Dickens left school in 1827 and obtained work as a solicitor's clerk. Two years later he became a legal reporter and then moved to parliamentary reporting. His initial writings on London life were brought together as *Sketches by Boz* (1836). In later life, he felt compelled to constantly revive his sense of place by walking the streets and alleys of the capital, and visiting its markets, public houses, workhouses, police courts and prisons. Throughout his works of

'imaginative vision' Dickens embraced the essential contradictions of 19th-century London, but always dwelt on the harsher side. Next to the wealth he set appalling poverty, for his characters' hope was frequently tinged with despair. Splendid public buildings contrasted with terrifying slums, all shrouded in the fog which seemed to emanate from a million coal fires, factory chimneys and steamboats on the river.

The 'Great Oven', the 'Great Wen', the 'Fever Pitch' and 'Babylon' were the images of London that Dickens evoked as the capital grew from a densely-packed, filthy town of some 1,400,000 souls when he first arrived to a sprawling city of 3,000,000 inhabitants, complete with railways, underground lines and a constellation of suburbs, when he died. The grinding and widespread poverty of Dickens' adolescent vision was to remain a stark reality for many Londoners well beyond the time of his death.

London locations mentioned

Titles are abbreviated as follows:

SB	Sketches by Boz
BH	Bleak House
BR	Barnaby Rudge
DC	David Copperfield
D&S	Dombey & Son
GE	Great Expectations
LD	Little Dorrit
MC	Martin Chuzzlewit
NN	Nicholas Nickleby
OCS	Old Curiosity Shop
OMF	Our Mutual Friend
OT	Oliver Twist
PP	Pickwick Papers

A selection of places and events (numbered on map):

1. Adelphi Terrace and Hotel. Favourite watering hole of Mr Pickwick and Mr Wardle – PP
2. Bleeding Heart Yard. Home to the Plornish family and Daniel Doyce's factory – LD
3. Bow Street Police Court. The Artful Dodger – OT – and Barnaby Rudge were taken to Bow Street Police Station.
4. Bridewell Workhouse, House of Correction – where Miss Miggs was chosen to be Female Turnkey for the County Bridewell – BR
5. City Rd. Windsor Terrace. Where Mr Micawber lived – DC
6. Clare Market, site of the Old Curiosity Shop and gin shops – B
7. Custom House. Peepy worked here – BH; also the late Mr Bardell – PP. Pip left his boat at a wharf near the Custom House – GE
8. Johnson's Beef House where David Copperfield ordered a small plate of beef.
9. Field Lane, Saffron Hill. Fagin's Den – OT
10. Fleet Prison. Where Mr Pickwick was imprisoned – PP
11. Garraway's, the famous city coffee house. Mr Flintwich was a customer here – LD. Also Nadgett, the enquiry agent – MC
12. Golden Square. David Copperfield and Martha found Little Emily here – DC
13. Gray's Inn. David Copperfield stayed in the Gray's Inn Coffee House when visiting Traddles – DC
14. Horn Coffee House, 29 Knightrider St. Where Mr Pickwick sent for a few bottles of wine while in Fleet Prison, to celebrate Mr Winkle's visit – PP
15. Jacob's Island (Folly Ditch). Home of Bill Sikes and where he was hanged whilst trying to make his escape – OT
16. Leadenhall Market. In the Green Dragon Sam wrote the famous 'Valentine' – PP; the offices of Dombey and Son were here.
17. Obelisk, generally called the 'obstacle'. David Copperfield had his trunk and half guinea stolen and had to start his journey to Dover on foot – DC
18. St. Mary Axe. This is the location of the pretty roof garden where Lizzie and Jenny Wren used to sit and talk above the premises of Pubsey & Co. – OMF
19. Gray's Inn, South Square. Mr Phunky had chambers here – PP. Traddles lived at no. 2 Holborn Court, nearby.

(map showing London and River Thames with numbered locations: Bishopsgate St, NN; Commercial Rd.; Lothbury; Guildhall, PP; Threadneedle Street, B; Aldgate Pump, D&S, NN; Mincing Lane, OMF; Eastern Counties Station; Great Tower St, BR; The Tower; OCS; OT, BR, MC; DC, OMF; London Bridge; Guy's Hospital, PP; ...rt Inn, PP; St. Katharine Dock; London Docks; Shadwell Basin; Lower Pool; Albion Dock; Canada Dock; River Thames; Bricklayers Arms Station; Rotherhithe New Rd.)

DICKENS' LIFE

- **places where Dickens lived** *(dates shown on map)*
- **a selection of events in Dickens' life:**

1. Craven St. Site of Hungerford Market where Dickens worked as a boy in Warren's Blacking House at Old Hungerford Stairs.
2. Gower St. North. Dickens' mother opened a school here attempting to delay payment of debts.
3. The King's Bench Debtors' prison where John Dickens was imprisoned for debt in 1824.
4. The New Marshalsea Debtors' Prison to which Dickens' father was then moved. To save money, his wife and younger children moved in with him.
5. Tenterden St. The Royal Academy of Music. Dickens' sister Fanny was a boarder here.
6. Wellington House Academy (corner of Granby St. and Hampstead Rd.). Dickens was sent here for two years after his father's release from prison.
7. Gray's Inn, Holborn. Dickens, aged 15, came to work here at the law firm of Ellis & Blackmore.
8. Chancery Lane where Dickens then worked for the Solicitor Charles Molloy.
9. Bell Yard, Carter Lane. Dickens rented an office here while a reporter in Doctors' Commons.
10. Essex St. Dickens attended the Unitarian Chapel here in 1842.
11. Dean St. Fanny Kelly's Theatre (now gone) owned and run by a retired actress; Dickens acted here in 1845.
12. Maiden Lane. Rules Restaurant was one of Dickens' favourite eating places.

Images of Dickens' London *The Marshalsea Prison (above right, c.1895) must have held grim associations for Dickens, since his father was incarcerated there for debt in 1823-4. Dickens was taken out of school and worked at a blacking factory at Hungerford Stairs, near Charing Cross (below left, 1823). He was to write powerfully about child exploit-ation in his novels. The criminal underworld was also an enduring obsession; at the beginning of* **Great Expectations,** *the hero, Pip, is surprised in a graveyard by Magwitch, who has escaped from the hulks (see below right, a prison ship at Deptford), where convicts awaiting transportation were held. Little of Dickens' London survives; the George Inn at Southwark (right, photographed in 1890) is an exception. Mentioned in* **Little Dorrit,** *it is the only sur-viving galleried inn in London.*

THEATRICAL LONDON

In medieval London players were restricted to performing in galleried innyards, bull-baiting arenas and the houses of the nobility. The City authorities feared that plays could corrupt morals and kept a close watch on entertainments. In 1574 a law was passed forbidding the construction of playhouses inside the City walls, and so London's first purpose-built theatres were located outside the City. They were constructed to retain many of the qualities of the old galleried inns. In 1642 the Puritans closed down all London's theatres. Dramatic life revived just before the Restoration, but the old unroofed playhouses did not reopen.

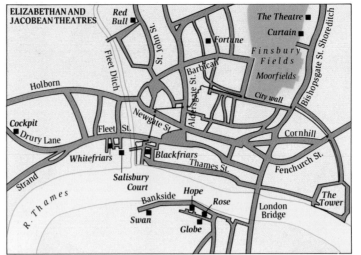

ELIZABETHAN AND JACOBEAN THEATRES

Shakespeare's London *The map (left) shows that Elizabethan and Jacobean playhouses were sited on the South Bank, which had housed entertainments since medieval times, in Shoreditch and off Fleet Street. Recent archaeological work has uncovered many details about the design of London's theatres in Shakespeare's day. The Curtain, London's first purpose-built theatre, was constructed in 1577, followed by the Rose on the South Bank in 1587. The Globe was built partly from materials salvaged from the Curtain, which had been demolished after objections from the Puritans.*

The Globe *The illustration (far right) shows an engraving by Wenceslas Hollar of the second Globe Theatre on Bankside. This was rebuilt in 1613 on the foundations of the original Globe Theatre, which had been constructed between 1598 and 1599.*

Shakespeare's workplace *The first Globe Theatre, in which Shakespeare performed, was destroyed by fire when a spark from a stage canon, used in a performance of* Henry VIII, *set light to the thatched roof.*

The new Globe Theatre *In 1993 construction of a new Globe Theatre began approximately 200 yards from the site of the original one. Like the first playhouse, known as Shakespeare's 'wooden O', the most recent version is polygonal and contains tiers of seating around an open yard for standing spectators. The roof is made of thatch and is open at the top, allowing natural daylight to illumine the action on stage. The photographs (above, left and right) show the new Globe as it appears from the outside, from the inside and under construction against the background of London's cityscape.*

ALHAMBRA Theatre of Varieties.

LEICESTER SQUARE, LONDON.

Royal patents for presenting spoken drama were granted to two companies by Charles II. Demand for theatre could not be satisfied by the patent houses in Drury Lane and Covent Garden alone, so 'minor' theatres operated outside the law until the monopoly was abolished in 1843. By 1850 London had two dozen theatres, clustered in Covent Garden, Drury Lane and the Strand.

Music-halls offered more robust and popular fare. Originally garish, brightly-lit halls often connected with pubs, they had become veritable palaces of entertainment by the end of the 19th century. Their heyday was not to last: London's first cinematograph show was given in 1895, and soon many theatres and music-halls were showing moving pictures. Nevertheless, the late 20th century has not been devoid of new theatres in London – two notable examples are the National Theatre and the Barbican in the City.

London's Theatreland *The Theatre Royal, Drury Lane (above left, 1812), was opened in 1663. It became the home of David Garrick, one of England's great Shakespearean actors, seen performing in an aquatint (above centre) dated 1808. The corniced facade of the Alhambra music hall at Leicester Square can be seen, (above right), while the painting (right, 1861), shows an early music-hall interior (the Pavilion). The map (below left) shows the location of music halls in Victorian London. Many of London's theatres are now closed, and the map (below right) shows London's lost theatres. Today's theatreland is largely concentrated in the West End (map, bottom left), although the National Theatre (bottom right) is part of a major arts complex on the South Bank.*

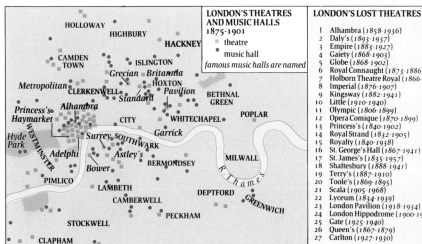

LONDON'S THEATRES AND MUSIC HALLS 1875-1901
- ■ theatre
- ● music hall
- *famous music halls are named*

LONDON'S LOST THEATRES

1 Alhambra (1858-1936)
2 Daly's (1893-1937)
3 Empire (1883-1927)
4 Gaiety (1868-1903)
5 Globe (1868-1902)
6 Royal Connaught (1873-1886)
7 Holborn Theatre Royal (1866-1880)
8 Imperial (1876-1907)
9 Kingsway (1882-1941)
10 Little (1910-1940)
11 Olympic (1806-1899)
12 Opera Comique (1870-1899)
13 Princess's (1840-1902)
14 Royal Strand (1832-1905)
15 Royalty (1840-1938)
16 St. George's Hall (1867-1941)
17 St. James's (1835-1957)
18 Shaftesbury (1888-1941)
19 Terry's (1887-1910)
20 Toole's (1869-1895)
21 Scala (1905-1968)
22 Lyceum (1834-1939)
23 London Pavilion (1918-1934)
24 London Hippodrome (1900-1982)
25 Gate (1925-1940)
26 Queen's (1867-1879)
27 Carlton (1927-1930)

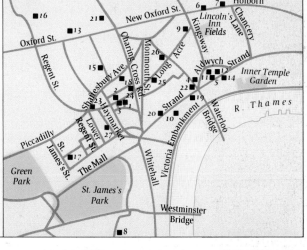

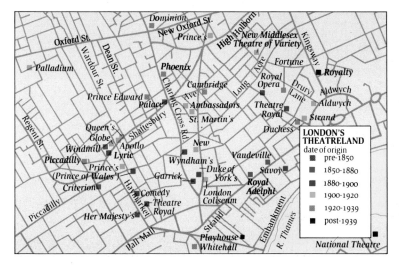

LONDON'S THEATRELAND
date of origin
- ■ pre-1850
- ■ 1850-1880
- ■ 1880-1900
- 1900-1920
- 1920-1939
- ■ post-1939

SHOPPING AND MARKETS

MARKETS

Great cities live by trade, and London is no exception. As the city grew and modes of transport improved, so it was supplied with goods from ever greater distances. In 1720, Daniel Defoe remarked that '... this whole kingdom, as well as the people, the land and even the sea in every part of it, are employed to supply London with provisions, fuel and timber'. As sailing ships gave way to steam, railways opened up hitherto unimaginable distances. The British Empire had reached its zenith and London's trade embraced the whole globe. Trade did not just mean goods; bonds, stocks and shares also played a role in the vast array of London's marketing.

The great Roman forum stood at the heart of *Londinium*, an emporium for goods brought by land and sea. Trade

The history of London's markets stretches back to Roman times. Indeed, many of the names of medieval markets (top left) are still familiar to a 20th-century Londoner – Cheapside, Smithfield, Leadenhall. Commodities, too, have been immortalised in street names such as Poultry, Milk Street and Honey Lane. The distribution of markets in Stuart London (centre left) reflects the spread of the city outside the medieval walls: Covent Garden, opened in 1670 but relocated to Battersea in 1973, was the first of many markets which were founded specifically to serve the residents of the new suburbs.

Markets in the 20th century London's markets today (bottom) present a rich and varied array. Although many of the old-established medieval markets, such as Billingsgate and Spitalfields, have been re-located in the outer periphery, medieval markets such as Leadenhall in the City, still survive. London's street markets capture a district's character in all its immediacy; markets such as Petticoat Lane and Brick Lane (to the east of the City) are both located in areas with a rich ethnic history – their Jewish and, more recently, Asian legacy.

The map shows the detailed range of specialities which characterise the street markets; flowers, antiques, furniture, second-hand clothes, tools, specialised foods, are all represented. Many residents of Greater London are within range of a street market (be it only a few fruit and vegetable stalls or a substantial street market selling a whole range of goods). Their vitality and variety are characteristics of the city.

Right *The Sunday street market at Petticoat Lane. A thriving old clothes market since the beginning of the 17th century, it is still a busy general market today.*

declined when the Saxons overwhelmed the city, but by the 7th century London was re-established and open markets were developed in Cheapside. As in other medieval cities, specialised traders clustered in distinctive quarters to defend their interests, and these local trades (bread, honey, milk, poultry) are all immortalised in the surrounding street names. Market activities were closely regulated – trading was permitted only in established marketplaces and during specified hours; few goods could be hawked openly in the streets. In time, more distant markets flourished at Spitalfields and in Southwark.

In 1666, the Great Fire dealt a mortal blow to some of the city's oldest markets, some of which were rebuilt as covered markets under royal charter (e.g. Leadenhall, Newgate, Billingsgate, the Stocks). At the same time, the fashion for suburban development was paralleled by the creation of new markets. Covent Garden (1670), built on the Earl of Bedford's land, was the first to be authorised; it was to be emulated by scores of other landowners during the next 200 years. The City's monopoly over all markets within a radius of six and two-third miles was widely challenged; many new markets developed unofficially.

With the advent of the railways, a number of wholesale food markets were opened near the main termini. The old sheds of Covent Garden had been replaced by an elegant market hall in 1830, Billingsgate and Leadenhall were rebuilt in the City, and handling of live cattle was transferred from Smithfield to the Caledonian cattle market in Islington (1855). It was closed during the Second World War. Over the past hundred years many of London's markets have been closed and others relocated (e.g. Billingsgate and Spitalfields) but eighty still survive in inner London and there are more in the suburbs and in surrounding towns.

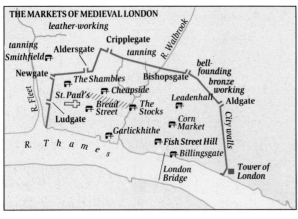

THE MARKETS OF MEDIEVAL LONDON

leather-working
tanning
Smithfield Aldersgate Cripplegate
Newgate tanning
The Shambles Bishopsgate
Cheapside bell-founding bronze working
St. Paul's
Bread Street Leadenhall Aldgate
The Stocks
Ludgate Corn Market
Garlickhithe
R. Thames Fish Street Hill
Billingsgate
London Bridge Tower of London

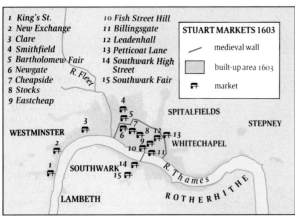

1 King's St.
2 New Exchange
3 Clare
4 Smithfield
5 Bartholomew Fair
6 Newgate
7 Cheapside
8 Stocks
9 Eastcheap
10 Fish Street Hill
11 Billingsgate
12 Leadenhall
13 Petticoat Lane
14 Southwark High Street
15 Southwark Fair

STUART MARKETS 1603
— medieval wall
▭ built-up area 1603
market

SPITALFIELDS STEPNEY
WESTMINSTER WHITECHAPEL
SOUTHWARK
R. Thames ROTHERHITHE
LAMBETH

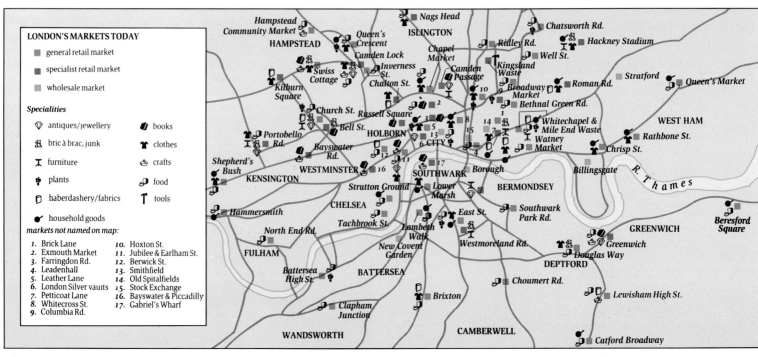

LONDON'S MARKETS TODAY
■ general retail market
■ specialist retail market
■ wholesale market

Specialities

♡ antiques/jewellery
🐚 bric à brac, junk
⊤ furniture
♣ plants
▯ haberdashery/fabrics
✿ household goods

📕 books
👕 clothes
🍴 crafts
🍴 food
⊤ tools

markets not named on map:

1. Brick Lane	10. Hoxton St.
2. Exmouth Market	11. Jubilee & Earlham St.
3. Farringdon Rd.	12. Berwick St.
4. Leadenhall	13. Smithfield
5. Leather Lane	14. Old Spitalfields
6. London Silver vaults	15. Stock Exchange
7. Petticoat Lane	16. Bayswater & Piccadilly
8. Whitecross St.	17. Gabriel's Wharf
9. Columbia Rd.	

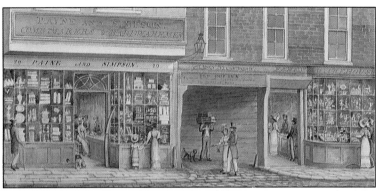

London's shopping history *From medieval times London was richly endowed with specialist shops. This view of the shopping street of Cornhill (early 18th century, above far right) clearly shows the elaborate overhanging signs, with which each individual shop advertised their wares. They were judged to be dangerous to passing pedestrians and were finally removed in 1762. 18th-century Londoners could be justly proud of their shops, with their enticing displays (above). A German visitor to London in 1786, Sophie v. la Roche, visited Oxford Street and remarked '... behind great glass windows absolutely everything one can think of is neatly, attractively displayed, and in such abundance of choice as to make one greedy ...'.*

Harrod's started life as a small grocer's shop in the Brompton Road, opened by H.C. Harrod, a tea retailer in 1853 (photograph far right 1891). Gradually, he acquired the surrounding terraces (map above right), until he had built up an 'island site', and the massive expansion of Harrod's store then began (diagram right). Harrod's now covers a 20-acre site.

19th-century Tobacco Dock is now a shopping village (illustration below), its vaults and upper levels converted into shops and boutiques. However, it ran into severe financial difficulties in 1990 as the dynamism of the Docklands revival declined.

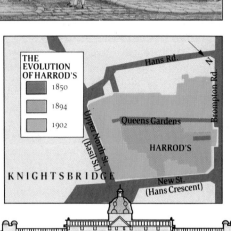

THE EVOLUTION OF HARROD'S

1850
1894
1902

Hans Rd.

Queens Gardens

HARROD'S

Brompton Rd.

Upper North St. (Basil St.)

KNIGHTSBRIDGE

New St. (Hans Crescent)

SHOPPING

London's earliest shops surrounded marketplaces in the walled city and were adjacent to unloading points along the Thames. That pattern held good for many centuries, but as London spread westward after 1650 so new clusters of shops developed to serve rich households living around the elegant squares and streets which were springing up all over the West End. Many of these shops were also dedicated to serving the Royal Court, and the King's move to St. James's Palace at the end of the 17th century (after Whitehall Palace had been destroyed by fire) was to have a profound effect on the surrounding neighbourhoods. Many royal warrant-holders set up business in the streets of St. James, proclaiming that they sold goods 'By appointment from' His or Her Majesty – many have retained both their warrants and

premises to this day. Over 800 royal warrant-holders are concentrated in the elite shopping districts of St. James and Mayfair. Many of the shops in this district have a long history; William Fortnum (a footman in the household of Queen Anne) and Hugh Mason set up a grocer's shop in 1707. Fortnum secured orders from the royal household and the business consequently thrived; to this day Fortnum and Mason's occupies the same site. A few yards along Piccadilly is Hatchard's bookshop, which opened in 1797, while Burlington Arcade was built in 1819 to provide an exclusive shopping parade, subject to strict regulations.

During the 19th century, the shopping districts of the West End began to radiate outwards, along Regent Street, Oxford Street and Tottenham Court Road. This was the century which saw the advent of department stores. Originally an American concept, these vast trading emporia, boasted (in the words of William Whiteley) 'We can supply anything from a pin to an elephant'. By the end of the century, department stores of ever-increasing size were to be found in the residential areas of Kensington, Brompton and Chelsea. While shops in the West End flourished, many in the City closed as its residential population fell from 128,000 (1801) to 27,000 (1901).

In the 1920s and 30s, the growth of American mass-consumerism, supported by advertising and the mass media, created an even greater demand. This was the era when shops such as Woolworth and Marks & Spencer began to colonise London's suburbs (*page 107*). Their presence in a district is an accurate index of a suburb's status as a regional shopping centre.

As the 20th century has progressed, fashions in shopping have changed. Some of the large department stores, founded during the optimistic and expansionist days of the British Empire, have fallen on hard times and been forced to close. Others, such as Whiteley's in Queensway, have been born again as shopping complexes, playing host to many separate retail outlets. Purpose-built shopping centres, with cinemas, restaurants and car parks, characterise the outer suburbs, while in the inner city a wide range of unlikely buildings (such as abandoned and neglected warehouses in Docklands) have been put to similar use.

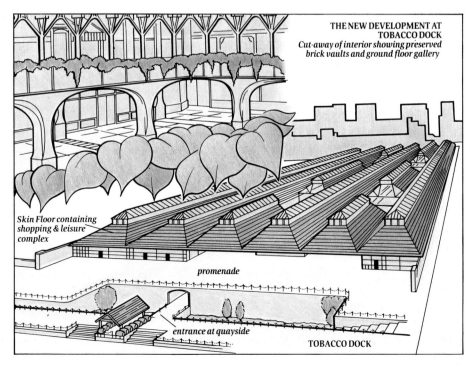

THE NEW DEVELOPMENT AT TOBACCO DOCK
Cut-away of interior showing preserved brick vaults and ground floor gallery

Skin Floor containing shopping & leisure complex

promenade

entrance at quayside

TOBACCO DOCK

CHAPTER 11
PLACES IN LONDON

THIS ATLAS is both a visual celebration of two thousand years of London's history and an inventory of economic and architectural achievement, which has been punctuated by phases of disaster and destruction as well as the poverty and deprivation which are the darker but inescapable sides of metropolitan life. The city is the sum total of its many neighbourhoods – some prestigious, others ordinary – which face the future with varying degrees of accumulated strength and potential. This final chapter focuses on a sample of these places. As the dual cores of old London, the City and Westminster stand proud in modern London and are obvious choices for our selection. So too is Docklands, which was so important to London's economy in the past and on which so much faith for London's future is pinned. 'Village London' evokes the historic settlements around which so many suburbs developed and much of the character of Greater London has been built. London's many bridges are of great visual and functional interest in their own right and remind us of how significant the Thames is in dividing Londoners' perceptions of their home town. Living 'north of the river' is very different from living 'south of the river', and few Londoners are equally familiar with both sides of the capital. London's universities and museums evoke the worldwide cultural importance of the city; its monuments and great cemeteries not only honour the dead, but also recall London's role as national capital and historic focus of a colonial empire. Perhaps less obvious is our choice of the hidden world beneath the surface of London and the many grand designs that were never realised to transform the capital. They contribute to the 'London that never was'. Equally surprising may be our selection of schemes that may fashion London's appearance and activities as it enters the third millennium.

As the 20th century draws to a close, London has lost most of its industry and now survives by providing services and managing international finance. If London, and the towns and villages in its commuting orbit, are to flourish in the years ahead, the city must respond to two challenges, one from within, the other from beyond. If it fails to do so then the future of the metropolitan region may be sombre. London's buildings and infrastructure are wearing badly. Much housing that had outlived its useful life was destroyed during the Second World War, and even more was erased in slum clearance schemes, but London still has a complex legacy of Victorian and Edwardian dwellings, inter-war suburban houses and post-war flats in need of expensive restoration or replacement. The surfaces of London's roads are pitted by repeated excavation and inadequate repair as old utilities are replaced and new ones, such as fibre-optics cables, are laid. Over 600,000 new holes are dug in London's streets each year. Many sewers were installed a century or more ago; frequent burst water mains reveal that many pipes need replacement. London's roads have to cope with lorries of ever increasing weight and dimension and are jammed with 2,200,000 cars each year. By 2001 the number may well rise to 2,800,000. The Underground system is often unreliable, overcrowded, dirty and very expensive. London's buses are little better as they crawl through what seems to be an unceasing rush hour that lasts from morning to night. Political geography is changing fast and London's vitality and financial supremacy cannot be taken for granted in the 'new Europe' that is emerging as a result of the Single European Market, the unification of Germany, the dramatic changes in Eastern Europe, the enlargement of the European Union, and the possible entry of states in Eastern Europe. Paris, Frankfurt and other European cities are planning to maximise their chances of capturing growing shares of business activity. Their success could be to the detriment of London unless effective action is taken to manage the city comprehensively, and meet the European challenge head on.

Paris provides the best example of a city that has been planned, managed and modernised over the past 25 years. No other city is promoting its future with such verve and commitment. If London is to compete with Paris and other European cities, it must be repaired, restored and reinvigorated. Its strengths and weaknesses must be evaluated and an integrated plan must be prepared, not just for the metropolis, but for the whole London region. Termination of the GLC (and Britain's other metropolitan authorities) in 1986 removed a politically controversial organisation, but also robbed the city of a London-wide framework for taking stock of problems. The City of London, plus 32 borough authorities with sharply different socio-economic and environmental characteristics and markedly divergent political sympathies, cannot pull together in the way necessary for London as a whole to flourish in the future.

What metropolitan development has occurred in recent years has been disjointed. Much faith has been pinned on redeveloping Docklands, not only as a mechanism for dealing with a major problem area, but also as the means of attracting corporate finance to the metropolis and answering an anticipated explosion in demand for office facilities. The area is served by a light railway that has been extended to Bank and to the Royal Docks. New highways link Docklands to other parts of the region, and the Jubilee Line extension offers a great improvement. The task of regenerating vast areas around the Royal Docks far downstream remains for the new millennium.

Other schemes are underway elsewhere in Greater London and more have been proposed: a fast rail link from Paddington to Heathrow has improved access to the airport; a CrossRail line tunnelled from beyond Liverpool Street to Paddington has been proposed to provide a fast east-west service with interchange facilities on several north-south Tube lines; rail services using the Channel Tunnel terminate at new facilities at Waterloo. The main terminal will, however, be at St Pancras, which will be served by a high-speed rail link through Kent. This will trim journeys to Paris and Brussels from three to two and a half hours. Unlike the North Circular Road, which is an identifiable highway being upgraded, the South Circular remains a bizarre string of suburban streets. Schemes for improving it and for installing motorways through congested suburbs have been shelved because of rising costs and widespread local opposition. The London Ring Main tunnel has been constructed to carry 285 million gallons of water at any one time from intake points to the south-west of London for distribution to many locations in the region. Some 120 feet beneath the surface, the 50-mile tunnel is deeper than most Tube lines and longer than the Channel Tunnel. There are proposals to increase its length early in the next century.

In 1993 the government launched a scheme to regenerate the East Thames Corridor, also known as the Thames Gateway. With a legacy of industrial decline and job losses, this area extends into south Essex and as far as the Isle of Sheppey in Kent. Its areas of derelict land offer great potential for accommodating new economic activities. Stations on the Channel Tunnel Rail Link will provide direct access to the European mainland; the Jubilee Line extension provides services to central London. However, the negative image of East London must be transformed if the area is to succeed in attracting the necessary volumes of investment for regeneration. After many delays and escalating costs, the major phases of the new British Library will be finished soon, enabling many scattered holdings to be consolidated. The fate of redundant railway land at King's Cross and other sites remains undecided. The banks of the Thames on the Isle of Dogs have been invaded by an extravaganza of post-modern construction. Office developments have enveloped Charing Cross station, Cannon Street and several other sites. Schemes for redeveloping the South Bank and land opposite the Tower of London will bring important changes to London's riverside in the years ahead. The former Bankside Power Station will house a major new art gallery, but the fate of the crumbling Battersea Power Station remains unclear. There is undoubted potential for redeveloping other areas in imaginative ways to enhance the sullied beauty of London's river.

In recent years many continental cities have been planned on a metropolitan scale and their economies are now forging ahead. This European experience suggests that London may be in peril without a grand design and an administrative mechanism to help implement it. The city runs the risk not only of lagging behind its European rivals, but also of repelling much-needed international investment which will be captured by Paris, Frankfurt or Berlin.

THE CITY

The City of London is both the oldest and in some ways the most modern part of the capital. Narrow streets, dating from medieval times and re-established after the Great Fire, contrast with broad, straight thoroughfares created in Victorian times or as part of the great rebuilding programme after the Second World War. Historic churches and St. Paul's cathedral still occupy their ancient sites, but underwent reconstruction in the 17th century and again in the 20th century. The Victorians transformed the Georgian City, with its small buildings and narrow streets, embarking on an ambitious scheme of road-widening and construction. Vast buildings, constructed in monumental style to house the banks, insurance companies and trading firms which managed the wealth of Victorian Britain and her Empire, are now being torn down and replaced by a new generation of office blocks designed to meet the requirements of computerised business transactions. New construction is drastically changing the face of the historic City – even offices built in the 1950s and 1960s are being demolished and replaced by striking buildings of post-modern design.

As offices and warehouses replaced homes in the City during the 19th century, so its residential population plummeted from 128,000 in 1851 to 9,000 in 1939 and a mere 4,000 in 1971. Since then, the figure has risen slightly (5,300) as a result of new blocks of flats being included in the Barbican development. In sharp contrast, about 300,000 people now commute into work in the 'Square Mile' each weekday.

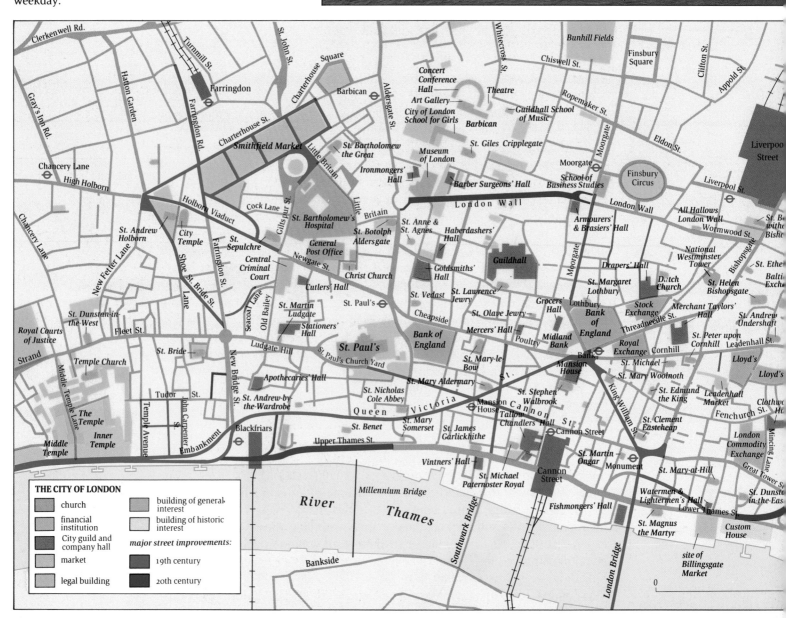

THE CITY OF LONDON

- church
- financial institution
- City guild and company hall
- market
- legal building
- building of general interest
- building of historic interest

major street improvements:
- 19th century
- 20th century

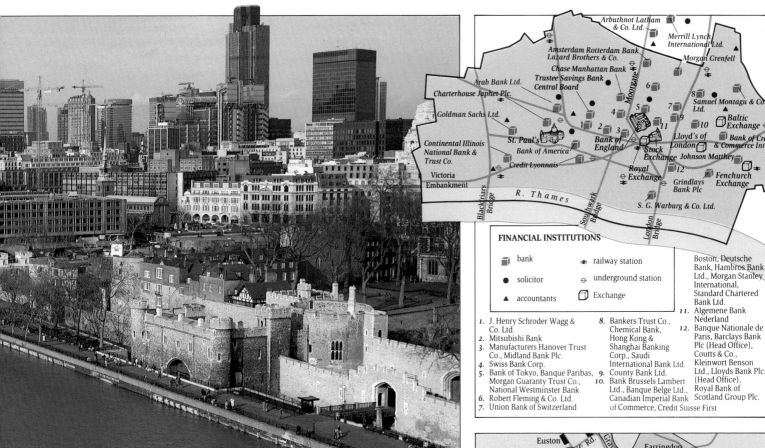

FINANCIAL INSTITUTIONS

⌂	bank	⊬	railway station
●	solicitor	⊖	underground station
▲	accountants	⬡	Exchange

1. J. Henry Schroder Wagg & Co. Ltd.
2. Mitsubishi Bank
3. Manufacturers Hanover Trust Co., Midland Bank Plc.
4. Swiss Bank Corp.
5. Bank of Tokyo, Banque Paribas, Morgan Guaranty Trust Co., National Westminster Bank
6. Robert Fleming & Co. Ltd.
7. Union Bank of Switzerland

8. Bankers Trust Co., Chemical Bank, Hong Kong & Shanghai Banking Corp., Saudi International Bank Ltd.
9. County Bank Ltd.
10. Bank Brussels Lambert Ltd., Banque Belge Ltd., Canadian Imperial Bank of Commerce, Credit Suisse First

Boston, Deutsche Bank, Hambros Bank Ltd., Morgan Stanley International, Standard Chartered Bank Ltd.
11. Algemene Bank Nederland
12. Banque Nationale de Paris, Barclays Bank Plc (Head Office), Coutts & Co., Kleinwort Benson Ltd., Lloyds Bank Plc. (Head Office), Royal Bank of Scotland Group Plc.

The modern City *The map (below left) shows today's City, whilst the photo (above) shows the buildings of many centuries. Clearly visible as a cluster of skyscrapers is the 'Square Mile', and the map (top right) identifies its leading financial institutions. The map (right) shows the pattern of office rents, which peak around the Bank of England. Rents have declined over the 1990s.*

Fleet Street *Close to the lawyers of The Temple and the clerics of St. Paul's, Fleet Street was the home of London's publishing trade for many centuries, and national newspapers were published there between 1702 and 1989 (map below). From that point, corporate mergers, new technology and working practices meant that editorial offices and, frequently, printing works began to relocate – in Docklands (map below right).*

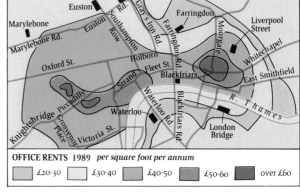

OFFICE RENTS 1989 *per square foot per annum*

£20-30	£30-40	£40-50	£50-60	over £60

MAIN NEWSPAPER SITES IN AND AROUND FLEET STREET PRE-1987

1. W H Smith (from 1920)
2. Newspaper House (Westminster Press)
3. Fleetway House (Amalgamated Press)
4. Northcliffe House (Associated Newspapers)
5. Temple House (Horace Marshall and Sons) National Press Agency Carmelite House
6. New Carmelite House

7. St. Bride Foundation Institute
8. Printing House Square
9. Bracken House
10. Sheffield Daily Telegraph
11. Liverpool Daily Post
12. Glasgow Herald
13. The Scotsman
14. Punch
15. Daily Chronicle & Lloyd's Weekly News
16. Birmingham Daily Post
17. Reuters & Press Association

NEWSPAPER MOVES SINCE 1987

● editorial   ▲ printing

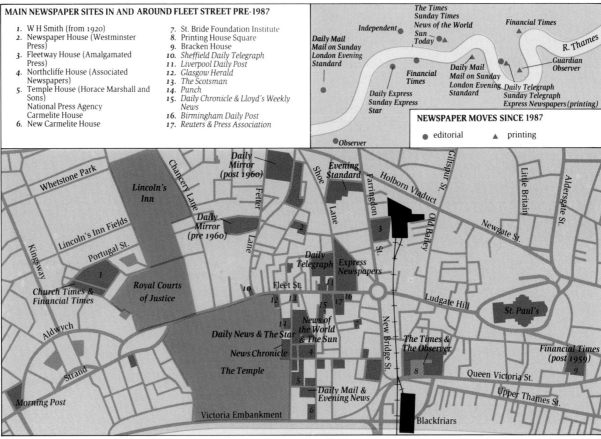

WESTMINSTER AND WHITEHALL

Almost a thousand years ago, the Saxon king Edward the Confessor chose to build an abbey and a royal palace on Thorney Island, and ever since, Westminster has been quite different in character from the commercial City. Westminster Abbey with its monastery was dedicated in December 1065, within days of Edward's death. William the Conqueror was crowned there in 1066 and almost all monarchs have followed that precedent. Henry III added a Lady Chapel (1220) and in 1245 rebuilt the abbey, inspired by the Gothic cathedrals at Amiens and Reims. Work was finished in 1269 and thereafter the abbey served as the burial place of kings up to the reign of George III. Rebuilding continued during the 14th and 15th centuries, with the magnificent Henry VII Chapel being completed in 1519. In 1540, following the dissolution of the monasteries, the Abbey's property and treasure were confiscated and the Benedictine monks expelled, but its fabric survived. Sir Christopher Wren and Nicholas Hawksmoor added two 225-feet towers that were completed in 1745 and substantial renovation was carried out in the 19th century.

William the Conqueror had Edward the Confessor's palace completely reconstructed, while his son, William Rufus, added Westminster Hall, the largest Norman hall in Europe. The building was remodelled in the late 14th century and has survived almost intact. Its splendid hammer-beam roof means that supporting piers are not required. Early parliaments met in the Abbey's Chapter House until 1529 when Henry VIII moved to Whitehall Palace. It was not until the 17th century that Parliament assembled in

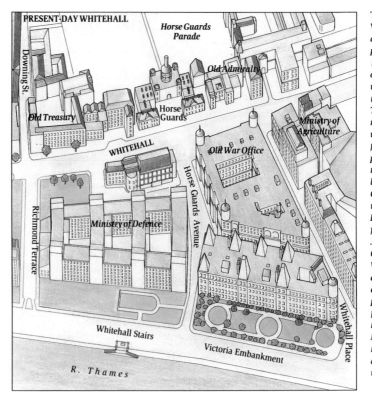

PRESENT-DAY WHITEHALL

The changing face of Whitehall Whitehall's ministries and official buildings span the period from the 17th century to the 1980s (map left). A reconstruction of the Palace of Whitehall as it would have appeared c.1695 (below) shows a marked contrast. The Banqueting Hall is all that remains of the vast Palace which was destroyed by fire in 1698. Just a few years earlier it was a maze of passages and yards traversed by two public highways running between Westminster and Charing Cross and down to the Thames. As well as the royal apartments, the Palace contained over 2,000 rooms for courtiers. To the west lay St. James's Park, redesigned for Charles II in the manner of Versailles, with shady avenues and an ornamental lake. 'Pelemele', a cross between croquet and golf, was played along the tree-lined walk in front of St. James's Palace and gave its name to Pall Mall and the Mall. Many plans were conceived to replace Whitehall Palace in an appropriately grand manner, but were never implemented.

Westminster Hall. Almost all of the old palace of Westminster was destroyed by fire in 1834, only the Jewel Tower and Westminster Hall survive. Charles Barry and Augustus Pugin won a competition to design the new parliament house, and combined a simple ground plan with an intricate Gothic design. The building that resulted is 940 feet long, has over 1,000 apartments and covers more than eight acres. On the south side the Victoria Tower rises to 336 feet but it is the smaller Clock Tower (316 feet), housing the clock and bell named Big Ben, that has made Westminster world-famous.

The third component of Westminster is Whitehall, which was once the site of a great palace. In the early 16th century, Cardinal Wolsey acquired York House and remodelled it in magnificent Tudor style. Henry VIII seized it in 1529, renaming it the Palace of Whitehall. Apartments in this, the largest palace in Europe, were reserved for visiting kings from Scotland prior to union in 1603. The name Great Scotland Yard evokes that function. Further embellishments were undertaken for the Stuart kings, including the classical Banqueting Hall (1619-22) designed by Inigo Jones. William III disliked

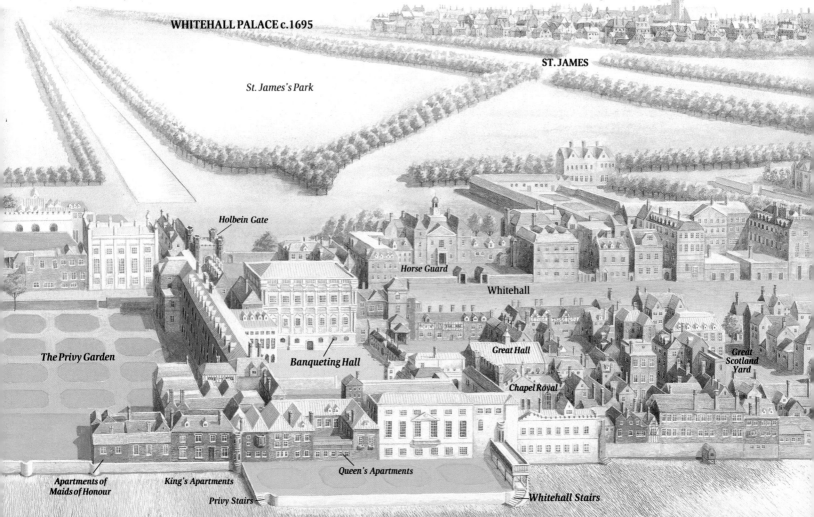

WHITEHALL PALACE c.1695

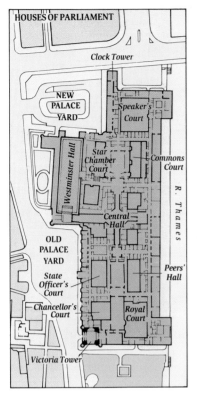

HOUSES OF PARLIAMENT

Clock Tower

NEW PALACE YARD

Speaker's Court

Westminster Hall

Star Chamber Court

Commons Court

R. Thames

Central Hall

OLD PALACE YARD

Peers' Hall

State Officer's Court

Chancellor's Court

Royal Court

Victoria Tower

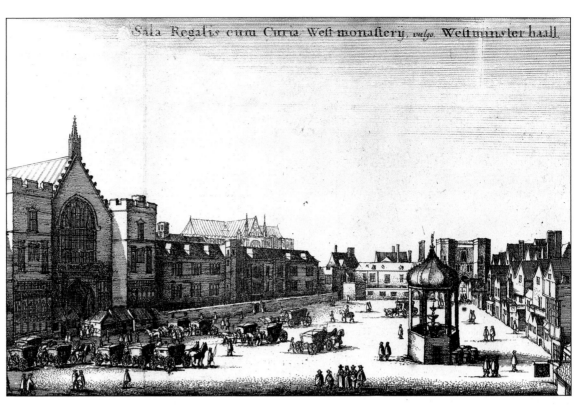

Sala Regalis cum Curia Westmonasterij, vulgo. Westminster haall.

Whitehall, preferring St. James's Palace and the clean air of Hampton Court, which was better for his asthma. In 1698 most of Whitehall Palace burned down; only the Banqueting Hall survived. The present appearance of Whitehall is the product of several centuries of building. Downing Street dates from the later 17th century and the Horse Guards from Georgian times. The Home Office, Foreign and Commonwealth Offices and the Treasury are Victorian (*page 95*). The massive Ministry of Defence was completed in 1959 and the Department of Health building as recently as the later 1980s.

The palace of Westminster was turned over to the State in 1529. From 1547, Parliament met in St. Stephen's Chapel and subsequently Westminster Hall. The Palace is shown in an engraving of the 1640s by Hollar (above). Virtually all the palace was burnt down in 1834, a fact proclaimed in a contemporary poster (right). The present Gothic revival building (plan top) was designed by Sir Charles Barry and Augustus Pugin (1837-60), and is depicted (below) in a painting by John Anderson, 1872.

DREADFUL FIRE!

VILLAGE LONDON
area built up by:

- 1813
- 1872
- 1897
- 1934
- present day
- parks and open spaces
- ○ village name in 1800

The urban sprawl *The map (above) shows the spread of London over the past two centuries. In 1800, the city was ringed with villages and hamlets set amidst meadows and market gardens. The vast majority of these villages were subsequently engulfed in suburbia. However, the remnants of old villages may still be traced in the form of street patterns, historic parish churches and houses, and fragments of village greens. The names of scores of London's villages survive on the name plates of railway and Underground stations.*

Chapter 11 Places in London
LONDON'S VILLAGES

London is richly endowed with neighbourhoods which, despite their absorption into the growing metropolis, have still retained their village character. The history of these villages varies greatly. Some, such as Shoreditch and Stepney, originated as agricultural settlements very close to the edge of the City, and soon became incorporated into the urban fabric. Others were at a greater distance from the City and remained free-standing settlements amidst the fields of Middlesex, Surrey, Kent or Essex until well into the 18th century and beyond. They were ultimately converted into suburbs by a number of factors:

turnpike roads brought Hampstead and Highgate into the parameters of 18th-century London, while 19th-century railway developments accounted for the absorption of Balham and Enfield. In the 20th century, Underground lines meant that even far-flung centres such as Edgware and Morden became part of Greater London. In addition, London has its share of planned 'villages' such as Hampstead Garden Suburb, which followed the opening of the Golders Green Underground station in 1907, and Bedford Park, which was described by John Betjeman as 'the most significant suburb built in the last century, probably in the western world'.

Tudor and Stuart monarchs had sought to contain the spread of London through legislation; James I feared that 'soon London will be all

England'. His attempts and those of successive administrations failed. The population of Middlesex, that most suburban of Home Counties, rose from a mere 70,900 living in hamlets, villages and market towns in 1801, to 792,000 in 1901 and an astonishing 950,000 just four years later. Not until the Green Belt legislation of 1938 and 1947 was London's suburban spread brought to a notable and controversial halt.

Villages such as Harefield are still distinctly recognisable as such, and not too much imagination is needed to identify the village structure of Blackheath, Clapham, Hampstead or Highgate. Traces, often more obscure, of London's historic villages abound elsewhere; beneath the fabric of present-day suburbia the structure of many of London's villages still survives.

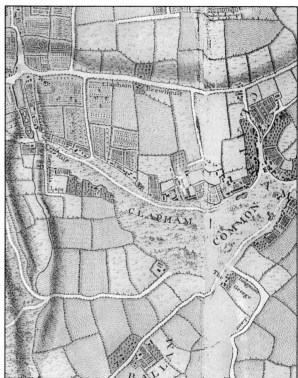

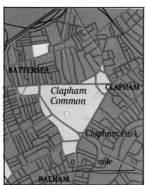

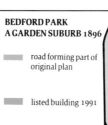

From villages to suburbs *The cartographer, John Rocque (1745), depicted Hampstead (above left) and Clapham (above right) when they were still largely rural, long before they were engulfed by bricks and mortar (maps centre). Constable's painting, 'Sir Richard Steele's cottage, Hampstead' c.1821 (above) shows a still rustic village street. The view of Holy Trinity Church, Clapham, 1845 (above right) is still much the same today.*

Bedford Park *This garden surburb (right) was laid out by Norman Shaw and other architects who designed the spacious, distinctive houses in the vernacular style of the 'Arts and Crafts' movement of the late 19th century. They were provided with their own church, inn, shops and tennis courts. The painting (below) depicts Tower House (right) and Queen Anne's Grove (1882), designed by Norman Shaw.*

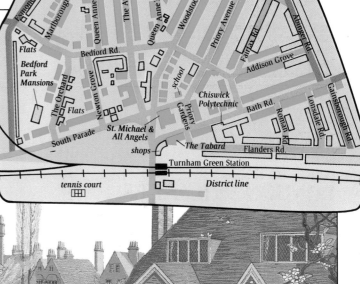

BEDFORD PARK
A GARDEN SUBURB 1896

road forming part of original plan

listed building 1991

0   yards   220

DOCKLANDS

Before the 1980s the vast complex of dock basins stretching for five miles downstream from the Tower of London formed unknown territory for most Londoners. The individual docks and warehouses were surrounded by high walls for protection against theft. Only dockers and seamen entered this enclosed world on which so much of the wealth of London and Britain had once depended. By the late 1960s the small upstream docks were closed, overtaken by fierce competition from European ports, poor labour relations and a lack of space for modern ships and cargo-handling techniques. After a decade of great uncertainty, the large Royal Docks were closed in 1981, and the basins at Tilbury – 26 miles downstream – were all that remained of the Port of London. As the docks closed, so their associated industries collapsed. No fewer than 18,000 dock-related jobs were lost between 1966 and 1981, when local unemployment stood at 24 per cent.

In 1969 a project was launched to revitalise St Katharine Dock (25 acres in size), situated just beyond the

Tower of London, by establishing a large hotel, a marina, a variety of housing and the World Trade Centre. In 1976 the Docklands Joint Committee, comprising representatives of local councils, presented the London Docklands Strategic Plan. It came at a most unpropitious time, at the height of a world economic crisis. The Plan sought to modernise the Docklands through more local authority housing and new industrial jobs. But private investment was sluggish, transport facilities outdated and large areas of derelict land remained. In 1981 the London Docklands Development Corporation (LDDC) was set up, and a year later an Enterprise Zone was created on the Isle of Dogs, enabling developers to enjoy rate and tax benefits.

The LDDC launched a vast regeneration programme, using public expenditure on new roads, sewers and public transport to attract market-led investment. A light railway was built – initially to link Docklands to Tower Hill. This now extends from Bank in the City to Beckton in the Royal Docks, and work has begun on extending the railway south of the river to Lewisham. In 1987 London City Airport was opened for business flights to

Canary Wharf (bottom right) *Early in 1988 the Canadian developers, Olympia and York, started work on the vast Canary Wharf scheme at the West India Docks. A 50-storey office tower, at One Canada Square, soaring to 800 ft and with 1.2 million sq. ft of office space, commands magnificent views of the dock basins and the Thames. The planners' dreams were challenged by the changing financial climate, fluctuating demand for office space and the terrorist bomb attack in 1996. Confidence has since returned and two new office towers are under construction: the 1.1 million sq. ft HSBC building and the 1.2 million sq. ft Citigroup building.*

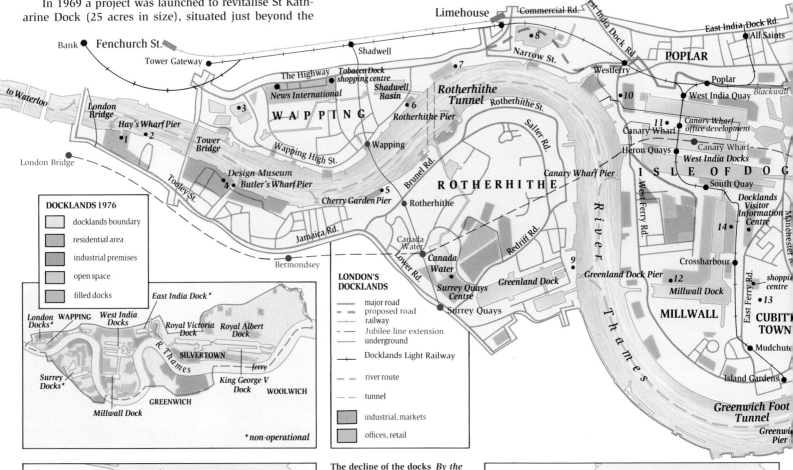

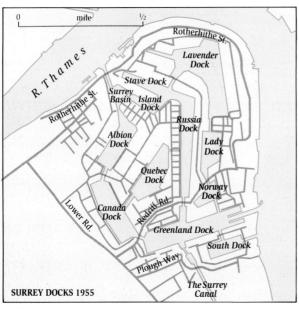

SURREY DOCKS 1955

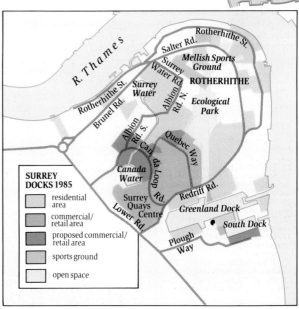

The decline of the docks *By the mid-1970s the smallest docks were no longer operational and the larger ones faced an uncertain future (map above left). Housing clustered outside their security walls, with industry still occupying important stretches of riverside. The Surrey Commercial Docks (map left) had flourished in the 19th century, importing timber from Scandinavia, grain and, later, dairy products. The docks were closed to traffic in 1970. Many basins were filled in and work started on landscaping and new roads. During the 1980s almost 5,000 dwellings were completed (map right), and many more have been built since. New industries moved in, including the Daily Mail's printing works. New open spaces on this site include a seven-acre ecological park.*

Europe. By 1996 £1,744 billion of state money had attracted a further £6,277 billion of private investment. New offices, housing, shopping centres and recreation facilities were constructed and land values soared. Incoming professionals and executives purchased new homes at prices beyond the reach of most East Enders. Many new jobs in Docklands involved transferring firms from elsewhere in London, bringing skilled workers with them, and the LDDC has been criticised for promoting outsiders' interests rather than concentrating on locals' needs for jobs and homes. The future of Docklands remains poised delicately on the assumption that London will continue to grow as the world's leading financial centre, creating more jobs, requiring more office space and attracting more international investment.

Transformation of Docklands (below) *The small upstream docks were among the first to be closed in the late 1960s. Now, following major projects at St Katharine Dock and Wapping, their transformation is complete. South of the river the Surrey Docks peninsula has experienced substantial regeneration. Work continues on the Isle of Dogs, where Canary Wharf towers above West India Docks, and new offices and homes are transforming the Millwall Dock. East of the River Lea, vast stretches of land and water in the Royal Docks are now being regenerated. London City Airport and new housing and shopping facilities at Beckton contribute to the revival of this area, which is now served by the Docklands Light Railway. The challenge of area regeneration will continue to be met in the 21st century.*

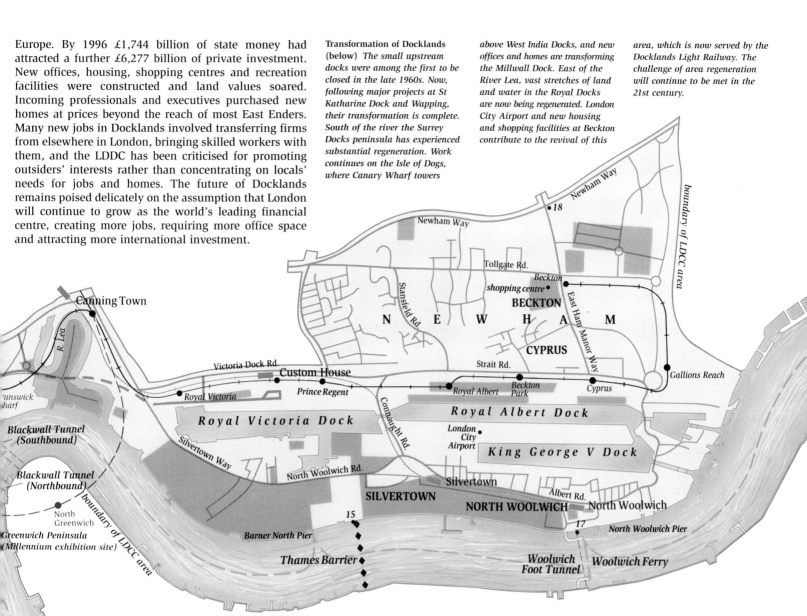

DOCKLANDS IN THE 90s: SITES OF INTEREST

1. Hay's Galleria, large complex of shops and eating places on the site of the old Hay's Wharf.
2. HMS Belfast, 1938 warship preserved as a floating museum.
3. St. Katharine Dock popular waterside attraction: new and old buildings, boats and yachts.
4. St. Saviour's Dock, conservation area includes fine 19th-century granaries and mills.
5. St. Mary's Church, Rotherhithe's fine Georgian village church.
6. The London Hydraulic Company Pumping Station – now closed, the 1890s buildings are to be a recording studio.
7. Free Trade Wharf, a large riverside redevelopment on the site of a once busy wharf, two original saltpetre warehouses remain.
8. Limehouse Basin, which linked the Port of London with England's inland waterway system.
9. Greenland Dock, built 1699, one of the few large areas of water remaining.
10. Cannon Workshops, built as stores and workshops, now used by small businesses.
11. Canary Wharf, largest commercial development in Europe, shops, office space, restaurants.
12. The Daily Telegraph printing plant.
13. Mudchute, now an urban farm, on site of silted mud dredged from Millwall Docks and grassed over.
14. The London Arena, 12,000 seat sports arena.
15. The Thames Barrier, London landmark, includes visitor centre.
16. London City Airport, flights for business sector to western European capital cities.
17. Railway Museum, in former North Woolwich Station built in 1847.
18. Mountaintop Ski Village, 'Beckton Alps', created out of industrial waste.

UNDERGROUND LONDON

Every Tube passenger hurtling between stations knows that there is a London under London. Nevertheless, a comprehensive map of the tunnels, pipes and cables beneath the capital has never been drawn up, so developers have to scrutinise many individual documents to avoid damage. Some of these hidden underground features date back centuries; many were installed after 1850. New rail tunnels, water mains and fibre-optic cables are being positioned now and more are planned.

Throughout the centuries, many tributaries of the Thames have been covered over and integrated into London's sewer system. The Walbrook, in the heart of the ancient city, was covered in the late 15th century. The same fate awaited the Fleet, whose polluted lower course was covered in 1760. Soon its upper stretches were also culverted and in 1855 were converted into a sewer. The culverted Westbourne passes through an aqueduct visible above Sloane Square Underground Station.

Streams and specially-dug ditches were London's earliest sewers, but could not cope with ever-increasing quantities of effluent. Cesspools harboured disease, while flushing lavatories, which started to be used in 1810, added to the pollution of the Thames. Urban cesspools were banned in 1848 and still greater amounts of waste were directed into the river. The Thames became so badly polluted that in 1858 MPs resolved to strengthen the powers of the newly-created Metropolitan Board of Works. By 1865, 70 miles of intercepting sewers had been installed to carry effluent to two major outfalls downstream of London. A network of drains serving every street and ultimately every building was installed as the built-up area expanded. London's sewer authorities are now faced with the expensive challenge of repairing and replacing the Victorian brickwork. As sewers crumble and collapse, foul water can stagnate, providing breeding grounds for armies of rats.

Work on railways under London began in 1860, when a 'cut and cover' tunnel was started beneath New Road from Paddington towards Moorgate. Trains started to run in 1863, linking the northern rail termini with the City. Other 'cut and cover' sections were dug to complete a misshapen 'Circle' in 1884. Smoke-consuming engines, where smoke was diverted into a tank behind the engine by means of an exhaust and released when trains emerged overground, operated until electrification in 1905. The remainder of the Underground system was made up by deep 'tube' tunnels, excavated by special machines. Separate companies constructed individual lines and operated electric trains. First came what is now part of the City branch of the Northern Line (1890), followed by the middle stretch of the Central Line (1900). The Waterloo and City Line ('the Drain') had been opened in 1898. Additional tube tunnels were excavated in the 1930s, playing a crucial role in London's suburban expansion. Yet more tunnels were dug for the Victoria Line (opened 1968) and the original Jubilee Line (completed 1979). The Jubilee Line extension from the West End to Docklands is due for completion in 1999. The Underground now runs over 100 miles of tunnels, though the total Underground network is considerably longer, since much runs on the surface.

Underground railways are not only used by the travelling public. For over 60 years the Post Office has operated its own underground electric railway to speed mail between sorting offices and main railway stations. During the Second World War, a three-mile section of the Central Line (Leytonstone to Gants Hill) served as an underground factory for aircraft components. Art treasures were stored in the Aldwych spur of the Piccadilly Line, and Down Street Station (closed in 1932) became a bunker for the War Cabinet. Platforms, shafts and even staircases in 79 tube stations provided makeshift night shelters for ordinary Londoners. In the early 1940s, eight deep tunnels

The construction of the first tunnel under the river – the Thames Tunnel – was greeted with some scepticism (cartoon, far right). Engineered by Marc Brunel, it took 18 years (1825-43) and cost £614,000 to complete. It never lived up to its promise as a pedestrian tunnel and was sold to the East London Railway Company.

The map (right) is the first edition of the Diagram of the London Underground drawn by Henry Beck and issued in January 1933. It shows the structure of the lines, but not their precise location or the relative distances between stations.

This cross-section (above) shows the station at Piccadilly Circus after it was extensively rebuilt in 1923. The complex network of tunnels, platforms, cables and – a recent innovation – escalators, was an engineering triumph.

The map (right) shows an underground view of the city. London's first underground railway, the Metropolitan (between Paddington and Farringdon) opened in 1863. Early lines are just beneath street level; later ones were excavated much deeper. The latest addition is the Jubilee Line extension through Waterloo and London Bridge to Docklands. Some underground stations have since closed down, becoming 'ghost stations'. Main sewers and underground rivers form other important features of subterranean London.

Some churches in London have crypts or vaults, many of which are architecturally fine. The one pictured (below near right) is St Mary Magdalene in Paddington, dating from 1895.

Many of London's sewers are excellent examples of Victorian engineering, such as the one pictured (below centre right). They are now over 100 years old and need constant maintenance and, in some instances, replacement.

The Post Office operates its own railway system beneath central London. Its fully automatic, driverless trains run between main sorting offices and railway terminals. The photograph (far right) shows mail being loaded.

were excavated beneath existing tube tracks, especially along the Northern Line. General Eisenhower used the Goodge Street tunnel as the HQ of America's wartime operations in Europe. In 1948 the deep tunnels at Clapham South provided temporary accommodation for 236 Jamaican immigrants who had sailed to Britain on *The Windrush*. Brixton was the nearest labour exchange and many settled nearby.

Subterranean London has also played a critical role in the city's defence. During the 1930s three subterranean civil defence 'citadels' were installed in north London and 'the Hole in the Ground' was excavated under the Treasury in Westminster to provide secure space for top politicians and civil servants. It is possible to visit the

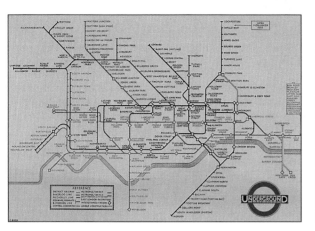

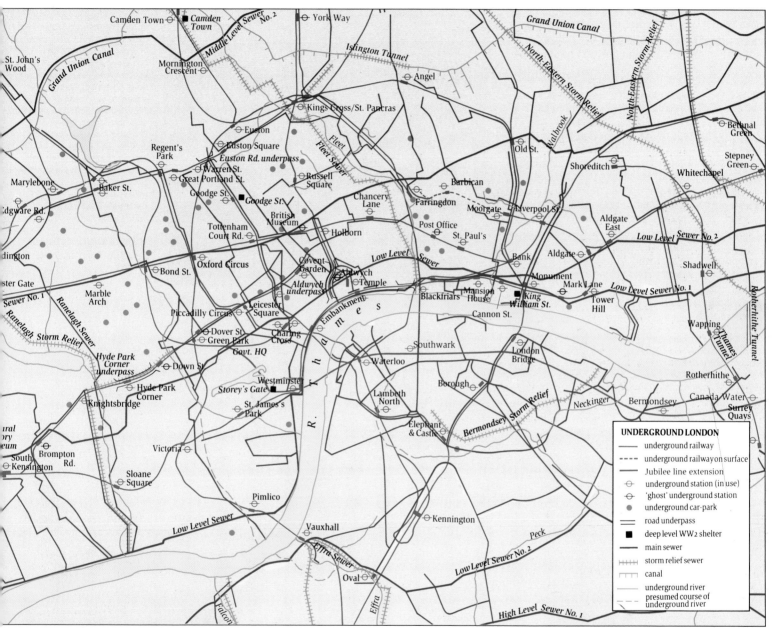

UNDERGROUND LONDON

——— underground railway
- - - - underground railway on surface
——— Jubilee line extension
⊖ underground station (in use)
⊖ 'ghost' underground station
● underground car-park
▬ road underpass
■ deep level WW2 shelter
——— main sewer
+++++ storm relief sewer
⊓⊓⊓ canal
——— underground river
- - - presumed course of underground river

Cabinet War Rooms, which accommodated Churchill and his Chiefs of Staff. Four subterranean fortresses were built more recently beneath central London to house vital government activities in event of nuclear war. The northern section of Kingsway Tram Tunnel has been equipped as an emergency control centre should London be faced with a major disaster.

The London 'Water Ring Main' has been excavated lower than the Underground and functions as a 50-mile aqueduct from Ashford-Sunbury in the south-west to Coppermills in the north-east. Located 120 feet below ground, it is designed to supply half the capital's water and plans are afoot for substantial extension early in the next century.

161

LONDON'S BRIDGES

Over two dozen bridges span the Thames in London, but only Tower Bridge lies downstream of the point selected by the Romans for siting their bridge almost 2,000 years ago. The original, wooden London bridge needed constant repair against the force of the tide and storms. In 1176 work started on a stone bridge which was completed in 1209, with a wooden drawbridge which could be raised on the north side to defend the City, and twenty arches. The bridge was surmounted by tightly-packed shops and houses – 198 of them by the mid-14th century. In places, the bridge thoroughfare was barely more than 9 feet wide. Not surprisingly, traffic congestion posed a serious problem and water-men were kept busy rowing travellers across the river.

For nearly 700 years London Bridge re-mained the only Thames crossing, but in 1729 a wooden bridge was constructed at Putney far to the west. In 1736 an Act of Parliament autho-rised a bridge to be built at Westminster, in the face of fierce opposition from boatmen and the Archbishop of Canterbury, who owned the only ferry large enough to convey horses and carts. Westminster Bridge was opened in 1750 and was followed by another stone bridge at Blackfriars in 1769. Both played a vital role in enabling urban development to extend on to the south bank of the Thames. Between 1758 and 1762, the houses were removed from London Bridge and the two central arches replaced by a single span to ease navigation. Four more bridges were built during the 18th century, two of wood (Kew, 1758-9 and Battersea, 1771-2) and two of stone (Richmond, 1774-7 and a second bridge at Kew, 1783-9). The first iron bridge came early in the 19th century (Vauxhall, 1811-16) and was soon followed by Waterloo (1811-17) and Southwark (1815-19), both designed by John Rennie, whose final task was to design a replacement for London Bridge.

The new century saw not only new styles of bridge but also new motives for construction. Suspension bridges were built for speculative purposes at Hammersmith (1825-7), to encour-age building on the south bank, and in central London, where the old Hungerford Bridge (1841-5) was designed to attract trade to Hungerford market. It survived for less than 20 years and the market site was redeveloped for Charing Cross Station. The first railway bridge was the Grosve-nor (1858-66), which brought the Brighton line to Victoria. Legislation was passed in 1877 which enabled the Metropolitan Board of Works to buy all of London's privately-constructed

London's Bridges *Old London Bridge was rebuilt in 1831, designed by John Rennie. It is pictured (left) in the late 19th century, congested with people and traffic. Westminster Bridge (top right, painted c. 1750) was opened in 1750, designed by Charles Labelye. A new technique was used, with caissons (large wooden boxes) being floated to where the piers were to be built, sunk and pumped dry to enable construction to take place inside them. Tower Bridge (1894) was designed to be in architectural harmony with the Tower of London. It is pictured (above right) under construction (c. 1889), and (far right) complete, with bascules raised (c. 1895) enabling tall ships to pass underneath. The Albert Bridge (right), 1873, designed by Roland Mason Ordish is one of several suspension bridges spanning the Thames.*

London Bridge *(reconstructed, below, as it might have appeared in the 16th century) was completed in 1209. The most extraordinary of the bridge buildings was the highly decorated Nonsuch House, brought in pre-fabricated sections from Holland and built in 1577. In stark contrast, traitors' heads rotted on pikes at the gateway to the Bridge.*

Kew Bridge 1759 (1903)

Kew Railway Bridge 1869

Hammersmith Bridge 1827 (1887)

Chiswick Bridge 1933

Barnes Railway Bridge 1849 (1891-5)

Richmond Lock Bridge 1894

Twickenham Bridge 1933

Richmond Railway Bridge 1848 (1908)

Richmond Bridge 1777 (1939)

Putney Bridge 1729 (1886)

Putney Railway Bridge 1889

Teddington Lock Bridge (1888-9)

bridges and abolish payment of tolls for their use. Previously only London and Westminster bridges had been toll-free. The pinnacle of achievement came with the construction of Tower Bridge (1885-94) which opened to enable tall shipping to reach the Pool of London. The story of London's bridges in the 20th century has been one of widening, rebuilding and replacement. John Rennie's London Bridge met a strange fate when it was replaced by the present bridge between 1968 and 1972 – it was sold for rebuilding at Lake Havasu City, Arizona.

Blackfriars Bridge 1769 (1869)

Alexandra Railway Bridge 1866 (widened 1890s)

Waterloo Bridge 1817 (1942-4)

Tower Bridge 1894

Hungerford (Railway) Bridge 1845 (1864)

Millennium Bridge 2000

London Bridge late 1st century AD (1973)

Southwark Bridge 1819 (1921)

Westminster Bridge 1750 (1862)

Lambeth Bridge 1862 (1932)

Grosvenor Railway Bridge 1858-66 (1963-7)

Chelsea Bridge 1858 (1937)

Albert Bridge 1873

Vauxhall Bridge 1816 (1906)

Battersea Bridge 1772 (1890)

THE BRIDGES OF LONDON

⌣ bridge

+++ railway bridge

where original bridge has been replaced, date of present bridge is shown in brackets

Battersea Railway Bridge 1863

Blackfriars Bridge *was designed by Robert Mylne, who completed the bridge in 1769. The picture (right) is taken from a stereoscopic card (1860).*

Wandsworth Bridge 1873 (1940)

Nonsuch House

LONDON BRIDGE

UNIVERSITIES AND MUSEUMS

London is one of the museum capitals of the world, with over 300 museums and galleries in the metropolitan area, visited 20 million times a year. They range in scale from national collections, such as the British Museum, to small specialist museums.

The British Museum was founded in 1753 and comprised three privately-owned public collections, bought by public lottery for £300,000 and kept in Montagu House, Bloomsbury. Other collections were incorporated over the ensuing half-century – Egyptian antiquities were acquired in the wake of the Napoleonic Wars and the still-contentious Elgin marbles soon afterwards. When, in 1823, George IV's library was added to the Collection, it became imperative to build new premises. A new building was designed by Sir Robert Smirke, and begun in 1823. Nevertheless, new acquisitions meant that space was still at a premium. This situation was alleviated when the Natural History collection was moved to South Kensington in 1881, housed in a building designed by Alfred Waterhouse. Some of the large profits made by the Great Exhibition of 1851 had been used to buy land in South Kensington, on which the Victoria and Albert, the Science and Natural History Museums were later built. Bloomsbury and South Kensington thus became the original focus of London's most famous collections. More recently, important local history collections in the suburbs have been opened to the public and new thematic museums have been created to celebrate design, the history of

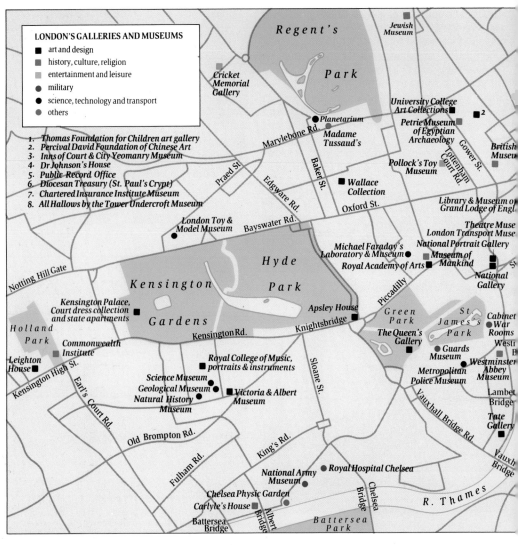

LONDON'S GALLERIES AND MUSEUMS
- ■ art and design
- ■ history, culture, religion
- ■ entertainment and leisure
- ● military
- ● science, technology and transport
- ● others

1. Thomas Foundation for Children art gallery
2. Percival David Foundation of Chinese Art
3. Inns of Court & City Yeomanry Museum
4. Dr Johnson's House
5. Public Record Office
6. Diocesan Treasury (St. Paul's Crypt)
7. Chartered Insurance Institute Museum
8. All Hallows by the Tower Undercroft Museum

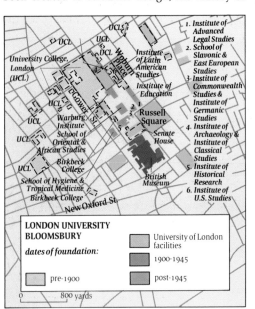

LONDON UNIVERSITY BLOOMSBURY

dates of foundation:
- ■ University of London facilities
- ■ 1900-1945
- ■ pre-1900
- ■ post-1945

0   800 yards

1. Institute of Advanced Legal Studies
2. School of Slavonic & East European Studies
3. Institute of Commonwealth Studies & Institute of Germanic Studies
4. Institute of Archaeology & Institute of Classical Studies
5. Institute of Historical Research
6. Institute of U.S. Studies

London's museums and galleries *The map (above) shows the wide range of galleries and museums within the capital.*

The academic quarter *in Bloomsbury (left) has the British Museum on its southern fringe and the University of London's Senate House (1936) soaring at its centre. The view, below, from the* **Illustrated London News** *(1933), shows 'A great temple of learning to arise in Bloomsbury', designed by the architect Charles Holden but never completed. The classical façade of the main building of University College was built 1827-29, following the design of William Wilkins, architect of the National Gallery. A macabre sight of University College is the clothed skeleton of philosopher Jeremy Bentham, one of its founding fathers, displayed in the main building.*

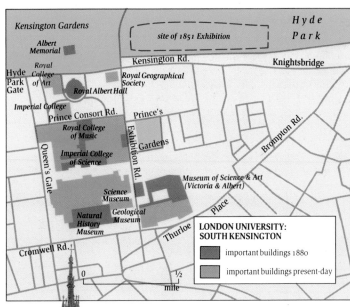

LONDON UNIVERSITY: SOUTH KENSINGTON
- ■ important buildings 1880
- ■ important buildings present-day

0   ½
         mile

South Kensington *The museums and colleges of South Kensington were built on property acquired after the Great Exhibition of 1851 (map above). Museums devoted to Natural History, Science, Geology, and Science and Art (the Victoria and Albert) are found in the southern section. At the centre is Imperial College, famed for science and technology; to the north the Albert Hall (1870) and elaborate Albert Memorial, 1876 (engraving, left) in honour of Queen Victoria's consort. The whole complex is clearly visible in the aerial photograph (right).*

London's Museums *The poster (left), designed by Edward Wadsworth in 1936, advertises the Imperial War Museum (1846), housed in the former Bethlehem Royal Hospital (Bedlam). Lord Leighton's house in Holland Park (1866), is now a museum. Its exotic Arab Hall (right, 1877-9), is based on drawings of Moorish Spain, its walls covered in tiles from Cairo, Damascus and Rhodes. The classical southern facade of the British Museum is depicted in a painting by Dudley Heath (below), 1890. Keats House (below right) Hampstead, was built 1815-16, Keats lived there 1818-21, and wrote Ode to a Nightingale there. Restored 1974-5, it is now a museum.*

London Transport, the moving image, theatre and, of course, London itself.

In the early 19th century, London contained the Inns of Court and historical medical schools but no actual university. In the 1820s, a group of politicians and scholars sought to establish a university which would charge moderate fees and be open to all, irrespective of race, creed or political belief. A site was purchased in Bloomsbury to house what is now University College (opened 1828). With the backing of the Archbishop of Canterbury, a rival university was opened at King's College in the Strand in 1831. In 1836 the University of London was set up as an examining and degree-conferring authority, a role it held until its reorganisation into a federal system of colleges in 1900. The University of London now has 50,000 undergraduates and 22,000 postgraduates. The capital also contains ten new universities. It also has the largest concentration of medical training in Britain.

The University of London *The organisation and geography of the federal University of London is highly complex (map below). A detailed map of the Bloomsbury focus of the University is shown (far left). Some colleges, such as University College and the London School of Economics, are tightly focused, but others occupy buildings in several parts of central and suburban London. Ten thousand students attend Brunel and City Universities, and a further 110,000 are enrolled at ten new universities (former polytechnics) in Greater London. Their premises are quite scattered, reflecting both the important expansion of higher education and the recent trend for institutional mergers.*

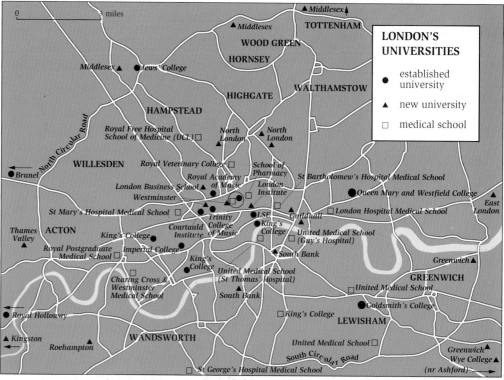

LONDON'S UNIVERSITIES

● established university
▲ new university
□ medical school

MONUMENTS AND CEMETERIES

CEMETERIES

Over 100 cemeteries are to be found within a 9-mile radius of Charing Cross, varying in size from one to 182 acres. Together, they occupy 3,000 acres, the size of a small London borough. In the early 19th century over 50,000 burials were taking place in London each year. London's ancient churchyards were filled to bursting, and many burials took place in scandalous conditions. Campaigners sought to establish cemeteries out in the surrounding countryside, and in 1832 the cholera epidemic finally demonstrated that new cemeteries were essential. Legislation was passed that very year and within a decade commercial companies had established a ring of seven metropolitan cemeteries. Further laws in 1850 enabled more cemeteries to be created.

LONDON'S CEMETERIES (date of opening shown in brackets)

✠ general (Church of England and others)

✝ Roman Catholic

✡ Jewish

1. Abney Park (1840)
2. Acton (1895)
3. Alperton (1917)
4. Barkingside (1923)
5. Barnes Common (1854)
6. Battersea New (1891)
7. Battersea St. Mary's (1860)
8. Beckenham (1877)
9. Bexleyheath (1876)
10. Brockley (1858)
11. Bromley (1905)
12. Brompton (1840)
13. Bunhill Fields (1665)
14. Camberwell (1856)
15. Camberwell New (1927)
16. Charlton (1855)
17. Chingford (1884)
18. Chiswick Old (1888)
19. Chiswick New (1933)
20. City of London (1856)
21. Croydon (1876)
22. Crystal Palace (1880)
23. Ealing and Old Brentford (1861)
24. Eastcote Lane (1900)
25. East London (1872)
26. East Sheen (1903)
27. Edmonton (1884)
28. Edmonton & Southgate (1880)
29. Eltham (1935)
30. Fulham (1865)
31. Golders Green (1902)
32. The Great Northern (1861)
33. Greenwich (1856)
34. Grove Park (1935)
35. Hammersmith (1869)
36. Hammersmith New (1926)
37. Hampstead (1876)
38. Harrow (1887)
39. Harrow (1899)
40. Hendon (1899)
41. Highgate (1839)
42. Isleworth (1879)
43. Alderney Road (1697)
44. Brady Street (1761)
45. East Ham (1919)
46. Fulham Road (1815)
47. Hoop Lane (1895)
48. Kingsbury Road (1840)
49. Lauriston Road (1788)
50. Montague Road (1884)
51. New Sephardi (1733)
52. Old Sephardi (1657)
53. Plashet Park (1896)
54. Rowan Road (1915)
55. West Ham (1857)
56. Willesden (1873)
57. Kensal Green (1832)
58. Kensington Hanwell (1855)
59. Kingston (1855)
60. Lambeth (1854)
61. Lee (1873)
62. Manor Park (1874)
63. Merton & Sutton (1947)
64. Mitcham, Church Road (1883)
65. Mitcham, London Road (1929)
66. Mortlake (1852)
67. North Sheen (1926)
68. South Metropolitan (1837)
69. Nunhead (1840)
70. Old Mortlake (1854)
71. Paddington, Mill Hill (1936)
72. Paddington, Willesden Lane (1855)
73. Pinner (1933)
74. Plaistow (1892)
75. Plumstead (1890)
76. Putney (1855)
77. Putney Vale (1891)
78. Queen's Road (1861)
79. Richmond (1853)
80. Roding Lane (1940)
81. Royal Hospital Chelsea (1692)
82. Royal Hospital Greenwich (1857)
83. St. Marylebone (1854)
84. St. Mary's (1858)
85. St. Pancras & Islington (1854)
86. St. Patrick's (1868)
87. St. Thomas's (1849)
88. Streatham (1892)
89. Streatham Park (1909)
90. Sutton (1889)
91. Teddington (1879)
92. Tottenham (1856)
93. Tottenham Park (1912)
94. Tower Hamlets (1841)
95. Twickenham (1868)
96. Walthamstow (1872)
97. Wandsworth (1878)
98. West Ham (1857)
99. Westminster (1854)
100. Willesden (1891)
101. Wimbledon (1896)
102. Woodgrange Park (1890)
103. Woolwich (1856)

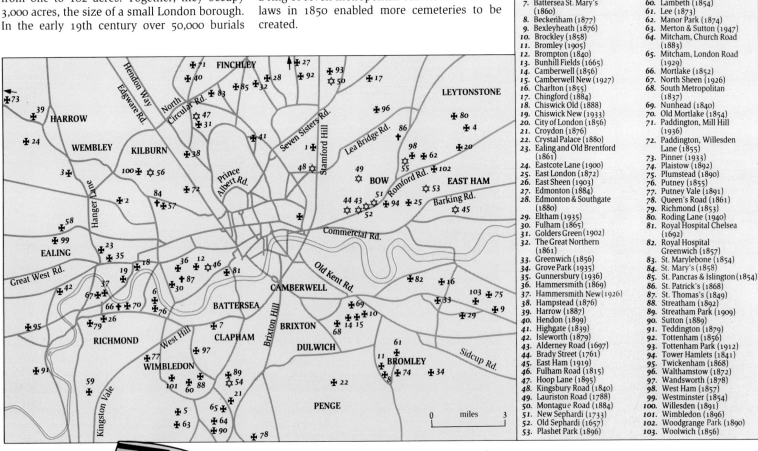

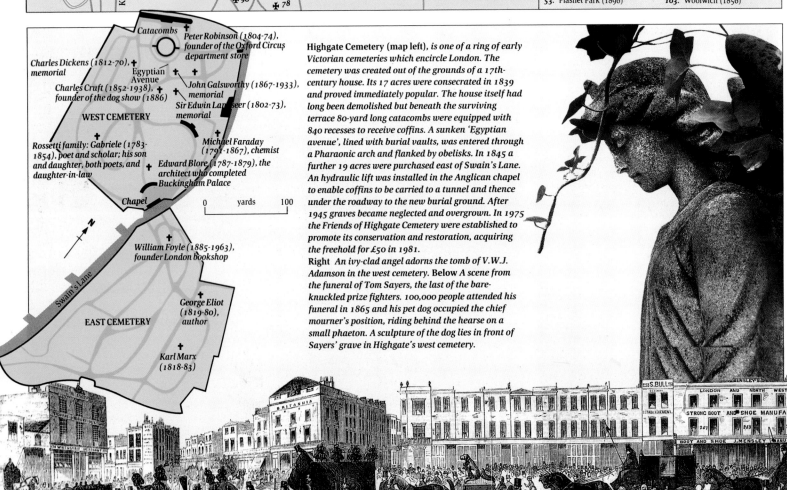

Highgate Cemetery (map left), *is one of a ring of early Victorian cemeteries which encircle London. The cemetery was created out of the grounds of a 17th-century house. Its 17 acres were consecrated in 1839 and proved immediately popular. The house itself had long been demolished but beneath the surviving terrace 80-yard long catacombs were equipped with 840 recesses to receive coffins. A sunken 'Egyptian avenue', lined with burial vaults, was entered through a Pharaonic arch and flanked by obelisks. In 1845 a further 19 acres were purchased east of Swain's Lane. An hydraulic lift was installed in the Anglican chapel to enable coffins to be carried to a tunnel and thence under the roadway to the new burial ground. After 1945 graves became neglected and overgrown. In 1975 the Friends of Highgate Cemetery were established to promote its conservation and restoration, acquiring the freehold for £50 in 1981.*

Right *An ivy-clad angel adorns the tomb of V.W.J. Adamson in the west cemetery.* **Below** *A scene from the funeral of Tom Sayers, the last of the bare-knuckled prize fighters. 100,000 people attended his funeral in 1865 and his pet dog occupied the chief mourner's position, riding behind the hearse on a small phaeton. A sculpture of the dog lies in front of Sayers' grave in Highgate's west cemetery.*

MONUMENTS

Central London is rich in monuments of all kinds and every suburb has at least one memorial recalling the sacrifices of war. Every British ruler since Elizabeth I (with the exception of Edward VIII) is on view in central London, together with politicians, soldiers, admirals, explorers, writers, social reformers and many others – no fewer than 370 statues are found on the outside of the Houses of Parliament alone. The most elaborate monument to an individual is surely the Albert Memorial, composed of granite, sandstone, limestone, slate, marble and mosaics and unveiled in 1876 after more than a decade of work. By contrast, plain Portland Stone is used in two of London's best-known monuments. In the City a 202-feet-high Doric column, the Monument, was designed by Sir Christopher Wren to commemorate the Great Fire. In Whitehall Sir Edward Lutyen's massive Cenotaph, constructed in 1920, honours all the nation's dead.

The oversize bronze statue of Sir Winston Churchill (1874-1965) half faces the House of Commons, (left). In Piccadilly Circus the aluminium statue of Eros (below) honours the philanthropy of the Seventh Earl of Shaftesbury (1801-85).

A stone replica of a howitzer surmounts the Royal Artillery Memorial at Hyde Park Corner (right), on which the names of 49,000 war dead are incised. London's monuments (map below) The greatest single concentration of monuments is in Westminster, especially close to Whitehall, Pall Mall and St James's Park, which formed the 'heart of the empire' in Victorian and Edwardian times. Other impressive clusters are in Chelsea, South Kensington, the City, along the Victoria Embankment and at Greenwich. The discreet Holocaust memorial garden is in Hyde Park.

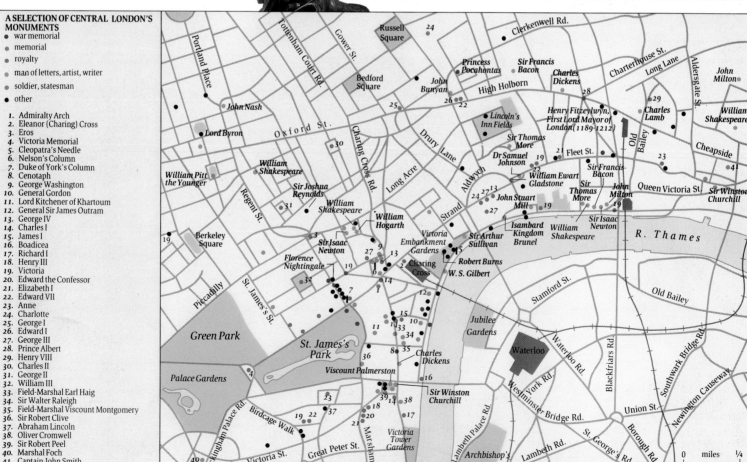

A SELECTION OF CENTRAL LONDON'S MONUMENTS

- ● war memorial
- ● memorial
- ● royalty
- ● man of letters, artist, writer
- ● soldier, statesman
- ● other

1. Admiralty Arch
2. Eleanor (Charing) Cross
3. Eros
4. Victoria Memorial
5. Cleopatra's Needle
6. Nelson's Column
7. Duke of York's Column
8. Cenotaph
9. George Washington
10. General Gordon
11. Lord Kitchener of Khartoum
12. General Sir James Outram
13. George IV
14. Charles I
15. James I
16. Boadicea
17. Richard I
18. Henry III
19. Victoria
20. Edward the Confessor
21. Elizabeth I
22. Edward VII
23. Anne
24. Charlotte
25. George I
26. Edward I
27. George III
28. Prince Albert
29. Henry VIII
30. Charles II
31. George II
32. William III
33. Field-Marshal Earl Haig
34. Sir Walter Raleigh
35. Field-Marshal Viscount Montgomery
36. Sir Robert Clive
37. Abraham Lincoln
38. Oliver Cromwell
39. Sir Robert Peel
40. Marshal Foch
41. Captain John Smith

167

LONDON THAT NEVER WAS

London's architectural history includes countless designs for projects that were never realised. Some were prepared after urban disasters – especially fires – others responded to the quest for grandeur, the need to provide new facilities, or the desire to make use of new technology. Many were speculative designs, others were commissioned by distinguished patrons. Some were works of vision and genius, others were simply banal. All shared a common fate – they never became reality.

The most famous item in this catalogue of urban imagination is Christopher Wren's reconstruction scheme following the Great Fire (1666) which envisaged a radically new approach to the remnants of the medieval City (*below*).

Grandiose visions An artist's impression (*below*) *of Christopher Wren's plan for rebuilding London after the Great Fire, drawn up within a week of the flames being controlled. Narrow streets and cramped courtyards were to be replaced by broad avenues radiating from piazzas and lined with buildings worthy of a great trading capital. Francis Goodwin designed a Grand National Cemetery* (*right*) *in 1824. Inspired by the architecture of Classical Greece, it was to occupy 150 acres and contain buildings modelled on ancient Athens. Distinguished citizens would be buried at its centre, the wealthy in a middle zone, and humbler folk on its margins.*

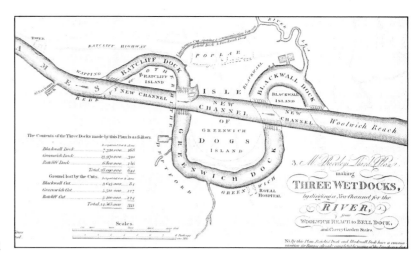

Unrealised schemes *William Reveley's (1796) scheme for straightening the Thames (right) entailed building a ¾-mile 'new channel', which would reduce sailing time as ships would no longer have to cope with adverse winds in Greenwich Reach. New docks and warehouses would improve berthing and cargo handling along the congested river. Many schemes to build elevated railways through the heart of London were proposed, including one to run the trains along the sides of buildings (below). However, cross-London links were achieved only when a number of companies built Underground lines by cut-and-cover methods or by excavating deep 'Tubes'.*

Rival projects were proposed by John Evelyn, who sketched a dozen interconnecting squares for his imaginary city and Robert Hooke, who designed a grid system of roads. Valentine Knight's plan included a canal running from the River Fleet to Billingsgate that would be lined with profit-making quays. Richard Newcourt wanted London greatly enlarged. Each scheme would have been costly and slow to implement. London had to recover fast as it needed to continue competing with its Dutch trading rivals. So the medieval street pattern re-surfaced, simply adorned with new buildings. It was almost three hundred years before the widespread devastation of the Second World War encouraged architectural imagination and inspiration to flourish once again on a vast scale in London.

Urban disasters offered the opportunity for enhancing as well as restoring the capital. Following the terrible fire of 1834 (*page 155*), almost a hundred projects were submitted for the new Houses of Parliament. Numerous speculative designs had been prepared long before then, including projects by Wren, William Kent (1733) and Sir John Soane (1779). Inigo Jones's scheme (1638) for reconstructing and enlarging Whitehall Palace would have given Charles I a grander residence than those of his rivals Philip IV in Madrid and Louis XIV in Paris, while Sir John Soane (1753-1837), designed a great palace for George IV in Green Park.

During the Victorian era, elegant but unrealised projects for churches, museums, exhibition halls, opera houses and parks proliferated, as did more mundane but arguably more essential schemes for cemeteries, hospitals, railways, sewers and much else besides. Together they make up the fascinating story of the London that never was.

LONDON AT THE MILLENNIUM

Cities throughout the world will commemorate the Millennium in their own ways, but London will be the main focus of the United Kingdom's year-long celebration. Over £4 billion has been invested in the capital's leisure industry, with 50 new hotels ready for the Year 2000. The Millennium Commission, funded from the National Lottery, has worked on schemes across the United Kingdom. Fourteen landmark projects and hundreds of regional or local projects have been devised, with the Millennium Experience providing the national focus. After fierce competition between cities, it was decided that Greenwich, in south-east London, should house the Experience which is expected to attract 12 million visitors from the UK and around the globe during 2000.

The choice was a logical one as Greenwich had housed the Royal Observatory, through which the meridian (north/south line) was aligned and from which longitude was measured, since 1675. Despite a growing need for a universal system, it was not until the Washington Conference of 1884 that the leading nations decided on a single prime meridian which would pass through Greenwich. Greenwich Mean Time became the global standard for measuring time. A brass bar in the courtyard of the old Royal Observatory marked 0° longitude. Accurate satellite technology has now redefined the true prime meridian 100 metres to the east.

In 1996 a site covering 120 hectares on the Greenwich peninsula, which lay beyond the LDDC zone of urban regeneration, was selected as the site of the Millennium Experience. It had housed one of the largest gasworks in Europe, as well as containing the southern entrance to the Blackwall road tunnel. The land had been polluted by industrial chemicals and required intensive restoration. The vast

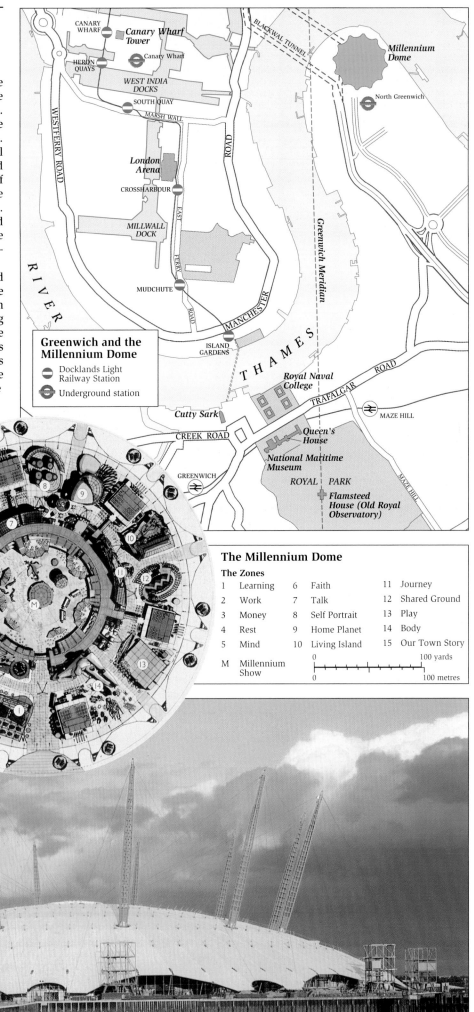

Greenwich and the Millennium Dome

- Docklands Light Railway Station
- Underground station

The Millennium Dome

The Zones

1	Learning	6	Faith	11	Journey
2	Work	7	Talk	12	Shared Ground
3	Money	8	Self Portrait	13	Play
4	Rest	9	Home Planet	14	Body
5	Mind	10	Living Island	15	Our Town Story

M Millennium Show

0   100 yards

0   100 metres

Greenwich has a distinguished *history as a royal residence and maritime base* (map left). *At the heart of the National Maritime Museum is the Queen's House (1635), designed by Inigo Jones. To the south lie the Royal Observatory and Park, while the historic tea-clipper –* the Cutty Sark *– lies close to the Thames. The former Royal Naval College will house part of the University of Greenwich. The Millennium Dome is in the midst of a Thames meander, north-east of Greenwich town centre. The extended Jubilee Line and the southern extension of the Docklands Light Railway enhance access to Greenwich from central London, Canary Wharf and the eastern suburbs.*

Millennium Dome was designed by Sir Richard Rogers. The first of its twelve steel masts was hoisted into place in October 1997. Then came the installation of the Dome's enormous translucent, glass-fibre canopy, with a circumference of 1km. The unique structure was first intended to contain an enclosed central area but this idea was replaced by an open arena surrounded by 14 segments. These house exhibits depicting life, work, and recreation in Britain, the diversity and challenges of the nation's environment and how Britain interacts with the rest of the world. State-of-the-art technology adds to the excitement of the displays and what may be learned from them.

The massive station at North Greenwich on the new Jubilee Line extension will provide the main form of public transport to the Exhibition. In addition, buses and boats will bring visitors from existing railway stations and park-and-ride sites. Early in 2000 the Docklands Light Railway extension to Lewisham will improve access to the historic town centre of Greenwich, whose Observatory, Royal Park and National Maritime Museum attract great numbers of visitors. Surrounding the impressive, but controversial, Dome are the Millennium Gardens and the Millennium River Walk offering views across the Thames. Immediately south-west of the Dome, the environmentally-friendly Millennium Village is being constructed to provide about a thousand homes plus community facilities and may well be expanded in the years ahead.

Millennium funds help finance other commemorative schemes in the capital. The Tate Gallery of Modern Art, partly funded by the Millennium Commission, will occupy the vast remodelled interior of Bankside power station opposite St Paul's Cathedral. This new gallery will house the Tate's collection of 20th-century art; the present building on Millbank will become the Tate Gallery of British Art. The Millennium Bridge will provide the first new Thames crossing in central London since Tower Bridge was opened in 1894. Due for completion in summer 2000, this pedestrian bridge will afford splendid views.

In Bloomsbury, the Great Court of the British Museum has been covered with a translucent roof and will accommodate an education centre and galleries for the museum's African collections. The round Reading Room, which used to form the heart of the British Library (now relocated to the Euston Road), will house an information room on the Museum's collections. The whole scheme is due for completion at the end of 2000. Millennium funds have also helped construct a new education centre at London Zoo, concentrating on the theme of conserving biological resources throughout the world.

Millennium celebrations will also focus on the largest Ferris wheel ever to be built which rises 135m from the South Bank almost opposite the Houses of Parliament. The 32 enclosed capsules of the 'London Eye' will afford views not only of London but as far out as Tunbridge Wells, Guildford, Windsor, Luton, Stansted and Rochester. Each ride will last about 30 minutes. At the end of five years, the wheel will be dismantled for erection at another site.

Millennium celebrations *will bring new activity and regeneration to the southern bank of the Thames – at Greenwich, in Southwark and on the South Bank. The giant Ferris wheel* (above), *known as the 'British Airways London Eye' was built horizontally on temporary islands in the Thames and then hoisted into position in Jubilee Gardens. The Millennium Bridge* (below), *due to open in 2000, will provide a direct pedestrian link between the City and the fast-developing cultural and leisure quarter in Southwark, site of the Globe Theatre and the new Tate Gallery of Modern Art. The massive structure of Bankside Power Station (designed by Sir Giles Scott but redundant for many years) has been transformed into the 'New Tate', with funding from the Millennium Commission, the London Borough of Southwark, and many other investors. A new glass structure spans the length of the building, adding two extra floors and providing natural light for the top galleries.*

Beating fierce competition *from other cities, Greenwich was selected in 1996 to host the nation's main Millennium commemoration. The vast Dome* (left), *evoking the Dome of Discovery at the Festival of Britain (1951), brings a dramatic new feature to the drab post-industrial riverside.*

Using funds *from the National Lottery, the Millennium Commission has supported hundreds of commemorative projects throughout the United Kingdom which promoted environmental sustainability, education, community activities and urban regeneration* (box below).

MILLENNIUM PROJECTS IN GREATER LONDON
British Museum Great Court Scheme: *transformation of the Inner court into a covered piazza*
London Zoo: *education centre*
Millennium Bridge: *new pedestrian bridge*
Tate Gallery of Modern Art: *a new national art gallery in the disused Bankside power station*
Southwark Cathedral: *visitor centre and grounds*
Hungerford Bridge: *creation of two new foot bridges*
Croydon Skyline: *new lighting to project large images in the town centre*
Mile End Park: *theme park along the Grand Union Canal*
Millennium Centre Eastbrookland Country Park: *environmental education centre, Dagenham*

LONDON OF THE FUTURE

London's history has been chequered by phases of growth interspersed with periods of decline. At the Millennium, the capital faces new challenges. During the 20th century, the metropolitan economy underwent profound de-industrialisation and it depends heavily on providing a great variety of services, from administration, higher education, tourism and insurance to high-level banking and commerce. London is a 'world city', comparable in financial strength with New York and Tokyo, but its role in the European Union – and the Euro-economy – is being challenged by Paris, Frankfurt and Berlin. Central London now heads the league of European regions in terms of wealth generation, but its ring of Victorian suburbs and areas in the East End are some of the most deprived localities in Britain, with high unemployment, low average incomes and worrying levels of educational achievement. The very rich and the very poor are co-existing, almost side by side.

In the present era of economic globalisation, London's main advantage continues to lie with the vast experience of its bankers, traders and financiers. However, the capital needs to modernise the facilities it provides, not only to satisfy international big business, but also the millions of ordinary Londoners who service its daily economy. In many respects London's history may be seen as a handicap for such a prosperous future – so much needs to be restored, renewed or replaced. Public transport is outdated, roads are congested, sewers and water supplies need repair or replacement, as do generations of London's housing.

As London's economy has changed over the past half century, so land has become available for redevelopment. Redundant railway land opened new opportunities at King's Cross and other locations. As riverside docks closed, the challenge of re-using the banks of the Thames in imaginative ways had to be faced. The upstream docks and the Isle of Dogs have already changed beyond recognition, but parts of the Royal Docks, former industrial sites and other areas of 'brown land' await regeneration.

Over the last quarter century *many of the City's Victorian and post-war office blocks have been torn down and replaced by vast and architecturally dramatic corporate headquarters. Viewed from across the Thames or from the air* (above), *the effect is both impressive and controversial. It expresses London's confidence and sustained dynamism as a global financial centre and world city for the new millennium.*

In the city centre, riverside development continues to pose social and architectural challenges. A new masterplan is being conceived to transform the 27-acre site of the South Bank Centre into a user-friendly cultural environment. Increasingly, the management of London needs to be implemented in a sustainable way, by reducing pollution of air and water, conserving and enhancing green spaces, and working to minimise environmental damage. It may be that the election of a Mayor for

The London Authority HQ *After long consultation it was decided to construct a new building to house the chamber and offices of London's future elected authority* (above). *Both the design and site of the new headquarters are controversial. The design has been compared with a giant eye or a massive headlamp. The site comprises vacant land immediately across the Thames from the Tower of London and close to London Bridge. This choice expresses faith in the future of London, while recognising the tremendous importance of its historic legacy.*

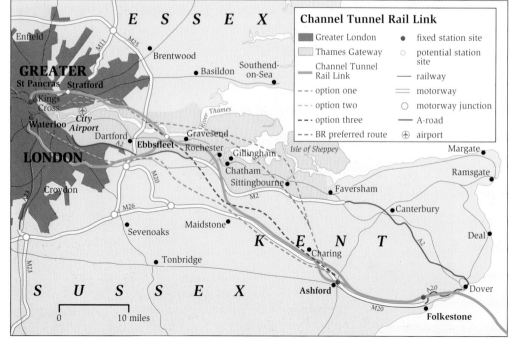

The Channel Tunnel Rail Link *The map* (left) *shows alternative routes that were proposed for the fast rail link. These options gave rise to much controversy among environmentalists and the residents of Kent. The chosen route will cross the Thames at Ebbsfleet, have a station at Stratford and terminate at St Pancras, where special platforms will be built for Eurostar trains. It is hoped that regeneration in the economically depressed Thames Gateway area to the east of London will be boosted by these new connections. Work on the 70 miles of track, one sixth of which will be underground will take an estimated six years to complete. Some Channel Tunnel services will continue to use the new terminal at Waterloo International.*

London (in 2000) will place sustainability more clearly on the political agenda.

Important investments in mass transit are being made but much more is needed to improve London's Underground and surface railways by modernising stations, trains and signalling systems, by maintaining existing tunnels and by providing new lines beneath the capital's crowded city centre. The Jubilee Line extension facilitates links between central London and Docklands. To the west, the direct rail link from Paddington provides an express service to Heathrow Airport. Further proposals for the 21st century include a north-south rail link (Thameslink 2000), a CrossRail link running east-west and the so-called Chelsea-Hackney line.

Waterloo International is the present terminal for Eurostar rail services from Paris and Brussels but St Pancras has been chosen as the final terminal and will eventually be served by a fast rail link to the Kent coast. Over 95 million passengers use London's five international airports, with over 58 million travelling through Heathrow and over 27 million using Gatwick. If a fifth terminal is built at Heathrow, the airport will be able to handle over 80 million passengers a year by 2015. London's roads form a patchwork of old and new. They are woefully inadequate to cope with the current volume of traffic and pose serious challenges for the new century. Controversial road-building schemes have been shelved because of high costs and environmental disruption. Campaigners have called for serious restrictions on the use of private cars in central London. Without doubt, existing roads need to be managed better and public transport to be greatly enhanced, not only for London's residents and workforce, but also for the almost 30 million tourists who visit the city each year.

The British Library *The book collections of Britain's major library have been moved form cramped premises within the British Museum to a splendid new library at St. Pancras (map right), which seats 1,200 and has sections dealing with humanities, rare books, music, manuscripts and science and technology. The famous Round Reading Room in the old library forms the heart of Norman Foster's British Museum Great Court Scheme (above). It has been covered by a vast glass-and-steel roof and will house reference books and databases. The general public will have access late in 2000.*

Located on a traffic roundabout at Waterloo, the new state-of-the-art IMAX cinema (below) seats almost 500 and shows 2D and 3D large-format films. This cylindrical building, enclosed in glass, forms a striking addition to the townscape and contributes to the regeneration of the area which is a gateway to London for many visitors. New luxury apartments and university facilities add to the transformation of this part of the South Bank.

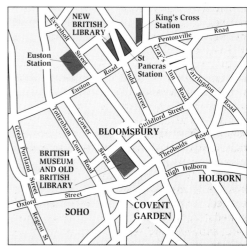

ETYMOLOGY OF LONDON PLACE NAMES

Not all London place-names are mysterious – **Highgate** actually does refer to a 'high gate' (probably a *toll-gate*) – but many are. London place-names have histories of different lengths. New ones are created from time to time; **Highgate's** history goes back to 1354; **Hendon's** ('high hill') to 959; **London's** to 115. Many changes can take place in the course of time: *Londinium* has become **London**; *Tidwulf's tree* has become **Elstree**; *St. Vedast's Lane* has become **Foster Lane**. In some cases, the cumulation of changes of form has led to amazing changes of meaning: today's **Cannon Street** began life as, in effect, 'candlemaker street'. Fortunately, the written records of these changes are copious enough to allow us to parade before you an ordered sequence of well-preserved fossils that preceded the place-name's latest form, like the well-attested evolution of the horse from *Eohippus* to *Red Rum*.

For reasons of space this appendix concentrates on the names of London's rivers and districts (or 'villages'), and the names of streets in its old City. They are listed in alphabetical order; after each map name comes its history given as clearly and succinctly as necessary. Each such history has one or more of the following parts in the following order:

Name /pronunciation (where noteworthy) / : 'meaning of the name as it might have appeared to those who coined it' [history of the forms of the name : analysis of the parts of the name (with their individual meanings) ; additional information] *cross-references to other relevant* **Names**.

Off-putting conventions have been avoided; **bold-face** means 'name listed in this appendix'; ? means 'perhaps'; ‹ means 'from'. The analysed parts of each name come from Old English (in use till about 1100) unless otherwise specified. Pronunciations are given in the version of the International Phonetic Alphabet used in the *Collins English Dictionary*; a key to these symbols is provided below.

The symbols used in the pronunciation transcriptions are those of the International Phonetic Alphabet. The following consonant symbols have their usual English values: *b, d, f, h, k, l, m, n, p, r, s, t, v, w, z.* The remaining symbols and their interpretations are listed in the tables below.

English Sounds

ɑ:	as in *father* ('fɑ:ðə), alms (ɑ:mz), *clerk* (klɑ:k), *heart* (hɑ:t), *sergeant* ('s:dʒənt)
æ	as in *act* (ækt), *Cædmon* ('kædmən), *plait* (plæt)
aɪ	as in *dive* (daɪv), *aisle* (aɪl), *guy* (gaɪ), *might* (maɪt), *rye* (raɪ)
aɪə	as in *fire* ('faɪə), *buyer* ('baɪə), *liar* ('lɪə), *tyre* ('taɪə)
aʊ	as in *out* (aʊt), *bough* (baʊ), *crowd* (kraʊd), *slouch* (slaʊtʃ)
aʊə	as in *flour* ('flaʊə), *cower* ('kaʊə), *flower* ('flaʊə), *sour* ('saʊə)
ɛ	as in *bet* (bɛt), *ate* (ɛt), *bury* ('bɛrɪ), *heifer* ('hɛfə), *said* (sɛd), *says* (sɛz)
eɪ	as in *paid* (peɪd), *day* (deɪ), *deign* (deɪn), *gauge* (geɪdʒ), *grey* (greɪ), *neigh* (neɪ)
ɛə	as in *bear* (bɛə), *dare* (dɛə), *prayer* (prɛə), *stairs* (stɛəz), *where* (wɛə)
g	as in *get* (gɛt), *give* (gɪv), *ghoul* (gu:l), *guard* (gɑ:d), *examine* (ɪg'zæmɪn)
ɪ	as in *pretty* ('prɪtɪ), *build* (bɪld), *busy* ('bɪzɪ), *nymph* (nɪmf), *pocket* ('pɒkɪt), *sieve* (sɪv), *women* ('wɪmɪn)
i:	as in *see* (si:), *aesthete* ('i:sθi:t), *evil* ('i:vᵊl), *magazine* (ˌmægə'zi:n), *receive* (rɪ'si:v), *siege* (si:dʒ)
ɪə	as in *fear* (fɪə), *beer* (bɪə), *mere* (mɪə), *tier* (tɪə)
j	as in *yes* (jɛs), *onion* ('ʌnjən), *vignette* (vɪ'njɛt)
ɒ	as in *pot* (pɒt), *botch* (bɒtʃ), *sorry* ('sɒrɪ)
əʊ	as in *note* (nəʊt), *beau* (bəʊ), *dough* (dəʊ), *hoe* (həʊ), *slow* (sləʊ), *yeoman* (jəʊmən)
ɔ:	as in *thaw* (θɔ:), *broad* (brɔ:d), *drawer* ('drɔ:ə), *fault* (fɔ:lt), *organ* ('ɔ:gən)
ɔɪ	as in *void* (vɔɪd), *boy* (bɔɪ), *destroy* (dɪ'strɔɪ)
ʊ	as in *pull* (pʊl), *good* (gʊd), *woman* ('wʊmən)
u:	as in *zoo* (zu:), *do* (du:), *queue* (kju:), *shoe* (ʃu:), *spew* (spju:) *true* (tru:), *you* (ju:)
ʊə	as in *poor* (pʊə), *skewer* ('sjʊə), *sure* (ʃʊə)
ə	as in *potter* ('pɒtə), *alone* (ə'ləʊn), *furious* (fjʊərɪəs), *nation* ('neɪʃən), *the* (ðə)
ɜ:	as in *fern* (fɜ:n), *burn* (bɜ:n), *fir* (fɜ:), *learn* (lɜ:n), *term* (tɜ:m), *worm* (wɜ:m)
ʌ	as in *cut* (kʌt), *flood* (flʌd), *rough* (rʌf), *son* (sʌn)
ʃ	as in *ship* (ʃɪp), *election* (ɪ'lɛkʃən), *machine* (mə'ʃɪ:n), *mission* ('mɪʃən), *pressure* ('prɛʃə), *schedule* ('ʃɛju:l), *sugar* ('jʊgə)
ʒ	as in *treasure* ('trɛʒə), *azure* ('æʒə), *closure* ('kləʊʒə), *evasion* (ɪ'veɪʒən)
tʃ	as in *chew* (tʃu:), *nature* ('neɪtʃə)
dʒ	as in *jaw* (dʒɔ:), *adjective* ('ædʒɪktɪv), *lodge* (lɒdʒ), *soldier* ('səʊldʒə), *usage* ('ju:sɪdʒ)
θ	as in *thin* (θɪn), *strength* (stɛŋθ), *three* (θrɪ:)
ð	as in *these* (ðɪ:), *bathe* (beɪð), *lather* ('lɑ:ðə)
ŋ	as in *sing* (sɪŋ), *finger* ('fɪŋgə), *sling* (slɪŋ)
ᵊ	indicates that the following consonant (*l* or *n*) is syllabic, as in *bundle* ('bʌndᵊl) and *button* ('bʌtᵊn)
x	as in Scottish *loch* (lɒx).
əɪ	as in Scottish *aye* (əɪ), *bile* (bəɪl), *byke* (bəɪk).

Length
The symbol : denotes length and is shown together with certain vowel symbols when the vowels are typically long.

Stress
Three grades of stress are shown in the transcription by the presence or absence of marks placed immediately *before* the affected syllable. Primary or strong stress is shown by ', while secondary or weak stress is shown by ˌ. Unstressed syllables are not marked. In *photographic* (ˌfəʊtə'græfɪk), for example, the first syllable carries secondary stress and the third primary stress, while the second and fourth are unstressed.

[In addition, the following special letters are used in the transcription of old manuscripts: *æ/Æ* (as above), *ð* (as above), þ (pronounced like *ð* or *θ* above), ȝ (like ʒ above in shape, but pronounced like *j* or *x* above or like the voiced equivalent of *x*). In earlier forms of a place name, ' at the end of a word shows that the scribe has omitted one or more letters; a bar over a letter (*ā*) shows that *m* or *n* has been omitted. In the analysis of place-name elements, such a bar over a vowel functions like ':' in the pronunciations: it shows that the vowel is pronounced long.]

A

Abchurch Lane [1291-92 (*Abbechirchelane*); by the church of *St. Mary Abchurch*]

Acton: 'oak-tree estate' [1181 (*Acton(e)*): *āc* 'oak' + *tūn* 'estate, home farm']

Acton Common ‹ **Acton**

Aldermanbury: 'alderman's manor; aldermen's manor' [c. 1130 (*Aldremanesburi*), 1336 (*Aldermannebury*): *(e)aldermann* 'elder, chief; alderman' + *burh* 'stronghold, fortified manor']

Aldersgate *see* **Aldersgate Street**

Aldersgate Street: 'street of Ealdred's gate' [1303 (*Aldresgatestrete*): *Ealdre(d)* (name) + *geat* 'gap; gate' + *strǣt* '(paved) road, street' (‹ Latin)]

Aldgate: 'ale gate' [c. 1095 (*Ealsegate*), 1275 (*Alegate*): *ealu* 'ale' + *geat* 'gap; gate'; City gate previously called *Æst Geat* 'east gate']

Aldgate Street: 'ale-gate street' [13th C. (*Alegatestrat*), c.1600 (*Aldgate street*); now **Aldgate**] ‹ **Aldgate**

The Artillery Ground [named officially in 1746; the 'Old Artillery Ground' near Spitalfields is mentioned in the diary of Samuel Pepys (1633-1703)]

Ashford: 'Eccel's ford' [1042-66 (*Echelsford*), 1062 (*Exforde*), 1488 (*Assheford*): *? Ecceles* (from name) + *ford* '(river-) ford']

B

The Bailey [c.1166 (*Bali*), 1298 (*le Bail*), 1431-32 (*la Baillye*), 1444-45 (*Old Bailey*): Middle English *bail, bailey* 'outermost wall or court (as of a castle), bailey' (‹ Old French); sited just west of the City wall; now *Old Bailey*]

Baker Street [after William *Baker*, 18th-century property-developer of land here that he acquired from W.H. Portman]

Balham /'bæləm/: 'Bælga's village; Bælga's river-bend land' [957 (*Bælgenham*), 1472 (*Balam*): *?Bælgen* (from name ?‹ *bealg* 'rounded') + *hām* 'settlement, village' or *hamm* 'river-bend land']

Banstead: 'bean-place' [675 (*Benstede*), 1062 (*Bænstede*), 1198 (*Banstede*): *bēan* 'bean' + *stede* 'place, stead']

Barbican Street [1348-49 (*Barbecanstret*), 1377 (*Barbycanstret*): Middle English *barbican* 'outer fortification, typically with a tower' (‹ Middle French ‹ Medieval Latin ‹ Persian) + *strǣt* '(paved) road, street' (‹ Latin)]

Barking: 'Berca's folk' [c.730 (*Bercingum*), 1086 (*Berchinges*): *Berc(a) (name) + -ing(as)* 'followers of']

Barnes: 'barns' [1086 (*Berne*), 1222 (*Bernes*), 1387 (*Barnes*): *bern* 'barn' (‹ *bere* 'barley' + *ærn* 'building for the stated purpose')]

Barnet /'bɑ:nɪt/: 'burnt (site)' [c. 1070 (*Barneto*): *bærnet* 'burnt'; area cleared for settlement by burning part of the former Middlesex and Hertfordshire woodland]

Basinger Lane, Basing Lane: 'Basing('s) lane' [1279-80 (*Basingelane*), 1324 (*Basingeslane*), 1544 (*Basinglane*): *Basing* (attested local surname, perhaps originally from *Basing* in Hampshire) + *lane* 'lane'; now part of **Cannon Street**] *compare* **Bassishaw Street**

Bassishaw Street: 'the Bas(s)ings' manor street; Bassishaw-ward street' [1279 ((the street of) *Basingeshawe*), c.1600 (*Bassings hall streete*): *Bas(s)ing(a)s* 'followers of Bas(sa)' (perhaps as in *Basing*stoke in Hampshire) + *haga* 'hedge; enclosure, property; town house' + *strǣt* '(paved) road, street' (‹ Latin); *Bassishaw* was also the name of a ward of the City of London; now *Basinghall Street*] *compare* **Basinger Lane**

Battersea: 'Beadurīc's (high) ground' [c.1050 (*Batrices ege*): *Batric* (name) + *ēg* 'high ground']

Battersea Park ‹ **Battersea**

Bayswater Road ‹ **Bayswater:** 'Bayard's watering place' [1380 (*Bayards Watering Place*): *?Baynardus*, henchman of William the Conqueror]

Bearward Lane, Berewards Lane: 'bear-keeper's lane' [1285 (*Berewardeslane*), 1417 (*Berwardeslane*): Middle English *bērewarde* 'bearward, bear-keeper' (‹ *?beraweard*) + *lane* 'lane']

Beckenham: 'Biohha's village' [862 (*Biohhahema mearc*), 1086 (*Bacheham*): *Biohha* (name) + *hām* 'settlement, village' (+ *mearc* 'boundary, landmark')]

Beddington: 'Beadda's folk's estate' [675 (*Bedintone*), c. 905 (*Beaddinctun*), 1229 (*Bedington*): *Bead(da)* (name) + *-ing* '-follower of' + *tūn* 'estate, home farm']

Bedfont: 'hollow spring' [1086 (*Bedefunt, Bedefunde*): *?byde(n), bede(n)* 'hollow (place), depression' + *funta* 'spring' (‹ Celtic ‹ Latin)]

Belleyetteres Lane: 'bell-maker's lane' [1306 (*Belleyetterslane*), 1421 (*Belleterlane*), 1667 (*Billiter Lane*); mainly *Billiter Lane* till the 19th century, but now **Billiter Street**] *see* **Billiter Street**

Bercher's Lane: 'barber's lane' [1193-95 (*Bercheuere lane*), 1260 (*Berchervereslane*), 1300 (*Berchenereslane*), 1372-73 *Berchereslane, Bercherlane*), 1386 (*Birchenlane*), 1493-94 (*Birchinlane*): ?Middle English *berdcherver(e)* 'barber' (‹ *?beardceorfere* ‹ *beard* 'beard' + *ceorfere* 'carver') + *lane* 'lane'; now *Birchin Lane*]

Bermondsey: 'Beornmund's (high) ground' [1086 (*Bermundeseye*): *Beornmund* (name) + *ēg* 'high ground']

Bethnal Green: 'Blitha's-corner green' [13th C. (*Blithehale*), 1443 (*Blethenalegreene*): *Blithe(n)* (from name) + *halh* 'corner/nook of land' + *grēne* '(village) green']

Beverley Brook ‹ **Beverley:** 'beaver rivulet' [693 (*Beferiþi*): *beofor* 'beaver' + *rīð* 'stream' + *-ig* (diminutive)]

Bexley: 'box(-tree) lea/grove' [c.780 (*Bixle*): *?byxe* (variant of *box* 'box-tree') + *lēah* 'grove; meadow, lea']

Bexleyheath ‹ **Bexley**

Billingsgate: 'Belling's gate' [c.1205 (*Bælȝesȝate*), c.1250 (*Bellinges-ȝate*): *Belling* (name) + *geat* 'gap; gate'; the *gate* presumably gave access to the Thames]

Billiter Street: 'bell-maker street' [1298 *Belȝeterslane*), 1349 *Belleȝeterestret*): Middle English *belleyetere* 'bell-founder' + *strǣt* '(paved) road, street' (‹ Latin)] *see also* **Belleyetteres Lane**

Bishopsgate *see* **Bishopsgate Street**

Bishopsgate Street: 'street of the bishop's gate' [1275 (*Bis(s)hopesgatestrete*): *biscop* 'bishop' (‹ Late Latin ‹ Greek *episcopos*) + *geat* 'gap; gate' + *strǣt* '(paved) road, street' (‹ Latin); a gate in the north-eastern part of the City wall was built by order of Erkenwald, Bishop of London 675-693]

Blackheath [1166 (*Blachedefeld*); from the dark colour of the peat thereon]

Blackheath Common ‹ **Blackheath**

Borough Road [*burh* 'stronghold, fortified manor'; from (the) *Borough* (of *Southwark*)] *see* **Southwark**

Bowyer Row [1359 (*Bowiarresrowe*): Middle English *bowiar* 'bowyer, (archery) bow-maker' + *rǣw* 'row' (of trees or houses); now *Ludgate Street* east of **Ludgate**]

Bread Street [1163-70 (*Bredstrate*): *brēad* 'bread' + *strǣt* '(paved) road, street' (‹ Latin); from the baking and selling of *bread* here]

Brent: 'Holy One' [c.974 (*Brægentan*): Latin *brigantiā* 'holy one' (‹ Celtic)]

Brentford: *(River) Brent ford'* [705 (*Bregunt ford*), 1222 (*Brainford*): **Brent** + *ford* '(river-) ford'] ‹ **Brent**

Bridge Street [1193-1212 (*vicus de ponte*), 1226-27 (*Brygestrate*); after **London Bridge;** later *Fish Street*; now *Fish Street Hill*]

Brixton: 'Brihtsige's stone' (1062 (*Brixges stane*), 1279 (*Brixistane*): *Brihtsige* (? name) + *stān* 'stone']

Broad Street [c. 1212 (*Bradestrate*), 1513 (*Brodestrete*):*brād* broad, wide' + *strǣt* '(paved) road, street' (‹ Latin)]

Bromley: 'broom (-plant) lea/grove' [862 (*Bromleag*): *brōm* 'broom' + *lēah* 'grove; meadow, lea']

Brompton: 'broom (-plant) estate' [1294 (*Brompton*): *brōm* 'broom' + *tūn* 'estate, home farm']

Buckingham Palace [?1825: originally *Buckingham House*; built 1703 for John Sheffield, Duke of *Buckingham*; remodelled about 1825 by Nash for George IV]

Bucklersbury: 'Bukerel's manor' [1343 (*Bokerellesbury*), 1477 (*Bokelersbury*): *Bukerel, Bucherel(l)* (attested surname) + *burh* 'stronghold, fortified manor']

Budget Row, Budge Row: 'budge-fur row' [1342 (*Bogerowe*), 1383-84 (*Bugerowe*), 1553-54 (*Bouge Rowe*), 1591 (*Budgerowe*): Middle English *bugee, boge* 'fur of lamb, rabbit or kid' (?‹ French) + Middle English *rowe* ‹ *rǣw* 'row' (of trees or houses); near the skinners' district]

Bull's Cross, Bulls Cross [1540 (*Bellyscrosse*), 16th C. (*Bulls Cross*): ?(Gilbert) *Bolle* (name attested locally in 1235) + *cros* 'cross' (‹ Old Irish ‹ Latin)]

Bunhill Fields: 'bone-hill fields' [1544 (*Bonhilles*), 1567 (*Bonhill Field*), 1799 (*Bunhill Fields Burying Ground*): *bān* 'bone' + *hyll* 'hill' + *feldas* 'fields'; perhaps from the transfer here of *bones* from St. Paul's charnel house for burial; also *Tindals Burying Ground* (1746)]

Bushy Park, Bushey Park (Middlesex) [1650 (*Bushie Park*), 1667 (*Bushey Park*): ?*busc* 'bush, shrub' + (ge)*hæg* 'enclosure']

Butchery [1349 (*Bocherie*)] *see* **Shambles**

C

Camberwell: [1086 (*Cambrewelle*): ? + *well(a)* 'well; spring']

Camden ‹ **Camden Town**

Camden Town [1795; after Charles Pratt, Earl *Camden* of *Camden* Place in Kent, local landowner]

Candlewick Street [1241 (*Kandelwiccestrate*), 14th C. (*Candelwikstrete*, etc.): ‹ *Candelwrichstrete* 'candlemaker street' (by folk etymology); now **Cannon Street**] *see* **Cannon Street**

Cannon Street: 'candlemaker street' [c. 1185 (*Candelwrichstrete*), 1480 (*Canyngesstrete*), 1664 (*Cannon-street*): ?*Candelwyrhta* 'candle-wright, chandler' + *strǣt* '(paved) road, street' (‹ Latin)] *compare* **Candlewick Street**

Canonbury: 'canon's manor' [1373 (*Canonesbury*): Middle English *canoun* 'canon' (‹ Old Northern French ‹ Latin) + *burh* 'stronghold, fortified manor'; after the *canons* of St. Bartholomew's in **Smithfield**, granted land here in 1253]

Carshalton /kɑːˈʃɔːl̩tᵊn; formerly keɪsˈhɔːtᵊn/: 'cress well-spring farm' [1086 (*Aultone*), 1235 (*Cresaulton*), 1279 (*Carshaulton*): *cærse* '(water-) cress' + *ǣwell, ǣwiell* 'spring; well-spring, (stream-) source' (‹ *ēa* 'stream' + *well(a), wiell(a)* 'well, spring') + *tūn* 'estate, home farm']

Carter Lane [1295 (*Carterestrate*), 1349 (*Carterestane*): Middle English *carter(e)* (‹ *crǣt* + *-ere*) + *strǣt* '(paved) road, street' (‹ Latin) or *lane* 'lane']

Caterham /ˈkeɪtərəm; formerly ˈkætərəm/: 'hill-fort village' [1179 (*Catheham*), 1200 (*Katerham*), 1372 (*Caterham*): ?Celtic *cater* 'hill-fort' (‹ Latin) + *hām* 'settlement, village']

Chancellor's Lane [1227 (*Newestrate*), 1278-79 (*Converslane*), 1338 (*Chauncellereslane*), 1454 (*Chauncerylane*): *Newestrate* 'new street'; *Converslane* 'converts' lane' (after the *Domus Conversorum* founded here in 1231-32 by Henry III for Jewish *converts* to Christianity); *Chauncellereslane, Chauncerylane* (from the subsequent use of the *Domus* as seat of the *Chancery* and perhaps office of the *Chancellor*); now *Chancery Lane*]

Charing Cross: 'turning-point cross' [c.1000 (*Cyrringe*), 1360 (*La Charryngcross*): *cierring* 'bend' (in a river or road) + *cros* 'cross' (‹ Old Irish ‹ Latin); from the *cross* put here in 1291 by Edward I to mark the last stage in the funeral procession of his first Queen, Eleanor of Castile, from Harby in Nottinghamshire to Westminster Abbey] *compare* **Waltham Cross**

Charlton: 'peasants' farmstead; peasants' (part of) estate' [1086 (*Cerletun*): *ceorla* '(free) peasants'' + *tūn* 'estate, home farm']

Charlton (Middlesex): 'Cēolrēd's folk's estate' [1086 (*Cerdentone*), 1550 (*Charleton*), 1594 (*Charleton al. Chertington*): ?*Cēolrēd* (name, with *d/l* confusion) + *-ing* '-follower of' + *tūn* 'estate, home farm']

Cheam /tʃiːm/: '(tree-) stump village; (tree-) stump river-bend land' [675 (*Cegeham*), 933 (*Cheham*): ?*cege* '(tree-) stump; underbrush' + *hām* 'settlement, village' or *hamm* 'river-bend land']

Cheapside: 'market-side' [1436 (*Chapeside*): *cēap* 'market; trade' + *sīde* 'side'; formerly *Westcheap*] *see also* **West Cheap**

Chelsea: 'chalk landing-place' [789 (*Celchyth*), 1214 (*Chelchee*): *cealc* 'chalk, lime; limestone' (‹ Latin) + *hȳð* 'landing place, harbour']

Chessington: 'Cissa's folk's hill' [1086 (*Cisendone*), 1279 (*Chessingdone*): *Cis(sa)* (name) + *-ing* '-follower of' + *dūn* 'hill']

Chicken Lane, Chick Lane [1181-89 (*Chikenslane*), 1540-41 (*Cheke Lane*): *cicen* 'chicken' + *lane* 'lane'; where *chickens* were reared and sold]

Chigwell: 'Cicca's folk's spring/well' [1086 (*Cingheuuella*), 1158 (*Chigwell*, etc.): *Cicc(a)* (name) + *-ing* '-follower of' + *wella, wiella* 'well; spring']

Chigwell Row ‹ **Chigwell**

Chingford: 'shingle-ford' [1086 (*Cingefort*): ?*cinge(l)* 'shingle' + *ford* '(river-) ford']

Chipping Barnet: 'market Barnet' [1329 (*Chepyng Barnet*), 1628 (*Chipping Barnet al. High Barnet*): *cēping, cī(e)ping* 'market' + **Barnet**; also *High Barnet*] ‹ **Barnet**

Chislehurst: 'gravel hill' [973 (*Cyselhyrst*), 1159 (*Chiselherst*): *cisel, ceosel* 'gravel; flint; pebble' + *hyrst, herst* 'hill(ock); copse']

Chiswell /ˈtʃɪzwel, -wəl/ **Street:** 'gravel street' [13th C. (*Chysel strate*), 1458 (*Cheselstrete*), late 16th C. (*Chiswell Street*): *cisel, ceosol* 'gravel; flint; pebble' + *strǣt* '(paved) road, street' (‹ Latin); perhaps from the character of the local soil]

Chiswick /ˈtʃɪzɪk/: 'cheese farm' [c.1000 (*Ceswican*): *ciese* 'cheese' + *wīc* 'farm or harbour of the stated kind']

City Road ‹ the City (of **London**)

Clapham: 'hill village' [c.880 (*Cloppaham*): ?*clop* (perhaps from a Germanic word cognate with *hill* and *culminate*) + *hām* 'settlement, village']

Clapham Common ‹ **Clapham**

Clapton: 'hill estate' [1339 (*Clopton*), 1593 (*Clapton*): ?*clop* (*see* **Clapham**) + *tūn* 'estate, home farm']

Clarence House [built 1825 for William IV when Duke of *Clarence*]

Clay Hill [1524 (*Clayhyll*): ?*Clay, Cleye* (local surnames) + *hyll* 'hill']

Clerkenwell /ˈklɑːkənwel/: 'clerics' well; scholars' well [c. 1150 (*Clerkenwell*): Middle English *clerken(e)* 'clerics', scholars' (‹ *clerc* ‹ Latin ‹ Greek) + *well(a)* 'well, spring']

Cock Lane [c.1200 (*Cockeslane*), 1543 (*Coklane*): *cocc* 'cock' + *lane* 'lane'; perhaps from *cock*fighting here]

Coleman Street: 'Coleman's street' [1181-83 (*Colemanestrate*): ?(St. Stephen) *Coleman* (variant name, recorded in 1276, of the former church of St. Stephen here) + *strǣt* '(paved) road, street' (‹ Latin)]

Cordwainer Street [1216-17 (*Corueiserestrate*), 1230 (*Cordwanere-strete*): Middle English *corviser, cordwaner* 'cordwainer: leather-merchant; shoemaker' (‹ Old French ‹ Old Spanish *cordovano* 'cordovan'); now *Bow Lane* and *Garlick Hill*]

Cornhill: 'hill where corn is grown; hill where corn is sold' [1055 (*Cornehulle*): *corn* 'corn' + *hyll* 'hill]

Counter's Creek: 'countess's creek' [after the *Countess* of Oxford, wife of the Earl of Oxford (after whom *Earl's Court* was named)]

Cowley: 'Cofa's lea/grove' [959 (*Cofenlea*), 1086 (*Covelie*), 1294 (*Cowelee*): *Cofen* (from name) + *lēah* 'grove; meadow, lea']

Cranford: 'crane-ford; heron-ford' [1086 (*Cranforde*): *cran* 'crane; heron' + *ford* '(river-) ford']

Crayford: '(River) Cray ford' [1322 (*Crainford*): Celtic *crai* ?'fresh, new; clean, pure' + *ford* '(river-) ford']

Cricklewood: 'crinkled wood' [1294 (*le Crikeldwode*), 1509 (*Crykyll Wood*): ?dialectal *crickled* 'bent' (perhaps related to *crinkled*) + *wudu 'wood']*

Cripplegate: 'low gate' [991-1002 (*Cripelesgate*): *crypel* (as in *cripple*, ‹ *crēopan* 'creep') + *geat* 'gap; gate'; in some dialects *cripplegate* still has a similar meaning]

Crooked Lane [1278 (*la Crokedelane*)]

Croydon: 'saffron valley' [809 (*Crogedene*): *croh* 'saffron + *denu* 'valley']

Crudrun Lane, Godrun Lane: 'Gōdrūn('s) lane' [1180-92 (*Godrun lane*), c. 1206-7 (*Godruneslane*), 1349 (*Gother lane*), 1472 (*Gutterlane*): *Gōdrūn, Guðrún* (woman's name) + *lane* 'lane'; now *Gutter Lane* (by folk etymology)]

Crutched Friars: 'be-crossed friars' [1405 (*le Crouchedfrerestrate*): Middle English *crutched* 'crossed' (‹ Latin) + Middle English *freres* 'friars' (‹ Old French ‹ Latin *fratres* 'brothers') (+ *strǣt* '(paved) road, street' ‹ Latin); after the insignia of the Augustinian *Friars* of the Holy *Cross*, whose house here was founded in 1298]

D

Dagenham /ˈdægənəm/: 'Dæcca's village' [695 (*Dǣccenham*), 1499 (*Dagnham*): *Dæccen* (from name) + *hām* 'settlement, village']

Dalston /ˈdɔːlstən/: 'Dēorla's estate' [1294 (*Derleston*), 1581 (*Darleston*): *Dēorla* (name) + *tūn* 'estate, home farm']

Dartford: '(River) Darent ford' [1086 (*Tarenteford*), 1089 (*Darenteford*), c. 1100 (*Derteford*): Celtic *Darent* 'oak river' + ford '(river-) ford']

Deptford /ˈdetfəd/: 'deep ford' [14th C. (*Depeford*): *deop* 'deep' + *ford* '(river-) ford']

Dicer's Lane, Dicer Lane: 'ditcher's lane' [1275 (*Dicereslane*, *Dikereslane*), 1411-12 (*Diserlane*): *dīc* 'ditch; dyke' + *-ere* '-er: - maker; -worker' + *lane* 'lane'; spellings with *Dicer-* and *Diser-* may show Norman-French influence; later *Rose Street*]

Dollis Brook ‹ **Dollis Farm** (Hendon): 'share of land' [? Middle English *dol* ‹ *dāl* 'share']

Dowgate: 'dove-gate' [1244 (*Douegat*), c. 1600 (*Dowgate*): ?*dūfe* 'dove, pigeon' + *geat* 'gap; gate'; a former watergate and wharf that gave its name to a street now called *Dowgate Hill*]

Down, Downe: 'hill' [1316 (*Doune*): *dūn* 'hill']

Dulwich /ˈdʌlɪdʒ, -ɪtʃ/: 'dill-marsh' [967 (*Dilwihs*), 1210-12 (*Dilewisse*): *dile* 'dill' (the herb) + *wisce* 'marshland' (not *wīc* as in **Greenwich**)]

Dulwich Common ‹ **Dulwich**

E

Ealing: 'Gilla's folk' [c. 700 (*Gillingas*), c. 1170 (*Yllinges*), 1553 (*Elyng*): *Gill(a)* (name) + *-ingas* '-followers of, -people of']

Earl's Sluice [after the first *Earl* of Gloucester, bastard son of Henry I (1086-1135) and lord of the manor of **Camberwell** and **Peckham**]

East Acton [1294 (*Estacton*): *ēast* 'east' + **Acton**] ‹ **Acton**

East Barnet [c. 1275 (*Est Barnet*): *ēast* 'east' + **Barnet**] ‹ **Barnet**

East Dulwich [1340 (*Est Dilewissh*): *ēast* 'east' + **Dulwich**] ‹ **Dulwich**

East Molesey ‹ **Molesey**

East Smithfield [1229 (*Estsmethefeld*), 1272 (*Est Smythefeld*): *ēast* 'east' + **Smithfield**] ‹ **Smithfield**

Eastcheap, East Cheap: 'east market' [c.1100 (*eastceape*): *ēast* 'east' + *cēap* 'market; trade'; former site of a meat-market]

Ebbgate: 'ebb-tide gate' [c. 1190 (*Ebbegate*): *ebb(a)* 'ebb, low tide' + *geat* 'gap; gate'; a former watergate; now *Swan Lane*]

Edgware: 'Ecgi's weir' [c. 975 (*Ǽcges Wer*): *Ǽcge* (name) + *wer* 'weir']

Edgware Road [1574 (*Edgware High Waie*)] ‹ **Edgware**

Edmonton: 'Ēadhelm('s) (folk's) estate' [1086 (*Adelmetone*), 1211 (*Edelmintone*), 1214 (*Edelmeston*), 1369 (*Edmenton*): *Ēadhelm* (name) (+ *-ing* '-follower of' or *-es* '-'s') + *tūn* 'estate, home farm']

Elbow Lane: 'old Bow Lane' [1343 (*Eldebowelane*), c. 1600 (*Elbow lane*): *eald* 'old' + *boga* '(archery) bow; arch, arched bridge' + *lane* 'lane'; *Elbow(e)* ‹ *Eldebowe* (by folk etymology, as the lane has a bend); now *College Street*]

Elstree: 'Tidwulf's tree' [785 (*Tiðulfes treow*), 1188 (*Tidulvestre*), 13th C. (*Idulfestre*), 1320 (*Idelstre*), 1487 (*Illestre*), 1598 (*Elstre*): *Tidwulf* (name) + *treow* 'tree']

Eltham /ˈeltəm/: 'Elta's village' [1086 (*Elteham*): *Elte* (from name) + *hām* 'settlement; village']

Enfield: 'Ē(a)na's field' [1086 (*Enefelde*), 13th C. (*Enesfeud*), 1293 (*Enfeld*): *Ē(a)na* (name) + *feld* 'field']

Enfield Highway [14th C. (*Alta Via*), 1610 (*the kings highe way leading from Waltham Cross toward London*)] ‹ **Enfield**

Epsom: 'Ebbi's village; Ebbi's river-bend land' [933 (*Ebesham*), 973 (*Ebbesham*), 1086 (*Evesham*), 1404 (*Epsam*): *Eb(bi)* + *hām* 'settlement, village' or *ham(m)* 'river-bend land']

Erith /ˈɪərɪθ/: 'gravel landing' [695 (*Earhyð*), 1610 (*Eryth*): *ēar* 'gravel; mud, earth' + *hyð* 'landing-place, harbour']

Esher /ˈiːʃə; formerly ˈeʃə/: 'ash (-tree) district' [1005 ((to) *Ǽscæron*), 1062 (*Esshere*): *æsc* 'ash(-tree) ' + *scearu* 'share, district; boundary']

Euston [after *Euston* in Suffolk, seat of the Duke of Grafton, lord of the manor of **Tottenham Court** when the *Euston* Road was built (1756-75)]

Ewell /ˈjuːəl/: 'well-spring' [675 (*Euuelle*), 1066 (*Ǽwelle*): *ǣwell, ǣwiell* 'spring; well-spring, (stream-) source' (‹ *ēa* 'stream' + *well(a), wiell(a)* 'well, spring')]

F

Faitour Lane: 'layabouts' lane' [1292 (*Faytureslane*), 1568 (*Feter Lane*): Middle English *faitour* 'vagrant; beggar' (‹ Anglo-French ‹ Old French ‹ Latin) + *lane* 'lane'; now *Fetter Lane*]

Falcon [(*Falcon Brook*): after the *Falcon* Inn, near Clapham Junction]

Farnborough: 'fern hill' [1180 (*Ferenberga*), 1242 (*Farnberg*): *fearn* 'fern, bracken' + *beorg* 'hill, mound; barrow']

Feltham /ˈfeltəm/: 'field village; mullein (-plant) village' [969 (*Feltham*), 1086 (*Feltehā*): *feld* 'field' or *felte* (name of various plants, such as *mullein, marjoram, couch grass*) + *hām* 'settlement, village']

Fenchurch Street [1377-78 (*Fancherchestret*), 1510 (*Fanchurche Strete*)] ‹ **Fenchurch:** 'fen church' [1337 (*Fanchurche*): *fenn* 'fen, marsh' + *cirice* 'church' (‹ Late Greek)]

Finameur Lane, Finamour Lane [1316 (*Fynamoureslane*), c. 1600 (*Finimore lane or fiue foote lane*): *Finamour* (attested local surname, of French origin) + *lane* 'lane'; later *Fye Foot Lane*]

Finchley: 'finch lea/grove' [c. 1208 (*Finchelee*): *finc* 'finch' + *lēah* 'grove; meadow, lea']

Finks Lane [1231-45 (*Finkeslane*), 1305-6 (*Fynghis Lane*), 1326 (*Fyncheslane*): *Fink* (attested local surname ‹ *finc* 'finch') + *lane* 'lane'; by the church of *St. Benet Fink* (same surname); now *Finch Lane*]

Finsbury: 'Fin's stronghold' [1231 (*Vinisbir'*), 1535 (*Fynnesbury*): *Fin* (‹ *Vin* name, perhaps Scandinavian) + *burh* 'stronghold, fortified manor']

Finsbury Fields ‹ **Finsbury**

Fleet Lane [1544 (*Fletelane*): *Flete* (name of river or person) + *lane* 'lane'] ‹ **River Fleet**

Fleet Street [1271-72 (*Fletestrete*): (River) *Flete* + *strǣt* '(paved) road, street' (‹ Latin)] ‹ **River Fleet**

Foot's Cray: 'Fot's (property by the River) Cray' [c. 1100 (*Fotescræi*), 1210 (*Fotescraye*): Fot *(name)* + *Cray* (river-name)] *see* **Crayford**

Foster (Vedast) Lane: 'St. Vedast's Lane' [1271 (*Seint uastes lane*), 1321 (*Seint Fastes lane*), 1337 (*Fasteslane*), 1359 (*Fasterslane*), 1428 (*Foster lane*); by the church of *St. Vedast*]

Friday Street [1138-60 (*Fridaiestraite*): *Frīgdæg* 'Friday' + *strǣt* '(paved) road, street' (‹ Latin); from a fish-market there on *Fridays* or a person called *Friday*]

Friern Barnet /ˌfraɪən ˈbɑːnɪt, ˌfriːən ˈbɑːnɪt/: 'friars' burnt (site)' [1235 (*le Bernet, la Barnate*), 1274 (*Frerennebarnethe, Frerebarnet*, etc.): Middle English *freren* 'friars'' (‹ Old French ‹ Latin) + *bærnet* 'burnt'; formerly held by the monastic order of Knights of St. John of Jerusalem] *see* **Barnet**

Fulham: 'Fulla's river-bend land' [c. 705 (*Fulanham*), c. 895 (*Fullanhamme*): *Ful(l)an* (from name) + *hamm* 'river-bend land']

G

Giltspur Street [1547 (*Gyltesporestrete*); the *gilt spur(s)* were made there or displayed on a local signboard; also *Knyghtryders Strete* (1547)]

Godliman Street [? after *Godalming* (in Surrey, formerly a leather-tanning centre); the Cordwainers' Hall was near this street] *compare* **Cordwainer Street.**

Golden Lane: 'Golda's lane' [1291-92 (*Goldinelane*), 1317 (*Goldenlane*): ? *Goldine* (from woman's name *Golda*, attested nearby in 1245) + *lane* 'lane']

Golders Green: 'Le Godere's green; Godyer's green' [1612 (*Golders Greene*), 1790 (*Groles or Godders Green*): (John le) *Godere* or (John) *Godyer* (names in 14th C. records) + *grēne* '(village) green'; the surname *Golder* is not attested]

Gracechurch Street: 'grass-church street' [1284 (*Garscherchestrate*), 1437 (*Gracechirche strete alias Graschirche strete*): *græs* 'grass' + *cirice* 'church' (‹ Late Greek) + *strǣt* '(paved) road, street' (‹ Latin); after the nearby church of *St. Benet Gracechurch*, perhaps thatch-roofed or grass-surrounded]

Gray's Inn Fields ‹ **Gray's Inn:** 'Gray's dwelling-place' [1396 (*Grays Inn in Holborne*): (Reginald de) *Gray* (died 1308; owner of the property) + *inn* 'dwelling, lodging'; formerly called *Portpoole Manor*]

Great Stanmore: 'great stony pool' [793 (*Stanmere*), 1392 (*Great(e) Stanmare*): Middle English *grēat* 'big' (‹ Old English *grēat* 'thick, bulky') + *stān* 'stone' + *mere* 'pool'; *Great* by contrast with *Little Stanmore* (*Whitchurch*)]

Green Hill Green, Greenhill Green (Middlesex, near Harrow) ‹ **Greenhill** [1334 (*Grenehulle*), 1563 (*Grenehill*), 1675 (*Green Hill*); ? after the local family *de Grenehulle, de Grenhulle* or *de Grenehill*]

The Green Park [because embellished with *green*ery (lawns and trees) rather than flowers]

Greenford: 'green (river-)ford' [845 (*Grenan Forda*): *grēne* 'green' + *ford* '(river-) ford']

Greenwich /ˈgrɪnɪdʒ, ˈgrenɪtʃ/: 'green harbour' [964 (*Grenewic*): *grēne* 'green' + *wīc* 'farm or harbour of the stated kind']

Gresham Street [1845; after the nearby *Gresham* College (founded 1579 by Sir Thomas Gresham); previously *Lad Lane* and *Cateaton Street*]

H

Hackney: 'Haca's (high) ground' [1198 (*Hakeneia*): *Hacan* (from name) + *ēg* 'high ground']

Hackney Marsh [1397 (*Hackenemershe*)] ‹ **Hackney**

Hackney Wick: 'Hackney farm' [1294 (*...atte Wyk in Hakeney*): **Hackney** + *wīc* 'farm or harbour of the stated kind'] ‹ **Hackney**

Hammersmith: 'hammer smithy' [1294 (*Hamersmyth'*): *hamor* 'hammer' + *smiððe* 'smithy']

Hampstead: 'settlement, homestead' [959 (*Hemstede*): *hām* 'settlement, village, home' + *stede* 'place, stead']

Hampstead Heath [1543 (*Hampstede Heth*)] ‹ **Hampstead**

Hampton: 'river-bend estate' [1086 (*Hammtone*): *hamm* 'river-bend land' + *tūn* 'estate, home farm']

Hampton Court: 'Hampton manor' [1476 (*Hampton Courte*): **Hampton** + Middle English *court(e)* 'manor' (‹ Old French)] ‹ **Hampton**

Hampton Court Park ‹ **Hampton Court**

Hanwell Common ‹ **Hanwell:** 'cock well' [959 (*Hanewelle*): *hana* 'cock(erel)' + *well(a)* 'well, spring']

Hanworth: 'cock's ward' [1086 (*Haneworde*), 1254 (*Hanesworth*): *hana* 'cock(erel)' (perhaps used as a personal name) + *worð* 'ward, enclosure']

Harefield /formerly ˈhɑːvəl/: 'army field' [1086 (*Herefelle*), 1206 *Herefeld*), 1223 (*Harefeld*): *here* 'army' + *feld* 'field']

Haringey: 'Haering's enclosure' [1201 (*Haringeie*)] *see* **Hornsey**

Harlington: 'Hygered's folk's estate' [831 (*Hygereding tun*), 1362 (*Herlyngdon*), 1521 (*Harlyngton*), 1535 (*Hardington al. Harlington*): *Hygered* (name, with *d/l* confusion) + *-ing* '-follower of' + *tun* 'estate, home farm']

Harmondsworth /formerly ˈhɑːmzwəːθ/: 'Heremōd's ward' [1086 (*Hermodesworde*), 1408 (*Hermesworth*): *Heremōd* (name) + *worð* 'ward, enclosure']

Harrow: 'heathen shrine' [767 (*Gumeninga Hergae*), 825 (*Hearge*), 1234 (*Herewes*), 1347 (*Harwo*), 1369 (*Harowe*), 1398 (*Harowe atte Hille*): (*Gumeninga* 'Guma's folk('s)'+) *hearg* 'heathen sacred place or temple' (+ *hyll* 'hill')]

Harrow on the Hill [1426 (*Harrowe on the Hill*)] ‹ **Harrow**

Hatton: 'heath estate' [1086 (*Hatone, Haitone*): *hǣð* 'heath' + *tūn* 'estate, home farm'; near **Hounslow** Heath]

Havering /ˈheɪvərɪŋ/: 'Hæfer's folk' [1086 (*Haueringas*): *Hæfer* (name) + *-ingas* '-followers of, -people of']

Haverstock /ˌhævəstɒk/ **Hill** ‹ **Haverstock:** 'cattle (live)stock; Stock's cattle' [1627 (*Haverstocke*): ?*aver(ia)* 'cattle' (which may have grazed there) (‹ Late Latin) + *stocc* 'livestock' or *Stock* (name of local family originally from *Stock* in Essex)]

Hayes: 'brushwood' [1177 (*Hesa*), 1610 (*Heys*): ?*hæse* 'brushwood(-land)']

Heathrow: 'heath-side row' [c.1410 (*La Hetherewe*): *hǣð* 'heath' + *rǣw* 'row' (of trees or houses)]

Hendon: 'high hill' [959 (*Hendun*): *hēan* (‹ *hēah* 'high') + *dūn* 'hill']

Heston /nonstandardly ˈhesən/: 'brushwood farm' [1123-33 (*Heston(e)*): ?*hæse* 'brushwood(-land)'+ *tūn* 'home farm, estate']

High Holborn: 'high (part of the River) Holborn' ‹ **Holborn**

Highbury: 'high stronghold' [1274 (*Neweton Barrewe*), c. 1375 (*Heybury*), 1548 (*Newington Barowe al. the manor of Highbury*): *Neweton/Newington* (as in **Stoke Newington**) + *Barrewe/Barowe* (after the landholder Thomas de *Barewe*) + *hēah* 'high' + *burh* 'stronghold, fortified manor'; called *high* because higher than **Canonbury** or *Barnsbury*]

Highgate: 'high (toll-)gate' [1354 (*Le Heighgate*): *hēah* 'high' + *geat* 'gap; gate']

Hillingdon: 'Hilda's hill' [c.1080 (*Hildendune*), 1086 (*Hillendone*), 1274 (*Hylingdon*): *Hilden* (from name, probably short for *Hild*(*rīc*), *Hild*(*wulf*), etc.) + *dūn* 'hill']

Hogsmill River: 'Hogs' mill river' [1638] ‹ **Hogs Mill** [1535 (*Hoggs Myll*), 1564 (*Hoggesmyll*): after (the family of John) *Hog* (attested 1179 in **Merton**)]

Holborn /ˈhəʊbən/: 'valley stream' [959 (*Holeborne*): *holh* 'valley, hollow' + *burna* 'stream, brook'; former name of (the upper part of) the **River Fleet**]

Holloway: 'valley way' [1307 (*Le Holeweye...*), 1553 (*Holowaye...*): *holh* 'valley, hollow' + *weg* 'road, way']

Homerton /ˈhɒmətən/: 'Hūnburh('s) estate' [1343 (*Humburton*), 1581 (*Hummerton*): *Hūnburh* (woman's name ? ‹ *hūne* 'horehound' + *burh* 'stronghold') + *tūn* 'estate, home farm']

Honey Lane [c.1200 (*Hunilane*), 1274-75 (*Honylane*): Middle English *hony* (‹ *hunig* 'honey') + *lane* 'lane'; from bee-keeping here]

Hook: 'hook (of land)' [1227 (*Hoke*), 1680 (*Hook*): *hōc* 'hook, bend, spit' (of land, as by a river)]

Hornsey: 'Haering's enclosure' [1201 (*Haringeie*), 1243 (*Haringesheye*), 1524 (*Harnesey*), 1564 (*Hornsey*): *Haering* (name) + (ge)*hæg* 'enclosure'; **Haringey** and *Harringay* have the same origin]

Hosier Lane [1328 (*Hosiereslane*), 1338 (*Hosierlane*)]

Houndsditch: 'dog's ditch' [1502 (*Hundesdich*), 1550 (*Houndesdyche*): *hund* 'dog, hound' (probably not a personal name as in **Hounslow**) + *dīc* 'ditch; dyke'; after the City of London ditch *Houndsditch* (1275 *Hondesdich*), into which dead dogs and rubbish were discarded]

Hounslow: 'Hund's mound; Hund's barrow' (1217 (*Hundeslawe*), 1341 (*Houndeslowe*): *Hund* (name) + *hlaw* 'mound, mount; burial mound']

Hoxton: 'Hoc's estate' [1086 (*Hochestone*): *Hoc* (name) + *tūn* 'estate, home farm']

Hubbard Lane [1231 (*vicus Sancti Andreæ*), 1252-65 (*venella Sancti Andree Hubert*); by the church of *St. Andrew Hubbard*; now **Philpot Lane**] *see* **Philpot Lane**

Hyde (Middlesex, near Hendon): 'hide of land' [1281 (*la Hyde*): *hīd* 'enough land to support a family' (generally believed to have been 120 acres)]

Hyde Park: 'park at Hyde (in the manor of Ebury)' [1204 (*Hida*), 1543 (*Hide Park*): *hīd* (as in **Hyde**) + Middle English *park* 'park' (‹ Old French)] *compare* **Hyde**

I

Ickenham: 'Ticca's village' [1086 (*Tichehā*), 1176 (*Tikeham*), 1203 (*Tikenham, Ikeham*), 1236 (*Ikenham*): *Ticcan* (from name, with loss of initial *T* through merger with final *t* of preceding *æt* 'at') + *hām* 'settlement, village']

Ilford: '(River) Hyle ford' [1086 (*Ilefort*), 1300 (*Hyleford*): *Hyle* '? trickling stream; ? still stream' (former name of the **River Roding**) + *ford* '(river-)ford'] *see also* **River Roding**

Isle of Dogs [1365 (*marsh of Stebenhithe*), 1593 (*Isle of doges ferm*), 1799 (*Poplar Marshes or Isle of Dogs*); of unknown origin, the present name is probably derogatory]

Isleworth /ˈaɪzᵊlwəːθ/: 'Gīslhere's ward' [695 (*Gislheresuuyrth*), 1231 (*Istleworth*): '*Gīslhere* (name) + *worð* 'ward, enclosure']

Islington: 'Gisla's hill' [c.1000 (*Gislandune*): *Gislan* (from name) + *dūn* 'hill']

Ivy Lane [13th C. (*Yvi lane*), 1280 (*Ivilane*); perhaps from *ivy* on nearby houses; previously *Alsies Lane*]

J

Jewry Street: 'street where Jews live' [1366 (*la Porejewerie*); also *Jewry*; previously **Poor Jewry**]

K

Kennington: 'Cæna's folk's estate' [1086 (*Chenintune*), 1263 (*Kenyngton*): *Cæn(a)* (name) + *-ing* '-follower of' + *tūn* 'estate, home farm']

Kensal Green: 'king's-wood green' [1253 (*Kingisholte*), 1367 (*Kyngesholt*), 1550 (*Kynsale Grene*): *cyning* 'king' + *holt* 'wood, holt; thicket' + *grēne* '(village) green']

Kensington: 'Cynesige's folk's estate' [1086 (*Chenist'*), 1235 (*Kensington*): *Cynes(ige)* (name) + *-ing* '-follower of' + *tūn* 'estate, home farm']

Kensington Gardens ‹ **Kensington Palace**

Kensington High Street ‹ **Kensington**

Kensington Palace ‹ **Kensington**

Kentish Town: 'Kentish estate; Le Kentiss(h)'s estate' [1208 (*Kentisston*), 1278 (*Le Kentesseton*), 1294 (*La Kentishton*), 1488 (*Kentisshtown*): *Kentish* (‹ *Kent* (place-name) or *Le Kentiss(h)* (name or nickname)) + *tūn* 'estate, home farm']

Keston: 'Cyssi('s) stone('s folk's boundary)' [862 (*Cystaninga mearc*), 1086 (*Chestan*), 1205 (*Kestan*): *Cyssi* (name) + *stān* 'stone' (+ *-inga* '-followers' of' + *mearc* 'boundary, boundary marker')]

Kew: 'quay on a neck of land' [1327 (*Cayho*): Middle English *kai* 'quay' (‹ Old French ‹ Celtic) + *hōh* 'neck/spur of land']

Kew Gardens ‹ **Kew**

Kidbrook: 'kite brook' [1202 (*Ketebroc*): *cēta* ‹ *cȳta* 'kite (bird)' + *brōc* 'brook']

Kilburn: 'royal stream; cows' stream; kiln stream' [c.1130 (*Cuneburna*), 1181 (*Keleburne*): *cyne-* 'royal' or *cȳna* 'cows' or *cyln* 'kiln' + *burna* 'stream, brook']

King Edward Street [1843: after *King Edward* (VI) (1537-53), who endowed Christ's Hospital school here; previously **Stinking Lane**] *see* **Stinking Lane**

King William Street [built 1829: after *King William* (IV) (1765-1837), who opened it]

Kings Cross: 'the King's cross-roads' [from a statue of *King* George IV at a cross-roads here from 1830 to 1845; formerly *Bradford (bridge), Battlebridge*]

Kings End [1550 (*Kings End*): owned then by *King's* College Cambridge, but perhaps already so called after the *King* family prominent locally since at least 1296]

Kingsbury: 'the King's stronghold' [1044 (*Kynges Byrig*): *cyning* 'king' + *burh* 'stronghold, fortified manor']

Kingsland: 'the King's land' [1395 (*Kyngeslond*)]

Kingston: 'the King's demesne' [838 (*Cyninges tun*): *cyning* 'king' + *tūn* 'estate, home farm']

Kingston Hill ‹ Kingston

Kingston upon Thames ‹ **Kingston, River Thames**

Knightrider Street: 'knight's street' [1322 (*Knyghtridestrete*): *cniht* 'youth; servant, soldier; knight' + *ridere* 'knight; rider' + *strǣt* '(paved) road, street' (‹ Latin); *knight* and *rider* overlapped in meaning for a time]

Kyroun Lane: 'Cynrūn('s) lane' [1259 (*Kyrunelane*), 1275 (*Kyroneslane*): ?*Cynrūn* (woman's name) + *lane* 'lane'; later *Maiden Lane*]

L

La Riole [1331 (*la Ryole, la Riole*), 1455-56 (*le Royall*): *La Ryole* (name of house here ‹ *La Réole* wine-exporting town in Bordeaux); later *Royal Street* (*Royal* ‹ *Réole* by folk etymology); now *College Hill*]

Ladle Lane [c.1300 (*Ladelane*): Middle English *ladel* 'ladle' (‹ *hlædel*) + *lane* 'lane'; *ladles* may well have been made here]

Laleham: 'withe village; withe river-bend land' [1042-66 (*Læleham*): *lǣl* 'twig, withe' + *hām* 'settlement, village' or *hamm* 'river-bend land']

Lambard's Hill: 'Lambert's rise' [1283 (*Lamberdeshul, Lambardeshull*), 1645 (*Lambert-Hill*), 1659-60 (*Lambeth Hill*): *Lambert, Lamberd* (surname) + *hyll* 'hill'; hill in such London street names means 'steep street'; now *Lambeth Hill* (*Lambeth* ‹ *Lambert* by folk etymology)]

Lambeth: 'lambs' landing-place' [1088 (*Lamhytha*), 1312 (*Lambhehithe*): *lamb* 'lamb' + *hȳð* 'landing-place, harbour']

Lambs Conduit Fields ‹ **Lamb's Conduit** [?1577: (William) *Lamb(e)* (name of the conduit's builder) + Middle English *conduit* 'aqueduct' (‹ Anglo-French ‹ Old French ‹ Medieval Latin]

Lampton: 'lamb farm' [1376 (*Lampton feld*), 1438 (*Lamtonfeld*), 1611 (*Lambton*), 1633 (*Lampton*): *lamb* 'lamb' (with *b* as *p* before *t*) + *tūn* 'estate, home farm' (+ *feld* 'field')]

Leadenhall Street: 'street of the lead(-roofed) hall' [1605 (*Leaden Hall Street*): after *Leadenhall* (a large local house)]

Lee: 'lea/grove' [1086 (*Lee*): *lēah* 'grove; meadow, lea']

Les(s)nes(s) Heath ‹ **Les(s)nes(s):** 'pasture promontory' [1086 (*Lesneis*): *lǣs* 'pasture; meadow' + *nǣs* 'headland, promontory, cape'; from its projection into the **Erith** Marshes]

Leveroune Lane: 'Lēofrūn's lane' [1233 (*Le Vrunelane*), 1331 (*Lyveroneslane*), 1353 (*Leverounelane*), 1604 (*Lither lane al. Liver lane*), 1682 (*Leather Lane*):? *Lēofrūn* (woman's name) + *lane* 'lane'; now *Leather Lane*]

Lewisham: 'Liof's village' [862 (*Liofshema mearc*), c.1060 (*Liofesham*): *Liof* (name) + *hām* 'settlement, village' (+ *mearc* 'boundary, landmark')]

Leyton: '(River) Lea demesne' [c.1050 (*Lugetune*): **Lea** + *tūn* 'estate, home farm'] ‹ **River Lea**

Leytonstone: 'Leyton (at the) stone' [1370 (*Leyton atte Stone*), 1426 (*Leyton Stone*): **Leyton** + *stān* 'stone' (from the reputed site here of a Roman milestone)] *see* **Leyton**

Lime Street: 'street where (quick)lime was burnt and sold' [1170-87 (*Limstrate*): *līm* '(quick)lime' + *strǣt* '(paved)road, street' ‹ Latin)]

Limehouse: '(quick)lime-kilns' [1367 (*Le Lymhostes*), 1547 (*Lymehouse*): *līm* '(quick)lime' + *āst* 'oast, kiln']

Lincoln's Inn Fields ‹ **Lincolns Inn:** 'Lincoln's dwelling-place' [1399 (*Lincolnesynne*): *Lincoln* (name) + *inn* 'dwelling, lodging'; after Thomas de *Lincoln*, owner of related property elsewhere, and Henry de Lacy, Earl of *Lincoln* (died 1311), lawyers' patron]

Little Britain [1329 (*Brettonstrete*), 1547 (*Britten Strete*), c. 1600 (*little Brittain streete*): after (Robert le) *Bretoun*, local landowner (attested 1274) (+ Middle English *strete* ‹ *strǣt* '(paved) road, street' (‹ Latin)]

Little Chelsea [1655 (*Little Chelcy*)] ‹ **Chelsea**

Liverpool Street [after R.B. Jenkinson, 2nd Earl of *Liverpool*, Prime Minister 1812-27]

Lombard Street [1318 (*Lumbardstret*): *Lumbard* 'Lombard' + *strǣt* '(paved) road, street' (‹ Latin); after the *Lombards* (north Italians, often early bankers) there]

London [115 (*Londinium*), 150 (*Londinion*), c.380 (*Lundinium*), 962 (*Lundene*), 12th C. (*Lundres*), 1205 (*Lundin*); origin unknown]

London Bridge [10th C. (*Lundene brigc*): *Lundene* 'London' + *brycg* 'bridge'] ‹ **London**

London Wall [1547 (*London Walle*); after the former city *wall;* previously also called *Babeloyne* 'Babylon' (1385-86)] ‹**London**

Long Ditton: 'long ditch estate' [1086 (*Ditune*), 1233 (*Ditton*), 1242 (*Longa Dittone*): *lang* 'long' + *dīc* 'ditch; dyke' + *tūn* 'estate, home farm']

Long Lane [1530 (*Long Lane*)]

Lothbury /'ləʊθbərɪ, -brɪ/: 'Lotha's manor' [1293 (*Lotheberi*): *Hloþa* (name) + *burh* 'stronghold, fortified manor']

Loughton /'laʊtᵊn/:'Luh(h)a's folk's estate' [1062 (*Lukintone*), 1200 (*Lucheton'*), 1331 (*Lughton*), 1338 (*Loughton*): *Luh(ha)* (name) + *-ing* '-follower of' + *tūn* 'estate, home farm']

Lower Clapton ‹ **Clapton**

Lower Norwood ‹ **Norwood**

Lower Sydenham ‹ **Sydenham**

Lower Tooting ‹ **Tooting**

Ludgate: 'postern gate; low gate' [1164-79 (*Ludgate*): *ludgeat* 'back gate' or 'low gate' (?‹ *lud-* ‹ *lūtan* 'bow (head), lower' + *geat* 'gap; gate')]

M

Malden /'mɔːldən/: 'cross hill' [1086 (*Meldon(e)*), 1225 (*Maldon*): *mǣl* 'sign; cross' + *dūn* 'hill']

Mark Lane: 'Martha's lane' [c.1200 (*Marthe-lane*), 1481 (*Markelane*): ?*Marthe* 'Martha' + *lane* 'lane'; formerly also *Mart Lane*]

Marlborough /,mɑːlbərə,-brə, îmɔːl-/ **House** [built 1709-11 by Wren for John Churchill (1650-1722), first Duke of *Marlborough*]

Mart Lane ‹ **Mark Lane**

Marylebone /'mærələbən, 'mærlə-/: '(St.) Mary's stream; (St.) Mary(')s) by the stream' [1453 (*Maryburne*): *Mary* (name) + French *le* 'the' (17th-C. insertion) + *burna* 'brook, stream'] *see also* **Tyburn Street**

Mayes Brook [after (Richard le) *May* (attested 1314) or his family]

Merton: 'pool(-side) estate' [967 (*Mertone*): *mere* 'pool' + *tūn* 'estate, home farm']

Mile End: 'mile's end' [1288 (*La Mile Ende*): *mīl* 'mile' (‹ Latin) + *ende* 'end, edge; district'; from its being a *mile* from **Aldgate**]

Milk Street [c.1140 (*Melecstrate*), 1153-67 (*Milkstrete*): *meoluc, milc* 'milk' + *strǣt* '(paved) road, street' (‹ Latin); from the selling of *milk*, and perhaps also the *milking* of cows, here]

Mill Hill [1547 (*Myllehill*): *myln* 'mill' (‹ Latin) + *hyll* 'hill']

Minchen Lane [1360 (*Mynchenelane*)] *see* **Mincing Lane**

Mincing Lane: 'nuns' lane' [12th C. (*M(e)ngenelane*): *mynecenu* 'nuns'' + *lane* 'lane'; formerly also **Minchen Lane**] *see also* **Minchen Lane**

Mitcham: 'big village' [1086 (*Michelham*), c.1150 (*Micham*): *micel* 'big, large, great' + *hām* 'settlement, village']

Molesey: 'Mūl's (high) ground' [675 (*Muleseg*), 967 (*Muleseye*): *Mūl* (name) + *ēg* 'high ground']

Monument Street [after its *Monument* (built 1671-7) commemorating the Great Fire of London (1666)]

Moor Fields: 'marsh fields' [*mōr* 'marsh, moor' + *feld* 'field; now *Moorfields*]

Moorgate: 'marsh gate' [*mōr* 'marsh, moor' + *geat* 'gap; gate']

Morden: 'marsh hill' [969 (*Mordune*), 1204 (*Moreden*): *mōr* 'marsh, moor' + *dūn* 'hill']

Moselle /məʊ'sɛl, -'zɛl (both according to the National Rivers Authority)/, **Moselle Brook** [previously *Campsborne* (1608)] ‹ **Muswell Hill**

Mottingham: 'Mod(da)'s folk's river-bend land' [973 (*Modinga hammes gemǣro*), 987 (*Modinga hæma mearc*), 1044 (*Modingeham*), 1206 (*Modingh'*): ?*Mod(da)* (name) + *-inge-, -inga-* '-followers' of' + *hamm* 'river-bend land']

Mugwell Street: 'Muc(c)a's well street' [c. 1200 (*Mukewellstrete*), 1279 (*Mogewelstrete*), 1544 (*Mugwellstrete*), c. 1600 (*Monkeswell Streete*): ?*Muc(c)a* (name) + *well(a)* 'well; spring' + *strǣt* '(paved) road, street' (‹ Latin); later called *Monkwell Street*, probably through folk etymology]

Muswell /îmʌzwəl; formerly îmʌzəl/ **Hill:** 'moss-spring hill' [c.1155 (*Mosewella*), 1535 (*Muswell*), 1631 (*Mussell Hill*): *mēos* 'moss' + *wella* 'well; spring' + *hyll* 'hill']

Mutton Brook: 'Mordin's brook' [1574 (*Mordins Brook*), 1819 (*Mutton Brook*): ?*Mordin* (name) + *brōc* 'brook']

N

Necklinger, Neckinger: 'noose; bend' [from (the Devil's) *Neckercher* 'the Devil's neckerchief' (former slang name for 'hangman's noose') or from the stream's sinuosity]

Needler Lane, Needler's Lane [1400-01 (*Nedlerslane*), 1403 (*Nedelerslane*), c. 1600 (*Needlers lane, Needlars lane*): Middle English *nedlere* 'needler: needle-maker; needle-seller' + *lane* 'lane'; now *Pancras Lane*]

New Bond Street [1732 (*New Bond Street*): after Sir Thomas *Bond*, developer of land in the area]

New Cross: 'new cross-roads' [? from the junction of the Old Kent Road with a road leading to **Dartford** in Kent]

New River [1625 (*the Newe River*); engineered 1609-13]

Newham [London borough comprising west **Barking, West Ham**, *East Ham* and **Woolwich** north of the Thames]

Newington (Middlesex): 'new estate' [1086 (*Neutone*), 1255 (*Newinthon*); another name for **Stoke Newington**] *see* **Stoke Newington**

Newington (Surrey): 'new estate' [c. 1200 (*Neuton*), 1258 (*Newenton*): *nīwe* 'new' + *tūn* 'estate, home farm'; *-ing-* probably from the *-an* of Old English "æt þǣm nīwan tūne" ('at the new estate')]

Newington Green [1480 (*Newyngtongrene*): **Newington** + *grēne* '(village) green'] ‹ **Newington** (Middlesex)

North End (Hampstead): 'the north end of Hampstead' [1741-45: *nor* 'north' + *ende* 'end, edge; district']

Northolt /'nɔːθəʊlt/: 'north corner' [960 ((*æt*) *norð healum*), 1086 (*Northala*), 1610 (*Northolt*): *norð* 'north' + *halh, healh* 'corner/nook of land'; contrasted with **Southall**]

Northwood [1435 (*Northwode*): *norð* 'north' + *wudu* 'wood'; north of **Ruislip**]

Norwood: 'north wood' [1176 (*Norwude*): *norð* 'north + *wudu* 'wood'; north of **Croydon**]

Norwood Common ‹ **Norwood**

O

Old Broad Street [eastern part of the former **Broad Street**; contrasted with the later *New Broad Street*] ‹ **Broad Street**

Old Change: 'old trading-place' [1297-98 (*Chaunge*), 1316-17 (*Eldechaunge*), 1393 (*Oldechaunge*): *eald* 'old' + ?Middle English *chaunge* 'change; merchants' meeting place' (‹ Anglo-French ‹ Old French ‹ Latin ‹ Celtic); this *change* may have been the royal mint formerly here]

Old Dean's Lane [1257 (*Eldedeneslane*), 1513 (*Eldens lane* alias *Warwik lane*): *eald* 'old' + Middle English *deen* 'dean' (‹ Middle French ‹ Late Latin) + *lane* 'lane'; after a former *Dean* of St. Paul's; now **Warwick Lane**]

Old Fish Street [1230-40 (*Westfihistrate*), 1272-73 (*Fihstrate*), 1293-94 (*Old Fistrete*); probably from a *fish*-market there; now **Knightrider Street**] *see also* **Knightrider Street**

Old Fish Street, Old Fish Street Hill [c. 1600 *Old Fishstreete Hill*); *hill* in such London street-names means 'steep street']

Old Ford: 'old (river-)ford' [1230 (*Eldefordmelne*), 1313 (*Oldeforde*): *eald* 'old' + *ford* 'river-ford' (+ *meln, myln* 'mill' ‹ Latin)]

Old Jewry: 'former Jewish quarter' [1327-28 (*la Oldeiuwerie*), 1336 (*la Elde Jurie*): (French *la* 'the' +) *eald* 'old' + Middle English *giwerie, juerie* 'Jews' territory, Jews' district' (‹ Anglo-French ‹ Old French); formerly *Colechurch Lane*]

Old Oak Common: 'old-grove common' [1380 (*Eldeholt*), c.1415 (*Oldeholte*), 1650 (*Common called Old Oake*): *eald* 'old' + *oak* (by folk etymology ‹ *holt* 'holt, grove, copse') + English *common* 'common land' (‹ Latin *commūnia*)]

Old Street: 'old road' [c.1200 (*Ealdestrate*): *eald* 'old' + *strǣt* '(paved) road, street' (‹ Latin)]

Orpington: 'Orped's folk's estate' [1042 (*Orpedingtun*), 1086 (*Orpinton*), 1207 (*Orpington'*): *Orped* (name) + *-ing* '-follower of' + *tūn* 'estate, home farm']

Osterley Park ‹ **Osterley:** 'sheep-fold lea/grove' [1274 (*Osterle*): *eowestre* 'sheep-fold' (‹ *eowu* 'ewe') + *lēah* 'grove; meadow, lea']

Oxford Street [1720; previously called *Tyburn Road, Road to Worcester, Road to Oxford*, etc.]

Oystergate [1259 (*Oystregate*); a former watergate where *oysters* may have been sold]

P

Paddington: 'Pad(d)a's folk's estate' [c. 1045 (*Padington*): *Pad(da)* (name) + *-ing* '-follower of' + *tūn* 'estate, home farm']

Pall Mall /'pæl'mæl/: 'mall or alley used for playing pall-mall' [1650 (*Pall Mall Walk*): *pall-mall* 'alley for pall-mall, a game of getting a ball through a raised ring by hitting it with a mallet' (‹ Middle French ‹ Italian)]

Pancras ‹ **St. Pancras**

Paternoster Row: 'rosary-makers' row' [1307 (*Paternosterstrete*), 1320-21 (*Paternoster Lane*), 1344 (*Paternosterowe*), 1374 (*Paternosterrowe*): *paternostre* 'pater-nosterer, rosary-maker (‹ Latin) + *rǣw* 'row' (of trees or houses)]

Peckham: 'peak village' [1086 (*Pecheham*), 1241 (*Peckham*): ?*pēac* 'peak, hill' + *hām* 'settlement, village']

Penge /pɛn(d)ʒ/: 'head wood, chief wood' [1067 (*Penceat*), 1206 (*Penge*), 1472 (*Pengewode*): ?Celtic *pen* 'head, top; chief' (as in *Pen(zance)*) + ?Celtic *cēt* 'wood']

Pentecost Lane [1280 (*Pentecostelane*), 1290 (*Pentecostes lane*): Middle English *Pentecoste* (Christian festival or man's Christian name) (‹ *pentecosten* ‹ Late Latin ‹ Greek) + *lane* 'lane']

Petersham Road ‹ **Petersham:** 'Peohtric's river-bend land' [675 (*Piterichesham*): *Peotric* (name) + *hamm* 'river-bend land']

Petty Wales: 'little Wales' [1298-99 (*petit Walles*), 1349 (*Pety Wales*): French *petit* 'little' + *Wal(l)es* 'Wales'; perhaps from Welsh people resident here]

Philpot Lane [1480-81 (*Philpot Lane*): after Sir John *Philpot*, Lord Mayor 1378-79; formerly probably **Hubbard Lane**] *see* **Hubbard Lane**

Piccadilly [? after *Pickadilly* (Hall) (a 17th-C. tailor's house nearby) ‹ *piccadil* 'border with a cut-out pattern, ornamenting especially the edge of a collar or ruff' (‹ French ?‹ Spanish)]

Pimlico [1630 (*Pimplico*), c.1743 (*Pimlico*): ? after *Pimlico* (Walk) in **Hoxton**, named after Ben *Pimlico*, 16th-C. innkeeper there]

Pinkwell [1754 (*Pinkwell*)]

Pinner: 'pin(-shaped) (river-)bank; Pinn(a's) river-bank' [1232 (*Pinnora*), 1332 (*Pinnere*): *pinn* 'pin, peg' or *Pinn* (name) + *ōra* 'bank, edge; slope' (‹ Latin)]

Pinner Green ‹ **Pinner**

Plaistow /'plɑːstəʊ, 'plæst-/: 'playing/sporting ground' [1278 (*Pleystowe*): *pleg* 'play' + *stōw* 'place']

Plaistow Levels: '? Plaistow level ground' ‹ **Plaistow**

Plumstead: 'plum(-tree) place' [c.965 (*Plumstede*): *plume* 'plum; plum-tree' + *stede* 'place, stead']

Ponders End [1593 (*Ponders ende*): *Ponder* (surname of local family) + *ende* 'end, edge; district'; on the **Enfield**/**Edmonton** border]

Pool River ‹ **Pool:** 'Pool of London' [1258 (*La Pole*): *pōl* 'pool']

Poor Jewry [1366 (*la Porejewerie*)] *see* **Jewry Street**

Poplar: 'poplar(-tree)' [1327 (*Popler*): Middle English *poplere* 'poplar' (‹ Middle French)]

Potter's Bar: 'Potter's Gate (to Enfield Chase)' [1509 (*Potterys Barre*), 1548 (*Potters Barre*): (*le*) *Pottere* (attested surname) + Middle English *barre* 'rod, bar; gate, barrier' (‹ Old French)]

Poultry: 'poultry-market' [1301 (*Poletria*), 1422 (*Pulterye*): Middle English *pultrie* 'poultry; poultry-market' (‹ Old French)]

Primrose Hill [1586 (*Prymrose Hill*); allegedly from the former profusion of *primrose(s)* there]

Pudding Lane (near Billingsgate): 'guts lane' [1360 (*Puddynglane*): Middle English *pudding* 'guts, entrails' + *lane* 'lane'; perhaps whence "the Butchers of Eastcheape" got rid of such parts of their animals]

Purley: 'pear-tree lea/grove' [1200 (*Pirlee*), 1220 (*Purle*): *pirige* 'pear-tree' (‹ *peru, pere* 'pear' ‹ Latin) + *lēah* 'grove; meadow, lea']

Putney: 'Put(t)a's landing' [1086 (*Putelei*), 1279 (*Puttenhuthe*): *Putten* (from name) + *hȳð* 'landing-place, harbour']

Putney Heath ‹ **Putney**

Pyl Brook ‹ **Pylford Bridge** [1548 (*Pillefordebrudge*)]

Pymme's Brook [after the family of Reginald *Pymme* of Edelmetone (**Edmonton**), attested locally since the 14th C.; formerly *Medeseye* (c. 1200)]

Q

Quaggy River: 'quagmire river; boggy river' [from its sluggish flow]

Queen Street [after Catherine of Braganza (1638-1705), *Queen* as wife of King Charles II; includes the former **Soper Lane**] *see* **Soper Lane**

Queen Victoria Street [opened 1871]

Queenhithe: 'Queen('s) dock' [898 (*Æðeredes hyd*): *Æðered* '?Ethelred, Alderman of Mercia' + *hȳð* 'landing-place, harbour'; now *Queenhithe Dock* perhaps because formerly owned by Isabella of Angoulême]

R

Radlett: 'cross-roads' [1453 (*Radelett*): ?*rād-(ge)lǣt(e)* 'road-junction' (‹ *rād* 'riding; road' + *(ge)lǣt(e)* 'junction of roads, cross-roads')]

Rainham (Essex): 'top-people's settlement' [1086 (*Renaham, Raineham*): ?*roeginga-ham* (as attested for *Rainham* in Kent) ‹ *roegingas* 'dominant folk' + *hām* 'settlement, village']

Red Cross Street [1275 (*Redecrochestrete*), 1341 (*Redecrouchestrete*), 1502 (*Redcrosse strete*): *rēad* 'red' + Middle English *crouche* 'cross' (‹ Latin) or *cros* 'cross' (‹ Old Irish ‹ Latin) + *strǣt* '(paved) road, street' (‹ Latin); perhaps after a local house or a boundary *cross*]

Redbridge [after a *red bridge* across the **River Roding** between **Wanstead** and **Ilford**]

Regent's Park [1817 (*The Regents Park*): after the Prince *Regent*, later George IV]

Richmond [1502 (*Richemont*): after Henry VII, Earl of *Richmond* in Yorkshire; formerly (*West*) *Sheen*]

Richmond Park ‹ **Richmond**

Richmond upon Thames ‹ **Richmond, River Thames**

River Brent ‹ **Brent**

River Ching [1562 (*the Boorne*), 1585 (*the Brook*): *burna* 'stream, brook'; *Ching* is a later back-formation from **Chingford**] *see* **Chingford**

River Colne ‹ **Colne** /kəʊn/: 'water stream' [1301 (*Collee*), 1351 (*Colne*): ? Celtic *colūn* 'water' + *ēa* 'stream'; found elsewhere too as a river-name]

River Crane [1825 (*Cran Brook*); formerly *Fishbourne*] ‹ **Cranford**

River Effra ‹ **Effra:** 'river-bank' [? *efre* 'bank' (of river); cognate with German *Ufer* 'bank, shore']

River Fleet ‹ **Fleet:** 'stream' [c. 1012 (*Fleta*): *fleot* 'inlet, stream']

River Gade ‹ **Gade:** 'Gǣte's stream' [1242 (*Gatesee*), 1349-96 (*Gateseye*), 1728 (*river Gade*): ?*Gǣtesēa* 'Gǣte's stream' (‹ *Gǣte* (name or nickname ‹ *gat* 'goat') + *ēa* 'stream')]

River Graveney: 'river by Tooting Graveney' [1272 (*Thoting Gravenel*): after (Richard de) *Gravenel*, lord of the manor of Lower/South **Tooting** in 1215 (whose family were perhaps from *Graveney* in Kent)]

River Lea Lea: 'bright' [895 (*Lygan*); perhaps related to *lēah* 'lea, meadow as light-filled place' (‹ Indo-European base ? *leuk, louk* 'light, brightness')]

River Mole [1214 (*aqua de Mulesia*), 1595 (*Moulsey River*)] ‹ **Molesey**

River Pinn [from **Pinner**] *see* **Pinner**

River Ravensbourne ‹ **Ravensbourne:** 'raven's stream' [1575 (*Ravensburn*): *hrǣfn* 'raven' + *burna* 'stream, brook']

River Roding [1576 (*Rodon*), 1622 (*Roding*); formerly *Hyle*] ‹ **Roding** (Essex): 'Rod(da)'s folk' [c. 1050 (*Rodinges*): *Rod(da)* (name) + *-ingas, -inges* '-followers of']

River Thames ‹ **Thames** /'tɛmz/: 'dark water' [?‹ Celtic base cognate with Sanskrit *tamasa-* 'dark']

River Wandle [from the *Wa(e)nd(e)l* of **Wandsworth**] *see* **Wandsworth**

River Wey ‹ **Wey:** 'flowing' [675 (*Waie*): ?‹ Indo-European *wegh* -(referring to motion, as in *wǣg* 'wave')]

Romford: 'ample ford' [1177 (*Romfort*), 1199 (*Rumford*): *rūm* 'roomy, spacious' + *ford* '(river-)ford']

Roper Lane: 'rope-maker's lane; Roper's lane' [1313 (*Ropereslane*): Middle English *roper* 'rope-maker' or *Roper(e)* (attested local surname) + *lane* 'lane'; now *Love Lane*]

The Ropery, Roper Street [1271 (*Roperestrete*), 1307 (*la Roperie*); after the *rope*-makers there; now **Thames Street**]

Rotherhithe: 'cattle landing-place' [c. 1105 (*Rederheia*), 1301 (*Rotherhethe*): *hrīðer* 'horned beast, ox; cattle' + *hȳð* 'landing-place, harbour']

Royal Botanic Gardens *see* **Kew Gardens**

Ruislip /'raɪslɪp/: 'rush(y) leap' [1086 (*Rislepe*), 1341 (*Ruysshlep*): *rysc* 'rush(-plant)' + ?*hlype* 'leap'; ? from a nearby crossing place of the **River Pinn**]

S

St. Botolph's Lane [1348-49 (*Seyntbotulfeslane*), 1432 (*Botulphesiane*), 1544 (*Botulphe Lane*); after the church of *St. Botulph Billingsgate*; now *Botolph Lane*]

St. Bride Street [? after *St. Bride*'s churchyard nearby]

St. Clement's Lane [1348 (*Seint Clementeslane*); by the church of *St. Clement Eastcheap*; now *Clement's Lane*]

St. Dunstans Lane [1329 (*Donstoneslane*), 1363 (*Seint Dunstoneslone*); by the church of *St. Dunstan in the East*; now *St. Dunstan's Hill*]

St. James ‹ **St. James's Palace**

St. James's Palace [built on the site of a leprosy hospital dedicated to *St.James* the Less]

St. James's Park ‹ **St. James's Palace**

St. Katharine's Dock [1422 (*Katerines Dokke*): *St. Katherine's* (from the name of a former local hospital founded 1148) + Middle English *dok* 'dock; wharf' (? ‹ Middle Dutch ‹ Latin)]

St. Laurence Lane [1320 (*Seint Laurencelane*), c. 1600 (*Poultney lane*); after the church of *St. Laurence Pountney*; now *Laurence Pountney Lane*]

St. Margaret Patten's Lane [1577 (*Rood Lane*): Middle English *rood* 'cross' (‹ *rōd*) + *lane* 'lane'; from a *rood* put before 1538 in the churchyard of *St. Margaret Pattens*, from which church comes the previous name of this lane now called *Rood Lane*]

St. Mary Axe: 'St. Mary at/of the Axe' [1275 (*strata Sancte Marie atte Ax*); after the church of *St. Mary Axe* (demolished 1561), housing the *axe* with which St. Ursula was said to have been martyred]

St. Mary (Axe) Street [now **St. Mary Axe**] *see* **St. Mary Axe**

St. Mary Cray: 'St. Mary('s church by the River) Cray' [1257 (*Creye Sancte Marie*)] *see* **Crayford**

St. Mary at Hill Lane [1275 (*venella Sancte Marie de la Hulle*), 1520-21 (*seint mary hill lane*); after the church of *St. Mary at Hill*]

St. Marylebone ‹ **Marylebone**

St. Martin Orgar Lane [1236-37 (*venella Sancti Martini*), c.1600 (*Saint Martins Orgar Lane*); after the church of *St. Martin Orgar; now Martin Lane*]

St. Martin's Le Grand: 'St. Martin the Great's' [1265 (*St. Martin le Grand*); after the former local church of *St. Martin le Grand*]

St. Michael's Lane [1303 (*Seint Micheleslane*); by the church of *St. Michael Crooked Lane*; later *Miles's Lane* (*Miles* ‹ *Michael*)]

St. Nicholas Lane [1381 (*Seint Nicholaslane*); by the church of *St Nicholas Acon*; now *Nicholas Lane*]

St. Pancras [1086 (*Sanctum Pancratiū*), 1588 (*Pankeridge al. St Pancras*); after *St. Pancras*, martyred under Diocletian (Roman emperor 284-305)]

St. Paul's Cathedral [built 1675-1710; replacing old *St. Paul's* (burnt 1666)]

St. Paul's Churchyard [after **St. Paul's Cathedral** or old *St. Paul's*]

St. Peter's Hill [1263 (*Venella sancti Petri*), 1378 (*Seint Petrelane*), 1564 (*Peter Lane*) c. 1600 (*Saint Peters Hill, Peter hill lane*); *hill* in such London street-names means 'steep street'; by the church of *St. Peter (the Little) Paul's Wharf*; formerly *Peter Lane*]

Salmon's Brook [1754: after (the family of John) *Salmon* (attested 1274 at **Edmonton**)]

Sanderstead /'sɑːndəstɛd/: 'sandy place' [c. 880 (*Sandenstede*), 1086 (*Sandestede*), 1221 (*Sanderstede*): ?*sanden* 'sandy' + *stede* 'place, stead']

Seacoal Lane: 'coal lane' [1253 (*sacolelane*), c. 1600 (*Seacole lane*): *sǣcol* 'coal, not charcoal' (delivered by *sea* or mined by the *sea*) + *lane* 'lane'; *coal* may have been delivered here from the **River Fleet**]

Sewardstone: 'Sigeweard's estate' [1176-90 (*Siwardeston'*): *Sigeweard* (name) + *tūn* 'estate, home farm']

Shambles, Butchery: 'meat-market; slaughterhouse' [1349 (*Bocherie*), 1425-26 (*Shameles*), 1530 (*le Fleshambles*): (*flǣsc* 'flesh, meat'+) Middle English *shambles* 'meat-market; slaughterhouse' (‹ Middle English *shamble* 'meat-seller's table' ‹ *sceamel* 'stool' ‹ Late Latin) or Middle English *butchery* 'slaughterhouse' (‹ Old French); now *Newgate Street*]

Shepherds Bush [1635 (*Sheppards Bush Green*), 1675 (*Shepperds Bush*)]

Shepperton: 'shepherd farm' [959 (*Scepertune*): ?*sceaphir(de), sceaphier(de)* 'shepherd' + *tūn* 'estate, home farm']

Shepperton Green [1754] ‹ **Shepperton**

Shitbourn Lane: 'shithouse lane' [1272-73 (*Shitteboruelane*), 1321 (*Shiteburghlane*), 1313 (*Shitebournelane*), 1467 (*Shirbouruelane* alias *Shetbouruelane*), 1540 (*Shirborne lane*), c. 1600 (*Sherborne lane*): Middle English *Shitebourne* 'shit stream' (by scribal error ‹ Middle English ? *shiteburgh* 'shithouse, privy' ‹ *scite* 'shit' + *burh, burg* 'stronghold, fortified manor') + *lane* 'lane'; now *Sherborne Lane* (by euphemism)]

Shoe Lane: 'shoe(-shaped) land; shelter-land' [1187-1216 (*Solande*), 1272 (*Sholand*), 1279 (*Sholane*): *scōh* 'shoe' or ?*scēo* 'shelter' + *land 'land']*

Shooters Hill: 'archer's slope' [1292 (*Shetereshelde*): Middle English *sheter* 'shooter, archer' (‹ ? *scēotere*) + *helde* 'slope']

Shoreditch: 'slope-ditch' [c. 1148 (*Soredich*): ? *scora* 'shore, bank; slope' + *dīc* 'ditch, trench; dyke'; the name's referent is obscure]

Sidcup: 'flat-top; camp-site top' [1254 (*Cetecopp'*), 1332 (*Sedecoppe*): ?*set* 'seat-shaped, flattened; camp' + *copp* 'hill-top']

Silk Stream: 'gully stream' [957 (*Sulh, Sulue, Sulc*), 13th C. (*Solke, Selke*): ?*sulh* 'plough; ?furrow' (? with *h* changed to *c* or *k* before *stream*)]

Sipson: 'Sibwine's estate' [13th C. (*Sibwineston*), 1391 (*Sibston*), 1638 (*Sipson*): *Sibwine* (name) + *tūn* 'estate, home farm']

Smallbury: 'narrow mound' [1436 (*Smalborow*), 1680 (*Smallbury Green*): *smæl* 'thin, narrow; small' + *beorg* 'hill, mound; barrow']

Smithfield: 'level field' [*smēðe* 'smooth, level' + *feld* 'field']

Soper Lane, Soper's Lane: 'soaper's lane' [c. 1246 (*Sopereslane*), 1282 (*Soperlane*), 1600 (*Sopers lane*): Middle English *sopere* 'soaper: soap-maker; soap-seller' (‹ *sāpe* + *-ere*) + *lane* 'lane'; now part of **Queen Street**] *see* **Queen Street**

South Mimms [1086 (*Mimes*), 1211 (*Mimmes*), 1253 (*Suthmimes*); *"South* in contrast to North Mimms in Hertfordshire....The name must remain an unsolved problem." – *The Place-Names of Middlesex*, p.76]

Southend [*sūð* 'south' + *ende* 'end, edge; district']

Southgate [1370 (*Suthgate, Southgate*): *sūð* 'south' + *geat* 'gap; gate'; by the *south gate* to **Enfield** Chase]

Southall: /'saʊθɔːl/: 'south corner' [1198 (*Suhauħ*), 1204 (*Sudhale*), 1261 (*Suthall(e)*): *suð* 'south' + *halh* 'corner/nook of land'; contrasted with **Northolt**]

Southwark /'sʌðək/: 'south fortress' [1086 (*Sudwearca*): *sūð* 'south' + (ge)*weorc* 'construction, fortification']

Spittlefields: 'hospital fields' [1561 (*Spyttlefeildes*): Middle English *spitel* 'hospital' (here, of St. Mary *Spital*, ‹ Medieval Latin) + *feld* 'field'; now *Spitalfields*]

Staines: 'stone(s)' [969 (*Stána*), 1086 (*Stanes*): *stān(as)* 'stone(s)'; probably after a nearby Roman mile*stone*]

Stamford Brook: 'stony ford brook' [1650 (*Stamford Brooke*) ‹ **Stamford:** 'stony ford' [1274 (*Staunford*): Middle English *stoon, ston, stan* ‹ *stān* 'stone' + *ford* '(river-)ford']

Stamford Hill: 'sand-ford hill' [1225 (*Sanford*), 1294 (*Saundfordhull*), 1675 (*Stamford Hill*)]

Stanmore [793 (*Stanmere*)] *see* **Great Stanmore**

Stanwell: 'stony spring' [1086 (*Stanwelle*): *stān* 'stone' + *well(a) 'well; spring']*

Stepney: 'Stybba's landing' [c. 1000 (*Stybbanhyþe*), 1542 (*Stebenheth al. Stepney*): *Stybban* (from name) + *hȳð* 'landing-place, harbour']

Stinking Lane [1228 (*Styngkynglane*); presumably from the *stink* of the nearby **Shambles**; now **King Edward Street**] *see* **King Edward Street**

Stockwell: 'tree-stump well' [1197 (*Stokewell*): *stocc* 'tree-trunk, stump, log' + *well(a)* 'well; spring]

Stoke Newington: 'log new estate' [1086 (*Neutone*), 1255 (*Newinthon*), 1274 (*Neweton Stoken, Stokeneweton*): *stocc* 'tree-trunk, stump, log' + *nīwe* 'new' + *-ing-* (perhaps as in **Newington** in Surrey) + *tūn* 'estate, home farm'] *see* **Newington** (Surrey)

The Strand: 'the bank (of the Thames)' [1185 (*Stronde*): *strand* 'shore, bank'; the Thames used to be wider]

Stratford: '(Roman-)road ford' [1066 (*Stratforde*): *strǣt* '(paved)road, street' (‹ Latin) + *ford* '(river-)ford'; where the old London-Colchester road crossed the **River Lea**]

Stratford-le-Bow: '(Roman-)road ford (at) the arched bridge' [1177 (*Stratford*), 1279 (*Stratford atte Bowe*), c. 1560 (*Stratford le Bow(e)*): *strǣt* '(paved) road, street' (‹ Latin) + *ford* '(river-)ford' (across the **River Lea**) + French *le* 'the' + *boga* '(archery) bow; arch, arched bridge'; now *Bow*]

Streatham /'strɛtəm/: '(Roman-)road village' [1086 (*Estreham*), 1175 (*Stratham*): *strǣt* '(paved) road, street' (‹ Latin) + *hām* 'settlement, village']

Sudbury: 'south stronghold' [1292 (*Suthbery*): *sūð* 'south' + *burh* 'stronghold, fortified manor']

Sunbury: 'Sunna's stronghold' [959 (*Sunnabyri*): *Sunn(a)* (name) + *burh* 'stronghold, fortified manor']

Surbiton: 'south grain-farm, south grange' [1179 (*Suberton*): *sū(ð)* 'south' + *bertūn* 'grain-farm; lord's grange' (‹ *bere* 'barley' + *tūn* 'estate, home farm')]

Sutton: 'south estate' [1181 (*Suthtona*): *sūð* 'south' + *tūn* 'estate, home farm']

Swithin Lane [1269-70 (*vicus Sancti Swithuni*), 1410-11 (*Seint Swithoneslane*), 1532 (*St. Swithens Lane*); after the church of *St. Swithin* in **Cannon Street**; now *St. Swithin's Lane*]

Sydenham /'sɪdᵊnəm/: 'Chippa's village' [1206 (*Chipenham*), 1315 (*Shippenham*), 1690 (*Sidenham*): *Syden* (from name ‹ *Shippen* ‹ *Chipen*) + *hām* 'settlement, village'; similar to *Chippenham* (Camb, Wilts), *Cippenham* (Bucks)]

Syvethe Lane, Syvthe Lane: 'chaff lane' [1258-59 (*Syvidlane*), 1322 (*Syvthelane*), 1533 (*Sedyng Lane*): *sifeða* 'siftings, chaff' + *lane* 'lane'; perhaps from threshing done nearby; now *Seething Lane*]

T

Teddington: 'Tuda's folk's estate' [969 (*Tudinton*), 1274 (*Tedinton*): *Tud(a)* (name) + *-ing* '-follower of' + *tūn* 'estate, home farm']

Thames Ditton: 'Thames ditch estate' [1005 (*Dictun*), 1235 (*Temes Ditton*): **(River) Thames** + *dīc* 'ditch; dyke' + *tūn* 'estate, home farm']

Thames Street [1222 (*vicus super Ripam Tamis*), 13th C. (*la rue de Thamise*), 1275 (*Tamisestrete*), 1308 (*Temestret*)] ‹ **River Thames**

Threadneedle Street: 'three-needle street; street where *threadneedle* is played' [1598 (*Three needle street*), 1616 (*Thred-needle-street*); ? from a local signboard or coat of arms displaying *three needles*, or from the children's game *threadneedle*]

Tilbury: 'Til(la's) stronghold' [c. 735 (*Tilaburg*), 1066-87 (*Tillabyri*), 1218 (*Tylleber, Tyllebery*): *Til(la)* (name) + *burh* 'stronghold, fortified manor']

Tooting: 'Tota's folk' [675 (*Totinge*): *Tot-* (from name) + *-ing* '-follower of']

Tooting Common ‹ **Tooting**

Tottenham: 'Tota's village' [1086 (*Toteham*), 1189 (*Totenham*), 1254 (*Tottenham*): *Tote(n)* (from name) + *hām* 'settlement, village']

Tottenham Court: 'Totta's-corner manor' [c. 1000 (*þottanheale*), 1083 (*Totenhala*), 1487 (*Totenhalecourt*), 1593 (*Totten Court*), 1741-45 (*Tottenham Court*): *Totten* (from name) + *halh, healh* 'corner/nook of land' (to *ham* influenced by **Tottenham**) + Middle English *court* 'manor' (‹ Old French)] *compare* **Tottenham**

Tottenham Court Road [1708 (*Tottenham Court Row*)] ‹ **Tottenham Court**

Tower Hamlets: 'hamlets near the Tower (of London)'

Tower Hill [after the adjacent *Tower* of London]

Turnagain Lane: 'blind alley' [1415 (*Turneageyne lane*): Middle English *turne-agayne lane* 'blind alley, cul-de-sac'; formerly *Wendageyneslane* (from 1293)]

Turnbaston Lane [1328 (*Tornebastonlane, Tornebastones-lane*), 1436 (*Turnebastlane*), 1568 (*Turnebaslane, Turnesbas-lane*): Middle English *?turnebaston* '?tollgate' (‹ Middle French) + *lane* 'lane'; now part of **Cannon Street**]

Turnham Green: 'round-village green; river-bend-land green' [c. 1235 (*Turneham*), 1396 (*Turnhamgrene*): *?trun, turn* 'circular, round' + *hām* 'settlement, village' or *hamm* 'river-bend land' + *grēne* '(village-) green'; near a big bend in the Thames]

Twickenham /formerly 'twɪtnəm/: 'Tuic(c)a's river-bend land' [704 (*Tuican hom, Tuiccanham*): *Tuic(c)an* (from name) + *hamm* 'river-bend land']

Twickenham Common ‹ **Twickenham**

Tyburn Street ‹ **Tyburn** /'taɪbən/: 'boundary brook' [959 (*Teobernan*), 1222 (*Tyburn'*): *?tēo* 'boundary' + *burna* 'stream, brook'; formerly marking the boundary of **Westminster** Abbey lands; sometimes also called *Marybourn*] *see also* **Marylebone**

U

Upminster: '(high-)up monastery' [1062 (*Upmynstre*): *upp* 'up' + *mynster* 'monastery, church']

Upper Clapton ‹ **Clapton**

Upper Norwood ‹ **Norwood**

Upper Sydenham ‹ **Sydenham**

Upper Tooting ‹ **Tooting**

Uxbridge: 'the Wixans' bridge' [c. 1145 (*Oxebruge, Wixebrug'*): *Wixan* (from tribal name) + *brycg* 'bridge' (over the **River Colne**)]

V

Vedast Lane *see* **Foster Lane**

Victoria [after Queen *Victoria* (1819-1901), reigned 1837-1901]

W

Walbrook: '(Celtic) Britons' brook' [1104 (*Walebroch*): *wal(h)* '(Celtic) stranger' + *brōc* 'brook']

Walham /ïwɔ:lhəm/ **Green:** 'de Wenden's Green' [1386 (*Wendenegrene*), 1615 (*Wandon's Green*), 1710 (*Wallam Green*), 1819 (*Walham Green*): *de Wenden(e), de Wanden(e)* (name of local family, perhaps originally from *Wendens* in Essex) + *grēne* '(village) green']

Waltham Abbey ‹ **Waltham** /'wɔ:lθəm; formerly 'wɔ:ltəm/: 'wood-land village; wood-land river-bend land' [1062 (*Waltham*): *wald, weald* 'wood-land, forest' + *hām* 'settlement, village' or *hamm* 'river-bend land']

Waltham Cross [**Waltham** + *cros* 'cross' (‹ Old Irish ‹ Latin); from the (Eleanor) *Cross* put here in 1291 by Edward I to mark the penultimate stage in the funeral procession of Eleanor of Castile] *see* **Waltham Abbey**; *compare* **Charing Cross**

Waltham Forest [London borough comprising **Walthamstow, Chingford** and **Leyton**]

Walthamstow /'wɔ:lθəmstəʊ; formerly 'wɔ:(l)təmstəʊ/ 'welcome place; Celts' place' [c. 1076 (*Wilcumestowe*), 1446 (*Walthamstowe*): *wilcume* 'welcome' or *walh* '(Celtic) stranger' + *stōw* 'place']

Walthamstow Mead: 'Walthamstow meadow' [**Walthamstow** + *mǣd* 'meadow'] ‹ **Walthamstow**

Walworth /'wɔ:lwəθ/: 'Celts' ward' [1006 (*Wealawyrð*), 1086 (*Waleorde*), 1196 (*Wallewurd*), 1354 (*Walworth*): *wealh, walh* '(Celtic) stranger' + *worð* 'ward, enclosure']

Wandsworth: 'Wændel's ward' [c. 1000 (*Wendleswurðe*): *Wænd(e)l* (name) + *worð* 'ward, enclosure']

Wanstead: '(wen-like) mound site' [c.1050 (*Wænstede*): *wænn* 'wen' + *stede* 'place, stead']

Wapping Dock ‹ **Wapping:** 'Wæppa's folk' [c.1220 (*Wapping'*), 1231 (*Wappinges*): *Wæpp(a)* (name) + *-ingas* '-followers of']

Warwick /ˌwɒrɪk/ **Lane** [1474-75 (*Werwyk Lane*); after the Earls of *Warwick*, who held property locally; formerly **Old Dean's Lane**]

Waterloo [after *Waterloo* in Belgium, site of the famous defeat of Napoleon in 1815]

Watford: 'hunters' ford' [944-46 (*Watford*), c. 1180 (*Wathford, Wathforda*): *wað* 'chase, hunting' + *ford* '(river-) ford']

Watling Street: 'prince's street' [c. 1213 (*Aphelingestrate*), 1289 (*Athelingstrate*), 1307 (*Watlingstrate*): *æðeling* 'prince; nobleman' (becoming *Watling* perhaps by folk etymology through similarity of sound to the Roman road *Watling* Street) + *strǣt* '(paved) road, street' (‹ Latin)]

Wealdstone Brook [1453 (*le Weldebroke*), 1548 (*Weyldbrooke*); also *Kenton Brook*; formerly *Lyddying (Water)*] ‹ **Wealdstone:** 'woodland (boundary-)stone' ‹ **Harrow Weald:** 'Harrow woodland' [1282 (*Weldewode*), 1388 (*Harewewelde*), 1603 (*Harrow weale*): **Harrow** + *weald* 'woodland, forest'] *see* **Harrow**

Weir /ïwɪə/ **Hall** [1086 (*Winehel(l)e*), 1198 (*Wylehale*), 1207-8 (*Wirhale, Wilehal'*), 1593 (*Wirehall, Wyerhall*): *Wylehale, Wyrhale* (attested surname)]

Welling [1362 (*Wellyngs*): *Welling* (surname of 14th-C. local landowners)]

Wembley: 'Wemba's lea/grove' [825 (*Wemba Lea*): *Wemba* (name) + *lēah* 'grove; meadow, lea']

West Bedfont [1086 (*Westbedefund*)] ‹ **Bedfont**

West Cheap: 'west market' [1304 (*Chepe*), 1249 (*Westchepe*): *west* 'west' (by contrast with **Eastcheap**) + *cēap* 'market; trade'; also *Cheap*; now **Cheapside**] *see* **Cheapside, Eastcheap**

West Drayton: 'west portage farm' [939 (*Drægton*), 1086 (*Draitone*), 1465 (*Westdrayton*): *west* 'west' (added perhaps to contrast with Ealing's *Drayton* (Green) + *dræg* 'drag; portage, slipway' (perhaps in reference to the adjacent **River Colne**) + *tūn* 'estate, home farm']

West End (Middlesex, near Northolt) [1274 (*(atte) Westende*), 1660 (*West End*)]

West End (Middlesex, near Pinner) [1448 (*le Westhend*)]

West Ham: 'west river-bend land' [958 (*Hamme*), 1186 (*Westhamma*): *west* 'west' + *hamm* 'river-bend land']

West Molesey [1200 (*Westmoleseie*)] ‹ **Molesey**

West Smithfield [*west* 'west' + **Smithfield**] ‹ **Smithfield**

West Thurrock [1219 (*West Turroc*, etc.): *west* 'west' + **Thurrock**] ‹ **Thurrock**: 'bilge; muck-heap' [1086 (*Turoc*): *þurruc* 'bilge; ship's bottom; muck-heap']

West Wickham: 'west farm village; west Romano-British-site village' [973 (*Wichamm*), 1086 (*Wicheham*), 1284 (*Westwycham*): *west* 'west' + *wīc-hām* 'farm village; village on Romano-British site']

West Wood Common [c. 1350 (*Westwode*)]

Westbourne [formerly *Knightsbridge Brook, Bayswater Rivulet*; flows into the Serpentine in Hyde Park] ‹ **Westbourne Green:** 'west-stream green; green west of the stream' [1222 (*Westeburne*), 1294 (*Westbourne*), 1548 (*Westborne Grene*): *west* 'west' + Middle English *bourne, burne* ‹ *burna* 'stream, brook' + *grēne* '(village) green']

Westminster: 'west monastery' [c.975 (*Westmynster*): *west* 'west' + *mynster* 'monastery, church'; previously *Thorney* (969)]

Whetstone [1417 (*Wheston*), 1437 (*Whetestonesstret*), 1492 (*Whetstone*): *hwetstān* 'whetstone']

White Chapel [1282 (*St Mary de Mattefelon*), 1340 (*Whitechapele by Algate*): *hwīt* 'white' + Middle English *chapel* 'chapel' (‹ Old French ‹ Late Latin); now *Whitechapel*]

White Cross Street [1226 (*Whitecruchestrete*), 1309-10 (*Whitecrouchestrate*), 1502 (*Whitecrosse Strete*): *hwīt* 'white' + Middle English *crouche* 'cross' (‹ Latin) or *cros* 'cross' (‹ Old Irish ‹ Latin) + *strǣt* '(paved) road, street (‹ Latin); after a local *white cross*]

Whitehall ‹ **Whitehall Palace** [1530 (*Whytehale*); from the name of the Lords' Chamber in the old Parliament; previously called *York Place* when the London residence of the Archbishops of York]

Willesden /'wɪlzdən/: 'spring's hill' [939 (*Wellesdune, Willesdone*), 1290 (*Willesden*): *well(a), wiell(a)* 'well; spring' + *dūn* 'hill']

Wimbledon: 'Wynman('s) hill; Winebeald('s) hill' [c. 950 (*Wunemannedune*), 1202 (*Wimeldon*), 13th C. (*Wymendon*), 1211 (*Wimbeldon, Wimbeldona*): *Wunemanne* (name) or *Winebeald* (name) + *dūn* 'hill']

Wimbledon Common ‹ **Wimbledon**

Winchmore Hill: 'Wynsige('s) boundary hill' [1319 (*Wynsemerhull*), 1543 (*Wynsmore hill*), 1586 (*Winchmore Hill*): *Winsige* (name) + (ge)*mǣre* 'boundary, border' + *hyll* 'hill'; near the southern boundary of **Edmonton** parish]

Wood End [1531 (*Wodehende*): *wudu* 'wood' + *ende* 'end, edge; district']

Wood Green [1502 (*Wodegrene*): *wudu* 'wood' + *grēne* '(village) green']

Wood Hall [1271 (*Wodehalle*), 1349 (*Wodhall*): *wudu* 'wood' + *hall* 'hall, manor']

Wood Street: 'street where wood was sold' [1156-57 (*Wodestrata*): *wudu* 'wood' + Latin *strāta* (source of *street*)]

Woodford: 'wood(-side) ford' [1062 (*Wodeforda*), 1225 (*Wudeforde*): *wudu* 'wood' + *ford* '(river-)ford']

Woodford Bridge [1238 ((Thomas de) *ponte de Wodeford*), 1429 (*Woodfordbrigge*): **Woodford** + *brycg* 'bridge'; name of **Woodford** east of the **River Roding**] ‹ **Woodford**

Woodford Wells [1285 ((William de) *fonte de Wodeford*)] ‹ **Woodford**

Woodruff Lane, Woodroffe Lane [1260 (*Woderouelane*), c. 1600 (*Woodroffe lane*): Middle English *woderove* 'woodruff (plant)' (‹ *wudurōfe*) + *lane* 'lane'; now probably *Cooper's Row*]

Woodside [1686 (*Woodside*): *wudu* 'wood' + *sīde* 'side']

Woolwich /'wʊlɪdʒ, -ɪtʃ/: 'wool harbour' [918 (*Uuluuich*): *wull* 'wool' + *wīc* 'farm or harbour of the stated kind']

Y

Yeading Brook ‹ **Yedding**

Yedding, Yeading /'jɛdɪŋ/: 'Geddi's folk' [757 (*Geddinges*), 1325 (*Yedding(g)s*), 1331 (*Yeddyng*): *Gedd(i)* (name) + *-(i)ngas* '-followers of']

BIBLIOGRAPHY

GENERAL AND 20TH CENTURY LONDON

The London Journal *is a scholarly periodical devoted to the history and contemporary life of London and its inhabitants.*

Abercrombie, P: Greater London Plan 1944, *HMSO 1945*
Abercrombie, P & Forshaw, JH: County of London Plan *London County Council 1943*
Ackroyd, P: Dicken's London: an imaginative vision *Headline 1987*
Aldous, T: Book of London's Villages *Secker & Warburg 1980*
Amery, C: Wren's London *Lennard, Luton 1988*
Anon: Report: improvements and town planning committee on the preliminary draft proposals for post-war reconstruction in the City of London *Batsford 1944*
Banks, FR: The Penguin Guide to London *Penguin 1958*
Barker, F: Edwardian London *Laurence King 1995*
Barker, F & Gay, J: Highgate Cemetery: Victorian Valhalla *John Murray 1984*
Barker, F & Hyde, R: London as it might have been *John Murray 1982*
Barker, F & Jackson, P: The History of London in Maps *Barrie & Jenkins 1990*
Barker, F & Jackson, P: London: 2000 years of a city and its people *Macmillan 1983*
Barker, TC & Robbins, LM: A History of London Transport *George Allen & Unwin 1975-6*
Barson, S & Saint, A: A Farewell to Fleet Street *Historic Buildings and Monuments Commission for England 1988*
Barton, NJ: The Lost Rivers of London *Phoenix House & Leicester University Press 1962*
Bell, WG: The Great Fire of London in 1666 *Lane 1920*
Betjeman, J: London's Historic Railway Stations *John Murray 1972*
Birdle, M & Hudson, D: The Future of London's Past *Rescue, Worcester 1973*
Bird, J: The Geography of the Port of London *Hutchinson 1957*
Bird, J: The Major Seaports of the United Kingdon *Hutchinson 1963*
Bolsterli, MJ: The Early Community at Bedford Park *Routledge & Kegan Paul 1977*
Breheny, MJ &Congdon, P: Growth and change in a core region: the case of South-East England *Pion, London 1989*
Brimblecome, P: The Big Smoke: a history of air pollution in London since medieval times *Methuen 1987*
Brownhill, S: Developing London's Docklands *Paul Chapman 1990*
Butler, T & Rustin, M (eds): Rising in the East? The Regeneration of East London *Lawrence & Wishart 1996*
Byron, A: London's Statues: a guide to London's outdoor statues and sculpture *Constable 1981*
Cady, M: The Book of London *Automobile Association, Basingstoke 1979*
Centre for Urban Studies: London, Aspects of Change *MacGibbon & Kee 1964*
Chandler, TJ: The Climate of London *Hutchinson 1965*
Charlton, J: The Tower of London: its buildings and institutions *HMSO 1978*
Clayton, KM [ed]: Guide to London Excursion: Twentieth International Geographical Congress *London school of Economics 1964*
Clayton, R [ed]: The Geography of Greater London *Philip 1964*
Clout, H [ed]: Changing London *University Tutorial Press, Cambridge 1978*
Clout, H & Wood PA [eds]: London: problems of change *Longman 1986*
Coppock, JT & Prince HC [eds]: Greater London *Faber 1964*
Corporation of the City of London: Reconstruction of the City of London *Batsford 1944*
Corporation of the City of London: The City of London: a record of destruction and survival *Architectural Press 1951*
Cox, J: London's East End: Life and Tradition *Weidenfeld & Nicolson 1994*

Creaton, H (ed): Bibliography of Printed Works on London History *Library Association 1994*
Croad, S: London's Bridges *Royal Commission on Historical Monuments England & HMSO 1983*
Crowe, A: The Parks and Woodlands of London *Fourth Estate 1987*
Dalzell, WR: The Shell Guide to the History of London *Michael Joseph 1981*
Darlington, I & Howgego, J: Printed Maps of London 1553-1850 *Philip 1964*
Davies, A: Literary London *Macmillan 1988*
Davis, T: John Nash: the Prince Regent's Architect *Country Life 1966*
Day, B: This Wooden 'O': Shakespeare's Globe Reborn *Oberon 1996*
Department of the Environment: Strategic Plan for the South East: review, government statement *HMSO 1978*
Department of the Environment: Thames Strategy *HMSO 1995*
Dolphin, P, Grant, E & Lewis, E: The London Region: an annotated geographical bibliography *Mansell 1981*
Donnison, D & Eversley DEC [eds]: London: urban patterns, problems and policies *Heinemann 1973*
Downes, K: Hawksmoor *Thames & Hudson 1970*
Dugdale, GS: Whitehall through the Centuries *Phoenix House 1950*
Dunning, JH & Morgan EV: An Ecomonic Study of the City of London *George Allen & Unwin 1971*
Dyson, T: The Medieval London Waterfront *Museum of London 1989*
Edwards, D & Pigram R: The Romance of Metroland *Midas, Tunbridge Wells 1979*
Edwards, D & Pigram R: London's Underground Suburbs *Boston Transport 1986*
Edwards, D & Pigram R: The Golden Years of the Metropolitan Railway and the Metro-land Dream *Bloomsbury Books 1988*
Feldman, D & Stedman Jones, G (eds): Metropolis London: Histories and Representations since 1800 *Routledge 1989*
Fiddes, A: The City of London: the historic square mile *Pevensey Press, Cambridge 1984*
Fishman, WJ, Breach, N & Hall, JM: East End and the Docklands *Duckworth 1990*
Fitter, RSR: London's Natural History *Collins 1945*
Fitzgibbon, C: The Blitz *MacDonald 1970*
Forman, C: Spitalfields: a battle for land *Hilary Shipman 1989*
Forshaw, A & Bergström, T: The Markets of London *Penguin 1983*
Forshaw, A & Bergström, T: The open spaces of London *Alison & Busby 1986*
Forshaw, A & Bergström, T: Smithfield: past and present *Hale 1990*
Foster, J: Docklands: cultures in conflict, worlds in collision *UCL Press 1999*
Fox, C (ed): London – World City 1800-1940 *Yale University Press, New Haven & London 1992*
Galinou, M [ed]: London's Pride: the glorious history of the capital's gardens *Anaya 1990*
Girouard, M: Sweetness and Light: the 'Queen Anne' Movement 1860-1900 *Clarendon Press, Oxford 1977*
Glanville, P: London in Maps *Connoisseur 1974*
Glanville, P: Tudor London *Museum of London 1979*
Gleichen: London's Open-air Statuary *Cedric Chivers Bath 1973*
Goode, D: Wild in Lodon *Michael Joseph 1986*
Grant, N: Village London: past and present *Pyramid Books 1990*
Grant, I & Maddren, N: The City at war *Jupiter 1975*
Gray, R: A History of London *Hutchinson 1978*
Green, DR: People of the Rookery: a pauper community in Victorian London *King's College, London 1986*
Greeves, IS: London Docks 1800-1900: a civil engineering history *Thomas Telford 1980*
Grimes, WF et al: Time on our side? *A survey of the archaeological needs of Greater London Department of Environment 1976*
Gwynn, RD: Huguenot Heritage: the history and contribution of the Huguenots in Britain *Routledge & Kegan Paul 1985*
Hall, JM: London: metropolis and region *Oxford University Press 1976*
Hall, PG: The Industries of London since 1861 *Hutchinson 1962*
Hall, PG: London 2000 *Faber 1963*
Hall, PG: Cities of Tomorrow *Basil Blackwell, Oxford 1988*
Hall, PG: London, 2001 *Unwin Hyman 1989*
Harrison, P: Inside the Inner City *Penguin 1983*
Harte, N: The University of London 1836-1986: an illustrated history *Athlone Press 1986*
Hawkins, R: Green London: a handbook *Sidgwick & Jackson 1987*
Hearsey, JEN: London and the Great Fire *John Murray 1965*
Henrey, R: London under fire 1940-45 *Dent 1969*

Hibbert, C: London : the biography of a city *Penguin 1980*
Hobhouse, H & Saunders A [eds]: Good and Proper Materials: the fabric of London since the Great Fire *London Topographical Society 1989*
Hobley, B: Roman and Saxon London: a reappraisal *Museum of London 1986*
Hoggart, K & Green, DR [eds]: London, a new metropolitan geography *Edward Arnold 1991*
Holden, CH & Holford, WG: The City of London: A Record of Destruction and Survival *Architectural Press 1951*
Humphries, S & Taylor J: The Making of Modern London 1945-1985 *Sidgwick & Hackson 1986*
Hunter, M & Thorne, R: Change at King's Cross: From 1800 to the Present *Historical Publications 1990*
Hyde, R: The A to Z of Georgian London *Harry Margary, Lympne Castle 1981*
Inwood, S: A History of London *Macmillan 1998*
Jackson, AA: London's Termini *David & Charles, Newton Abbot 1969*
Jackson, JA: Semi-detached London *George Allen & Unwin 1973*
Jenkins S: Landlords to London: the story of a capital and its growth *Constable 1975*
Jenner, M: London Heritage. The Changing Style of a City *Michael Joseph, 1988*
Johnson, D: The City ablaze: the second great fire of London 29 December 1940 *William Kimber 1980*
Jones, E & Sinclair DJ [eds]: Atlas of London and the London Region *Pergamon, Oxford 1968*
Jones, LR: The Geography of London River *Methuen 1931*
Kiek, J: Everybody's Historic London *Quiller Press 1984*
King, AD: Global cities: post-imperialism and the internationalization of London *Routledge 1990*
Leapman, M [ed]: The Book Of London *Weidenfeld & Nicolson 1989*
Lloyd, D [ed]: Save the City: a conservation study of the City of London *Society for the Protection of Ancient Buildings 1976*
Lobel, M [ed]: The British Atlas of Historic Towns Volume III. The City of London from prehistoric times to c.1520 *Oxford Univeristy Press and Historic Towns Trust 1989*
London County Council: County of London Plan *Macmillan 1943*
London County Council: The Administrative County of London Development Plan, 3 volumes *LCC 1951*
London County Council: The Youngest County *LCC 1951*
London County Council: The County Planning Report *LCC 1960*
London Transport: Planning London's Transport *London Transport 1995*
Luckin, B: Pollution and Control: a social history of the Thames in the nineteenth century *Adam Hilger, Bristol 1986*
Mack, J & Humpries S: The Making of Modern london 1939-1945: London at war *Sidgwick & Jackson 1985*
Mander, R & Mitchenson, J: The Theatres of London *Rupert Hart-Davis 1961*
Maré, E de: London's River: past, present and future *Reinhardt 1958*
Martin, JE: Greater London: an industrial geography *Bell 1966*
Matheson, J & Holding, A (eds): Focus on London 99 *The Stationery Office 1999*
McAuley, I: Guide to Ethnic London *Immel 1993*
McRae, H & Cairncross, F: Capital City: London as a financial centre *Methuen 1991*
Meller, H: London Cemeteries *Avebury, Aldershot 1981*
Merrifield, R: The Archaeology of London *Heinemann 1975*
Merriman, N (ed): The Peopling of London *Museum of London 1993*
Milne, G: The Great Fire of London *Historical Publications 1986*
Ministry of Housing and Local Government: The South East Study *HMSO 1964*
Ministry of Town and Country Planning: The Greater London Plan *HMSO 1945*
Morris, J: Londinium: London in the Roman Empire *Weidenfeld & Nicolson 1982*
Munby, DL: Industry and Planning in Stepney *Oxford university Press 1951*
Munton, RJC: London's Green Belt: containment in practice *Allen & Unwin 1983*
Naib, Al SK: London's Dockland: past, present and future *Thames & Hudson 1990*
Olding, S: Exploring Museums: London *HMSO 1989*
Olsen, DJ: Town Planning in London *Yale Universiry Press 1964*
Olsen, DJ: The Growth of Victorian London *Batsford 1976*
Ormsby, H: London and the Thames *Sifton Praed 1924*
Phillips, H: Mid-Georgian London *Collins 1964*
Phillips, H: The Thames about 1750 *Collins 1951*
Piper, D: London: an illustrated companion guide *Collins 1980*
Plummer, B & Shewan, D: City Gardens. An Open Spaces Survey in the City of London *Belhaven 1992*

Porter, R: London: A Social History *Hamish Hamilton 1994*
Porter, S: The Great Fire of London *Sutton, Stroud 1996*
Power, MJ: John Stow and his London *Journal of Historical Geography, 11, 1985, 1-20*
Prockter, A & Taylor, R: The A to Z of Elizabethan London *Harry Margary, Lympne Castle 1979*
Pudney, J: Crossing London's river *Dent 1972*
Pudney, J: London's Docks *Thames & Hudson 1975*
Rasmussen, SE: London: the unique city *Jonathan Cape 1937*
Robinson, E: London, illustrated geological walks 2 volumes *Scottish Academic Press, Edinburgh 1984-5*
Saint, A (ed): Politics and th People of London: The London County Council, 1889-1965 *Hambledon Press 1989*
Saint, A & Darley, G: The Chronicles of London *Weidenfeld & Nicolson 1994*
Saunders, A: Regent's Park *David & Charles, Newton Abbot 1969*
Schofield, J & Dyson, T: Archaeology of the City of London *City of London Archaeological Trust 1980*
Schubert, D & Sutcliffe, A: The 'Haussmannization' of London?: The Planning and Construction of Kingsway-Aldwych *Planning Perspectives, 11, 1996, pp115-144*
Seaborne, M: Photographers' London 1839-1994 *Museum of London 1995*
Shepherd, J, Westway, J & Lee, T: A Social Atlas of London *Oxford University Press 1974*
Sheppard, F: London, a History *Oxford University Press 1998*
Simmie, J (ed): Planning London *UCL Press 1994*
Smith, DH: The Industries of Greater London *PS King 1933*
South East Economic Planning Council: A Strategy for the South East *HMSO 1967*
South East Joint Planning Team: Strategic Plan for the South East *HMSO 1970*
South East Joint Planning Team: Strategic Plan for the South East 1976 Review *HMSO 1976*
Stamp, G: The changing metropolis: earliest photographs of London 1839-79 *Penguin 1986*
Stedman Jones, G: Outcast London: a study in the relationship between classes in Victorian society *Oxford University Press 1971*
Tames, R: City of London Past *Historical Publications, New Barnet 1995*
Thomas, D: London's Green Belt *Faber 1970*
Thompson, FML: Hampstead: building a borough 1650-1964 *Routledge & Kegan Paul 1974*
Thompson, FML [ed]: The Rise of Suburbia *Leicester University Press 1982*
Thompson, FML [ed]: the University of London and the World of Learning 1836-1986 *Hambledon Press 1990*
Thurston, H: Royal Parks for the People *David & Charles, Newton Abbot 1974*
Tindall, G: The Fields Beneath *Temple Smith 1980*
Townsend, P: Poverty and Labour in London *Low Pay Unit 1987*
Trench, R & Hillman, E: London under London: a subterranean guide *John Murray 1984*
Wales, HRH Prince of: A Vision of Britain: a personal view of architecture *Doubleday 1989*
Wallace, D: London: the Circle Line Guide *Penguin 1990*
Wallmann, S: Living in South London: perspectives on Battersea 1871-1981 *Gower, Farnborough 1982*
Warner, M: The Image of London: views by travellers and emigrés 1550-1920 *Trefoil 1987*
Webb, E: Literary London *Spellmount, Tunbridge Wells 1990*
Weightman, G: Bright Lights, Big City: London Entertainment, 1830-1950 *Collins & Brown, 1992*
Weightman, G: London River: the Thames story *Collins & Brown 1990*
Weightman, G & Humphries, S: The Making of Modern London 1914-1939 *Sidgwick & Jackson 1984*
Weinreb, B & Hibbert, C [eds]: The London Encyclopaedia *Macmillan 1983*
Wheeler, A: The Tidal Thames: the history of a river and its fishes *Routledge & Kegan Paul 1979*
White, HP: A Regional History of the Railways of great Britain Volume III Greater London *David & Charles, Newton Abbot 1963*
Williamson, E. & Pevsner, N: London Docklands, An architectural guide *Penguin 1998*
Wilmott, P & Young, M: Family and Class in a London Suburb *Routledge & Kegan Paul 1960*
Wilson, D: The Tower of London *Constable 1978*
Yelling, JA: Slums and Redevelopment: Policy and Practice in England, 1918-45, with Particular Reference to London *UCL Press 1992*
Young, K & Garside, P: Metropolitan London: politics and urban change 1837-1981 *Edward Arnold 1982*
Young, K & Wilmott, P: Family and Kinship in East London *Routledge & Kegan Paul 1957*
Ziegler, P: London at War 1935-45 *Sinclair Stevenson 1995*

HISTORICAL LONDON

CHAPTER 1: LAND UNDER LONDON

Merriman, N: Prehistoric London *HMSO 1990*

CHAPTER 2: ROMAN LONDON

Chapman, H, Hall, J & Marsh, G: The London Wall Walk *Museum of London 1985*
Hall, J & Merrifield, R: Roman London *HMSO 1986*
Marsden, P: Roman London *Thames & Hudson 1980*
Merrifield, R: London – City of the Romans *Batsford 1983*
Milne, G (ed): From Roman Basilica to Medieval London *HMSO 1992*
Milne, G: The Port of Roman London *Batsford 1985*
Milne, G: Londinium – Map and Guide to Roman London *Ordnance Survey 1983*
Milne, G: Roman London *Batsford 1995*

CHAPTER 3: SAXON AND NORMAN LONDON

Bailey K: "The Middle Saxons" in S Bassett [ed] The Origins of Anglo-Saxon Kingdoms *Leicester University Press, London 1989*
Brooke, C & Keir, G: London 800-1216: the shaping of a city *Secker & Warburg, London 1975*
Clark, J: Saxon and Norman London *HMSO, London 1989*
Cowie, R & Whytehead, R: Lundenwic: the archaeological evidence for middle Saxon London *Antiquity 63 1989-706-18*
Darby, HC & Campbell, EMJ [eds]: The Domesday Geography of South-East England *Cambridge University Press 1962*
Horsman, V, Milne, C & Milne, G: Aspects of Saxo-Norman London: 1 Building and Street Development *London & Middlesex Archaeological Society Special Paper 11 1988*
Vince, A: Saxon London: an archaeological investigation *Batsford 1990*
Yorke, B: Kings and Kingdoms of Early Anglo-Saxon England *BA Seaby, London 1990*

CHAPTER 4: MEDIEVAL LONDON

Barron, CM: Richard Whittington, in Studies in London History: Essays presented to PE Jones *[eds] A Hollaender and W Kellaway Hodder & Stoughton 1969*
Bolton, JL: The medieval English economy 1150-1500 *Dent, Rowman & Littlefield 1980*
Holt, R & Rosser, G [eds]: The medieval town, a reader in English urban history *Longman, Harlow 1990*
Reynolds, S: An introduction to the history of English medieval towns *Clarendon Press, Oxford 1977*
Rosser, AG: Medieval Westminster 1200-1540 *Clarendon Press, Oxford, 1989*
Schofield, J: Medieval London Houses *Yale University Press, New Haven & London 1995*
Sharpe, RR: London and the Kingdom *Longmans, Green & Co 1895*
Thomson, JAF [ed]: Towns and townspeople in the fifteenth century *Alan Sutton 1988*
Thrupp, SL: The Merchant Class of Medieval London *University of Michigan, Ann Arbor 1948*
Unwin, G: The Guilds and Companies of London *Methuen 1908*
Victoria County History of London: ed W Page: Ecclesiastical history and religious houses *Archibald Constable 1908*
Williams, G: London: from Commune to Capital *Athlone Press, University of London 1963*

CHAPTER 5: TUDOR AND STUART LONDON

Beier, AL & Finlay R [eds]: London 1500-1700: The Making of the Metropolis *Longman 1986*
Brett-James, NG: The growth of Stuart London *Allen & Unwin 1935*
Brigden, S: London and the Reformation *Clarendon Press, Oxford 1989*
Cobb, G: London City Churches *Batsford 1977*
Harding, VA: The Population of London 1550-1700: A Review of the Published Evidence *London Journal, 15, 1990, pp111-128*
Reddaway, TF: The Rebuilding of London after the Great Fire *Edward Arnold 1940*
Schofield, J: The building of London from the Conquest to the Great Fire *Colonnade 1984*
Weinstein, S: Tudor London *HMSO 1994*

CHAPTER 6: GEORGIAN LONDON

Bull, GBG: Thomas Milne's land utilization map of the London area in 1800 *Geographical Journal, 122, 1956*
George, D: London, Life in the Eighteenth Century *Routledge & Kegan Paul 1951*
Johnson, N: Eighteenth-Century London *HMSO 1991*
Laxton, P: The A to Z of Regency London *London Topographical Society and Harry Margary, Lympne Castle 1985*

Ogburn, M: Spaces of Modernity. London's Geographies 1680-1780 *Guilford Press 1999*
Rocque, J: A plan of the cities of London and Westminster from an actual survey in 1746 *London Topographical Society and Harry Margary, Lympne Castle 1971*
Rudé, G: Hanoverian London 1714-1808 *Secker & Warburg 1971*
Schwarz, L: London in the Age of Industralisation: Entrepreneurs, Labour Force and Living Conditions, 1700-1850 *Cambridge University Press 1992*
Spate, OHK: The growth of London, AD 1660-1800, in [ed] Darby, HC, An Historical Geography of England before 1800 *Cambridge University Press 1936*
Summerson, J: Georgian London *Barrie & Jenkins 1978*
Summerson, J: The Life and Work of John Nash *George Allen & Unwin 1980*

CHAPTER 7: VICTORIAN LONDON

Adburgham, A: Shopping in style: London from the Restoration to Edwardian elegance *Thames & Hudson 1979*
Barker, T: Moving millions: a pictorial history of London Transport *London Transport Museum 1990*
Bennett, Arnold: Riceyman Steps *1923*
Davis, J: Reforming London: The London Government Problem, 1855-1900 *Clarendon Press, Oxford 1988*
Dyos, HJ: Victorian suburb: a study of the growth of Camberwell *Leicester University Press 1961*
Dyos, HJ & Wolff, M [eds]: The Victorian city: images and realities *Routledge & Kegan Paul 1973*
Fishman, WJ: East End 1888 *Duckworth 1988*
Fried, A & Elman, R [eds]: Charles Booth's London *Hutchinson 1969*
Gissing, George: The Nether World *1889*
Green, DR: From Artisans to Paupers: Economic Change and Poverty in London, 1790-1870 *Scolar Press 1995*
Grossmith, G & W: The Diary of a Nobody *1892*
Hollingshead, J: Ragged London in 1861 *Dent Everyman reprinted in 1986*
Hyde, R [ed]: The A to Z of Victorian London *London Topographical Society and Harry Margary, Lympne Castle 1987*
Keating, P [ed]: Into Unknown England 1866-1913: selections from the social explorers *Fontana 1976*
Kynaston, D: The City of London, vol. 1: A World of its Own, 1815-90 *Chatto & Windus 1994*
Kynaston, D: The City of London, vol. 2: Golden Years, 1890-1914 *Chatto & Windus 1995*
Mayhew, H: London Labour and the London *Poor Penguin 1985*
Morrison, Arthur: A Child of the Jago *1896*
Olsen, DJ: The growth of Victorian London *Penguin 1979*
Olsen, DJ: Town planning in London: the eighteenth and nineteenth centuries *Yale University Press 1982*
Port, MH: Imperial London: Civil Government Building in London, 1851-1915 *Yale University Press, New Haven & London, 1995*
Kynaston, D: The City of London, vol. 1: A World of its Own, 1815-90 *Chatto & Windus 1994*
Sheppard, F: London 1808-1870: The Infernal Wen *Secker & Warburg 1971*
Stedman Jones, G: Outcast London *Penguin 1984*
Walkowitz, JR: City of Dreadful Delight: Narratives of Sexual Danger in Late Victorian London *Virago 1992*
Weightman, G & Humphries, S: The making of modern London 1815-1914 *Sidgwick & Jackson 1983*
White, J: Rothschild Buildings: life in an East End tenement block 1887-1920 *Routledge & Kegan Paul 1980*
Winter, J: London's Teeming Streets, 1830-1914 *Routledge 1993*
Wohl, AS: The eternal slum: housing and social policy in Victorian London *Edward Arnold 1977*
Yelling, JA: Slums and slum clearance in Victorian London *Allen & Unwin 1986*

PLACE NAME HISTORIES

Ekwall, E: English River names *Clarendon Press, Oxford 1928*
Ekwall, E: Street names of the City of London *Clarendon Press, Oxford 1954*
English Place Name Society: Middlesex, Essex, Hertfordshire, Surrey, English Place name Elements *Cambridge University Press*
Field, J: Place names of Greater London *Batsford 1986*
Pointon, GE [ed]: BBC Pronouncing Dictionary of British Names *Oxford University Press 1990*
Rossiter, S [ed]: Blue Guides: *Ernest Benn Ltd, London 1965*
Wallenberg, JK: Kentish Place Names *Uppsala 1931*
Wallenberg, JK: Place names of Kent *Uppsala 1934*
Wells, JC: Longman Pronunciation Dictionary *Harlow 1934*
Wittich, J: Discovering London Street Names *Shire Publications Ltd. Princes Risborough 1990*

PICTURE ACKNOWLEDGEMENTS

All the illustrations in this book are the work of Ralph Orme, Swanston Graphics. The publishers would like to thank the following museums, publishers and picture agencies for permission to base illustrations on their photographs or to reproduce them. Where there is no such acknowledgement, we have been unable to trace the source or the illustration is a composition by our illustrators and contributors.

p.14   *(centre top and background to page)* Guildhall Library, Corporation of London

p.15   *(centre top)* Guildhall Library, Corporation of London/Bridgeman Art Library, London; *(centre right)* Science & Society Photo Library; *(bottom right)* National Remote Sensing Centre Ltd

p.20   *(top)* Museum of London Archaeology Service; *(centre)* The British Museum; *(bottom left)* The Museum of London

p.21   *(top)* The Museum of London; *(bottom)* Dr Pamela Greenwood & Newham Museum Service

p.23   *(centre right)* Times Newspapers; *(bottom right)* The Museum of London.

p.24   The Museum of London

p.25   The Museum of London

p.26   The Museum of London

p.27   The Museum of London

p.32   The Museum of London

p.34   The Museum of London

p.35   The Museum of London

p.36   The Museum of London

p.37   The British Library

pp.38-9   Michael Holford

p.40   The Museum of London

p.41   The Museum of London

p.42   *(centre)* Michael Holford; *(bottom left)* The Museum of London

p.44   Guildhall Library, Corporation of London/ Bridgeman Art Library, London

p.45   The Museum of London

p.46   The British Library

p.47   The Museum of London

p.48   *(top)* Greater London Records Office; *(centre)* The British Museum

p.50   The British Museum

p.51   *(inset)* Rector and Churchwardens of the United Parishes of SS Magnus-the-Martyr, St Margaret, New Fish Street and St Michael, Crooked Lane *(right)* Bridgeman Art Library, London

pp.54-5   Society of Antiquaries

p.55   *(top)* The Mansell Collection

p.56   The Museum of London

p.58-9   The Mansell Collection

p.64   The Museum of London

p.65   The Museum of London

p.66   *(top and centre)* The Museum of London;

pp.66-7   *(centre)* National Monuments Record; *(bottom)* The Museum of London

p.67   *(right)* The Museum of London

p.68   The Museum of London

p.69   *(top)* The Museum of London/National Monuments Record; *(centre)* Guildhall Library, Corporation of London; *(bottom right)* The Museum of London

pp.70-1   Reproduced by Gracious Permission of Her Majesty the Queen

p.72   The Museum of London

p.73   The Mansell Collection

pp.74-5   The British Library

p.75   Guildhall Library, Corporation of London

p.76   The Museum of London

p.77   *(left, top right and bottom right)* The Museum of London; *(centre)* The Governor and Company of the Bank of England;

p.79   *(top)* Guildhall Library, Corporation of London; *(centre right)* The Museum of London

p.80   The Museum of London

p.81   The Museum of London

p.82   *(bottom)* Guildhall Library, Corporation of London/Bridgeman Art Library, London

pp.82-3   Guildhall Library, Corporation of London/Bridgeman Art Library, London

p.83   *(right)* Guildhall Library, Corporation of London; *(right)* National Maritime Museum, Greenwich

p.84   The Museum of London

p.85   The Dickens House, London

pp.86-7   London Transport Museum

pp.88-9   The Museum of London

p.89   *(centre)* Royal Institute of British Architects; *(right)* Private Collection (Mark Girouard)

p.90   *(top)* The Museum of London; *(bottom)* Courtauld Institute Galleries, London (Courtauld Collection)

p.91   *(centre right)* Vestry House Museum; *(bottom right)* The Museum of London

p.92   *(right)* Bancroft Road L.H. Library

pp.92-3   Museum in Docklands project

p.94   *(top)* The Museum of London; *(left)* Architectural Review/A. Acland; *(centre)* Illustrated London News; *(bottom right)* Courtesy of the Trustees of the V&A

p.95   *(top right)* The Museum of London; *(top left)* The Mansell Collection, courtesy of the Trustees of the V&A

p.97   *(centre)* The Salvation Army; *(right)* Andres Press Agency; *(bottom left)* Punch

p.98   *(top)* Illustrated London News; *(bottom left)* The British Library

p.99   Illustrated London News

p.100   *(top)* The Museum of London; *(top right)* Kensington and Chelsea Public Libraries; *(left)* Hulton-Deutsch Collection; *(bottom left)* Guildhall Library, Corporation of London

p.101   *(top right and right)* Greater London Records Office; *(bottom left)* Guildhall Library, Corporation of London

p.102   *(bottom left)* Punch; *(bottom right)* Southwark Arts Libraries

p.104   *(left)* Guildhall Library, Corporation of London; *(bottom)* Illustrated London News

p.105   *(left)* Guildhall LIbrary, Corporation of London/ Bridgeman Art Library, London; *(right)* Illustrated London News

p.106   Private Collection

p.107   *(top)* Selfridge's; *(bottom left)* Marks & Spencer

p.108   London Transport Museum

p.109   Royal Institute of British Architects

p.110   Popperfoto

p.112   London Borough of Camden Local History Library

p.113   London Transport Museum

p.114   *(right and bottom right)* Hulton-Deutsch Collection; *(bottom left)* Greater London Records Office

p.115   *(top left and bottom left)* London Transport Museum; *(right)* Hulton-Deutsch Collection; *(bottom right)* Greater London Records Office

p.117   *(top right)* C.E.G.B.; *(centre right and below right)* Gunnersbury Park Museum

p.118   *(left)* Popperfoto; *(bottom centre)* London Topographical Society; *(above right and bottom right)* The Times

p.119   London Transport Museum

p.120   Associated Press

p.121   Tony Stone Worldwide

p.122   Aerofilms

p.124   The Times

p.125   Alan J. Millard

p.126   *(centre)* Hulton-Deutsch Collection

pp.126-7   Hulton-Deutsch Collection

p.127   *(centre)* The London Journal Trust; *(right)* National Magazine Company Ltd.

p.129   Aerofilms

p.132   *(centre)* New Zealand House

pp.132-3   *(above)* National Westminster Bank; *(bottom)* Architectural Association/Valerie Bennett

p.133   *(top right)* Architectural Association/Valerie Bennett; *(right and centre right)* Architectural Association/Jane Beckett

p.134   *(right)* Punch; *(bottom left)* The Museum of London

p.135   *(left)* Punch; *(top right)* The Times

p.136   *(centre)* Illustrated London News; *(bottom right)* The Museum of London

p.137   *(bottom left)* East London Mosque; *(bottom centre)* Michael Nicholson/Corbis; *(right)* Popperfoto

p.138   *(centre top and bottom left)* The Museum of London; *(centre below)* Nick Daley

p.139   *(top right and left)* The Museum of London; *(bottom right)* Guildhall Library, Corporation of London

p.140   *(top)* Westminster City Libraries; *(centre, bottom left and bottom right)* The Museum of London; *(bottom centre)* Illustrated London News

p.141   London Transport Museum

p.142   *(top left)* Guildhall Library, Corporation of London/Bridgeman Art Library, London; *(top centre and below left)* The Museum of London

p.143   *(top right)* Aerofilms; *(below right)* London Transport Museum

pp.144-5   The Dickens House, London

p.146   *(centre, bottom left and bottom right)* Richard Kalina/The International Shakespeare Globe Centre Ltd; *(top right)* Guildhall Library, Corporation of London

p.147   *(top left and top centre)* The Museum of London; *(top right)* Victoria and Albert Museum; *(centre left)* Diana Howard: London Theatres and Musical Halls 1850-1950; *(bottom right)* National Theatre

p.148   John Topham Picture Library

p.149   *(top left and top right)* The Museum of London; *(below right)* Harrods

pp.152-3   Telegraph Colour Library

p.155   *(top right and bottom)* The Museum of London; *(centre right)* Westminster City Libraries

p.157   *(top, centre right and bottom)* The Museum of London; *(centre left)* Paul Mellon Collection

p.159   Canary Wharf

p.160   *(top)* London Transport Museum; *(bottom right)* Architectural Association/Canon Parsons

p.161   *(top left)* Bridgeman Art Library, London; *(top right and bottom left)* London Transport Museum; *(bottom right)* The Post Office

p.162   Hulton-Deutsch Collection

p.163   *(top)* By Gracious Permission of Her Majesty the Queen; *(centre and bottom centre)* The Museum of London

p.164   *(bottom left)* Illustrated London News

p.165   *(top left)* London Transport Museum; *(top right)* Architectural Association/ E. Hurwicz; *(centre right)* The Museum of London; *(below right)* Keith Wynn/Photocraft Hampstead; *(bottom left)* Aerofilms

p.166   *(bottom)* Westminster City Libraries; *(above right)* John Gay

p.167   *(left and right)* The Times; *(centre)* Royal Festival Hall

pp.168-9   *(top)* Guildhall Library, Corporation of London; *(bottom)* Paul Draper/Sunday Times

p.169   Guildhall Library, Corporation of London

p.170   *(centre and bottom)* Imagenet/The Millennium Experience

p.171   *(top)* Ove Arup & Partners: photo by Nick Wood/ Hayes Davidson; *(bottom)* Tate Gallery/Hayes Davidson

p.172   *(top)* Chorley Handford; *(centre right)* Foster and Partners: photo by Nigel Young

p.173   *(top right)* The British Museum; *(bottom)* BFI/CDP

Endpapers   The Mansell Collection

INDEX

This index lists: (1) all districts, place localities, estates, parks, open spaces and rivers within Greater London and surrounding counties named in the Atlas; (2) streets of historical importance, especially those on maps of the City of London; (3) all buildings and institutions named in the Atlas.

It also serves as an index to the many subjects and themes covered in the Atlas, e.g. air raids, banks, cemeteries, etc. The names of people mentioned in the text or commemorated by the statues listed on page 167 are not indexed. Streets and buildings with the City (of London) are so indicated. Places and buildings outside the City are located by reference to the first element of the postal area within which they lie, e.g. e, ec (outside the City), n, se, sw, w and wc. Places outside the London postal area are located by county, e.g. Berkshire, Essex, Kent, Surrey, etc. More precise locations are given in
a few cases, e.g. South Bank, Westminster.

Apart from early chapters in the Atlas which show Latin and Anglo-Saxon names, and a few unavoidable historical spellings, all names and spellings in the Atlas are given in their modern English form. Variant forms of the names of churches (symbolised †) are indicated by brackets or by the equation of different forms.

The following abbreviations have been used:
a/c also called
f/c formerly called
form. formerly
f/s formerly spelled
mod. modern
n/c now called
n/s now spelled
OE Old English

Abbey Mills E industrial complex 93; pumping station 104
Abchurch Lane CITY Medieval London 53; 59
Abney Park N 166
Acton W agriculture 78; growth 88, 112; trams 91; railway and Underground stations 113; bombing 119; village in 1800 156; cemetery 166
Acton Common W agriculture 78
Acton Town W Underground station 113
Adam and Eve Camden Town coaching inn 82
Adam and Eve Gardens NW pleasure gardens 142
Addington Surrey geology 18
Addiscombe Surrey trams 91
Addiscombe Road Surrey railway station 91
Adelmetone *see* Edmonton
Adelphi WC 76
Adelphi Music Hall W 147
Adelphi Terrace WC Dickens 144
Admiralty SW 76, 95, 154
Admiralty Arch WC 95, 132, 167
agriculture 78-9
air raids 118-19
airports City Airport 159
Akeman Street SW Saxon London 42
Albert Bridge SW 163
Albert Memorial SW 164
Albion Dock SE 92, 158
Alderney Road Cemetery E 166
Aldersgate CITY Roman gate 29, 31, 34; ward of Medieval city 52; Tudor London 60, 61; industry 116
Aldersgate Street CITY Medieval London 49; Tudor London 58, 60, 61
Aldgate CITY Roman gate 29, 31, 34; ward of Medieval city 52; Tudor London 59, 60, 61; tailoring industry 92; Underground station 153, 161
Aldgate East E Underground station 113, 153, 161
Aldgate High Street E Tudor London 59
Aldgate Pump E Dickens 145
Aldgate Street CITY Medieval London 49, 53
Aldwych WC Saxon London 41; Underground station 161
Aldwych Theatre WC 147
Aldwych Underpass WC 161
Alexandra Palace N railway station 113
Alexandra Park N 141
Alexandra Railway Bridge CITY 163
Alhambra Music Hall WC 147
All Hallows Barking (*n/c* All Hallows by the Tower) CITY † Saxon church 41, 59; Medieval church 50; 76; Undercroft Museum 165
All Hallows Bread Street CITY † Medieval church 50
All Hallows Grasschurch (*n/c* All Hallows Gracechurch Street) CITY † Medieval church 50; Great Fire 69; 76
All Hallows the Great CITY † Medieval church 50
All Hallows Honey Lane CITY † Medieval church 50
All Hallows the Little CITY † Medieval church 50
All Hallows Staining CITY † Medieval church 50; 15th-century tower 59
All Hallows by the Tower *see* All Hallows Barking
All Hallows on the Wall CITY † Medieval church 50
All Saints E railway station 158
All Souls W † 76, 77
Alperton Middlesex cemetery 166
Alteham *see* Eltham
Ambassadors Theatre WC 147
Amersham Buckinghamshire geology 18
Angel N trams 91; gentrification 127; Underground station 161
Angel Road N railway station 113
Angell Estate SE 75
Antwerp N Belgium trade with London 53
Apollo Gardens SE pleasure gardens 142
Apollo Theatre W 147
Apothecaries' Hall CITY 58, 76, 152
Archbishop's Park SW 167
architecture Georgian 76-7
Armourers' and Braziers' Hall CITY 62, 152
Arnold Circus E 129
Arsenal N football ground 143
Artillery E Huguenot church 136
Artillery Ground E 79
Arundel House WC 61
Arundel Stairs WC 23
Ashford Kent commuting 126; South-East Regional Plan 128
Ashford Middlesex village in 1800 156
Ashstead Surrey Roman building 28
Assembly House Highgate pleasure gardens 142
Astley's Music Hall SE 147
Athenaeum Club W 76
Audley Estate W 75
Augustinian Friars, House of CITY 51
Austin Friars CITY † 59, 60
Avery Hill Kent park 141
Aylesbury Buckinghamshire geology 18

Bacheham *see* Beckenham
Back Lane E 17th-century development 65
Bagnigge Wells N pleasure gardens 142
Bagshot Surrey growth 112
Bailey, The *see* Old Bailey
Baker Street W bazaar 107; railway station 113; Underground station 113, 161
Bakers Arms E trams 91
Bakers' Hall CITY 62
Bakewell Hall CITY 61
Balham SW comparative poverty 102; sewer 104; railway and Underground stations 91, 113; suburban development 157
Balham Hill Road SW shops 106
Baltic Exchange CITY 152
Bangladeshi population 137
Bank CITY Underground station 161
Bank of England CITY 77, 152; museum 165
Bankruptcy Court CITY 99
banks CITY 152-3
Bankside SE 17th century 64
Bankside Gallery SE 165
Banqueting Hall Whitehall 95, 154
Banstead Surrey geology 18; growth 112; bombing 119; village in 1800 156
Barber Surgeons' Hall CITY 152
Barbican CITY city wall 59; Underground station 152, 161
Barbican Art Gallery CITY 165
Barbican Centre CITY 152
Barbican Street CITY Medieval London 49
Barking Essex geology 18; flooding risk 23; Domesday Book Berchingas 43; growth 89, 112; railway station 91; 113; suburban expansion 114; bombing 119; Greater London Development Plan 129; open spaces 141; village in 1800 156; *see also* Berecingas
Barkingside Essex railway station 113; cemetery 166
Barklies Inn CITY 61
Barley Mow N pleasure gardens 142
Barnard's Hall WC 58
Barnard's Inn CITY 58, 99
Barner North Pier E 159
Barnes SW railway station 91; growth 112; village in 1800 156
Barnes Common SW 141; cemetery 166
Barnes Railway Bridge SW 162, 163
Barnet Hertfordshire geology 19; growth 88; suburban expansion 114; immigrant population 137; open spaces 141
Barnsbury N gentrification 127; Jewish settlement 136
Barons Court SW Underground station 160
Bartholomew Fair CITY market 148
Bartholomew's Hospital Estate E 75
Basildon Essex new town 125
Basinger Lane CITY Medieval London 53
Bassishaw CITY ward of Medieval city 52
Bassishaw Street CITY Medieval London 49
Battersea SW agriculture 79; trams 91; factories 93; cholera epidemics 96; housing 101; comparative poverty 103, 104; population decline 12; industry 116; power station 117; employment 117; bombing 118, 119; village in 1800 156; suburban development 157; cemetery 166
Battersea Bridge SW 163
Battersea High Street SW market 148
Battersea Park SW 139, 141
Battersea Park Road SW housing 100-101; railway station 113
Battersea Railway Bridge SW 163
Battle Bridge Estate NW 75
Battle Bridge Gate Islington tollgate 82
Baynard's Castle CITY 43, 58, 61
Bayswater W 88, 89; comparative poverty 102; Jewish settlement 136; Underground station 160
Bayswater and Piccadilly Market W 148
Bayswater Road W 83; housing 100-101; market 148
Bayswater Tea Gardens W pleasure gardens 142
Bear Quay CITY market 79
Bearward Lane CITY Medieval London 53
Beckenham Kent geology 19; Domesday Book Bacheham 43; growth 88, 112; railway station 113; bombing 119; village in 1800 156; cemetery 166
Beckenham Place Park Kent 141
Beckton E prehistoric site 20, 21; railway station 93, 113; shopping centre 159
Beckton Crossness E pumping station 104
Beckton Park E railway station 159
Beddington Surrey Domesday Book Beddintone 43
Bedfont Middlesex village in 1800 156
Bedford (Duke of) Estate NW 75
Bedford Park W garden suburb 157
Bedford Square WC buildings 133
Bedlam *see* Imperial War Museum
Belgrave Hospital for Children SW 96
Bell Street W market 148
Belsize House NW pleasure gardens 142
Belsize Park NW suburban development 157
Belvedere House SE pleasure gardens 142
Belvedere Tea Gardens N pleasure gardens 142
Bercher's Lane CITY Medieval London 53
Berchingas *see* Barking
Berecingas tribal people of East Saxon kingdom 40
Beresford Square SE market 148
Bergen C Norway trade with London 52
Berkeley Estate E 75
Berkeley Square E 75
Berkshire commuting population 112, 126; Green Belt and new towns 125; main roads into London 127; South-East Regional Plan 128
Bermondsey SE Saxon find 41; growth 74, 89; estates 75; market 79; housing 100, 101; comparative poverty 103; local government district 1855 104; licensed premises 103; market 106; population decline 112; industry 116; employment 117; bombing 118, 119; Irish population 137; village in 1800 156
Bermondsey Abbey SE 43
Bermondsey Spa Gardens SE pleasure gardens 142
Bermondsey Storm Relief SE sewer 161
Berners Estate 74
Berwick Street WC market 79, 148
Bethlehem Lunatic Hospital (*a/c* Bethleham Royal Hospital) (Bedlam) SE 76, 77; *see also* Imperial War Museum
Bethnal Green E 17th-century development 64; factories 93; housing 101; licensed premises 103; local government district 1855 104; market 106; population decline 112; railway station 113; poverty 115; industry 116; bombing 118, 119; Jewish settlement 136; immigrant population 137; village in 1800 156; Underground station 161
Bethnal Green Museum of Childhood E 165
Bethnal Green Road E market 148
Beverley Brook SW 19, 78; sewer 104
Bexley Kent geology 19; growth 112; bombing 119; immigrant population 137; open spaces 139; village in 1800 156
Bexleyheath Kent railway station 91; Greater London Development Plan 129; cemetery 166
Billericay Essex growth 112
Billingsgate CITY Roman baths 29; Medieval London 49, 53; ward of Medieval city 52; Tudor London 59, 60, 61; 17th century 64; market 79, 148, 152
Billiter Lane CITY Medieval London 49
Birka S Sweden trade with London 42
Birkbeck College WC 164
Bishops Park SW 141
Bishops Stortford Essex commuting 126
Bishopsgate CITY Roman gate 29, 31, 34; ward of Medieval City 52; Tudor London 59, 60, 61; railway station 91, 113; bombing 118
Bishopsgate Street CITY Medieval London 49, 53
Black Cap Camden Town coaching inn 82
Black Friars CITY † 60
Black Lyon Stairs SW 23
Black Prince SE pleasure gardens 142
Black Queen, Shacklewell N pleasure gardens 142
Blackfriars Bridge 163
Blackfriars Railway Bridge 163
Blackfriars Road SE 83; market 106
Blackfriars Station CITY commuter traffic 126; 152, 153; Underground station 161
Blackfriars Theatre CITY 146
Blackheath SE geology 18; common and market gardens 78; 79; growth 89; railway station 91; 113; comparative poverty 103; park 141; open spaces 141; rugby ground 143
Blacksmiths' Hall CITY 62
Blackwall E railway station 113; industry 116
Blackwall Reach Thames 93
Blackwall Tunnel SE 105, 159
Bleeding Heart Yard SE Dickens 145
Bletchingley Surrey Roman building 28
Blitz, The 118-19
Bloomsbury WC furniture industry 92; factories 93; county court 99; comparative poverty 103; Jewish settlement 136; university 164, 165
bombing Zeppelin raids 118; World War II 118-119
Bond Street W Underground station 161
Boodles Club W 76
Bordeaux SW France trade with London 52
Borough SE Underground station 161
Borough High Street SE 83; bombing 118
Borough Market SE 79, 148
Borough Road SE 83; railway station 113
Boundary Street E housing development 101
Bourne & Hollingsworth W department store 107
Bournemouth Hampshire (now Dorset) South-East Regional Plan 128
Bow E factories 93; county court 99; industry 116
Bow Creek E 81, 93
Bow Road E shops 107
Bow Street WC police court 99; Dickens 144

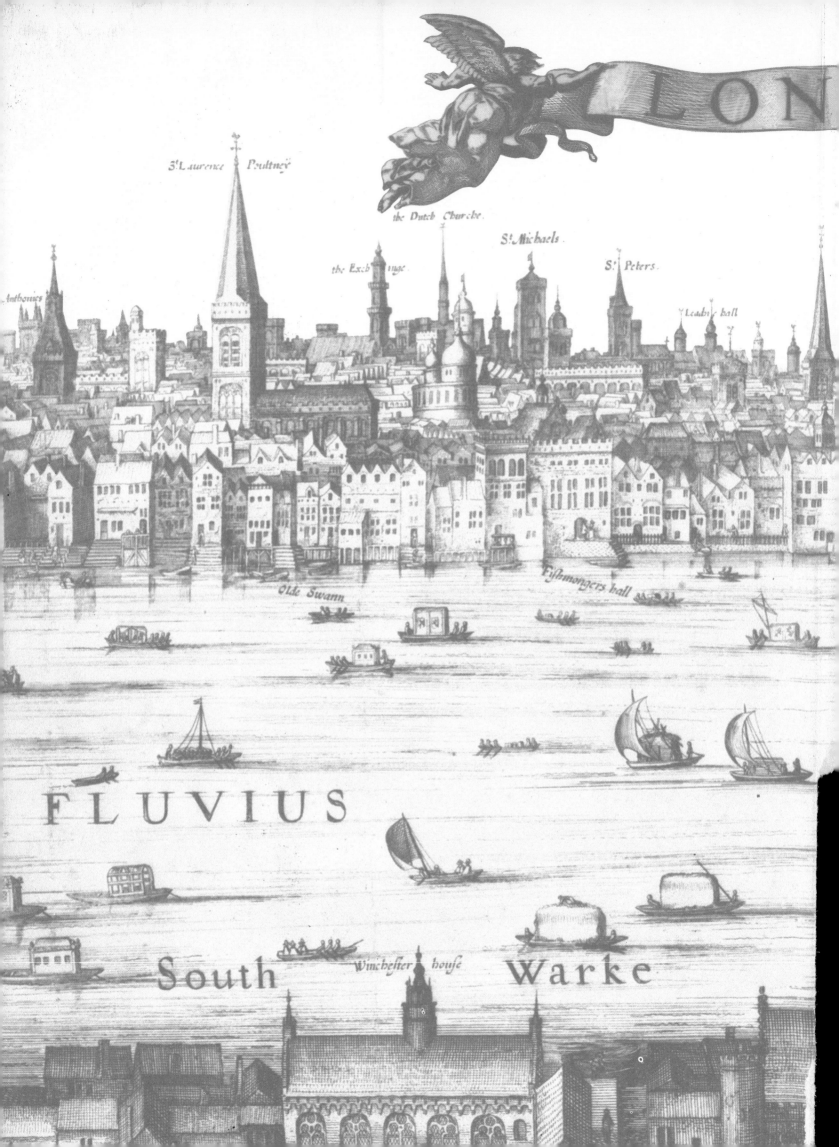

St. Laurence Poultney

the Dutch Churche.

St. Michaels.

the Exchange.

St. Peters.

Anthonies

Leaden hall

Olde Swann

Fishmongers hall

FLUVIUS

South

Winchester house

Warke

LON